Estimation and Classification of Reserves of Crude Oil, Natural Gas, and Condensate

Chapman Cronquist
Oil & Gas Consultant
Houston, Texas

Henry L. Doherty Memorial Fund of AIME
Society of Petroleum Engineers
Richardson, TX USA

SPE BOOK SERIES

Book Editor
Arlie M. Skov

SPE Books Committee 2001

Author

Chapman Cronquist is a registered Professional Engineer in Texas with over 40 years of oil and gas reservoir engineering experience, both domestic and international, and in both operations and research. During 1985-1987, he chaired an SPE Task Force on Reserves Definitions and was an SPE Distinguished Lecturer on Reserves Definitions during 1987-1988. He was elected a Distinguished Member of SPE in 1988. Chap has written papers and/or made presentations on waterflooding, PVT behavior, improved oil recovery, gas reservoir performance, basic and advanced reservoir engineering, probabilistic methods, and reserves estimation. He has a BS in geology from Rensselaer Polytechnic Inst. and a MS in petroleum and natural gas engineering from Pennsylvania State U.

Disclaimer

This book was prepared by members of the Society of Petroleum Engineers and their well-qualified colleagues from material published in the recognized technical literature and from their own individual experience and expertise. While the material presented is believed to be based on sound technical knowledge, neither the Society of Petroleum Engineers nor any of the authors or editors herein provide a warranty either expressed or implied in its application. Correspondingly, the discussion of materials, methods, or techniques that may be covered by letters patents implies no freedom to use such materials, methods, or techniques without permission through appropriate licensing. Nothing described within this book should be construed to lessen the need to apply sound engineering judgment nor to carefully apply accepted engineering practices in the design, implementation, or application of the techniques described herein.

ISBN 978-1-55563-090-4
ISBN 978-1-61399-225-8 (Digital)

Society of Petroleum Engineers
222 Palisades Creek Drive
Richardson, TX 75080-2040 USA
http://store.spe.org/
books@spe.org
1.972.952.9393

FOREWORD

Among the many varied and multidisciplinary tasks a petroleum engineer undertakes in the practice of his profession, none are individually more important than the estimation and classification of oil and gas reserves. That is the subject of this remarkably comprehensive book by my friend, Chap Cronquist.

The determination of reserves in petroleum engineering differs in an important manner from the problems typically addressed by the more conventional engineering disciplines, such as civil, electrical, or mechanical. In these, the physical properties of the materials of interest—and the laws that govern their behavior—are either well established or can be measured directly. The petroleum engineer, however, must deal with inferred properties, and with the infinite variability of Mother Nature.

The subsurface "boxes" that trap hydrocarbons are naturally occurring geological features of a complex nature that cannot easily be predicted or quantified. The characteristics, size, and continuity of these "boxes" must largely be inferred, using petrophysical, geological, and geophysical data, relying heavily upon the petroleum engineer's experience, judgment, intuition, and his or her ability to appropriately employ innovation and analogy. To estimate reserves, the petroleum engineer must also use an impressive array of more conventional engineering and mathematical tools, including probability theory. These procedures are admirably detailed in this book.

The value of this book is not limited to practicing petroleum engineers. It will also serve as a valuable resource, reference, and educational tool to all who have a need to understand the significance of oil and gas reserves estimates and their inherent uncertainty. This includes not only the staff and management of oil and gas companies, but also the many governmental entities who are either charged with regulating oil and gas activities or who benefit from them, as well as financial institutions, attorneys, accountants, politicians, and consultants. Reserves estimates carry substantial economic and political weight. Truly, the professionals entrusted with the responsibility of estimating and classifying oil and gas reserves provide a valuable service to society, with wide-ranging implications that profoundly impact the global economy.

Arlie M. Skov
Editor and SPE President 1991

DEDICATION

This work is dedicated to my son, Neil,

born of love 30 October 1962,

died by his own hand 8 September 1994.

Fly on, free spirit.

PREFACE

Since the late 1950's and early 1960's, when the major works on reserves estimation were published, significant progress has been made in characterizing the geology of oil and gas reservoirs, in understanding how reservoir heterogeneity impacts well and reservoir performance, and in mathematically simulating well and reservoir behavior with progressively more sophisticated computer models.

Computer simulation methods to estimate reserves, however, are data and manpower intensive. For many older reservoirs, and for many small reservoirs being developed today by small companies and independents, neither the data nor the manpower required for this technology is available. Even in highly sophisticated operations, reserves estimates frequently must be made before sufficient data can be collected or before there is sufficient time for detailed analysis. In many cases the engineer must rely on empirical methods and rules of thumb to make preliminary reserves estimates, pending collection of additional data and observation of well performance. These methods, and an extensive set of references, are included in this work to assist the practicing engineer with this aspect of reserves estimating.

Many of the publications cited, especially those needed to estimate PVT properties, are not readily available at remote locations. Accordingly, graphs and charts routinely used to estimate such properties are included in Appendix E. In addition, the correlations from which the charts were developed are included for PC-proficient engineers.

The emphasis here is on estimating reserves attributable to primary production mechanisms. Appendix F, however, is an overview summary of estimating reserves attributable to improved recovery methods. Detailed discussion of these methods is beyond the scope of this work, and references to more extensive treatments are provided.

Estimation of reserves of natural gas from coal deposits, coalbed methane, is beyond the scope of this work. Several references are provided; however, this technology has developed rapidly over the last several years and improvements continue to be made. Thus, the reader is advised to review the current literature on this topic and consult with recognized experts.

Reserves estimation is needed for many purposes, including financial reporting to the investing public and asset management by operators. Each entity in the industry tends to establish slightly different guidelines, the totality of which cannot be summarized in a work of this type. In the final analysis, however, "it all starts at the wellhead," and the emphasis here is so focused.

This material should not be considered a "cookbook." Reserves estimation is quite subjective; it requires considerable experience and a high degree of professional skill and judgment. No matter how copious the basic data, in the final analysis, the reliability of any reserves estimate depends on the experience and integrity of those responsible.

Reserves estimation should be synergetic and draw on the skills of many specialists, including geophysicists, geologists, log analysts, production engineers, reservoir engineers, statisticians, and economists. The reserves estimator cannot master all these skills, but should be sufficiently knowledgeable to meld individual contributions into a consistent whole. Just as the reserves estimator cannot master all the contributing elements, so this work cannot present all the details of the individual specialties.

The work is sufficiently complete that it may be used as a working reference for reserves estimation by experienced reservoir engineers. If used as a text, students should be versed in basic reservoir engineering and have a good understanding of probability and statistics.

ACKNOWLEDGMENTS

This work has its origins in *Reserves Estimation*, a softbound book prepared by the author for IHRDC during 1990 as part of their *Video Library for Exploration and Production Specialists*. IHRDC's courtesy in assigning their copyright to the Society of Petroleum Engineers for that part of this work is appreciated.

This work, however, is a significant expansion over the IHRDC book and contains much new material. Since 1990, many significant papers regarding reserves estimation have been published, the results of which have been included in appropriate sections of this work. The section on material balance analysis of volumetric gas reservoirs has been expanded to include sour gas. There is an appendix that covers correlations to estimate the PVT properties of reservoir fluids. The section on classification of reserves has been expanded to include definitions from many of the major producing areas of the world. The most significant development since 1990, however, was the 1997 joint adoption of new reserves definitions by the Society of Petroleum Engineers and the World Petroleum Congresses. These definitions, which provide for either deterministic or probabilistic procedures to estimate and classify reserves, led to a major chapter on the subject. In preparation of this chapter, the author has benefited from discussions with members of the SPE Standing Committee on Reserves and with members of the SPE Subcommittee involved in the preparation of a Supplement to the 1997 SPE-WPC definitions.

Much of the material in this work was used by the author in presenting a course on reserves estimation in the Graduate Program in Petroleum Engineering at the University of Houston. Sincere appreciation is expressed to the students for persevering during this baptism of fire, for helping to find the inevitable typos, and for sharing their experience.

Appreciation is extended to industry reviewers, including Victor Alcobia, Ed Capen, DeAnn Craig, Amiel David, Ian Dunderdale, Arlen Edgar, Harold Gilliland, Ron Harrell, Harry Jung, Claude McMichael, Bill Miller, Forest Mintz, Noel Rietman, Pete Rose, Jim Ross, Wim Swinkels, Robert Thompson, and Ronald Watson, who reviewed some or all of the work at various stages and made many helpful suggestions that contributed to the final product. Special thanks are reserved for Arlie Skov, the Editor of this work, who waded diligently through numerous pages of material and shared generously his expertise and many years of experience.

Much of the material compiled in this work has been taken from industry publications, especially those by the AAPG, the SPE, and Gulf Publishing Company. Appreciation is expressed to numerous authors for their efforts and for sharing their experience. A diligent effort has been made to cite all sources.

While pleased to acknowledge the help received in the preparation of this work, the author recognizes "the buck stops here" and accepts responsibility for errors or omissions herein.

As with all things, improvement is possible. Your comments would be welcome.

Chap Cronquist
Houston, Texas
Spring 2001
CHAPCRONQUIST@AOL.COM

CONTENTS

1. INTRODUCTION

2. ANALOGY METHODS

4. MATERIAL BALANCE METHODS

5. PERFORMANCE/DECLINE TREND ANALYSIS

6. SPECIAL PROBLEMS IN RESERVE ESTIMATION AND CLASSIFICATION

7. FIELD EXAMPLES

8. RESERVES ANALYSIS, FIELD DEVELOPMENT, AND PERFORMANCE MONITORING

9. PROBABILISTIC ESTIMATION AND CLASSIFICATION OF RESERVES

APPENDICES

A. PETROLEUM RESERVES DEFINITIONS OF SOCIETY OF PETROLEUM ENGINEERS AND WORLD PETROLEUM CONGRESSES

B. INTERPRETIVE COMMENTS REGARDING SPE-WPC [1997] RESERVES DEFINITIONS

C. COMPARISON OF SPE-WPC AND U.S. SEC DEFINITIONS FOR PROVED RESERVES

D. STANDARDS PERTAINING TO THE ESTIMATING AND AUDITING OF OIL AND GAS RESERVES INFORMATION

E. EMPIRICAL EQUATIONS AND CORRELATION CHARTS TO ESTIMATE PVT PROPERTIES OF RESERVOIR FLUIDS

J. METHOD TO CALCULATE THE COMPOSITION AND SOME PROPERTIES OF RESERVOIR GAS FROM THE COMPOSITION OF SEPARATOR FLUIDS

NOTATION AND ABBREVIATIONS

REFERENCES 380

AUTHOR INDEX 405

SUBJECT INDEX 412

ONLINE RESOURCES 417

Estimation and Classification of Reserves of Crude Oil, Natural Gas, and Condensate

CHAPTER 1

INTRODUCTION

1.1 Rationale for Reserves Estimation

The term "reserves" means different things to different groups. To the banking industry, reserves are the amount of capital retained to meet probable future monetary demands. To the oil and gas industry, reserves are the amount of crude oil, natural gas, and associated substances[1] that can be produced profitably in the future from subsurface reservoirs.[2]

1.1.1 Reasons for Reserves Estimates

Estimates of oil and/or gas reserves are required for different purposes by different entities involved in the industry and at different stages in the maturity of the property[3] for which the estimate is made. Entities concerned with oil and/or gas reserves include:

- companies and individuals responsible for exploration, development, and operation of oil and/or gas properties,
- buyers, sellers, and evaluators of interests in oil and/or gas properties,
- banks and other financial institutions involved in financing exploration, development, or purchase of oil and/or gas properties,
- regulatory, taxation, and other agencies with authority over operators of oil and/or gas properties,
- governmental agencies responsible for the planning and development of national energy policies,
- investors in companies involved in exploration and production of oil and/or gas,
- owners of mineral interests—e.g., individuals, companies, governments—and
- parties to a dispute or arbitration involving reserves.

Depending on the nature of their involvement in the industry, each of these entities may have different requirements for reserve estimates and, accordingly, tend to establish different guidelines regarding reporting of such estimates. In 1997, in an effort to standardize reserves reporting, Society of Petroleum Engineers and World Petroleum Congresses (SPE-WPC), acting jointly, promulgated a set of definitions, which are reproduced in **Appendix A.** These

---

[1] The term "associated substances" includes solution gas, condensate, natural gas liquids, and sulfur. However, methods for estimating reserves of natural gas liquids and sulfur are beyond the scope of this work. A publication from the Petroleum Society-CIM [1994] includes a brief discussion of estimation and classification of reserves of these two products.

[2] This simple definition is discussed further in Sec 1.3.16.

[3] The term "property" reflects mineral law in the U.S. In countries where private ownership of minerals is not recognized, comparable terms might include "concession area," "entity," "reservoir," or "field."

definitions, which are discussed in more detail in Sec. 1.3.5, include proved, probable and possible reserves. Interpretive comments—by the author—regarding these definitions are provided in **Appendix B**.

The U.S. Securities and Exchange Commission (U.S. SEC) requires reporting proved reserves—only—and has promulgated detailed guidelines for such reporting, which are compared—in part—with SPE-WPC [1997] definitions in **Appendix C**. Additional discussion regarding these definitions is provided in Sec. 1.3.6.

Depending on the scope of their operation, operators of oil and/or gas properties require estimates of reserves and judgments about the degree of uncertainty in those estimates at various stages of exploration, development, and production. Of concern are:

- potential resources of crude oil or natural gas on undrilled prospects,
- volume and degree of uncertainty of reserves on prospects being developed,
- sizing and design of equipment to process reserves and transport them to market, and
- opportunities for additional profit from incremental reserves that might be attributable to stimulation of producing wells, infill drilling, equipment modifications or additions, or improved recovery projects.

1.1.2 Uncertainties in Reserves Estimates

Estimates of oil and/or gas reserves are inherently uncertain.[4] Reserves of oil and/or gas are the *commercially* recoverable portion of accumulations of fluid hydrocarbons in the pore spaces of rocks buried in the subsurface, the extent and nature of which cannot be determined with a high degree of precision. Recovery of oil and/or gas from subsurface reservoirs is controlled to a significant degree by the heterogeneities of the reservoir rock and by the type of reservoir drive mechanism.[5] Neither of these factors can be determined with a reasonable degree of certainty until after an accumulation has been developed and placed on production. DeSorcy [1979] has discussed the source and probable magnitude of many of the errors associated with estimates of oil and/or gas reserves.

In addition to the foregoing uncertainties, which are physical in nature, there are commercial uncertainties. Recovery of oil and/or gas from a subsurface reservoir is controlled, in the long run by costs to acquire rights to exploit a tract or concession; costs to explore the tract; costs of associated dry holes; costs to develop the accumulation; costs to produce, treat, and move oil and/or gas to market; and the market value of the quantities sold. Typically, these costs are incurred over a period of many years, with significant expenditures frequently being required years before any income is realized.

The commercial environment in which oil and/or gas reserves must be estimated and the associated uncertainties are no less important than the physical environment. In today's oil and gas industry, the commercial environment may be subject to more uncertainty than the physical environment. The degree of uncertainty in estimates of reserves depends mainly on:

- the degree of geologic complexity,
- maturity of the property,

---

[4] Because of this, estimates of reserves should always be qualified to indicate the degree of uncertainty. The most commonly used such qualifiers are "proved," "probable," and "possible," a point discussed in more detail in Sec. 1.3.

[5] Depending on circumstances, the economic scenario and operating environment may have an even greater influence than these factors on the amounts of oil and/or gas potentially recoverable from an accumulation.

- the quality and quantity of geologic[6] and engineering data,
- the operating environment, and
- the skill, experience and integrity of the estimators.

Each of these factors is discussed briefly in the following paragraphs.

Geologic Complexity

The degree of geologic complexity in a group of properties may vary widely. At one extreme, the properties may be in an area characterized by low structural relief, little or no faulting, no unconformities, and having oil and/or gas reservoirs in the same general type of depositional unit. The lower Tuscaloosa (Cretaceous) trend in southwestern Mississippi (southeast U.S.) is an example. At the other extreme, the properties may be in an area characterized by extensive faulting, numerous unconformities, multiple depositional units and complex reservoir fluids. The U.S. overthrust belt and the North Sea are examples.

Maturity

The maturity of an oil and/or gas reservoir may be described in terms of three stages of development and production.

1. Geologic delineation/reservoir characterization, which includes discovery of the oil and/or gas accumulation and the period required to: (a) delineate the major geologic features controlling the extent of the accumulation and (b) determine the characteristics of the reservoir rock/fluid systems.
2. Reservoir optimization, which includes the period of additional development, sustained production, and reservoir surveillance required to: (a) determine reservoir drive mechanisms and probable recovery efficiencies, (b) establish optimum well spacing and production policy, and (c) implement improved recovery projects, if needed to increase commercial recovery of oil and/or gas.
3. Settled production, which includes the period during which wells, including those which have responded to fluid injection (if any), have developed performance/production trends that may be analyzed to estimate reserves.

Quality and Quantity of Data

Minimum data necessary to estimate reserves with a reasonable degree of confidence vary widely from one field to the next and depend (in part) on the geology and maturity of the field and on the geology, maturity, and drive mechanism of individual reservoirs. For a field monitored from discovery and producing by primary reservoir drive mechanisms, acquisition of most or all of the following data is recommended:[7]

- concession and/or production-sharing agreement and/or joint operating agreement, as appropriate to the property,
- contracts for gathering and/or processing and/or for sales of production,
- sufficient seismic data to identify and delineate major geologic features and probable reservoir limits,

---

[6] The term "geologic" includes "geophysical," as discussed in Appendix B.

[7] The operator of a property typically has considerable discretion in obtaining geologic and engineering data. Depending on circumstances, some of the data gathered by an operator may not be readily available to third parties who must make independent estimates of reserves for various purposes.

- subsurface control (wells) sufficient to delineate major structural and stratigraphic features, fluid contacts, and reservoir limits,

- drillstem, wireline, or cased-hole formation tests on significant zones not initially placed on production,

- full-hole cores in key wells in all major reservoirs,

- sufficient wireline logs[8] to identify all zones potentially productive of oil or gas and to characterize the lithology, net thickness, porosity, saturation, petroleum type, and probable productivity of each zone,

- wireline formation tests to determine fluid gradients, to locate water levels, and to determine degree of intra- and interzone pressure communication,

- sidewall samples to supplement full-hole cores and to help resolve log interpretation problems,

- initial potential tests as each well zone is placed on production that include production rates of oil (or condensate), gas, and water; wellhead pressure; choke size; specific gravity of produced fluids; and salinity of produced water, if any,

- bottomhole transient pressure test on each well zone placed on production, preferably during the initial test period,

- samples of reservoir fluids from all major reservoirs to define composition, saturation pressure, and other physical properties, with sufficient samples to ensure representative data and to identify possible spatial variations in fluid properties,

- depending on the nature of the reservoir rock/fluid system and drive mechanism, special core analysis data, including water/gas, water/oil, or gas/oil relative permeability, capillary pressure data, pore volume compressibility vs. net overburden pressure, etc.,

- monthly production of oil, gas, and water and production method for each well,

- monthly potential tests on each well, including production rates of oil (or condensate), gas, and water; wellhead pressure; choke size; and artificial lift (if any) data,

- bottomhole transient pressure tests, as warranted by apparent changes in individual well performance,

- sufficient historical (static) bottomhole pressure data to determine reservoir drive mechanisms and to identify possible discontinuities between wells completed in the same reservoir,

- historical data on all downhole remedial work, including stimulation; same-zone or new-zone recompletions; and/or installation of or change to artificial lift equipment,

- in multizone fields, through-casing monitoring to detect possible drainage of behind-pipe zones, especially in waterdrive fields,

---

[8] The selection of a cost-effective suite of wireline logs is quite subjective and typically will vary from one operator to the next in the same geologic trend. In wildcat areas, where the petrophysics of the objective section usually are unknown, the selection of a cost effective log suite typically will be by trial and error. The problem is particularly acute in gas-prone sections [Schlumberger Well Surveying Corp. 1991].

- historical operating data for each well, including revisions to downhole pumping and surface processing equipment that have affected production rate of oil, gas, or water,

- current sales prices and sufficient historical operating cost data to facilitate estimating average current and future costs and economic limit,

- costs to drill and complete wells and install treating and processing facilities, and

- operator's plans (if any) to drill or work over wells or to modify or augment production equipment, and historical, current, or anticipated future restrictions on processing or transportation facilities, or on market conditions.

In reservoirs with improved recovery projects, this list should be expanded to include surveillance of monthly-injected volumes and pressures from each injection well. Data requirements for improved recovery projects tend to be method- and project-specific. Talash [1988] discussed data requirements for waterflood projects. The Natl. Petroleum Council [NPC 1984] published an extensive bibliography of papers on improved recovery methods that can provide insight regarding data requirements for such methods.

Operating Environment

The operating environment of an oil and/or gas property is one of the major factors controlling the costs of developing and operating the property. These costs and the market price of the oil, gas, and condensate have a direct impact on the minimum size of a commercially exploitable accumulation and, thus, whether any portion of the accumulation may be classified as "reserves," as defined in Sec. 1.3.

In the North Sea, for example, in 1988 an accumulation containing 30 million bbl of oil in place was considered marginal [Horne *et al.* 1988]. It was a candidate for commercial exploitation at that time because it was near an existing production platform and could be produced from satellite facilities.[9] In contrast, in west Texas (an area with an extensive infrastructure and relatively low operating costs), accumulations less than 1% of that size are commercially exploitable.[10]

Skill, Experience, and Integrity of the Estimators

Even a casual reader of industry journals and related news media cannot help but notice reports of significant reductions in reserves attributed to various firms. Recently, for example, the stock of Hondo Oil (in the U.S.) was delisted from the American Stock Exchange because "proved reserves...in its Annual Report were overstated...(owing to)...steep decline...in the rate of production at its Columbian wells" [*Wall Street Journal* 1998]. Reportedly, Hondo's proved reserves of natural gas had to be reduced from 52.5 to 0.7 MMScf.[11] In the absence of any published data, one can only guess why such a decline had not been anticipated. A reduction in proved reserves of this magnitude, however, certainly calls into question the skills of Hondo's reserves estimators.

---

[9] In 1996, the U.K. Dept. of Trade and Industry estimated the minimum commercial size for offshore—i.e., North Sea—prospects to be 15 million bbl.

[10] In still further contrast, it is noted that, in the mid 1970's, BP discovered a field west of the Shetland Islands that was estimated to contain 3 to 4 billion bbl of heavy oil. It was considered noncommercial because of the remote location, depressed prices, and high costs associated with production of heavy crude.

[11] The reduction in ultimate recovery was less than implied by this statement because there was no account of production during the 7 months between the two estimates.

In 1979, SPE published *Standards Pertaining to the Estimating and Auditing of Oil and Gas Reserves Information*, **Appendix D**, in which minimum standards for education and experience of estimators were recommended. Such standards are no less valid today, if not somewhat understated. Since those standards were promulgated, advances in technology for reservoir characterization and simulation mandate a team approach for reserve estimation and classification for major projects. Such an approach has been adopted by many major companies and consulting firms. All too often, however, such approaches become secondary to time and monetary constraints, and other pressures.

1.2 Definitions of Petroleum Fluids

The following definitions, which are consistent with those of the U.S. Energy Information Admin. [1996], are adopted, in part, from Martinez [1987]. Additional discussion regarding definitions of these fluids is provided by Moses [1990], McCain [1990], and Barrufet [1998].

1.2.1 Petroleum

Petroleum is "a general term to apply to all naturally occurring mixtures of predominantly hydrocarbons"[12] [Martinez 1987]. Petroleum includes natural gas, crude oil, and natural bitumen. (The reserve estimation techniques discussed in this work do not include techniques for estimating reserves of natural bitumen.)

1.2.2 Crude Oil

"Crude oil is the portion of petroleum that exists in the liquid phase in natural underground reservoirs and remains liquid at atmospheric conditions of temperature and pressure" [Martinez 1987]. Crude oil may contain small amounts of nonhydrocarbons produced with the liquids. Crude oil has a viscosity equal to or less than 10,000 cp at initial reservoir temperature and atmospheric pressure, on a gas-free basis. (This is approximately the viscosity of 8° API crude oil.)

Crude oil may be subclassified as: *extra heavy* (stock-tank gravity less than 10° API), *heavy* (stock-tank gravity 10 to 22.3° API), *medium* (stock-tank gravity 22.3 to 31.1° API), and *light* (stock-tank gravity greater than 31.1° API) [Martinez 1987]. The upper limit for "heavy" crude is somewhat arbitrary, and other classifications have been proposed that set a stock-tank gravity of 25° API as the upper limit.

Crude oil also has been classified as *naphthenic, mixed base, or paraffinic*. This classification is based on a characterization factor determined from density and atmospheric boiling-point curves. As discussed in **Appendix E**, the crude oil characterization factor may influence the bubblepoint pressure of the corresponding reservoir oil. This classification scheme, however, is less useful for reservoir engineering than a scheme based on API gravity, which is used in many empirical correlations that may be used to estimate properties of crude oil (Appendix E).

1.2.3 Natural Gas

"Natural gas is the portion of petroleum that exists either in the gaseous phase, or is in solution in crude oil, in natural underground reservoirs, and that is gaseous at atmospheric

---

[12] This definition of "petroleum" conflicts with the Greek origin of this word—i.e., petra (rock) + oleum (oil)—and definitions found in most modern dictionaries, which define petroleum as "an oily, thick, flammable, usually dark colored liquid...that is a mixture of various hydrocarbons."

pressure and temperature" [Martinez *et al.* 1987]. Natural gas may include amounts of nonhydrocarbons.[13]

Natural gas may be subclassified as *associated* or *nonassociated* gas.[14] Associated natural gas is found in contact with, or dissolved in, crude oil in a natural underground reservoir. Nonassociated natural gas is found in a natural underground reservoir that does not contain crude oil. (Many gas reservoirs classified as "nonassociated" contain oil columns that cannot support a statutory[15] oil well completion.)

1.2.4 Solution Gas

Solution gas, also (incorrectly) called casinghead gas,[16] is natural gas that is dissolved in reservoir oil under initial reservoir conditions of pressure and temperature and that is liberated from solution by reduction in pressure and temperature as the oil is produced through surface gas/oil separation equipment.

1.2.5 Lease Condensate

Lease condensate, also called condensate or distillate, is a petroleum liquid consisting mostly of pentanes and heavier that is in the gas (vapor) phase under initial reservoir conditions and that condenses to the liquid phase when the gas is produced through surface separation equipment operating under ambient conditions on a lease.[17] In some operations, lease condensate may be recovered from oil well gas.

1.2.6 Plant Products

Plant products are petroleum liquids and associated products that have been extracted from lease gas at a central plant,[18] generally utilizing absorption and/or refrigeration processes.[19] Plant products typically include ethane, propane, butane, and heavier components referred to as natural gasoline. Plant products may also include sulfur and helium or other valuable materials associated with petroleum.

In general, a complete reserve estimate for a property should account for all the above substances or products.

1.3 Classifications of Reserves

"Then you should say what you mean," the March Hare went on. "I do," Alice hastily replied, "At least —at least I mean what I say—that's the same thing, you know."

Lewis Carroll, *Alice in Wonderland*

---

[13] After natural gas has been processed through surface separation and dehydration equipment and otherwise prepared for sale, it may be called "sales gas" or "residue gas." Care must be taken in estimating reserves of this substance to account for volume losses of natural gas attributed to this processing and to lease usage— e.g., fuel [Spencer 2001].

[14] Natural gas also has been classified as associated (free) natural gas, associated (dissolved) natural gas, and nonassociated (free) natural gas.

[15] Regulatory agencies in the U.S. and Canada, for example, classify producing wells as either "gas wells" or "oil wells," usually based on the well's initial producing GOR. In such jurisdictions, the classification of a well may be considered in establishing well spacing and maximum efficient rate (MER). If the oil column is too thin to support production with a GOR below the statutory limit for an oil well, wells would have to be drilled on gas well spacing and classified accordingly.

[16] Casinghead gas is gas produced from an oil well, which may include both solution gas and gas cap gas.

[17] The definition of condensate has been the source of considerable controversy among OPEC countries. Less restrictive definitions, which make it difficult to distinguish between condensate and light oil, have been proposed.

[18] Interests in plants typically differ from interests in leases (or units) sourcing the plant, which must be accounted for in estimates of net reserves of plant liquids attributable to a specific leasehold interest.

[19] Such products, together with lease condensate, are called "natural gas liquids."

1.3.1 Background

Over the years, there have been numerous classifications of oil and/or gas reserves promulgated by various individuals and organizations. The principle purposes of a classification system are: (1) to promote uniformity and (2) to qualify the degree of risk for each class of reserve.[20] Frequently, such classifications have used the terms *reserves* and *resources* interchangeably and, in some cases, in different contexts, which has caused confusion.

1.3.2 Procedures To Estimate and Classify Reserves[21]

Procedures[22] to estimate and classify reserves (ECR) have been described as *deterministic* or *probabilistic*. Deterministic procedures typically use the "best estimate" of the "true value" of each relevant input parameter in a data set to calculate, for each data set, a single "best estimate" of reserves. Such reserves, or a part of such reserves, may be classified as proved, probable, and/or possible, based on the engineer's judgment and relevant guidelines regarding the probability of actually producing such reserves.

In contrast, probabilistic procedures typically use the full range of potential values of each relevant input parameter in a data set to calculate, for each data set, a set of reserve estimates that reflect the frequency distributions of the basic data. Such reserves may be classified as proved, probable, and/or possible, depending on the cumulative frequency distribution of the set of reserve estimates and relevant guidelines. Procedures for making such estimates are discussed in Chap. 9.

While it has been recognized that there is always some degree of uncertainty in estimating reserves, most of the published literature on reserve estimating has focused on deterministic methods. Probabilistic calculations, however, have received continuing attention in the geologic literature in the evaluation of exploration "plays" [Newendorp 1975, Megill 1979, Capen 1992, and Otis and Schneidermann 1997].

With the advent of the North Sea discoveries in the late 1960's and early 1970's, probabilistic calculations were applied with increasing frequency to estimating and classifying reserves in these fields, which led Keith *et al.* [1986] to propose adoption of a system of reserve estimation and classification based on these methods.

Probabilistic calculations are considered appropriate for many scenarios; however, there are many situations where deterministic calculations may be more appropriate. Accordingly, in 1997 Society of Petroleum Engineers (SPE) and World Petroleum Congresses (WPC) promulgated a set of definitions for reserves that recognized both probabilistic and deterministic procedures.

The methods discussed in Chaps. 2 through 5 cover deterministic procedures because they form the basis for any discussion of probabilistic calculations. Probabilistic procedures are discussed in Chap. 9.

---

[20] It seems apparent that these purposes have yet to be achieved, given the large number of extant definitions (outlined in Secs. 1.3.3 thru 1.3.12) and the intense dialog within the industry regarding deterministic vs. probabilistic procedures for estimating and classifying reserves.

[21] This topic is discussed in more detail in Chap. 9.

[22] The term "procedure" is used in this work to describe deterministic and probabilistic calculations to avoid confusion with the term "method," historically used in phrases like "analogy method" and "volumetric method," etc. Thus, either deterministic or probabilistic procedures may be used with the volumetric method–or for any other method–for ECR.

1.3.3 Lahee's Classification

Among the earliest classifications of reserves were those advocated by Lahee [1945] at the 1944 meeting of the American Petroleum Institute (API) and in 1955 at the World Petroleum Congresses (WPC). As shown in **Table 1.1**, Lahee's 1955 system classified reserves as *proved, probable, possible,* or *hypothetical*. From this table, it may be noted that the term "reserves" was used to include oil and/or gas *yet to be discovered*. (for example, Clause "C.2" in the definitions) As discussed below, current practice generally is to restrict usage of the term reserves to the petroleum in *discovered* accumulations that is commercially recoverable.[23] Depending on circumstances, however, possible reserves might be assigned to untested prospects in a trend of structures proved productive in the objective horizon.

TABLE 1.1—LAHEE'S 1955 CLASSIFICATION OF RESERVES

A. Proved
   1. Drilled
   2. Undrilled
B. Probable
   1. Undeveloped part of pool
   2. Secondary recovery
   3. Behind pipe
C. Possible
   1. In productive areas
      (a) Shallower or deeper pools within field limits
      (b) Pools outside field limits
   2. Areas not now producing but which are geologically similar to producing areas in the region
D. Hypothetical - from non producing regions underlain by sediments that produce in other areas

1.3.4 The "McKelvey Box"

In most western countries, the basis for current terminology for the classification of reserves and resources of *all* minerals, including oil and/or gas, is based on work by McKelvey [1972], a former Director of the U.S. Geological Survey (USGS). As shown by **Table 1.2**, which has been called a "McKelvey box," *resources* (total minerals initially in place) are assigned two attributes: (a) degree of geologic assurance and (b) feasibility of economic—i.e., commercial—recovery.

Where there is a high degree of geologic assurance regarding the resource, it is classified as "identified," or discovered. Where there is a low degree of geologic assurance the resource is classified as "undiscovered."

Reserves are that part of the resource base that has been discovered—they are in *known* accumulations—and for which there is a high feasibility of commercial recovery. The classifications jointly promulgated by SPE and WPC in 1997, which are discussed in Sec. 1.3.5, are identified in bold face type in Table 1.2. The terms "probable" and "possible" imply increasing degrees of geologic and economic uncertainty, as indicated by the position of these two terms in this table.

---

[23] Unfortunately, the term "reserves" frequently is used by the geological community to mean petroleum initially in place. One sees the term "in-place reserves" in the literature of the American Assn. of Petroleum Geologists, for example, which usually is intended to mean petroleum initially in place.

The term "recoverable resources" is comparable to the reservoir engineer's term "ultimate recovery." The term "ultimate recovery" does not imply whether these volumes are considered proved, probable, or possible reserves. Reserves are time-dependent; thus, when reporting reserves, the "as of" date always should be stated. Otherwise, the term initial reserves or ultimate recovery should be used, as appropriate. The term *recoverable* reserves is redundant; the term reserves *implies* recoverable. The term *in place* reserves is ambiguous; the preferred term is in-place *resources.*

TABLE 1.2—USGS CLASSIFICATION OF RESOURCES (THE MCKELVEY BOX) ILLUSTRATING RELATIONSHIP WITH SPE/WPC [1997] CLASSIFICATIONS OF RESERVES

	<----------------------- T O T A L R E S O U R C E S -------------->				
	I D E N T I F I E D (DISCOVERED)			U N D I S C O V E R E D	
	DEMONSTRATED (SAMPLED)			HYPOTHETICAL (KNOWN DISTRICTS)	SPECULATIVE (UNDISCOVERED DISTRICTS)
	MEASURED	INDICATED	INFERRED		
E C O N O M I C	R E S E R V E S PROVED UNPROVED PROBABLE POSSIBLE				
MARGINALLY ECONOMIC	". . . now border on being economically producible." [McKelvey 1972]				
SUB-ECONOMIC	". . . require substantially higher price or major cost reducing advance in technology to be economically producible." [McKelvey 1972]				

INCREASING COMMERCIALITY - - >

<-------------------- I N C R E A S I N G   G E O L O G I C   A S S U R A N C E

Terminology for oil and gas resources that is preferred here is illustrated by **Fig. 1.1,** which is adapted from a comparable figure used by U.S. DOE/EIA (EIA).

1.3.5 SPE-WPC

In 1997, SPE-WPC promulgated definitions for proved, probable, and possible reserves of "petroleum," which are reproduced in Appendix A.[24] The term "petroleum" was introduced in these definitions in lieu of the term "crude oil, condensate, natural gas, natural gas liquids, and associated substances," which was used in SPE 1987 definitions.[25]

Table 1.3 summarizes the possible combinations of reserve classification, development status, and producing category, which are implied by SPE-WPC [1997] definitions. Footnotes to this table summarize applications recommended here.

---

[24] During 1998, Pemex, the Mexican national oil company, adopted these definitions, which resulted in a reduction of their proved oil reserves (as of 1 January 1999) from 45 to 24.7 billion bbl [Petzet 1999].

[25] As provided by the SPE [1987], "Associated substances include sulfur, helium and carbon dioxide, which may be associated with naturally occurring petroleum deposits, and which may be of sufficient commercial value to warrant extraction and sale as separate products." In the author's view, the (tacit) omission of such substances from the 1997 SPE-WPC definitions should not preclude application of these definitions to estimating and classifying reserves of these substances.

The degree of uncertainty in reserve classifications[26] also may be qualified by a combination of terms to describe development status and producing category. For example, the phrase "proved, developed, producing" usually implies less uncertainty in a reserve estimate than the phrase "proved, behind pipe (not producing)."

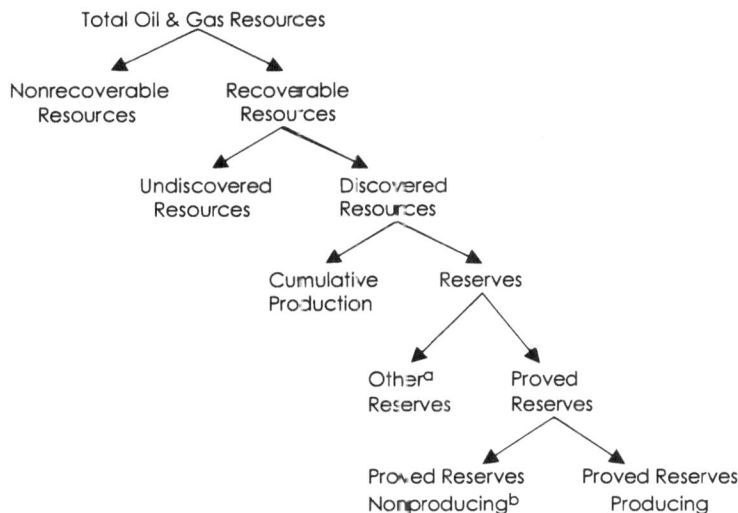

a. In reporting to EIA, "other" reserves include only "indicated additional" reserves that do not meet the criteria necessary to be classified as proved.
b. Nonproducing reserves include those that did not produce during the prior reporting year.

Fig. 1.1—Terminology for oil and gas resources [after EIA, 1996].

The term "producing" identifies a well category to which reserves are attributed. As shown by **Fig. 1.2**, a producing well might be assigned both proved and proved plus probable reserves, depending on the degree of uncertainty in the projection of the observed production trend at the date of the estimate. In this context, it is noted that describing a *reservoir* as "proved to be productive," does not necessarily imply that all the reserves attributed that reservoir are classified as "proved."

McKelvey's [1972] scheme recognizes uncertainty in geologic definition and feasibility of commerciality. This scheme does not provide a convenient method to recognize *engineering* uncertainty—e.g., recovery efficiency. SPE-WPC [1997] and other classification schemes use the term "probable" to imply either, or both, geologic and engineering uncertainty.[27]

Recognizing this problem, the internal classification scheme of some firms recognizes geologically proved resources[28]—oil and/or gas proved to be initially in place—and allocates

---

[26] The discussion here reflects so-called deterministic procedures to classify reserves, which have been widely used for onshore reservoirs. Such reservoirs frequently are developed competitively, and reserves are assigned to individual wells. Thus, development status and producing category become important considerations. Please see Chap. 9 for further discussion of uncertainty in reserve classification.

[27] It is observed here that, depending on circumstances, the degree of geologic uncertainty typically is qualified by subjective judgments regarding (for example) presence of hydrocarbons or reservoir continuity; the degree of engineering uncertainty, in contrast, may be amenable to quantification by statistical analysis. This point is discussed further in Chap. 9.

[28] This terminology might cause confusion because there is no definition of "proved resources" in the public domain.

TABLE 1.3—CLASSIFICATION, RECOVERY METHOD, DEVELOPMENT STATUS AND PRODUCING CATEGORY FOR RESERVES

CLASSIFICATION	RECOVERY METHOD							
	PRIMARY RECOVERY				IMPROVED RECOVERY			
	DEVELOPED			UNDEVELOPED	DEVELOPED			UNDEVELOPED(g)
	PRODUCING	NON-PRODUCING			PROD	NON-PROD		
		SHUT-IN	BEHIND PIPE (d)			SI	BP	
PROVED		(b)	(e)	(j)	(f)	(h)	(i)	
UNPROVED								
Probable	(a)	(c)						
Possible								

a- May be attributed to alternate interpretation of well performance, more efficient reservoir drive mechanism than anticipated for estimate of proved reserves, and/or anticipated installation of pump or compressor, etc.; should not be assigned unless there are also proved reserves assigned to same entity.

b- Generally assigned if well has been flow tested at commercial rates and—at the date of the estimate—is waiting on sales contract, waiting on completion, waiting on facilities, etc.

c- May be assigned in cases when well has been flow tested at marginal rates and is shut-in pending stimulation or installation of pump or compressor and such operations generally have been successful in establishing commercial flow rates in analogous situations.

d- Assign only if there is a reasonable expectation reserve will be produced through existing production casing and within less than about 20 years; otherwise, reserve status should be "undeveloped."

e- Assign only if zone has been flow tested at commercial rates, or it is analogous to other zones that have been flow tested at commercial rates or to other areas of the same reservoir which have produced at commercial rates.

f- Assign only after there has been evidence of response to fluid injection, which may include pressure and/or production response.

g- Generally assigned in cases where injection wells and/or production wells to complete pattern have not been drilled and/or improved recovery facilities have not been installed.

h- May be assigned to previously operational project which has been shut-in for operational reasons or project reassessment.

i- May be assigned if reserves meet other criteria for proved classification and existing injection wells and facilities will be used for project.

j- Assign for operator only if in drilling budget and on drilling schedule; assign for non-operator only if AFE has been approved.

Footnotes reflect the views of the author and do not necessarily represent an official position of either SPE or WPC

*part of this resource to proved reserves and part to probable or possible reserves, as appropriate, considering technical and/or economic uncertainty.

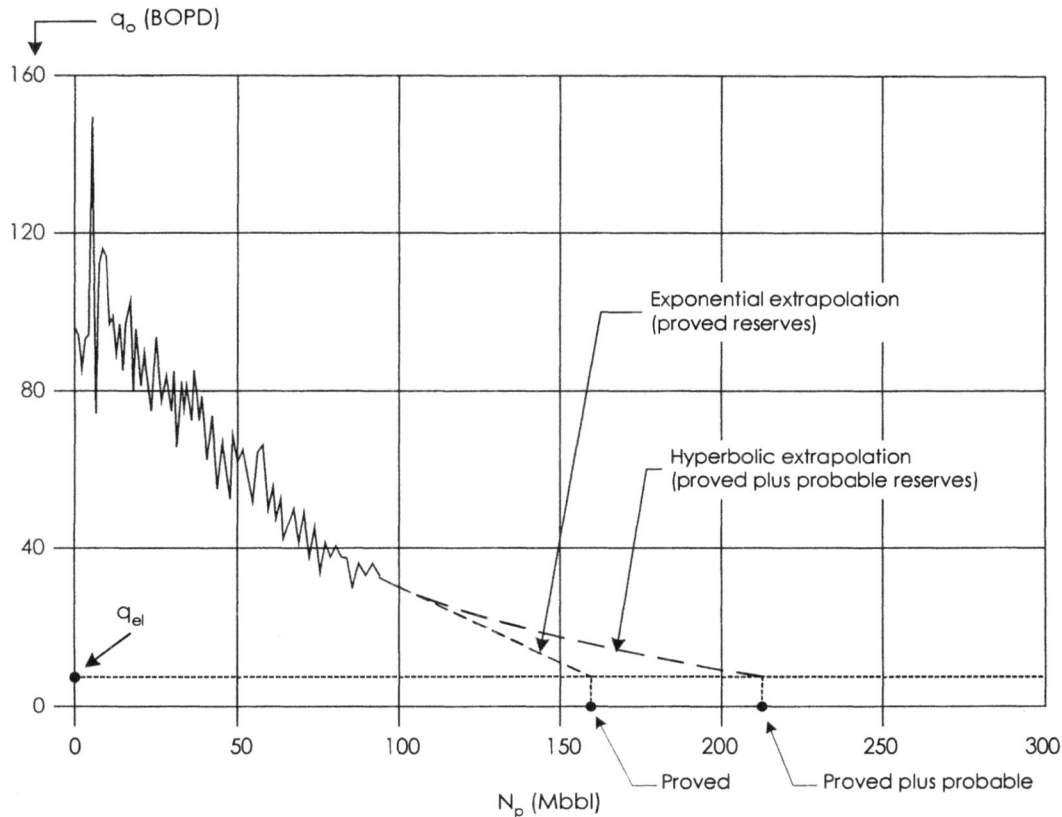

Fig. 1.2—Assignment of proved and proved plus possible reserves to a producing well.

1.3.6 U.S. Securities and Exchange Commission (U.S. SEC)

U.S. SEC also has promulgated definitions for reserves [U.S. Federal Securities Laws 1982]. These definitions, however, are for proved reserves only because U.S. SEC does not require public companies to report other classes of reserves. In Appendix C, key phrases in current U.S. SEC definitions are compared with those in SPE-WPC [1997] definitions for proved reserves. In most cases, the definitions are comparable. However, for reserves attributable to improved recovery methods, current U.S. SEC definitions are more restrictive than those of SPE-WPC [1997]. U.S. SEC definitions require that an improved recovery technique must be proved effective in the "same reservoir." SPE-WPC [1997] definitions, however, require "successful testing" in the reservoir or "one in the immediate area with similar rock and fluid properties."

1.3.7 Canada

Petroleum Soc. of Canadian Inst. of Mining, Metallurgy and Petroleum (PS-CIM)

In 1993 PS-CIM, released a Special Report, "Definitions and Guidelines for Classification of Oil and Gas Reserves" [DeSorcy *et al.* 1993]. PS-CIM recognized "proved," "probable" and "possible" reserves. The 1993 PS-CIM definitions,[29] however, differ from those of SPE-

---

[29] Reportedly, these definitions are not in widespread use in Canada.

WPC [1997] in that they assign probability levels to each of the three classifications,[30] as follows:

- Proved reserves: "generally an 80% or greater probability that at least the estimated quantity will be recovered."

- Probable reserves: "40 to 80% probability that the estimated quantity will be recovered."

- Possible reserves: "10 to 40% probability that the estimated quantity will be recovered."

Despite the reference to probability levels in the 1993 CIM definitions of reserves, as of early 1997, major consulting firms in Canada qualify their estimates of proved reserves with such terms as "reasonable certainty" or "high degree of certainty."

Alberta Energy Resources Conservation Board

Alberta's Energy Resource Conservation Board (ERCB)[31] historically has reported "established reserves," which are defined as: "Those reserves recoverable under current technology and present and anticipated economic conditions, specifically proved by drilling, testing or production; plus that judgment portion of contiguous recoverable reserves that are interpreted from geologic, geophysical or similar information, with reasonable certainty to exist" [ERCB 1987].

The classification "established reserves" historically used by the Alberta ERCB and other agencies "typically comprises proved reserves plus one-half probable reserves" [DeSorcy *et al.* 1993].

Ontario Securities Commission and Canadian Provincial Securities Admin.

The Ontario Securities Commission [OSC] and the Canadian Provincial Securities Admin. have promulgated definitions for proved and for probable additional reserves that are essentially identical.[32] Also promulgated was a set of guidelines regarding the qualifications of the reserve estimator and required supporting documents.

Proved reserves are defined as being "estimated as recoverable under current technology and existing economic conditions, from that portion of a reservoir which can be reasonably evaluated as economically productive on the basis of analysis of drilling, geological, geophysical and engineering data, including the reserves to be obtained by enhanced recovery processes demonstrated to be economic and technically successful in the subject reservoir" [OSC 1992].

Probable additional reserves are differentiated from proved reserves in that the above described data "does not demonstrate (the reserves) to be proved under current technology and existing economic conditions but where such analysis suggests the likelihood of their existence and future recovery" [OSC 1992].

Probability levels for the above two classifications are not specified, but reserve estimators are required to state "the degree of risk assigned to values, particularly of probable additional reserves including a statement...disclosing the approximate amount...by which the volume of probable reserves...have been reduced to allow for the risk associated with obtaining production from probable reserves" [OSC 1992].

---

[30] In June 1999, PS-CIM promulgated industry definitions that provided for 90, 50, and 10% probability for proved, proved plus probable, and proved plus probable plus possible reserves, respectively.

[31] In 1995 the Alberta ERCB was merged with other agencies and became part of the Alberta Energy and Utilities Board.

[32] It is noted here that the term "reasonable certainty" does not appear in either of these definitions.

1.3.8 Australia

Australian Stock Exchange

In May 1995, the Australian Petroleum Production and Exploration Assn. (APPEA) approved mandatory guidelines for companies reporting oil and/or gas reserves to the Australian Stock Exchange. These guidelines were intended to be "read in conjunction with the Definitions for Oil and Gas Reserves...(of)...the Society of Petroleum Engineers (May 1987)." [33] Among other considerations, they provide clarification regarding treatment of fuel usage and flare losses, stating "Reserves should generally be quoted nett (sic) of fuel and flare losses" and "Gas...after processing to sales specification...after removal of inert substances."

APPEA [1995] guidelines specify "probability of exceedance" of 90% and 50% to define proved reserves and proved plus probable reserves, respectively. Probabilities for proved plus probable plus possible reserves were not specified. As stated by APPEA [1995], the inclusion of probability levels "does not imply a preference for either probabilistic or deterministic techniques."

The APPEA [1995] defined "contingent reserves" as those "quantities of hydrocarbon estimated in discovered reservoirs that may be recoverable...but...the requirements of technical and economic recovery were not satisfied." In this context it was noted that "Until technical and economic recovery is established the use of the term 'Reserve' should be avoided."

Western Australia

Operators in the state of Western Australia are requested, where feasible, to follow SPE guidelines in reporting reserves to the Dept. of Minerals and Energy. Reportedly, however, "reserve estimates for all fields (are) based on a Monte Carlo type probability estimate (>90% and >50% probability of recovery)."[34] In fields where economic development is not currently planned or where reserves are not sufficiently well-defined to classify as proved/probable, the estimated quantities are classified as "possible resources." The probability of recovery of such estimates (reportedly) ranges from 10 to 50%.[35]

1.3.9 Former Soviet Union (FSU)

Classifications used by the FSU differ from the foregoing, as they appear to reflect the status of exploration and development of the resource rather than the risk associated with the reserve estimate or the commercial feasibility of exploitation.[36] In the last 10 years or so, there have been numerous publications regarding FSU reserve[37] classifications, each somewhat different from the others. The definitions here have been summarized from several sources [Semenovich *et al.* 1976, Connelly and Krug 1992, Grace *et al.* 1993, Nemchenko *et al.* 1995]:

- Class "A:" reserves attributed to reservoirs (or areas of reservoirs) being produced and which have been fully characterized regarding rock/fluid properties and drive

---

[33] Although not so stated, it is presumed that this recommendation will carry forward to the 1997 SPE-WPC definitions.

[34] Graham Oakley, informal memorandum, 3 Sep 1996.

[35] Op. cit.

[36] Reportedly [Pearson 1997], the need to consider economic factors for outside financing has led to inclusion of such factors on a case by case basis by Russian analysts in estimating reserves.

[37] In some published discussions of the FSU classifications, the term "reserves" apparently refers to in-place volumes; the term "recoverable reserves" refers to "the portion of reserves which can be recovered with economic profits when using a rational approved technology of production and meeting environmental protection requirements" [Nemchenko *et al.* 1995].

mechanism (recovery efficiency); calculations expected to reflect no more than about 10 % uncertainty.

- Class "B:" reserves attributed to reservoirs (or areas of reservoirs) proved commercially productive by flow tests at different levels and by favorable log/core data and for which a development plan has been approved; drive mechanism (recovery efficiency) may be estimated; calculations expected to reflect no more than about 15 % uncertainty.

- Class "C1:" reserves attributed to reservoirs (or areas of reservoirs) where some wells are proved commercially productive and others are characterized by favorable log and core data; also includes areas of reservoirs adjacent to areas in which Class "A" or "B" reserves have been assigned; Class C1 reserves serve as the basis for outstep drilling; calculations expected to reflect no more than about 50% uncertainty.

- Class "C2:" reserves attributed to untested zones or fault blocks adjacent to or above reservoirs with a higher classification.

- Class "C3:" prospective *resources* of oil and/or gas in identified prospects approved for drilling and situated within an oil and/or gas region, and in developed fields in horizons untested by drilling, but proved to be productive in other fields.[38]

- Class "D1:" forecasted *resources* of oil and/or gas in trends or regions where commercial oil or gas fields have been discovered.

- Class "D2:" forecasted *resources* of oil and/or gas in regions or basins where commercial oil and/or gas fields have not been discovered but which appear to be favorable, based on similarities to known productive regions or basins.

In **Table 1.4**, FSU classifications (boldface) are compared to SPE-WPC[1997] definitions.

1.3.10 Peoples Republic of China (PRC)

The National Mineral Reserve Committee of the PRC (NMRC-PRC) identifies reserves under the "State Plan" and reserves not under the "State Plan," the former being reserves that can yield "social economical returns under currently available technical and economical conditions." Reserves not under the "State Plan" can be shifted to the former category if warranted by improvements in economic conditions and/or technical developments.[39]

NMRC-PRC provides guidelines regarding general standards for commercial flow rates for oil wells. These rates range from 0.3 metric ton per day (mtn/D) per well for reservoirs less than 500 m [1,640 ft] deep to 10 mtn/D per well for reservoirs more than 4000 m [13,124 ft] deep. (For 30° API oil, these rates would correspond to 2.2 and 71.7 BOPD/well, respectively.) General standards also are provided for gas wells, with the observation that both such standards are affected by prices and other factors and may be adjusted by NMRC-PRC.

Rather than providing definitions like SPE-WPC [1997], for example, NMRC-PRC relates reserve classifications to the stage of exploration, appraisal, and development (somewhat like the FSU system), as outlined in the paragraphs following:

---

[38] To avoid possible misinterpretation of earlier publications, the reader is alerted that the Semenovich *et al.* [1976] publication did not recognize Class "C3" and included reserves attributed to both tested and untested structures in Class "C2."

[39] This material (believed to be authoritative) was taken from a document titled "Petroleum Reserve Standard," which is an English translation prepared by Dr. Qin Luo of *Standards Drafted by the Office of Professional Committee for Petroleum and Natural Gas of the NMRC.*

Proved reserves:

- may be calculated when appraisal drilling is complete or nearly complete and form the basis for a development plan, a scenario in which it is expected that the reservoir limits, petrophysical properties, fluid content, and reserves can be determined reliably from data gathered during appraisal drilling,
- may also be assigned to a limited area around a discovery well where commercial flow has been established, as defined above,
- may also be assigned to small accumulations where the appraisal well becomes the development well, and
- are classified depending on the extent of appraisal and development drilling and the complexity of the reservoir, as follows:

   Developed proved reserves (Class I) are attributed to reservoirs where development drilling is complete and facilities are installed,

   Under-development proved reserves (Class II) are attributed to reservoirs where appraisal drilling has been completed and form the basis for a development plan; relative error of the calculated reserves should be within 20%.

TABLE 1.4—USGS CLASSIFICATION OF RESOURCES (THE MCKELVEY BOX) COMPARING SPE-WPC [1997] AND FSU CLASSIFICATIONS OF RESERVES AND RESOURCES

		TOTAL RESOURCES			
	IDENTIFIED (DISCOVERED)			UNDISCOVERED	
	DEMONSTRATED (SAMPLED)			HYPOTHETICAL (KNOWN DISTRICTS)	SPECULATIVE (UNDISCOVERED DISTRICTS)
	MEASURED	INDICATED	INFERRED		
ECONOMIC	RESERVES — PROVED UNPROVED — A B C1 C2 C3 — PROBABLE — POSSIBLE			D1	D2
MARGINALLY ECONOMIC	"...now border on being economically producible." [McKelvey 1972]				
SUB-ECONOMIC	"...require substantially higher price or major cost reducing advance in technology to be economically producible." [McKelvey 1972]				

INCREASING COMMERCIALITY ------>

<-------------------- INCREASING GEOLOGIC ASSURANCE

Generally proved reserves (Class III):

- are assigned to reservoirs with any combination of complex faulting, multiple horizons, complex lithology, or natural fractures, with relative error of the calculated reserves within 30%, and

- are the basis for carrying out a "successive exploration and development" program, during which the development wells function as "exploration" wells to acquire additional reservoir information and, after 3 years, may be upgraded to developed reserves.

Probable reserves:

- are calculated during the drilling of a few appraisal wells after commerciality has been established by wildcat drilling, with relative error of the calculated reserves within 50%,

- form the basis for additional appraisal drilling and establishment of a development plan,

- may also be assigned in the case where appraisal drilling has been suspended due to marginal results, and

- should, in general, comprise no more than 30% of proved plus probable reserves in a given area.

Possible reserves:

- are assigned to areas where wildcat drilling has established flow or found shows and are used for planning appraisal drilling, and

- may be assigned to potential reservoirs in geological zones similar to those proved by wildcat drilling or to nonobjective zones with good shows but which have not been tested by appraisal drilling.

In addition to the general guidelines summarized here, the definitions provide detailed specifications regarding the amount and type of seismic, log, core, and laboratory data for each classification. For example, for proved reserves cores should have been taken in "at least one third of the appraisal wells...(with)...core length more than 30% of the total thickness of the oil zone...(with)...core recovery more than 90%."

1.3.11 Norway

Norwegian Petroleum Directorate (NPD)

The NPD classification scheme for resources and reserves[40] is based on the stage of appraisal, development planning, and/or production [Norwegian Petroleum Directorate 1997], as follows:

- Class 1 - *reserves* in production.[41]

- Class 2 - *reserves* with an approved development plan.

- Class 3 - recoverable *resources* in a late planning phase (approval by authorities of a plan of development and operation (PDO—or actual production if PDO approved—within 2 years).

- Class 4 - recoverable *resources* in an early planning phase (PDO approval within 10 years).

- Class 5 - recoverable *resources* that may be developed in the long term.

---

[40] In the NPD system, reserves are defined as *initially* recoverable quantities, which differs from the 1997 SPE-WPC system in which reserves are *remaining* commercially recoverable quantities.

[41] Class 0 is initial reserves where production has ceased and equal actual production to date of cessation.

- Class 6 - recoverable *resources* where development is not very likely; previously designated "small technical discovery."
- Class 7 - *resources* in new discoveries for which the evaluation is not complete.
- Class 8 - *resources* from possible future measures to increase the recovery factor.
- Class 9 - *resources* in prospects; the estimates shall be given a probability of discovery.
- Class 10 - *resources* in leads; the estimates shall be given a probability of discovery.
- Class 11 - unmapped *resources*; estimated by statistical analysis of geologic plays.

Each field may have resources/reserves in one or more different classes. For each classification, "low," "base," and "high" estimates are made of recoverable volumes. Use of stochastic methods is not mandated for such estimates. However, the following guidelines are provided by NPD:

- Low estimate - probabilities shall be reported for the recovery of an amount equal to or greater than the estimate—e.g., P90 or P80.
- Base estimate - probability of recovery not stipulated, except that if stochastic methods are used, the estimate should correspond to the mean.
- High estimate - probabilities shall be reported for the recovery of an amount equal to or greater than the estimate—e.g., P10 or P20.

If a field contains more than one deposit,[42] an estimate is required for each deposit.

Statoil

Statoil, the Norwegian "state oil company," classifies resources as either undiscovered or discovered [Abrahamson *et al.* 1993, Skaar *et al.* 1996]. Reserves are considered part of the discovered resources that are expected to be recovered profitably. Reserves are classified as follows:

- P90 (low estimate): "the resource volume of which there is a 90% certainty that the final outcome will be larger or equal."
- Expectation (mean estimate): "the expected final recoverable resource volume."
- P10 (high estimate): "the resource volume of which there is only 10% certainty that the final outcome will be larger or equal."

The classification scheme avoids the terms "proved," "probable," and "possible," as they are considered "dubious."

The Statoil scheme focuses on "segments," which are defined as "the smallest geological unit of a petroleum accumulation used in the resource classification." A segment is "the part of a prospect that will be tested by one single well, that is, the area to which the result of the well can be extrapolated with a reasonable degree of certainty." Statoil apparently considers that "reasonable certainty" regarding the existence of (discovered) resources should be established on a segment-by-segment basis—i.e., by appraisal drilling of each segment. In Statoil's view, the segment concept "helps planning the number of wells required to prove and delineate a (potential) discovery."

---

[42] NPD's definition of "deposit" appears to be consistent with that for a "reservoir"—i.e., a petroleum accumulation limited by stratigraphic and/or structural boundaries and fluid contacts that is in pressure communication.

1.3.12 United Kingdom (U.K.)

London Stock Exchange (LSE)

Regarding mineral, oil, and natural gas companies, the "Listing Rules" of the LSE recognize two classifications of reserves: (a) proven and (b) probable [U.K. FSA 2000].

For mineral companies primarily involved in the extraction of oil and/or gas resources, proven reserves means "those reserves which on the available evidence and taking into account technical and economic factors have a better than 90% chance of being produced."

For such companies, probable reserves means "those reserves which are not yet 'proven' but which on the available evidence and taking into account technical and economic factors have a better than 50% chance of being produced."

Companies involved only in exploration for oil and/or gas and not in commercial production are not eligible for listing on the LSE. For new listings, LSE rules specify that: (a) proven reserves must be sufficient to sustain commercial production for at least the 2 years subsequent to initiating production and (b) the aggregate value of proven plus probable reserves must not be less than 50% of the expected aggregate market value of the company's equity share capital immediately following admission to listing.

The listing application must include a report by a "competent person,"[43] and must include:

- geological characteristics, including porosity, permeability, net pay, and anticipated recovery mechanism for reserves,

- description of resources (if any) other than proved and probable reserves,

- production schedule, and

- net present value of reserves (with proven and probable reserves analyzed separately) together with assumptions regarding discount rates and economic conditions.

Oil Industry Accounting Committee

In December 1987, the Oil Industry Accounting Committee promulgated mutually exclusive definitions for "commercial reserves" that, at a company's option, may be either:[44] (a) *proven and probable* reserves, which were defined as "estimated quantities...recoverable...from known reservoirs...considered commercially producible...(with)...a 50 percent statistical probability that the actual quantity...recoverable...will be more than the amount estimated...the equivalent statistical probability for the proven component...(is)...90 percent," or (b) *proved developed and undeveloped* reserves, which were defined as "estimated quantities...(expected)...with reasonable certainty to be recoverable...under existing economic and operating conditions."

The difference between these two definitions is significant:

- Under definition (a), "commercially producible" implies the following:
  -management intent of developing and producing,
  -a reasonable assessment of the future economics of such production,

---

[43] The qualifications of a "competent person," definition of the term "aggregate value," and detailed requirements for the "competent persons" report are provided in "The Listing Rules," to which the reader is referred [UK-FSA 2000].

[44] In January 2000, the Oil Industry Accounting Committee promulgated "SORP- Statement of Recommended Practice" in which previous SORP's and guidance notes were combined into a single document. In the words of the committee "No significant new guidance has been introduced in this consolidation process." (The OIAC maintains a website at www.oiac.co.uk.)

-a reasonable expectation that there is a market for all or substantially all the
expected hydrocarbon production, and

-evidence that the necessary production, transmission, and transportation facilities
are available or can be made available.

- Under definition (b), commerciality is defined by "prices and costs as at the date the
estimate is made."

Department of Trade and Industry (DTI)

The DTI recognizes three classifications: (a) discovered recoverable reserves,
(b) undiscovered recoverable reserves,[45] and (c) potential additional reserves. For discovered
recoverable reserves, the following definitions are utilized:

- Proven: "reserves which on the available evidence are virtually certain to be
technically and economically producible—i.e., having a better than a 90% chance
of being produced."

- Probable: "reserves which are not yet proven but which are estimated to have a
better than a 50% chance of being technically and economically producible."

- Possible: "reserves which at present cannot be regarded as 'probable' but which are
estimated to have a significant but less than a 50% chance of being technically and
economically producible."

"Potential additional reserves" are those attributed to discoveries about which there is
insufficient information or which fail to meet the technical and economic criteria for
inclusion as possible reserves.

"Undiscovered reserves" are attributed to mapped, undrilled prospects and are based on
Monte Carlo analyses of volumetric parameters for each prospect maintained in a DTI
database.

1.3.13 Other Classifications

As shown by **Table 1.5**, most classifications of reserves reported at the WPC in 1987[46]
appear to be based on the McKelvey [1972] system and use the terms "proved," "probable,"
and "possible" to identify progressively more uncertainty. As indicated by the footnotes to
Table 1.5, some countries—e.g., Austria, Brazil, and The Netherlands—assign probabilities
to the three classifications. The probability levels, however, differ from one country to the
next. In several countries, different entities apparently use slightly different classification
methods. Since the 1987 WPC, some entities may have developed revised classification
schemes; thus, the information in Table 1.5 should be considered a "snapshot" in time.

1.3.14 Use of Geophysical Data

From a review of the definitions for reserves here and in Appendices B and C, it is apparent
that for reservoir volumes to be classified as "proved," they must be in reasonable proximity
to potentially productive wells in the same reservoir. The utilization of only geophysical
data to define reservoir limits and classify reserves at locations in areas not delineated by
drilling is controversial. This historical heavy reliance on subsurface (well) data to the
exclusion of geophysical data in such areas may not be consistent with continuing advances
in geophysical methods of reservoir definition. Recognizing this problem, Schlagenhauf and

---

[45] DTI usage of the term "reserves" is not consistent with usage elsewhere regarding undiscovered resources.
[46] A comparable table has not appeared in subsequent WPC proceedings.

TABLE 1.5—COMPARISON OF RESERVES CLASSIFICATION SYSTEMS*

Country or Organization	DISCOVERED			UNDISCOVERED
		UNPROVED		
	PROVED	PROBABLE	POSSIBLE	POTENTIAL RECOVERY
Australia** A (1a)	Proved and probable		Possible	
" B (1b)	P1***	P2*** (Incl P1)	P3*** (Incl P1 + P2)	
" C (1c)	Identified & economic	Demonstrated	Inferred	Hypothetical + Speculative
Austria (2)	Proved	Probable	Possible	Potential
Brazil (3)	Proved	Probable	Possible	Potential
Canada** (4)	Proved (Established)	Probable (Best current)	Possible (Upside)	Potential resources (Future discoveries)
Denmark A (5a)	Proved	Unproved		Potential
" B (5b)	Proved	Probable	Possible	
Ecuador A (6a)	Proved	Probable	Possible	
" B (6b)	Proved	Probable	Possible, geologically and geophysically	
France (7)	Proved	Probable	Possible	Hypothetical
Germany	Proved	Probable	Technically & geologically indicated	Speculative
India (8)	Proved (A + B)	Probable (C1)	(C2)	(D1 + D2)
Iran (9)	Proved	Probable	Possible + Speculative	
Malaysia	Proved	Probable	Possible	
Mexico	Proved	Probable		Potential
Netherlands (10)	Proved	Unproved		Potential
P.R. China** (11)	Proved	Probable	Possible	Resources
Spain	Proved	Probable	Possible	
U.K.** (12)	Proved	Probable	Possible	
U.N. 1977 (13)	Proved	Incremental less certain	Incremental least certain	Undiscovered
U.N. 1979 (14)	r - 1 - E	r - 1 - S + r - 1 - M	r - 2	r - 3
U.S.A. (15)	Proved or Measured	Probable or Indicated	Possible or Inferred	Hypothetical + Speculative
F.S.U.** (16)	A + B + part C1	Part C1	Part C1 + part C2	Part C2 + D1+ D2
Venezuela (17)	Proved	Semi-proved	Not proved	
W. E. C. (18)	Proved	Additional in proven oil provinces		Additional in other areas

*   After World Petroleum Congresses (1987)

**   Please see Sec. 1.3.7, 1.3.8, 1.3.9, 1.3.10, and 1.3.12 for additional discussion regarding these countries.

*** Preferred notation is 1P, 2P, and 3P, respectively. (Please see **Notation**)

TABLE 1.5—CONTINUED

(1a)   Bureau of Mineral Resources

(1b)   Australian Minerals and Energy Council system indicates categories as: P1 (proved), known with 93 % certainty; P2 (proved + probable), 60 % probability; P3 (proved + probable + possible), 5 % probability

(1c)   National Energy Advisory Committee adopted the 'McKelvey' system, but uses the word 'resources' and subdivides into recoverable, economic, sub-economic, (para-marginal and sub-marginal), identified (demonstrated and inferred), and undiscovered (hypothetical and speculative)

(2)   Proved is defined as having more than 90 % certainty, probable as 50-90 % certainty, and possible as less than 50 % certainty

(3)   Proved reserves are those delineated by producing wells and geological barriers. Probable and possible are estimated by a statistical approach.

(4)   Established, best current estimate and upside are used by some and correlate, respectively, to 90 %, 50 %, and 10 % probability.

(5a)   Producing companies preference

(5b)   Government preference

(6a)   Ministry for Natural Resources and Energy treats primary reserves as probable and possible and, separately, reserves from secondary recovery.

(6b)   The State Oil Company (CEPE) considers the proved and probable as being the total reserves, but allows possible reserves as those indicated by only geological and geophysical methods.

(7)   Comite des Techniciens, Chambre Syndicate de la Recherche et de la Production du Petrole et du Gas Naturel

(8)   Reserves are only calculated for categories A, B, and C1. For oil in place there is an additional category, C2, and a category called jabalance, for discovered, non-economic oil. For prognosticated reserves there are two categories, D1 and D2. D1 reserves are estimated for rock units with proved oil and gas potential within the limits of the basin. D2 reserves are calculated in rock units with proved oil and gas potential in areas with similar geologic framework. D2 reserves also are calculated for other rock units not previously considered in a proved petroliferous basin

(9)   Iran proposes that a probabilistic approach should define the classes given.

(10)   The system is based on a probabilistic approach whereby the total reserves correspond with the expectation of reserves using a Monte Carlo technique. Of the total reserves the amount that has a 90 % chance of being exceeded is called proved and the remainder is called unproved.

(11)   Proved reserves are made up of developed, undeveloped, and 'basically proved' reserves. 'Basically proved' reserves are those in geologically complex reservoirs, which will take considerable time to drill and prove up.

(12)   Proved reserves are virtually certain; probable reserves have above 50 % certainty; possible reserves, less than 50 % certainty.

(13)   UN Expert Group Report in *Natural Resources Forum* Vol. 1, No. 4, July 1977

(14)   *The International Classification of Mineral Resources*, March 1979

(15)   USGS Circular 831

(16)   Semenovich, et al. (1976)

(17)   Ministry of Energy and Mines, Regulations of October 1966

(18)   World Energy Conference (1980). Details unavailable.

Jaynes [1995] proposed revising the 1987 SPE definitions[47] to allow use of 3D seismic to delineate both proved and probable reserves in areas beyond direct offsets. Their paper includes guidelines to ensure "geophysical due diligence."

Utilization of borehole-calibrated 3D seismic data for reservoir definition is practiced widely. Thus, reserve estimators may have to exercise considerable judgment in classifying reserves in newly developing fields where modern geophysical methods have been incorporated in the subsurface interpretation. Additional comments regarding the application of geophysical data to reserve estimation have been provided by Robertson [2001].

1.3.15 Mining Industry Definitions

Given the apparently common origins for definitions of reserves for both the petroleum and mining industries, this section is provided to illustrate current views by the latter and to identify differences between the two industries.

Council for Mining and Metallurgical Institutions

Early in 1996, the Council for Mining and Metallurgical Institutions (CMMI) proposed a standardized classification of resources and reserves for international use [Riddler 1996]. The proposed scheme seems to be based, in large part, on that proposed by McKelvey [1972], which was discussed in Sec. 1.3.4.

The CMMI proposal classified mineral *resources* as "measured," "indicated," or "inferred," with definitions for these terms similar to those proposed by McKelvey [1972]:

- *Measured mineral resources* were classified as "that part of a mineral resource which has been explored, sampled and tested through appropriate exploration techniques at locations such as outcrops, trenches, pits, workings and drill holes which are spaced closely enough to confirm geological continuity and from which collection of detailed reliable data allows tonnage/volume, densities, size, shape, physical characteristics, quality and mineral content to be estimated with a high level of certainty. This category requires a high level of confidence in and understanding of the geology and controls of the occurrence." (Author's underline)

- *Indicated mineral resources* were differentiated from measured mineral resources by "locations...too widely spaced or inappropriately spaced to confirm geologic continuity but which are spaced closely enough to assume geological continuity and from which collection of reliable data allows...(estimates) with a reasonable level of confidence, but not a high degree of certainty."

- *Inferred mineral resources* are based on data similar to that required for measured and indicated mineral resources, but which are "limited or of uncertain quality and reliability...and mineral content can be estimated with a low degree of certainty and low level of confidence."

Mineral *reserves* were classified as "part of a measured or indicated mineral resource...on which appropriate assessments have been carried out to demonstrate at the time of reporting...that could justify exploitation under realistically assumed technical and economic conditions."

---

[47] The 1997 SPE-WPC definitions do not directly address the utilization of geophysical data in areas not delineated by drilling, and comments by cited authors might be considered in applying these definitions.

CMMI also proposed that "The term mineral reserve need not necessarily signify that exploitation facilities are in place or operative nor that all government approvals have been received provided that there are reasonable expectations of such approval."

The CMMI proposal defined *proved* mineral reserves as "part of a measured mineral resource on which detailed technical and economic studies have been carried out to demonstrate, at the time of reporting, that it can justify exploitation under specific technical and economic conditions."

Probable mineral reserves were defined as "part of a measured or indicated mineral resource on which sufficient technical and economic studies have been carried out to demonstrate, at the time of reporting, that it can justify exploitation under appropriate technical and economic conditions."

The CMMI proposal did not recognize possible mineral resources or possible reserves.

Mining Industry Definitions Compared to Those for the Petroleum Industry

Regarding the CMMI proposals, the following observations are made:
(a) The phrase "specific technical and economic conditions" used in the definition of proved mineral reserves seems less restrictive that the term "current economic conditions" used in the 1997 SPE-WPC definitions of proved reserves.
(b) The apparent lack of any provisions to describe the status of development or production seems less restrictive than the usual practice in the oil and gas industry.
(c) The use of probabilistic calculations to determine levels of confidence is not mentioned.

Regarding the two industries, it is noted that the degree of uncertainty in the evaluation of all subsurface mineral deposits, both solid and fluid, is strongly influenced by the geologic setting of such deposits. In establishing degrees of uncertainty, mining industry definitions typically distinguish between "sample" spacing, which is either: (a) close enough to confirm geologic continuity—the scenario for measured (proved) resources—or (b) too wide or inappropriate, such that geologic continuity may only be assumed—the scenario for indicated (probable) resources.

Qualifications regarding sample (well) spacing and geologic continuity are not stated succinctly in petroleum industry definitions. They must be inferred from interpretation of terms like offset and outstep location in the context of geologic setting and drive mechanism.

1.3.16 What Are "Reserves?"

Despite the publication of thousands of words on the subject, a globally acceptable answer to this question is not readily forthcoming. One of the major issues is interpretation of the phrase "commercially recoverable," which is part of most definitions of reserves. The multiyear time frame of exploration and production mandates a long-term view of "commerciality." Over the years, however, short-term downturns in wellhead prices have led to production operations with negative cash flow. [48]

Under current U.S. SEC regulations, such production, *at the date of the estimate,* would not be considered "reserves." For various reasons, however, major companies, typically with other sources of income, may continue operations of such marginal operations, despite short-

---

[48] Under such a scenario, the validity of the operator's ease may be questioned. The habendum clause in many leases provides for continuance "so long as oil or gas shall be produced in paying quantities." Legal scholars, however, comment that such clauses are interpreted in the light of "normal conditions and not of depressed nor unusual economic conditions that might of themselves make a lease unprofitable." Reportedly, "any production, whether or not in paying quantities, is ...sufficient under the Louisiana prescription statute" [Sullivan 1955].

term losses. Shut down of steamdrive operations, for example, may prove to be more costly in the long term. Also, the production may be needed to meet refinery demand.

As a practical matter, the definition of reserves is dependent on the purpose of the estimate and/or the guidelines of the agency for which the estimate is made. For example, reserve estimates prepared under strict interpretation of current U.S. SEC guidelines must exclude volumes when current economic conditions result in production at a loss. If however, the operator can demonstrate efforts to reduce costs to eliminate such losses, exceptions might be made. Reserve estimates prepared under guidelines[49] that allow "reasonably projected physical and economic conditions" might, depending on circumstances, include such production. In general, so long as production is being sold in a bona fide market, it seems the production should be considered reserves.

1.4 Reserve Estimates—How Accurate?

As noted by Arps [1956], the methods used to estimate reserves and the accuracy of the result depend on the type, amount, and quality of geologic and engineering data available, all of which generally will depend on the maturity of development and production of the property.[50] **Fig. 1.3** illustrates schematically the relationships between:

- the maturity of a property,
- the general methods used to estimate reserves for the property, and
- the range of uncertainty in the estimate of ultimate recovery.

Historically, Fig. 1.3 has been interpreted to infer that the estimate of *ultimate recovery* (initial reserves) for a property should become more accurate with maturity. While this seems intuitively obvious and might be true for single wells or single reservoirs, it might not generally be true for complex, multireservoir fields. Also, it has been observed that estimates of (remaining) *reserves* may not become more accurate with maturity. These points are discussed briefly in the following paragraphs.

Regarding estimates of (remaining) reserves, Hefner and Thompson [1996] presented the results of a study in which 12 industry experts were asked individually to estimate reserves for a group of five producing wells at four different stages of maturity. The purpose of the study was to compare deterministic and probabilistic estimates.

With the permission of the principal author, Cronquist [1996] examined another aspect of the data [Hefner 1992] developed during the study. It was observed that, for each of the wells examined in that study, estimates of (remaining) reserves did not converge on actual (remaining) reserves with maturity. It was also observed that, considering the estimates in the aggregate, there was no statistically valid relation between the accuracy of the estimates and the maturity of the wells.

Regarding such estimates on a global scale, a major international company provided data for each of 30 fields showing the degree of uncertainty in their estimates of remaining reserves and the percent of initial reserves that had been produced from each of these fields.[51]

---

[49] California State Board of Equalization, Rule 468.

[50] As used here, the term "property" may refer to a single well, an aggregate of wells producing from a single reservoir, or a complex field with numerous reservoirs.

[51] David Morgan, personal communication, 1995.

The degree of uncertainty in the company's estimate of remaining reserves for each field was expressed as "kappa," defined as:

KAPPA = (LOG NORMAL STANDARD DEVIATION)/(REMAINING MEAN RESERVES)

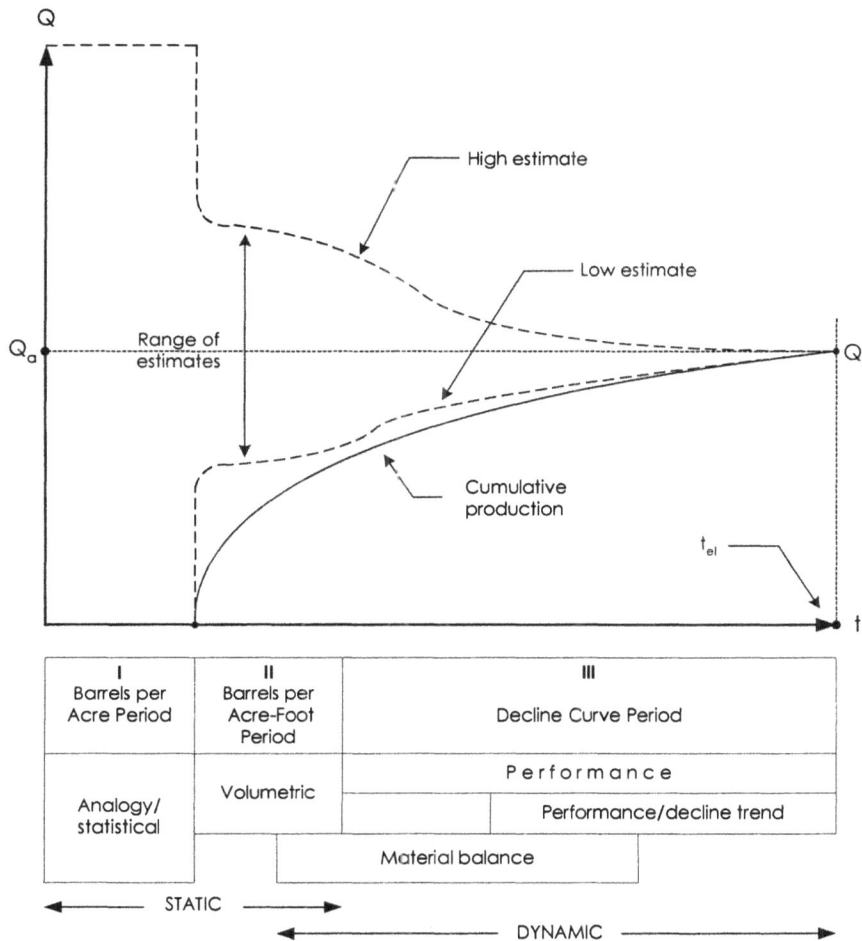

Fig. 1.3—Periods in estimating ultimate recovery, Q, and reserves, illustrating how reserves estimating methodology and range of recovery estimates depend on maturity of reservoir [after Arps 1956].

Another measure of the degree of uncertainty—developed here from other provided data—is the ratio of the company's high estimate (P10) to the low estimate (P90),[52] identified here as "P-Rat". On **Fig 1.4,** both "kappa" and "P-Rat" are plotted vs. percent initial reserves produced at the time of the estimate. It seems apparent that, for this group of fields, the degree of uncertainty did not change with maturity.[53] In fact, the trend lines on Fig 1.4 suggest the degree of uncertainty may have increased with maturity!

---

[52] Such estimation procedures are discussed in Chap. 9.

[53] This analysis is a snapshot in time for a group of fields and does not address the question of how the degree of uncertainty changes with maturity for a given field.

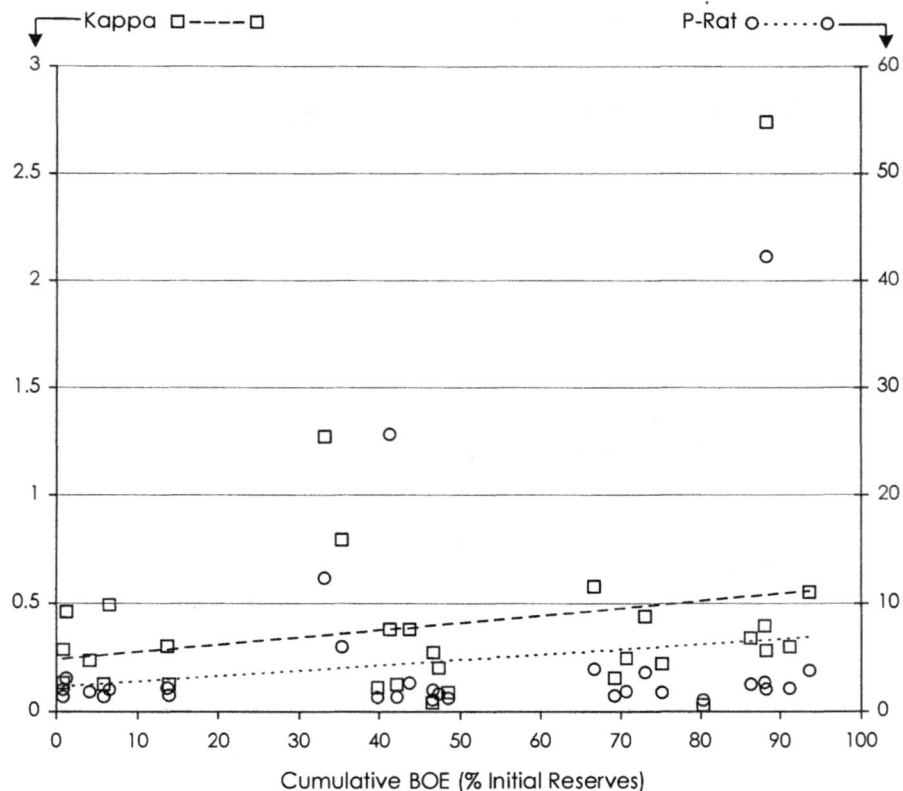

Fig. 1.4—Kappa and P-Rat vs. fraction of initial reserves produced [after Morgan 1995].

Regarding estimates of ultimate recovery (initial reserves), Spears and Dromgoole [1992], in a study of 25 fields in the North Sea, observed significant revisions within 4 years after the fields were sanctioned for development. Such revisions, made after a period of delineation drilling, are contrary to the scenario implied by Fig. 1.3. In general, the overestimates were attributed to the geologic setting being more complex than mapped after the delineation phase.

1.5 Methods To Estimate Reserves

As shown by Fig. 1.3, methods to estimate reserves may be categorized [Arps 1956] as:

I.   *analogy/statistical*— the "Barrels per Acre Period," or Period I,

II. *volumetric* —the "Barrels per Acre Foot Period," or Period II, and

III. *performance*—the "Decline Curve Period," or Period III.

The first two methods, analogy/statistical and volumetric, have been called "static" because they are based on geologic and engineering data before sustained production begins.

The third method, performance, has been called "dynamic" because it is based on historical observations of pressure and produced volumes after sustained production begins.

Static methods do not involve analysis of historical rate/time trends, although such trends might be developed from the reserve estimate by the engineer to compute discounted future net income. In this sense, such methods are considered here to be "volumetrics driven." In contrast, dynamic methods, especially performance/decline trend methods, tend to involve analysis of observed rate/time trends for the estimation of (remaining) reserves based on calculation of the economic limit. In this sense, such methods are considered here to be "economics driven."

Each of the methods, described briefly below, is discussed in detail in Sec. 2 through 5.

1.5.1 Analogy/Statistical Methods

Analogy/statistical methods typically are used to estimate ultimate recovery for *undrilled prospects* and to supplement volumetric methods of estimating reserves in the early stages of development and production. In some areas, only analogy/statistical methods are feasible until clearly defined performance trends are observed. In addition, such methods may be used to estimate reserves for *undrilled tracts* in partially developed reservoirs, or in (proved) productive trends.

The methodology is based on the assumption that the analogous reservoir or well is comparable to the subject reservoir or well, regarding those aspects that control ultimate recovery of oil and/or gas. The weakness of the method is that the validity of this assumption cannot be determined until subject reservoir or well has been on production long enough to estimate reserves based on volumetrics and/or performance.

1.5.2 Volumetric Methods

Volumetric methods are used when subsurface data from wells and appropriately calibrated seismic data are sufficient for structural and isopachous mapping of the objective field or reservoir. Objectives of this mapping include: (a) estimation of oil and/or gas initially in place and (b) identification of additional areas for development. The fraction of oil and/or gas initially in place that is commercially recoverable may be estimated using a combination of analogy and analytical methods.[54] Analytical methods to estimate recovery efficiency are discussed in Sec. 3.7.

In a relatively uncomplicated geologic setting, it may be possible to make a reasonably accurate volumetric estimate of oil and/or gas initially in place with relatively sparse subsurface control. In contrast, in a complex geologic setting—i.e., one characterized by extensive faulting and/or complex stratigraphy—it may not be possible to make accurate volumetric estimates until the field is almost completely developed, *if ever*.

1.5.3 Performance Methods

Performance methods are used after a field, reservoir, or well has been on sustained production long enough to develop a trend of pressure and/or production data that can be analyzed, usually mathematically, to estimate oil and/or gas initially in place and/or future production. The analysis may involve: (a) material balance calculations, (b) computer simulation, and/or (c) "fitting" historical trends of production rates, water/oil ratio, gas/oil ratio, water/gas ratio, condensate/gas ratio, and/or pressure.

Material Balance

The term "material balance" generally refers to a computational procedure in which the fluid properties and pressure-performance history of a *reservoir* are averaged, thereby treating the reservoir as a "tank." Material balance equations are formulated to calculate volumes of oil

---

[54] For major prospects, computer simulation may be used at this stage to estimate reserves and future production rates. Simulation at this stage is considered an extension of the volumetric method, rather than a "performance method," as discussed in the next section.

and/or gas initially in place. These methods may be used when there are sufficient historical reservoir pressure and production data to perform reliable calculations.[55]

For reliable material balance calculations, the reservoir should have reached semisteady-state conditions—i.e., pressure transients should have affected the entire hydrocarbon fluid system. Depending on reservoir fluid and rock properties and on reservoir drive mechanism, this could require cumulative production of as much as 5 to 10% of the oil and/or gas initially in place.

Reliable application of this method requires accurate historical production data for *all* fluids (oil, gas, and water), accurate historical (static) bottomhole pressure data, and PVT data representative of *initial* reservoir conditions. If computer simulation is being considered, historical production and bottomhole pressure data may be required for *each well* in the reservoir, depending on the degree of reservoir complexity and the purpose of the study.

For gas reservoirs producing by pressure depletion, a graphical form of the material balance equation may be used to estimate gas initially in place *and* reserves. For oil reservoirs, the material balance equation may be used to estimate oil and/or gas initially in place and probable drive mechanism. To estimate *reserves* from oil reservoirs, however, the material balance equation must be used in a form that includes the appropriate relative permeability data in a predictive mode adapted for computer simulation. Otherwise, recovery efficiencies may be estimated by analogy.

Computer Simulation

Computer simulation may be considered a form of material balance analysis in which the spatial distribution of rock, fluid, and rock/fluid properties in the reservoir are represented in a computer model by a grid system, or a set of interconnected "tanks." The computer model may be used to calculate oil and/or gas initially in place, to match observed pressure-performance history, and to forecast future production rates. The grid system may be configured to simulate the well or reservoir (or sections of the reservoir) under study. For some applications, the reservoir (or well) model may be coupled with models to simulate production facilities.

In high-permeability reservoirs, where reservoir pressure does not exhibit large areal variations, "zero dimensional" or "tank" model material balance calculations usually will be acceptable. In low-permeability reservoirs, where reservoir pressure may exhibit large areal variations, or in geologically complex reservoirs, it may be necessary to use one or more multidimensional reservoir simulation models.

Performance/Decline Trend Analysis

Performance/decline trend analysis generally is used in the mature stage of production from a reservoir, after reasonably well-defined historical trends have been observed in the performance of the wells completed in the reservoir. The analysis procedure may be considered to have two aspects: (a) analysis of trends of "performance indicators" and (b) analysis of declining trends in production rate.

The term "decline trend analysis" refers to analysis of declining trends of the production rates of oil or gas—the principal products of oil or gas wells, respectively—vs. time or vs. cumulative production to estimate reserves. Depending on the reservoir drive mechanism and

---

[55] Although historically considered a "method" to estimate reserves, material balance calculations are used to estimate oil and/or gas initially in place and, except for volumetric gas reservoirs, must be supplemented by other procedures to estimate recovery efficiency and reserves.

operating practices, however, it may be possible to estimate reserves of individual wells before there is a decline in the production rate of the principal product. This procedure is called here "performance trend analysis."

For example, in many bottomwater-drive oil reservoirs, especially those with "thin" oil columns, wells typically begin producing water early in life. Depending on pump capacity, oil production from these wells may be maintained at more or less constant rates for a substantial part of productive life. Frequently, a semilog plot of the fraction of oil in the total liquid production vs. cumulative oil for individual wells will be linear. Extrapolation of the fraction of oil in total liquid to an economic limit may be used to estimate reserves for these wells. Other examples are discussed in Sec. 5.2.

The trend analysis procedure is based on the assumption that those factors that control the fitted trend will continue in the future. The significance of this assumption is discussed in Sec. 5.1.

1.5.4 Combination Methods

Time and data constraints permitting, more than one method should be used to estimate reserves. Typically, in the early stages of development and production, reserves will be estimated using a combination of analogy and volumetric methods. In some areas, it may be feasible to utilize seismic data to help determine reservoir or field size before there are sufficient well data to prepare more reliable geologic maps. As development continues and the early wells begin to develop pressure/production trends, reserves for those wells may be estimated using performance methods or performance/decline trend analysis. Reserves for undrilled tracts in a developing area may be estimated by analogy with older wells in the same or similar reservoirs in the field.

1.5.5 Reconciliation Between Methods

As noted previously, methods used to estimate reserves before initiation of sustained production are considered *static* methods. Typically, such methods yield results significantly different from the *dynamic* methods used after initiation of sustained production.[56] Depending on when they are used, analogy, statistical, volumetric, and computer simulation methods might be considered static methods. Material balance, computer simulation (with history match) and performance methods typically are considered dynamic methods.

Field case histories have attested repeatedly that well and reservoir performance typically differ significantly from that estimated prior to initiating sustained production [Buchanan and Hoogteyling 1979, Cronquist 1984, Hazeu *et al.* 1988, Markum *et al.* 1978, and Van Rijswijk *et al.* 1981].

In the North Sea Thistle field, for example, Nadir and Hay [1978] reported that the first seven development wells encountered reservoir conditions in excellent agreement with conditions predicted by the operator's predevelopment reservoir model. This model had been developed from seismic mapping, geologic modeling, and five widely spaced appraisal wells. After initiating sustained production, however, wellhead pressures of wells in the crestal area of the field declined faster than had been anticipated, and "it became obvious that...production rate...would be significantly below expectation" [Nadir 1980].[57]

---

[56] As noted by Dake [1994], the time delay between discovery and initiation of sustained production tends to be longer for fields where there is no production infrastructure in place than for fields which can readily be placed on production shortly after discovery. The former category would include remote or far offshore locations, while the latter would include near offshore or onshore locations that can readily be tied into an existing production infrastructure.

[57] Shortly before initiating sustained production, the operator increased estimated ultimate recovery from this field from

Continuous monitoring of well and reservoir performance throughout reservoir life and reconciliation of differences between static and dynamic reserve estimates is an *essential element of good reservoir management.* For additional discussion of this topic, please see Secs. 5.1.5 and 8.5.

---

approximately 50 million tonnes to almost 80 million tonnes. In subsequent years, however, the estimate was gradually reduced to approximately 55 million tonnes [Corrigan 1993]. 1 tonne of crude oil equals approximately 7.5 bbl.

CHAPTER 2

ANALOGY METHODS

2.1 Background

As noted in Sec. 1.4.1, analogy methods generally are used to estimate ultimate recovery—or unit recovery factors[1] —of oil and gas: (a) for a prospect before it is drilled and (b) to supplement volumetric/analytical methods during the early stages of development and production of such prospects. Analogy methods also may be used to estimate reserves for wells planned for outstep or for infill locations in proven areas.

In some scenarios, analogy may be the *only* feasible method until there are sufficient pressure and/or production data for a reliable analysis of performance. Such scenarios include areas of widely spaced development, where subsurface information may be too sparse to facilitate reliable volumetric mapping, and reservoirs where log, core, and/or test data are insufficient for reliable characterization.

Two broad categories of analogy methods are discussed in Sec. 2.2 and 2.3, analytical and statistical, respectively.

Irrespective of the method used, however, analogy and subject[2] *reservoirs* should be similar regarding:

- structural configuration,
- lithology and depositional environment of the reservoir rock,
- nature and degree of principal heterogeneity,
- average net thickness and ratio of net-to-gross pay,
- petrophysics of the rock/fluid system,
- initial pressure and temperature,
- reservoir fluid properties and drive mechanism,
- spatial relationship between free gas, oil, and aquifer at initial conditions—i.e., "stacked" or "en echelon"[3]—and well spacing.

---

[1] Such factors might include: (a) barrels of oil (or standard cubic feet of gas) per acre, per foot of pay, or per acre-foot; (b) percent recovery of oil (or gas); or (c) basic rock/fluid parameters like porosity, water saturation, formation volume factor, etc.

[2] In this context, the terms "analogy" and "subject" are synonyms for the terms "prototype" and "model," respectively, which are used interchangeably in this work.

[3] The term "stacked" is defined here as oil *completely* overlain by free gas and *completely* underlain by water. In this case, potential reservoir drive mechanisms are bottomwater drive and/or gas-cap expansion, accompanied by gas and/or water coning. In this situation, the ratio of vertical to horizontal permeability would have a significant influence on well performance and recovery efficiency. The term "en echelon" is defined here as oil *partially* overlain by free gas and *partially* underlain by water. In this case, potential reservoir drive mechanisms are edgewater drive and/or gas-cap expansion, accompanied by gas over-running and, possibly, by water under-running. In this situation, the ratio of vertical to horizontal permeability would tend to be less significant.

Seldom, if ever, will all these requirements be met, and adjustments usually must be made to compensate for the differences. Judicious adjustments from prototype to model require considerable local knowledge and reservoir engineering experience.

2.2 Analytical

Analytical analogy methods include using recovery *factors*—e.g., stock-tank barrels of oil per acre foot of reservoir—or recovery *efficiencies*—i.e., percent recovery—observed in analogous reservoirs to estimate oil and/or gas recovery from reservoirs being studied.

In addition, or as modifiers, to these aggregate factors, basic rock and/or fluid parameters—e.g., porosity, water saturation, and formation volume factor—may be used. For example, the recovery factor, F_R, observed in a prototype might be adjusted as follows:

$$[F_R]_M = [F_R]_P [\phi S_h]_M / [\phi S_h]_P ,$$

where the subscripts "M" and "P" denote model and prototype, respectively, S_h is initial hydrocarbon saturation, and ϕ is porosity.[4] In many areas, estimates of rock properties (porosity, initial water saturation, and net pay) from wireline logs are subject to considerable uncertainty. In the absence of core data or definitive formation tests, analogy may be the only method initially available to estimate reserves.

Such analogies may be drawn from mature reservoirs in comparable geologic and engineering settings. Several examples are provided for mature areas in the U.S.

2.2.1 U.S. Regional Recovery Efficiencies

From data published by the American Petroleum Institute [API 1980],[5] *aggregate* recovery efficiency[6] was estimated here for groups of oil reservoirs in various regions in the U.S., **Table 2.1**. (Comparable data on natural gas reservoirs in the U.S. have not been published.) From this table, the following observations are offered:

- Comparing north and south Louisiana, for example, note the influence of lithology and drive mechanism on recovery efficiency. The estimated average recovery efficiency for North Louisiana carbonate reservoirs is about what might be expected for these reservoirs from theoretical calculations of solution-gas-drive recovery efficiency for "average" carbonates discussed in Sec. 3.7.2.

- Comparing Oklahoma with Pennsylvania, for example, note the significant difference in waterflood recovery efficiency, which probably results from the significantly poorer quality rock in the Pennsylvania waterfloods compared to that in Oklahoma.

- Comparing south Louisiana with the Texas Gulf Coast, for example, note there is reasonable agreement in waterdrive recovery efficiencies. The small difference in

---

[4] This simple adjustment tacitly assumes that recovery factor is proportional to the ϕS_h product. Depending on circumstances, a more complex factor might be more appropriate. For example, consideration might be given to using the parametric groups introduced in Eqs. 3.16 and 3.18.

[5] Given the date of this reference, readers may question the validity of these estimates. The API [1980] data generally were compiled from mature fields, and actual recovery efficiencies are anticipated to be reasonably close to those estimated.

[6] The reader is cautioned that these are averages of many fields in the region indicated. Data reported by the API were insufficient for statistical analysis, but significant variations are expected from one project to the next in the same region. Thus, these data should be considered guidelines for preliminary estimates only and should not replace more site-specific data.

the average probably is attributable to slightly poorer quality sands in the Texas Gulf Coast compared to those in the Louisiana Gulf Coast.

TABLE 2.1—ESTIMATED ULTIMATE RECOVERY EFFICIENCY OF OIL FROM DIFFERENT LITHOLOGIES AND DRIVE MECHANISMS FROM SEVERAL AREAS IN THE U.S.

State or Region	Dominant Lithology	Main Drive Mechanism	Recovery Efficiency[a]
North Louisiana	Carbonate[b]	Pressure depletion	.15
South Louisiana	Sandstone[c]	Water drive	.51[d]
Oklahoma	Sandstone[e]	Water flood	.34
Pennsylvania	Sandstone[f]	Water flood	.20
Permian Basin	Carbonate[g]	Water flood	.39[h]
Texas Gulf Coast	Sandstone[i]	Water drive	.47[j]
Texas Panhandle	Carbonates[k]	Pressure depletion	.23
West Virginia	Sandstone[l]	Pressure depletion	.21

a. Calculated from data published by the API [1980], Table IV. Almost all fields included produce "light" oil.
b. Mostly lower Cretaceous-Jurassic, with porosity in the range 5 to 20%; permeability 1 to 500 md.
c. Chiefly Miocene, with porosity in the range 15 to 35%; permeability, 50 to 2500 md.
d. Estimated ultimate recovery efficiency from a few, high quality, giant fields in the range 56 to 66%.
e. Mostly Pennsylvanian, with porosity in the range 5 to 20%; permeability, 5 to 100 md.
f. Mostly poor quality Devonian graywackes.
g. Chiefly Permian shelf carbonates with porosity in the range 5 to 20%; permeability, 1 to 100 md.
h. Estimated ultimate recovery efficiency from some of the better floods in the range 50 to 60%.
i. Chiefly lower Cenozoic, with porosity in the range 15 to 35%; permeability, 50 to 2500 md.
j. Estimated ultimate recovery efficiency from several high quality, giant fields in the range 52 to 61%, with a few approaching 70 to 74%.
k. Mostly upper Paleozoic, with porosity in the range 5 to 15%; permeability, 1 to 100 md.
l. Middle Paleozoic, with porosity in the range 5 to 15%; permeability, 1 to 100 md.

From a review of these data—admittedly limited—it seems apparent that variations in aggregate oil recovery efficiency between regions are attributable principally to differences in reservoir rock/fluid systems and to differences in reservoir drive mechanisms. The Soc. of Petroleum Evaluation Engineers [1998] has provided useful summary discussions of regional variations in rock quality and reservoir drive mechanism for various areas in the U.S. in addition to those summarized in Table 2.1.

In 1984 the API published estimates of oil initially in place (OIP) and primary recovery efficiency for 533 reservoirs in the U.S. (see Sec. 3.7.2). Also published were estimates of primary and waterflood recovery efficiency for 230 reservoirs in the U.S., which are discussed in **Appendix F.**

2.2.2 Other Recovery Efficiency Data

Estimates like the foregoing may be developed for other areas from detailed data published annually by most state oil and gas regulatory commissions in the U.S., by the Alberta Energy Resource Conservation Board (ERCB),[7] by the Saskatchewan Dept. of Energy and Mines, various regional geological societies, and other such agencies. Pending development of specific data for the reservoir in question, this type of regional data may be useful in making *preliminary* estimates of reserves.

2.3 Statistical

Statistical analogy methods include using per-well recoveries of oil and/or gas from analogous wells in the same producing trend or in analogous geologic settings to estimate recoveries from wells being studied.

Two types of statistical analogy methods are discussed in Sec. 2.3.1 and 2.3.2: (1) iso-ultimate recovery maps[8] and (2) analysis of observed frequency distributions of ultimate recovery.

When analogy methods are applied to estimate reserves for individual *wells,* analogy and subject well should be similar regarding:

- well completion, including type of stimulation (if any),
- production method,
- initial absolute open flow potential for gas wells,
- initial potential and/or productivity index for oil wells, and
- economic limit.

2.3.1 Iso-Ultimate Recovery Maps

In some producing areas, ultimate recovery of oil or gas from individual wells may be controlled by geologic trends. These trends might include, for example, depositional environment, intensity of fracturing, or degree of diagenesis. In such cases, a so-called "iso-ultimate recovery" map may be made by posting and contouring estimated ultimate recovery from individual wells in the area of interest. Subject to several cautions, discussed below, such a map may be used to estimate reserves for undrilled tracts. Chen *et al.* [1986] provided an example of an iso-recovery map for the fractured, low-permeability Austin chalk trend in central Texas.

Exercise caution in the application of this technique! Factors other than geologic may control ultimate recovery of oil or gas. Different completion and stimulation procedures may result in different ultimate recoveries from individual wells. For example, Fast *et al.* [1977] reported an apparent dependency between size of fracture treatment and ultimate gas recovery from wells completed in the "J" Sand in the Denver-Julesberg basin of western U.S. Similar

---

[7] In 1995 the Alberta ERCB was merged with other agencies and became part of the Alberta Energy and Utilities Board.
[8] Such maps are considered graphical statistical analyses of observed data.

observations were made by Blauer *et al.* [1992] regarding the Spraberry (oil) Trend in the Permian Basin in the southwestern U.S. Thus, before relying on iso-ultimate recovery maps, it is good practice to determine whether there is a statistically valid relation between completion or stimulation method, or another parameter, and ultimate recovery.

Wells in the area of interest may be capable of draining more than a "spacing unit."[9] In this case, wells adjacent to undrilled tracts and those placed on production early in reservoir life may exhibit higher ultimate recoveries than wells in central locations and those placed on production late in reservoir life. These possibilities should be investigated before using an iso-ultimate recovery map to estimate reserves for undrilled tracts.

2.3.2 Log-Normal Distribution of Reserves

Arps and Roberts [1958], Kaufman [1963], and others noted that, in a given geologic setting, a log-normal distribution is a reasonable *approximation*[10] to the frequency distribution of "field size"—i.e., to the ultimate recoveries of oil or gas from those fields. As discussed in Chap. 9, the frequency distribution of other geologic and/or engineering parameters also may be *approximated* with log-normal distributions. As discussed in the ensuing paragraphs, these observations may be used to estimate reserves for a prospect on a geologic trend, or reserves for wells planned for undrilled tracts in partially developed areas.

Estimates for Fields

Table 2.2 shows 28 fields in the Leduc formation in the Rimbey-Meadowbrook reef trend in Alberta (Canada) and estimated ultimate oil recovery (initial reserves) attributed to them [Lee and Wang 1986].

The list is ordered by discovery date. To test for log normality, these data are ordered by field size—estimated ultimate oil recovery—and plotted on log-normal graph paper in **Fig. 2.1**. It may be observed that these data follow a linear trend with a "correlation coefficient" of 0.96, which indicates the distribution is approximately log normal.[11] Lee and Wang [1986] provide no information regarding a possible correlation of initial reserves with other factors that may be related to initial reserves—e.g., area of seismic closure, nature of seismic anomaly, or structural relief.

Given this information, the "best estimate" for the ultimate recovery from the *next* prospect along this trend, given a discovery, is the *median* of the distribution (the value at the 50th percentile, which corresponds to "0" standard deviations on the "x" axis of Fig 2.1). As indicated by the dashed line the median estimate is approximately 11 MM bbl,[12] which is between 14.2 and 8.6 MM bbl, the field sizes represented by Sylvan Lake and St. Albert B in Table 2.2. Other interpretations of this type of data to estimate proved, probable, and possible reserves for prospects in proven trends are discussed in Chap. 9.

---

[9] In the U.S. and Canada, regulatory authorities historically have assigned allowable tract sizes or well spacing. In many jurisdictions in the U.S., for example, a "spacing unit" for an oil well would be a 40-acre tract of land overlying the reservoir.

[10] Statements by some authors that reserves *should* be log-normally distributed are not consistent with studies that suggest that distributions other than log normal might also be reasonable approximations to the state of nature.

[11] This analysis was performed using STATISTIX™ (www.statistix.com). Other such programs are available; no endorsement is intended. The correlation coefficient is the Shapiro-Francia statistic, which was computed using the Wilk-Shapiro procedure to test the distribution of the logarithms of the ultimate recoveries in Table 2.2 for "normality." The mechanics of the process are described here for readers who may not have access to this computer software.

[12] As with most interpretations of statistical data, beauty is in the eye of the beholder! Use of the phrase "best estimate" is consistent with typical interpretations of this type of analysis. Users desiring a higher confidence level might prefer to consider making an estimate at the 80th (or higher) percentile. Irrespective of the interpretation, such estimates, strictly speaking, cannot be considered "reserves" until a discovery has been made in the trend being analyzed.

TABLE 2.2—ESTIMATED ULTIMATE OIL RECOVERY AND DISCOVERY DATE, LEDUC POOLS OF THE RIMBEY-MEADOWBROOK REEF TREND, ALBERTA, CANADA

Pool	Discovery Date*	Ultimate Recovery (MM bbl oil)
Leduc-Woodbend A	Feb 47	351.5
Redwater	Jul 48	1295.4
Golden Spike A	Sep 49	290.8
Leduc-Woodbend G	May 50	1.1
Golden Spike B	Jul 50	4.0
Acheson	Jul 50	142.9
Golden Spike C	Feb 51	2.6
Wizard Lake A	Mar 51	366.7
Glen Park	Jul 51	27.6
Leduc-Woodbend D	Aug 51	15.0
Bonnie Glen	Sep 51	765.9
Leduc-Woodbend E	Mar 52	2.0
Leduc-Woodbend D	Apr 52	0.7
Westerose	May 52	169.6
Leduc-Woodbend C	Jun 52	0.9
St. Albert B	Oct 52	8.6
Farydell-Bon Accord	Dec 52	14.7
Homeglen-Rimbey	Jan 53	111.0
Leduc-Woodbend F	Feb 53	6.6
Yekau Lake A	Jun 55	4.7
Morinville A	Aug 55	0.6
St. Albert A	Jul 56	20.1
Morinville B	Jun 60	14.7
Sylvan Lake	Oct 61	14.2
Morinville C	Oct 63	3.4
Wizard Lake B	Apr 64	1.0
Lanaway	Jun 64	2.2
Yekau Lake B	Jul 67	0.3

* Spud date of discovery well.

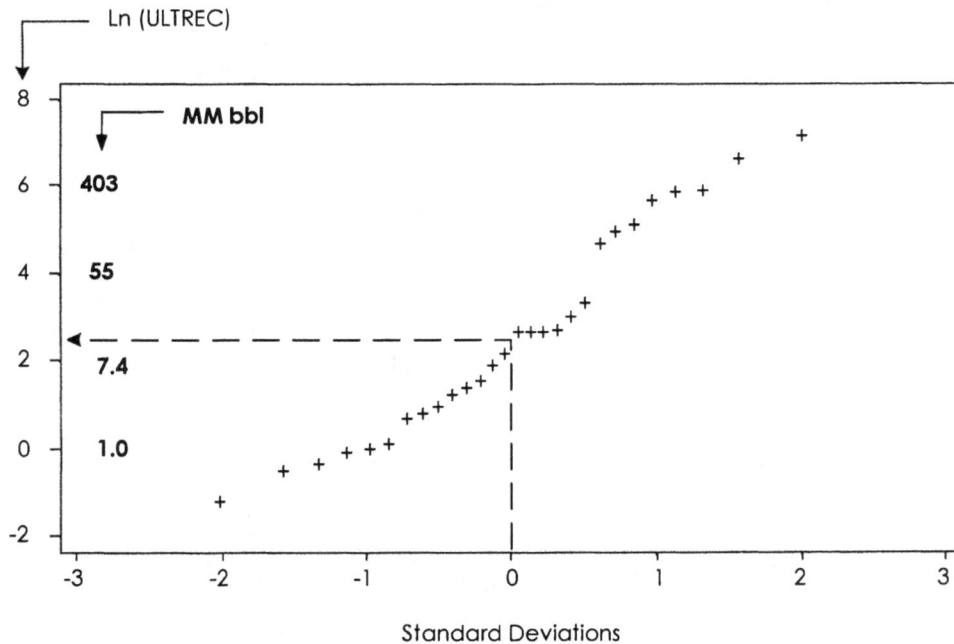

Fig. 2.1—Frequency distribution of ultimate recovery from 28 oilfields in the Rimbey-Meadowbrook reef trend, Alberta, Canada [after Lee and Wang 1986].

As noted by Lee and Wang [1986], other interpretations of these data may be made, by using a successive sampling model.[13] This procedure is beyond the scope of this work, and interested readers are referred to those authors.

For a discussion of the "pros" and "cons" of this procedure, see Newendorp [1975]. Also, Pruit [1978] has provided a discussion of statistical relations between reserves and various geologic factors in the Denver basin of western U.S.

Estimates for Wells

The technique discussed in the previous section also may be used to estimate ultimate recovery for individual wells on undeveloped tracts in partially developed areas. Before using this technique, however, it should be determined if per-well reserves are truly "random," and are not related to some geologic or operating parameter, as discussed in Sec. 2.4.

Table 2.3 lists estimated ultimate oil production (initial reserves) for each of 40 wells producing from the Green River formation in the Uinta basin of Utah (U.S.).

There is no statistically valid correlation between individual well reserve estimates and net pay or "frac" treatment or any other such parameter in this area. Using the same procedure as previously, it was determined that these data can be approximated by a log-normal distribution; the correlation coefficient is 0.98. The *median* value of this distribution is 38 Mbbl. *Given a discovery*, this is the "best estimate" for ultimate recovery for each of the next few wells drilled in this area.[14]

If per-well initial reserves are approximately log-normally distributed, as in this example, use of the *arithmetic average* to estimate ultimate recovery for the next few wells to be drilled will be biased on the high side.[15]

As drilling and production continue, the analysis should be updated as reserves for existing and new wells are revised. Interpretations of these data to estimate proved, probable, and possible reserves for wells in proven areas are discussed in Sec. 9.8.

The random distribution of individual well reserves has been observed frequently by the author. For example, such distributions have been observed in the Grimes gas field, Sacramento Valley, California, and in the Morrow gas trend, southeastern New Mexico. In both of these cases, there was no statistically valid correlation between the ultimate gas production for the individual wells in these areas and net pay, "frac" treatment, initial potential, or any other parameter for these same wells. Comparable unpublished observations have been made by others for Devonian gas wells in the Appalachians (eastern U.S.) and elsewhere.

2.4 Variations in Operating Practice

Factors such as well spacing, completion technique, operating cost, and operating procedure may have a significant influence on ultimate recovery of oil and/or gas. If there are significant differences in these factors from one property to the next, caution should be exercised in making comparisons or statistical analyses of wells between such properties.

---

[13] The observant reader might note an apparent correlation of LN(ULTREC) with discovery sequence. The regression coefficient, however, is only 0.3, which is too small to support this hypothesis.
[14] The phrase "best estimate" is used here in the same context as in the previous section.
[15] The comments in Note 12 are also relevant to this analysis.

TABLE 2.3—ESTIMATED ULTIMATE OIL RECOVERY FROM 40 WELLS, GREEN RIVER FORMATION, UINTA BASIN, UTAH, U.S.

Lease	Well	Ultimate Recovery (M bbl)
Federal "C"	3-15	33.3
Federal "B"	2-14	35.6
Federal "D"	2-22	27.8
Federal-State "A"	2-11	28.9
Federal "B"	3-14	20.6
Federal "C"	4-15	19.4
Federal-State "A"	3-11	53.3
Federal "D"	3-22	77.8
Federal-State "A"	3-12	83.3
Jones-Federal "B"	5-21	83.4
State-Federal "B"	2-02	50.0
Federal "A"	2-10	38.9
Federal-State "A"	4-11	12.2
Federal "B"	5-14	150.0
Federal "C"	6-15	138.9
Federal-State "A"	6-11	50.1
Federal "A"	4-10	77.8
State-Federal "B"	3-02	94.4
Smith-Federal "A"	1-21	83.3
State-Federal "B"	5-02	13.9
Federal "A"	3-10	22.2
Smith-Federal "A"	4-21	49.9
Smith-Federal "A"	3-21	77.8
Federal "E"	1-16	111.1
Federal "E"	3-16	16.7
Federal "B"	6-14	20.6
Jones-Federal "B"	2-21	22.2
Federal "C"	8-15	30.1
Federal-State "A"	1-12	33.3
Federal "A"	6-10	36.1
Smith	5-16	66.7
Federal "E"	9-16	111.1
Federal "B"	7-02	17.8
Smith	2-16	22.2
Federal "E"	7-16	30.0
Jones	2-28	37.8
Federal "B"	8-14	46.1
Federal "A"	8-10	133.3
Jones-Federal "A"	1-29	38.9
Federal "E"	8-16	27.8

For example, in waterfloods in heterogeneous, stratified carbonates (such as those in the U.S. Permian Basin) well spacing has a demonstrated effect on ultimate recovery [Ghauri *et al.* 1974, George and Stiles 1978]. It is suggested that, in similar geologic settings, well spacing should *always* be considered in making comparisons between otherwise similar fields.

In this type of reservoir, high-water-cut wells should be kept "pumped off" so that water cannot backflow into zones still producing oil. For various reasons, this practice is not universally followed, and significant differences in recovery efficiency may occur between different operators in the same area.

In reservoirs with anisotropic permeability distribution or preferential fracture orientation, waterflood response may be influenced by pattern orientation [Guidroz 1967, Beliveau *et al.* 1993].

In low-permeability reservoirs, the method of stimulation may influence ultimate recovery. Thus, in reviewing production data from possible analogous fields or reservoirs, it is good practice to attempt to determine the nature of development, completion, and operating practices that may have influenced performance.

Other factors being more or less the same, in areas characterized by low operating costs and closely spaced wells, recovery efficiency of oil and/or gas generally will be greater than in areas characterized by high operating costs and widely spaced wells. Most of the statistical data on recovery efficiency in the U.S. has been developed from operations with onshore properties, which historically have had relatively low operating costs and closely spaced wells. These conditions are not typical of those in other areas of the world, especially offshore areas. Thus, caution should be exercised in using recovery efficiencies determined from reservoirs in the U.S. to estimate recovery efficiencies from otherwise analogous reservoirs elsewhere.

CHAPTER 3

VOLUMETRIC METHODS

3.1 Background

Volumetric methods to estimate reserves generally are used early in the life of a reservoir, before there are sufficient production and/or pressure data to use the performance method, and for behind-pipe zones,[1] which may not be placed on production until the current completion zone is abandoned.

Although it is the most widely used method to estimate reserves, the volumetric method may be subject to considerable uncertainty, depending on the geologic setting and the amount and quality of geologic and engineering data. Thus, good practice mandates checking reserves estimated using volumetric methods against well and reservoir performance at the earliest practical stage of production.

The discussion in this section tacitly assumes the use of so-called "deterministic" calculations, whereby a best estimate is made of the value of each parameter used in the calculations, resulting in a single best estimate for reserves for each data set. As discussed in Chap. 9, probabilistic calculations of reserves typically are based on the same equations as those used for deterministic calculations. Probabilistic calculations, however, treat each parameter as a random variable and result in a range of estimates of reserves for each data set.

Regarding estimates of reserves made early in field or reservoir life, Arps [1956] observed that reserves for "very prolific fields...have been generally underestimated, while (reserves for) the poorer ones...are usually overestimated."

For offshore fields, substantial expenditure typically must be made for construction of one or more drilling and production platforms and production facilities prior to initiating development and production. In such cases, significant underestimates of reserves or the nature of the production may lead to significant undersizing of platform(s) and/or misdesign of facilities, resulting in suboptimal financial realization.

In one of the earliest published discussions of the volumetric method, Deussen [1936] estimated that, through 1930, oil recovery factors of 477 and 767 bbl/acre-ft had been observed in the Goose Creek and Orange fields, respectively. These fields are located on the Texas Gulf coast, U.S.

In *principle,* the volumetric method involves calculating: (a) the amount of oil and gas initially in place by a combination of volumetric (geologic) mapping, petrophysical analysis, and reservoir engineering and (b) the fractions of oil, gas, and associated products initially in

---

[1] Depending on the amount of subsurface data available, it may not be feasible to map all behind-pipe zones, and analogy methods may have to be used to estimate such reserves.

place that are expected to be recovered commercially—i.e., the recovery efficiencies—using analytical methods and/or analogy.

In *practice,* however, the volumetric method typically involves: (a) assuming the drainage volume to be approximately equal to the developed acreage times the average net pay[2] (or some fraction thereof), based on experience and (b) estimating the recovery factor (barrels of oil, or cubic feet of gas, per acre foot) by analogy with similar reservoirs in the area, with reserves being estimated as the recovery factor times the drainage volume.

The procedure used generally will depend on the type of structure, on the trapping mechanism, and on the amount of subsurface geologic and engineering data available *at the time the estimate must be made.* As Havlena [1966] observed "it is impossible to prepare a 'universal cookbook' that satisfies all conditions; each reservoir and even each well presents a specific problem."

Estimating the percent of oil and gas initially in place that can be expected to be recovered commercially usually involves: (a) assuming the probable drive mechanism, generally by analogy with similar reservoirs in the same producing trend and (b) estimating the recovery efficiency attributable to that drive mechanism, considering the rock and fluid properties of the reservoir, reservoir heterogeneity, well spacing, well position, reservoir geometry, and other factors that control recovery efficiency.

DeSorcy [1979] provided broad ranges of recovery efficiencies observed for various types of reservoirs and primary drive mechanisms, as follows:

Drive Mechanism	Expected Recovery (%)		Remarks*
	Oil Reservoirs	Gas Reservoirs	
Oil expansion**	2 to 5		Higher recoveries reported.
Gas expansion		70 to 95	As low as 30% in low-permeability reservoirs; lower in tight gas reservoirs.
Solution gas	10 to 30		
Gas cap	20 to 50		Higher recovery efficiency generally associated with effective gravity segregation.
Water drive	25 to 50	45 to 70	As low as 10% for thin oil columns; occasionally as high as 70%, or higher.#
Gravity segregation	30 to 70		

* These remarks include DeSorcy's [1979] plus the author's.
** Includes effects of rock compressibility.
# In the East Texas field (U.S.), for example, oil recovery efficiency is anticipated to be 90% of OIP. However, a recovery efficiency this high is very unusual and is not likely to be observed elsewhere.

Such broad ranges provide only guidelines. Estimates for a specific reservoir should fall in a narrower range, with consideration of pertinent fluid, rock, and rock/fluid properties,

---

[2] Estimation of net pay is discussed in Sec. 3.6.3.

principal heterogeneities, well spacing, probable drive mechanism, and experience with analogous reservoirs.

Procedures to estimate oil and gas initially in place and recovery efficiency are discussed in the following sections.

3.2 Equations to Estimate Oil, Gas, and Condensate Initially in Place

Volumetric methods to estimate oil, gas (both free and dissolved), and condensate initially in place are discussed in this section. Material balance methods to estimate oil and free gas initially[3] in place are discussed in Chap. 4.

The development of this section tacitly assumes the phases (liquid or gas) in the reservoir at initial conditions can be determined, based on subsurface information (logging, RFT fluid samples, and pressure gradients) and well tests. For reservoirs containing fluids at near-critical conditions, however, such determinations may be subject to considerable uncertainty. Sec. 3.7.1 includes a brief discussion regarding identification of reservoir fluids.

3.2.1 Oil Reservoirs

For an oil reservoir, or for the oil column of an oil reservoir with a gas cap, oil initially in place (OIP) may be calculated as

$$N_i = 7,758 \phi_o(1 - S_{wo})A_o h_{no}/B_{oi}, \dotfill \text{(3.1a)}$$

where[4] N_i = oil initially in place, STB,
           7,758   = barrels in an acre foot,[5]
           ϕ_o = average porosity in the oil zone, fraction,
           S_{wo} = average water saturation in the oil zone, fraction,
           A_o = area of the oil zone, acres,
           h_{no} = average net oil pay, feet, and
           B_{oi} = average initial formation volume factor, RB/STB or RV/SV.[6]

For some applications, Eq. 3.1a may be written as

$$N_i = 7,758 \phi_o(1 - S_{wo})V_{to}R_{ngo}/B_{oi}, \dotfill \text{(3.1b)}$$

where V_{to} = gross volume of initially oil-bearing rock, acre-ft, and
        R_{ngo} = average net-to-gross ratio in initially oil-bearing zone.

Depending on the application, the estimated volume of oil initially in place usually is rounded to the nearest thousand stock-tank barrels (MSTB).[7]

---

[3] Rather than the historically used term "original," the term "initial" is used here, being more appropriate as a time descriptor and consistent with the subscript i used to denote initial conditions in SPE [1993] notation.

[4] Notation is consistent with SPE [1993] standard, but is introduced here and in subsequent sections for reader convenience and to identify the proper units.

[5] Constant is not exact.

[6] As discussed in Sec. 3.4.2, Well Test and PVT Data, for volatile oils the initial formation volume factor is highly dependent on surface gas/oil separation equipment. Also, for oils that are undersaturated at initial conditions, B_{oi} will be less than B_{ob} estimated from empirical correlations based on initial producing GOR.

[7] Given the uncertainties in most oilfield data, three significant figures are the maximum that should appear in estimates of oil and gas initially in place and of reserves.

If the units for area and net pay are hectares and meters, respectively, the constant in Eq. 3.1a becomes 1.0, and the units for OIP are $10^4 \, m^3$.

Porosity, water saturation, and oil formation volume factor should be *volume*-weighted averages in the *oil* zone. Sources of and methods of analyzing these parameters are discussed in Secs. 3.3 and 3.4.

If there is a significant oil/water transition zone initially present, special procedures may be required to determine the volume-weighted average initial water saturation. Please refer to Sec. 3.6.6 for discussion of this point.

Solution gas dissolved in the oil at initial reservoir conditions may be calculated as

$$G_{Si} = N_i R_{si}, \dotfill (3.2)$$

where G_{Si} = solution gas initially in place, scf, and
$\quad\quad\quad R_{si}$ = average initial solution gas/oil ratio, scf/STB.

Depending on the application, the estimated volume of solution gas initially in place usually is rounded to the nearest thousand standard cubic feet (Mscf).

In some areas, GOR units are cubic meters of gas per cubic meter of oil (m^3/m^3). The conversion[8] is 1 scf/STB = 5.614 $m^3/m.^3$

3.2.2 Gas Reservoirs

For a nonassociated gas reservoir, or for a gas cap, free gas[9] initially in place may be calculated as

$$G_{Fi} = 43{,}560\phi_g(1 - S_{wg})A_g h_{ng}/B_{gi}, \dotfill (3.3a)$$

where G_{Fi} = free gas initially in place, scf,
$\quad\quad$ 43,560 = cubic feet in an acre foot,[10]
$\quad\quad\quad \phi_g$ = average porosity in the free gas zone, fraction,
$\quad\quad\quad S_{wg}$ = average water saturation in the free gas zone, fraction,
$\quad\quad\quad A_g$ = area of gas cap or gas reservoir, acres,
$\quad\quad\quad h_{ng}$ = average net thickness of gas cap or gas reservoir, feet, and
$\quad\quad\quad B_{gi}$ = average initial formation volume factor, Rcf/scf or RV/SV.[11]

For some applications, Eq. 3.3a may be written as

$$N_i = 43{,}560\phi_g(1 - S_{wg})V_{tg}R_{ngg}/B_{gi}, \dotfill (3.3b)$$

where V_{tg} = gross volume of initially gas-bearing rock, acre-ft, and
$\quad\quad\quad R_{ngg}$ = average net-to-gross ratio in initially gas-bearing zone.

---

[8] Constant is not exact.
[9] Such gas usually is not marketable until it has been processed to meet pipeline specifications by removal of nonhydrocarbon components and condensate, after which some fraction may be considered reserves.
[10] Constant is exact.
[11] B_{gi} is the reciprocal of E, which frequently is used in gas engineering; units for E are scf/Rcf.

Depending on the application, the estimated volume of free gas initially in place usually is rounded to the nearest thousand standard cubic feet (Mscf).

If the units for area and net pay are hectares and meters, respectively, the constant in Eq. 3.3a becomes 1.0, and the units for GIP are $10^4 \, m^3$.

Porosity, water saturation, and initial formation volume factor should be *volume*-weighted averages in the *gas cap* or *gas reservoir*. Sources of and methods of analyzing these parameters are discussed in Secs. 3.3 and 3.4.

Condensate (also called distillate) in the vapor phase at initial reservoir conditions (but measured as a liquid at surface conditions) may be calculated as

$$C_i = G_{Fi}R_{ci}, \dots\dots\dots\dots\dots\dots\dots\dots\dots\dots\dots\dots\dots\dots\dots\dots\dots\dots\dots\dots\dots\dots\dots (3.4)$$

where C_i = condensate (distillate) initially in place, STB,
         R_{ci} = initial condensate/gas ratio (CGR), STB condensate/MMscf,[12] and
         G_{Fi} = free gas initially in place, MMscf.

Depending on the application, the estimated volume of condensate initially in place usually is rounded to the nearest thousand stock-tank barrels (MSTB).

3.3 Sources of Data

Eqs. 3.1 through 3.4 require three general types of data: petrophysical data, fluid data, and volumetric data. The sources of these data are noted briefly below; analysis is discussed in Sec. 3.4.

3.3.1 Petrophysical Data[13]

Petrophysical data needed for Eqs. 3.1 and 3.3 include average water saturation, S_w,[14] and average porosity, ϕ. Water saturation may be estimated from log analysis, capillary pressure data, analysis of cores taken using oil-based mud,[15] or relative permeability data. Depending on circumstances, all these data may be necessary to ensure a reliable estimate.

Petrophysical analysis is a vitally important part of volumetric mapping and estimation of oil and gas initially in place. Detailed discussion of this aspect of reserve estimation, however, warrants a separate book. It would be presumptuous to attempt to summarize in this work the publications of others in this specialty [Asquith 1982, Dewan 1983, Dresser Atlas 1982, Ellis 1987, Hilchie 1982, Jorden and Campbell 1986, Schlumberger 1987, Timur 1987, and Tixier 1987]. Thus, this section is limited to a few brief comments.

For mature producing areas of the world, long practice has led to the establishment of generally accepted procedures for: (a) the selection of diagnostic suites of logs and (b) the procedures to analyze these logs. While there are similarities between some producing areas, no two are identical. Typically, each producing area presents a different set of problems in

---

[12] As discussed in Sec. 3.4.2 under Well Test and PVT Data, the initial CGR may be dependent on surface separation equipment, especially for rich gases. Also, "raw" condensate produced through surface equipment may have to be blended (stabilized) to reduce the boiling point to meet pipeline specifications. Appropriate adjustment might have to be made to separator volumes before estimating reserves of such stabilized condensate.

[13] A probabilistic view of petrophysical analysis is presented in Sec. 9.9.2.

[14] The term "connate" or "interstitial" typically is used interchangeably with the term "average" to qualify water saturation. Unfortunately, however, the term "irreducible" also is used in this context, which is not correct. The term "irreducible" should be used only when describing immobile water saturation, as determined by relative permeability or capillary pressure data.

[15] If cores are taken with oil-based mud, a refined oil and minimum chemical additives in the mud will help minimize wettability changes. Otherwise, the oleic filtrate from such muds may alter the native wettability, possibly reducing the interstitial water saturation to a very low value.

log analysis. Thus, site-specific knowledge is essential for reserve estimation using such analyses. In addition to company proprietary publications, much of the industry's experience in formation evaluation in different operating areas has been summarized and published by leading service companies. For example, from time to time, Schlumberger Well Surveying Co. has published well logging monographs on various areas of the world.[16]

Use caution in applying results of computer analyses of well logs provided by service companies. In comparing the results of analyses made by their company's computer programs with those made by three leading service companies, Peeters and Visser [1991] observed that "Results differed by up to 35% in equivalent hydrocarbon columns." Their comparison was made using a simple logging suite run over gas- and water-bearing intervals of a shaly sandstone in a North Sea well.

Error or truth should not be attributed to any of the participants in this study; rather, there are different views regarding the analysis of the test section. It is noted here that the section evaluated had average core porosity of 14%, compared to average log-derived porosity of 14 to 20%. None of the participants in the project were provided with core data; thus, one might reasonably question whether the test simulated a real situation. It is here speculated that even greater differences would be observed in a rock section with more complex mineralogy, like that observed in the Green River basin (U.S.), for example.

Measurement and interpretation of capillary pressure data are discussed by Amyx *et al.* [1960] and Keelan [1977], among others. Historically, such data have been obtained at ambient conditions. However, in analyzing sandstone samples ranging from well consolidated to friable, Ajufo [1993] reported significant differences in capillary properties determined at overburden pressure compared to those determined at ambient conditions. As might be expected: (a) larger differences were observed for samples with poor sorting and/or less cementation and (b) at overburden pressure, the P_c vs. S_w curves shifted towards higher irreducible water saturation.

Porosity may be determined from core and/or log analysis. Core data, however, are essential to characterizing subsurface reservoirs, and consideration should be given to obtaining high-quality core data from any reasonably large accumulation. In this context, Archer and Wall [1986] observed that "A target of 30% cored wells will provide reasonable reservoir control on all but the most complex geology—the coring should be 75% complete before the reservoir has 50% of its wells." Full-hole cores are preferred over sidewall samples. It is recognized, however, that it may not be feasible to obtain full-hole cores from all, if any, wells, and reliance may have to be placed on sidewall samples. If both full-hole cores and sidewall samples are obtained from the same interval, attempts should be made to correlate data obtained from sidewall samples with that obtained from full-hole cores.

Core handling and analysis procedures have been discussed by others [Amyx *et al.* 1960 and Keelan 1977, 1982]. The reader is cautioned that, depending on circumstances and the objectives of the analyses, different procedures may be used in the laboratory to analyze cores. The results of these procedures, even using the same core sample, may not agree. Of vital importance is an understanding of the procedures and the limitations and potential source of error in using such procedures.

Whenever feasible, petrophysical calculations using models considered appropriate for one log suite should be cross-checked against those made with other models and log suites. Mathematical models and log suites appropriate for analysis of water-bearing zones may not

---

[16] Neither endorsement of Schlumberger Well Survey Co. nor exclusion of others is intended by this reference.

necessarily be appropriate for analysis of oil and/or gas-bearing zones. The results of petrophysical analysis of a logged interval should be consistent with analysis of flow tests and cores, if there are any through the same interval.

In situations where data are not available, reference may be made to tabulations of reservoir data published by the Society of Petroleum Engineers [Jenkins 1987]. Such data can serve as *guidelines*, pending obtaining site-specific data.

3.3.2 Fluid Data

Fluid data needed for Eqs. 3.1 through 3.4 include initial formation volume factor, B_{oi} or B_{gi}, and initial solution gas/oil ratio (GOR), R_{si}, or initial CGR, R_{ci}.[17]

Formation volume factors may be estimated with a reasonable degree of accuracy using empirical correlations with data from well tests at initial conditions. It is good practice to cross-check such estimates using laboratory PVT data from reservoir fluids in the general area whenever such data are available. All such estimates, empirical or laboratory, should be checked against subsequent production and bottomhole pressure data. If the producing GOR increases more rapidly than anticipated, suggesting the presence of free gas, estimates of initial bubblepoint pressure might be re-examined.

Each well completed in a reservoir should be flow tested under *stabilized conditions* for at least 24 hours as soon as feasible after the well has "cleaned up." For each stabilized flow period, data should include:

- gas and liquid production rates from each stage of separation,
- pressures and temperatures for each stage of separation,
- specific gravity of first- and second-stage separator gas, stock-tank vapors, and stock-tank liquid, and
- flowing wellhead pressure and choke size or, if producing by artificial lift, description of equipment and operating details.

If water is produced during such a test (after recovery of any completion fluids), samples should be obtained for laboratory analysis. Comprehensive discussions regarding well testing are provided by Kimmel and Dalati [1987], Smith [1987], and Dake [1994].

For gas wells, it is common practice in many areas to obtain samples of gas and liquid from the first-stage separator for compositional analysis and the stabilized shutin tubing pressure (SITP) at the conclusion of the test period.

Pending obtaining PVT samples, data obtained from oil well tests at initial reservoir conditions may be used to estimate bubblepoint pressure and initial formation volume factor using empirical correlations provided in Appendix E.[18]

Regarding the bubblepoint pressure estimated from correlations, there are three possibilities:

- The estimated bubblepoint pressure may be significantly less than the static reservoir pressure, in which case the reservoir oil may be treated as an undersaturated system and a correction must be made to the estimated formation volume factor, B_{ob}, to estimate B_{oi}, as discussed in Appendix E.

---

[17] The term "condensate/gas ratio" or CGR is used here for *gas* wells, in preference to the frequently used term "gas/oil ratio," which is misleading and should be used only for *oil* wells.

[18] It is recognized that equipment is available for on-site determination of PVT properties. However, such equipment is not readily available at remote locations and is not routinely used by small operators.

- The estimated bubblepoint pressure may be approximately equal to the static reservoir pressure, in which case the reservoir oil may be treated as a saturated system, and the estimated formation volume factor, B_{ob}, may be considered equal to B_{oi}.

- The estimated bubblepoint pressure may be significantly greater than the static reservoir pressure, in which case it must be concluded that the well produced free gas during the test, the test GOR is not representative of the reservoir oil, and the formation volume factor estimated from correlations must be considered invalid.

Additional comments regarding these matters are provided in Sec. 3.4.2 under Well Test and PVT Data.

Regarding gas reservoirs, there are no simple correlations to estimate dewpoint pressure and formation volume factor from well test data like there are for oil reservoirs. As discussed in Appendix E, the Nemeth-Kennedy [1967] correlation may be used to estimate the dewpoint pressure with reasonable accuracy. This correlation requires the composition of the reservoir gas, which can be calculated from the composition of the first-stage separator fluids, as shown in Appendix E.

The gas formation factor may be estimated from correlations presented in Appendix E, but the composition or the specific gravity of the reservoir gas must be estimated to use these correlations.

If the test GOR is in the range 3,000 to 3,500 scf/STB, a reservoir fluid near the critical point should be suspected; extreme care should be taken to get reliable tests at initial reservoir conditions. Caution should be exercised in sampling such reservoirs, as cases have been reported where fluid composition varied from undersaturated oil near the base of the accumulation to retrograde gas at the top, with no apparent gas/oil contact [Neveux et al. 1988, Cazier et al. 1995]. McCain [1992] discussed the problems with getting representative samples from retrograde gas reservoirs initially at saturation pressure.

3.3.3 Volumetric Data

Volumetric data needed for Eqs. 3 1 and 3.3 include reservoir area and average net pay. Early in the development of a field or reservoir, before there are enough subsurface geologic data for mapping, reserves may be estimated on a per-well basis.[19] *Reservoir* area in Eq. 3.1 or 3.3 becomes *drainage* area estimated for the well or wells in question. Drainage area usually is estimated by analogy with wells with similar rock/fluid properties and producing with the same reservoir mechanism. Drainage volume usually is calculated as estimated drainage area times the true vertical net pay logged in the well or some fraction of the net pay, depending on circumstances.

As development proceeds, and sufficient subsurface geologic data become available, structure and isopachous maps should be made on all key horizons. The procedures are discussed in Sec. 3.6.

3.4 Analysis of Data

All raw data from both field and laboratory should be analyzed carefully and cross-checked for reasonableness and consistency before being used for engineering calculations. For

---

[19] On "seismically defined" prospects, after calibration of the predrilling seismic interpretation, preliminary geophysical maps may be made of reservoir structure and/or productive limits. Depending on the indicated field size and geologic complexity, additional seismic data may be acquired in the course of mapping and developing the accumulation.

example, it is good practice to check laboratory PVT data against empirical correlations and against field tests. Discrepancies are not uncommon and may be caused by several factors, including erroneous measurements, transposition errors, and/or calculation errors.

All laboratory-determined petrophysical data should be reviewed carefully and cross-checked for consistency against results of other methods for determining critical parameters. For example, it is good practice to check irreducible water saturation determined from capillary pressure tests against that determined from relative permeability tests on the same or comparable rock sample(s).

In 1988, the Soc. of Core Analysts conducted an Electrical Resistivity Study that involved analyses of selected rock samples by 25 laboratories. Reporting the results, Sprunt et al. [1990] observed that the laboratories used "a wide variety of techniques (and) obtained a wide variety of results for brine resistivity measurements, and core saturation (n) and cementation exponent (m) determination." The range of variation in m and n within one standard deviation[20] of the assumed "true" values could lead to errors in computed water saturation of approximately 5 to 10 saturation units; at a 95% confidence level (two standard deviations), the potential errors are doubled [Sprunt et al. 1990].

Muskat [1949], Amyx et al. [1960], Havlena [1966 and 1968], Archer and Wall [1986], and PS-CIM [1994] provided comprehensive discussions of data analysis for general reservoir engineering. Procedures for data analysis, however, typically are site-specific and problem oriented. Thus, the methodology for characterizing a bottomwater drive, heavy oil reservoir in fractured carbonates will be substantially different from the methodology for characterizing a retrograde gas reservoir in a friable sandstone. There are, however, broad principles and caveats common to characterizing many accumulations. For example, Amaefule et al. [1993] provided a comprehensive discussion of a methodology to identify "flow zones" that apparently has been successful in a number of geologic settings throughout the world. Also, discussions regarding integration of capillary pressure data with log and core analysis have been provided by Alger et al. [1989], Hawkins et al. [1993], Hawkins [1994], and Haynes [1995].

3.4.1 Statistical Analysis

Law [1944] and Bulnes [1946] were among the earliest investigators to note that porosity and permeability of reservoir rocks tend to display certain statistical distributions. For example, porosity in both sandstones and carbonates frequently (although not always) may be approximated with a *normal,* or Gaussian, distribution. Holtz [1993] and Capen [1993] identified situations where a log-normal distribution was a better representation of the data.[21] The implications of these observations are discussed more fully in Chap. 9.

Permeability tends to exhibit a log-normal distribution—the logarithm of permeability tends to be normally distributed. In progressively lower-porosity rocks, the variance exhibited by permeability tends to be progressively larger.

Determination of the statistical distribution of porosity and permeability is key to characterization of the reservoir in question and to the proper treatment of other reservoir properties with which porosity and permeability may be correlated. In addition, statistical analysis may suggest the presence of more than one type of depositional unit in the reservoir [Bennion and Griffiths 1966, Testerman 1962, Sneider et al. 1977, Amaefule et al. 1988, and

---

[20] "within one standard deviation " would encompass approximately two-thirds of the data set.
[21] Statistical analysis of data sets, however, is not without ambiguity. For example, depending on the statistical test applied, one may draw different conclusions regarding the nature of a distribution [Murtha 1993].

Amaefule *et al.* 1993]. Such occurrences may have an important bearing on volumetric mapping and estimating reserves.

Average water saturation, S_w, is one of the parameters needed for volumetric calculations of oil and gas initially in place with Eqs. 3.1 and 3.3. If there is not a significant oil/water transition zone, the average water saturation in the reservoir may be *approximately* equal to the irreducible water saturation, S_{wi}. However, if there is a significant oil/water transition zone, consideration should be given to using the procedures discussed in Sec. 3.6.6.

In some reservoirs, determination of S_w from log measurements may be subject to considerable uncertainty. Reliable analysis may be precluded by uncertainties in the saturation "model" and/or uncertainties in the saturation or resistivity exponents and/or the resistivity of interstitial water.[22] These points are discussed further in Sec. 9.9.2.

Data from core analyses, including capillary pressure and relative permeability data, might have to be used to estimate S_w. Depending on rock texture, there might be statistically valid correlations between S_{wi} and one of several different rock parameters, including porosity, permeability, or the square root of permeability divided by porosity. The reader is referred to Jensen *et al.* [1997] for guidelines regarding statistical analysis.

If porosity is normally distributed, the best value to use for these correlations is the arithmetic average porosity; if permeability is log-normally distributed, the best value to use is the median permeability [Warren and Price 1961]. Cronquist [1984] reported there was better agreement between volumetric and material balance estimates of gas initially in place in a very heterogeneous Miocene sand gas reservoir by using the geometric mean average water saturation rather than the arithmetic average water saturation.

In some cases, it may be necessary to use capillary pressure curves to help estimate irreducible water saturation. The Leverett "J" function [Leverett 1941] historically has been used to relate P_c, S_w, ϕ, and k. Due to wide variations in rock texture, however, this procedure may not be satisfactory. In this situation, one solution is to select the capillary pressure curve from rock samples that have porosity and permeability approximately equal to the median porosity and median permeability for the reservoir or rock unit being investigated. If samples with such properties are not available, one might use procedures discussed by Amyx *et al.* [1960] that recently were reviewed by Pletcher [1994].

Irrespective of the procedures used to determine these parameters, estimates for S_w and ϕ should be weighted by rock volume before being used in Eqs. 3.1 and 3.3. Typically, S_w and ϕ will be determined for each well zone on an interval-by-interval—e.g., foot-by-foot—basis. If, at the time estimates must be made, the wells are on approximately uniform spacing, weighting on a footage basis may be adequate. Thus, weighted average porosity for each well-zone would be

$$\phi_W = \Sigma(\phi_1 h_1 + \phi_2 h_2 + \ldots + \phi_N h_N)/\Sigma(h_1 + h_2 + \ldots + h_N) , \quad\quad\quad (3.5\text{a})$$

and the weighted average hydrocarbon saturation for each well-zone would be

$$(1 - S_w)_W = \Sigma[(1 - S_{w1})\phi_1 h_1 + (1 - S_{w2})\phi_2 h_2 + \ldots + (1 - S_{wN})\phi_N h_N]/$$

$$\Sigma(\phi_1 h_1 + \phi_2 h_2 + \ldots + \phi_N h_N) , \quad\quad\quad (3.5\text{b})$$

---

[22] A general saturation "model," for example, is $S_w = [aR_w/R_t\phi^m]^{1/n}$, where notation is SPE [1993] standard.

where ϕ_n = porosity for the nth net pay interval in the well zone,

$(1 - S_{wn})$ = hydrocarbon saturation for the nth net pay interval,

$\phi_n h_n$ = porosity-thickness product for the nth net-pay interval,

N = total number of net pay intervals in the well zone, and

$\Sigma(h_1 + h_2 + \ldots + h_N)$ = sum of net pay in all intervals in the well zone.

For uniformly spaced wells, reservoir average porosity may be determined as

$$\phi_R = \Sigma(\phi_{w1}\, h_{w1} + \phi_{w2}\, h_{w2} + \ldots + \phi_{wM}\, h_{wM})/$$

$$\Sigma(h_{w1} + h_{w2} + \ldots + h_{wM}) , \quad\ldots\ldots\ldots\ldots\ldots\ldots\ldots\ldots\ldots\ldots\ldots\ldots\ldots\ldots\ldots (3.6a)$$

and reservoir average hydrocarbon saturation would be

$$(1 - S_w)_R = \Sigma[(1 - S_{ww1})\phi_{w1}h_{w1} + (1 - S_{ww2})\phi_{w2}h_{w2} + \ldots + (1 - S_{wwN})\phi_{wM}h_{wM}]/$$

$$\Sigma(\phi_{w1}h_{w1} + \phi_{w2}h_{w2} + \ldots + \phi_{wM}h_{wM}) , \quad\ldots\ldots\ldots\ldots\ldots\ldots\ldots\ldots\ldots\ldots\ldots (3.6b)$$

where ϕ_{wm} = average porosity for the mth well,

S_{wwm} = average water saturation for the mth well,

h_m = net pay in the mth well,

M = number of wells in the reservoir,

ϕ_R = average porosity for the reservoir, and

$(1 - S_w)_R$ = average hydrocarbon saturation for the reservoir.

If wells are not approximately uniformly spaced, it might be preferable to: (a) calculate the footage-weighted porosity and hydrocarbon saturation for each well, as in Eqs 3.5a and 3.5b, (b) calculate $\phi_{wm} (1 - S_w)_{wm} h_m$ for each well, (c) contour these values, and (d) planimeter the resulting map to determine the reservoir hydrocarbon pore volume.

3.4.2 Adjustment/Correlation of Data

Petrophysical and fluid data from *all* sources should be compared and cross-checked for consistency. Several aspects of this process for porosity, water saturation, PVT, and well test data are discussed in the following paragraphs. Complete discussions have been provided by others [Amyx *et al.* 1960, Havlena 1966, Keelan 1977, Muskat 1949,[23] Smith *et al.* 1992, and Timmerman 1982].

Porosity

Depending on lithology, wellbore conditions, and logging device(s), porosities indicated by wireline logs may or may not agree with those determined from core analysis. The problem is especially severe in loosely consolidated sands, shaly sands, and fractured, vugular carbonates. Under these conditions, log response may be unreliable, core recovery frequently will be incomplete, and correlation of core porosity with log response may be poor.

---

[23] Despite the "antiquity" of this reference, Morris Muskat's insight is no less valid today than it was more than 50 years ago. His book remains a veritable bible of reservoir engineering and is highly recommended reading.

Core depths, which are determined from drillpipe measurements, seldom agree with wireline log depths, and appropriate adjustments must be made before direct comparison is possible. A core gamma log usually will facilitate correlation between core depths and log depths and help to identify intervals where cores were not recovered. Accepted procedure is to correct core depths to wireline log depths and attempt to correlate log response with core porosity at the log-adjusted depth. Frequently, this will result in a poor correlation, especially in laminated sandstones and heterogeneous carbonates.

One procedure that may improve the correlation is to develop a running average of core porosity over several feet, double weighting the central values. An averaging process may come closer to the averaging process inherent in most porosity logging tools, which integrate formation response over several feet. The procedure used to develop a running average of core data should be compatible with logging tool response; consultation with logging company engineers is recommended.

An alternative procedure to improve the correlation between log response and core porosity has been discussed by Havlena [1966]. This procedure, called a forced fit, treats core porosity and log response as separate sets of spatially random variables, not necessarily dependent on log or core depth. For each well zone of interest, core porosity and log porosity are ordered separately in ascending order, regardless of the core or log depth. The sets of values, in ascending order, are plotted, and a least-squares fit is used to determine the relation between log and core porosity.

Usually, porosity from full-hole cores is considered the standard against which log response is calibrated. However, in highly fractured or vugular carbonates, neither core nor log data may provide representative porosity data. In addition, core analysis of loose, unconsolidated sands[24] may result in abnormally high porosity values, even though the cores are taken with a rubber sleeve and compacted before the analysis [Elkins 1972].

Well Test and PVT Data

For black oils, initial GOR's determined from well tests (those at initial reservoir pressure) should be in reasonable agreement with GOR's determined from laboratory flash liberation tests. If there is not reasonable agreement, consideration should be given to retesting the well and/or reviewing the laboratory data. Additional points regarding well tests and PVT data were in Sec. 3.3.2.

For progressively more volatile oils, the laboratory flash liberation tests might not replicate field separator conditions, and adjustments might have to be made. In any event, the laboratory formation volume factor at the bubblepoint, B_{ob}, and the corresponding GOR, R_{sb}, should be adjusted to compensate for separator conditions in the field, before being used in Eqs. 3.1 and 3.2. The procedure for making these adjustments is discussed by Amyx *et al.* [1960], Moses [1986], and McCain [1990].

If bottomhole samples were used for PVT analysis, there is a possibility of nonrepresentative sampling, especially if the reservoir has low permeability and was at or below the initial bubblepoint pressure at the time of sampling. GOR's from initial well tests may be significantly different from those determined from laboratory separator tests. Recommended procedure is to: (a) rely on empirical correlations based on data from initial well tests to estimate parameters needed in Eqs. 3.1 and 3.2, pending obtaining representative

---

[24] Mattax *et al.* [1975] discussed methods for core analyses of friable and unconsolidated sands. However, caution should be exercised in utilizing core data from such sediments, as petrophysical properties may have been altered by the coring/sampling process.

samples or (b) adjust the sample data to reflect the correct initial bubblepoint pressure, following the procedure discussed by McCain [1990].

When material balance or reservoir simulation calculations are made, the PVT data should be smoothed using the Y and the ΔV functions [Amyx *et al.* 1960]. The Y function is used to smooth the total relative volume data, B_t; the ΔV function, the relative oil volume data, B_o. These smoothing procedures should be performed by the laboratory doing the PVT studies. There may be circumstances, however, where these procedures are not utilized or obscure poor-quality data. Thus, the raw data should always be reviewed carefully.

3.5 Initial Reservoir Conditions

The PVT terms in Eqs. 3.1 through 3.4 should represent fluid properties at reservoir temperature and initial reservoir pressure. Highly accurate determination of reservoir temperature and initial reservoir pressure may not be critical for volumetric calculations, given the uncertainties in other factors. However, accurate determination of reservoir temperature may be critical for PVT analyses, especially for near-critical fluids.

For most reservoir engineering calculations, pressure and temperature data are adjusted to the depth that corresponds to the centroid of the principal reservoir fluid, oil or free gas. For oil reservoirs with significant gas caps, however, it may be appropriate to adjust such data to the depth of the initial gas/oil contact. For reservoirs producing with significant water influx, where material balance calculations may be appropriate, pressures should be adjusted to the depth of the initial water level for proper treatment of water influx equations.

3.5.1 Pressure

Accurate determination of initial reservoir pressure is critical if material balance methods are to be used to estimate oil or free gas initially in place.[25] In addition, for an oil reservoir, comparison of initial pressure with bubblepoint pressure can provide useful information regarding whether an initial gas cap is a reasonable possibility.

As noted in Sec. 6.5, accurate determination of initial reservoir pressure is especially important in geopressured gas reservoirs. Also, as discussed in **Appendix G**, determination of aquifer pressure and aquifer gradient may be important in estimating initial hydrocarbon/water contacts. In this context, the reader is referred to Dake [1994].

Initial reservoir pressure measured in new discoveries should be checked against local pressure gradient information. Subnormal initial pressure might be indicative of partial drainage from other wells or reservoirs producing from the same formation and may establish a need for early pressure maintenance. In a classic study of aquifer interference in the East Texas basin (U.S.), it was determined that production from the East Texas field had reduced initial reservoir pressure in the Hawkins field, which was 14 miles west from the East Texas field, by 280 psi [Bell and Shepard 1951, Rumble *et al.* 1951].

In certain areas, however, subnormal[26] reservoir pressures may be attributed to the geologic setting rather than to partial drainage. For example, several geologic basins in North America have gas-saturated sections that exhibit subnormal initial reservoir pressure [Davis 1984]. Such areas include Elmworth and Medicine Hat fields (Alberta, Canada) and Eastern Ohio and Wyoming (U.S.). Gas accumulations in these areas typically occur in low-porosity, low-permeability rocks and do not have an associated downdip aquifer. For example, in the

---

[25] As discussed in Sec. 1.1.2 under Quality and Quantity of Data, and in Sec. 4.1.2 under Data Required, reliable application of the material balance method mandates historical bottomhole pressure data in addition to accurate initial pressure.

[26] In this context, the term "subnormal" refers to situations where the initial pressure gradient is less than about 0.4 psi/ft.

Elmworth field, which produces from a depth of approximately 6,500 ft, gas productive rocks have porosity ranging from 4.5 to 12.5%. Permeability ranges from 0.5 to 150 md. Initial pressure gradients in this field range from 0.30 to 0.33 psi/ft.

The methodology of measuring and interpreting bottomhole pressure data to estimate static reservoir pressure has been discussed by Matthews and Russell [1967], Earlougher [1977], and Plisga [1987]. Detailed discussion of these procedures is beyond the scope of this work.

3.5.2 Temperature

Accurate determination of reservoir temperature is important because laboratory PVT data should be measured at reservoir temperatures. In addition, accurate reservoir temperature is necessary for calculations of the performance of gas reservoirs. It is good practice to determine the initial temperature/depth profile in each producing well in a field using a continuous recording thermometer of the type described by Plisga [1987]. Such baseline data can be useful in diagnosing well problems.

Caution should be exercised in using bottomhole temperatures observed during openhole logging of deep wells, as these temperatures will have been reduced below normal formation temperature by circulation of the drilling fluid. Depending on circumstances, reductions on the order of 10 to 20°F may be observed.

3.5.3 Regional Correlations

In the early stages of development and production of a field or reservoir, measured bottomhole pressure and temperature data may be too sparse to provide a reliable estimate of initial conditions. In this case, regional correlations may be helpful.

Initial Reservoir Pressure

Typically, correlations[27] for initial formation pressure vs. depth are of the following form:

$$p_i \approx g_P D + p_r \dots\dots\dots\dots\dots \tag{3.7a}$$

or

$$p_i \approx g_P D_{ss} + p_r \, ,\dots\dots\dots\dots\dots \tag{3.7b}$$

where p_i = initial pressure, psi,
       g_P = pressure gradient, psi/ft,
       D = depth, feet,
       D_{ss} = depth subsea, feet, and
       p_r = reference pressure at $D = 0$, or at $D_{ss} = 0$, psi.

In many geologic settings, initial pressure gradients vary from about 0.44 to 0.46 psi/ft, which are attributed to an overlying column of fresh to salt-saturated water. However, in some areas (the North Sea, for example) initial gradients may be slightly greater, which are attributed to partial isolation caused by faulting and/or stratigraphy. In other areas, (the U.S. Gulf Coast and offshore Nigeria, for example) initial gradients may be substantially higher,[28] which may be attributed to undercompaction.

---

[27] The notation ≈ is used throughout this work, rather than =, to designate equations derived from correlations.
[28] Geopressures, the term to describe such gradients, are discussed in Sec. 6.5.

Determination of initial pressure in the aquifer is an important aspect of reservoir characterization [Dake 1994]. For example, such pressures may be used to estimate the vertical position of the initial free water level, as discussed in Appendix G.

Formation Temperature

Correlations for formation temperature are functionally similar to those for initial reservoir pressure:

$$T_f \approx g_T D + T_r \dotfill \text{(3.8a)}$$

or

$$T_f \approx g_T D_{ss} + T_r \, , \dotfill \text{(3.8b)}$$

where T_f = formation temperature, degrees,
    g_T = thermal gradient, degrees/foot,
    D, D_{ss} = as previously defined, and
    T_r = reference temperature at $D = 0$, or at $D_{ss} = 0$, degrees.

Correlations for initial reservoir pressure and formation temperature for several regions in the world are listed in **Appendix H.** These correlations were compiled from numerous sources and there is no consistency regarding units (psia vs psig) or reference depths (D vs. D_{ss}). Size of databases and regression coefficients are rarely reported. Accordingly, reservoir temperature and initial pressure estimated using these correlations should be considered preliminary and not a substitute for actual measurements.

Simple, linear correlations of reservoir temperature and initial pressure vs. depth may not be valid in areas where initial reservoir conditions may be influenced by tectonic stress, undercompaction, or major unconformities. In these settings, there may be significant departures from a linear trend of reservoir temperature and initial pressure versus depth. Such departures have been observed, for example, offshore Texas-Louisiana [Dickinson 1953], the California Coast Ranges [Berry 1973], offshore Nova Scotia [Mudford and Best 1989], Prudhoe Bay,[29] and the Delaware basin in west Texas-New Mexico [Luo et al. 1994].

3.6 Volumetric Mapping

3.6.1 Types of Traps

The degree of uncertainty in a volumetric estimate of oil and gas initially in place is influenced more by the degree of geologic complexity of the reservoir and by the amount and quality of geologic data[30] than by almost any other combination of factors.

For example, regarding the Leman field in the North Sea, one of the world's largest gas fields, Craig et al. [1977] commented "an accurate volumetric determination of the gas in place is impossible because of seismic interpretation problems and the limited well control that results from the cluster-type development...the situation is aggravated further by a complex fault system."

---

[29] Arlie Skov, personal communication, 2000.
[30] Geologic data includes geophysical data.

In general, three broad categories of reservoirs may be identified: (a) those controlled by structure, (b) those controlled by stratigraphy, and (c) those controlled by a combination of structure and stratigraphy. Reservoirs on simple anticlinal closures, uncomplicated by faulting and containing laterally continuous porosity are relatively easy to map. The Santa Fe Springs field in California and the Abqaiq pool in Saudi Arabia are examples of simple structural traps. Volumetric mapping of these types of fields is subject to the least uncertainty. At the other extreme are stratigraphically complex reservoirs on highly folded and faulted structures. The Painter Reservoir field in the U.S. western overthrust belt and the Maloosa field in Italy are examples of complex structural traps. These types of fields are extremely difficult to map, and reserve estimates based on volumetrics are subject to substantial uncertainty.

Engineers responsible for estimating reserves should become familiar with the geologic setting in which the reserves occur and with the stratigraphic and structural complications associated with that setting. Reservoir geology impacts all phases of reservoir management, including drilling and completion techniques, well testing, well and reservoir performance, reserve estimation, and improved recovery techniques. Levorsen [1967] and Lowell [1985] provided treatments of petroleum geology that are recommended reading. In addition, the American Assn. of Petroleum Geologists (AAPG) publishes, on a continuing basis, geologic field case histories worldwide. Some of these publications are listed in the References.

3.6.2 Volumetric Mapping of Reservoirs

Volumetric mapping of reservoirs is highly specialized and, depending on circumstances, may require close collaboration between geophysicists, sedimentologists, production geologists, petrophysicists, and reservoir engineers. Thus, this section should not be considered any more than an overview of the topic. Depending on circumstances, it may be cost-effective to use computer technology. Because of rapid changes in this area, however, any reference here would be out of date, and consultation with current users and with software vendor's technical representatives is recommended.

After a discovery well and one or more confirmation wells have been drilled on a prospect, seismic data (if available) should be integrated with subsurface (well) data in the preparation of structure maps on key horizons. As the field or reservoir is developed, core, log, and well test information should be integrated into the mapping process to delineate gas/oil, gas/water, or oil/water contacts on each reservoir, as applicable.

In structurally controlled accumulations with mappable fluid contacts, a procedure first described by Wharton [1948] may be appropriate. Tearpock and Bischke [1991] provide details of this procedure. As discussed by these authors, gross and net sand isopach[31] maps should be made for each productive horizon. Fluid/fluid contacts should be transferred to the net sand isopach for each reservoir and net hydrocarbon isopachs should be prepared. PC programs to perform many of these tasks are commercially available. Contouring algorithms, vary, however, and may lead to interpretations that are not consistent with the structural/stratigraphic setting of the mapped accumulation. Thus, such programs should be used with caution.[32]

---

[31] The term "net sand isopach" refers to an isopach of net porous interval, either sand or carbonate, that is judged to be productive of fluids (either hydrocarbons or brine) at commercial rates.

[32] For some tips to help identify incorrect mapping, readers are referred to "Quick Look Techniques" at www.scacompanies.com (June 2000), a Web site maintained by Subsurface Consultants & Assocs. L.L.C. No endorsement is intended by this reference.

Depending on the purpose for mapping, or on the anticipated reservoir drive mechanism, different initial oil/water "contacts" may be considered. As illustrated by **Fig. 3.1**, there are several possibilities:

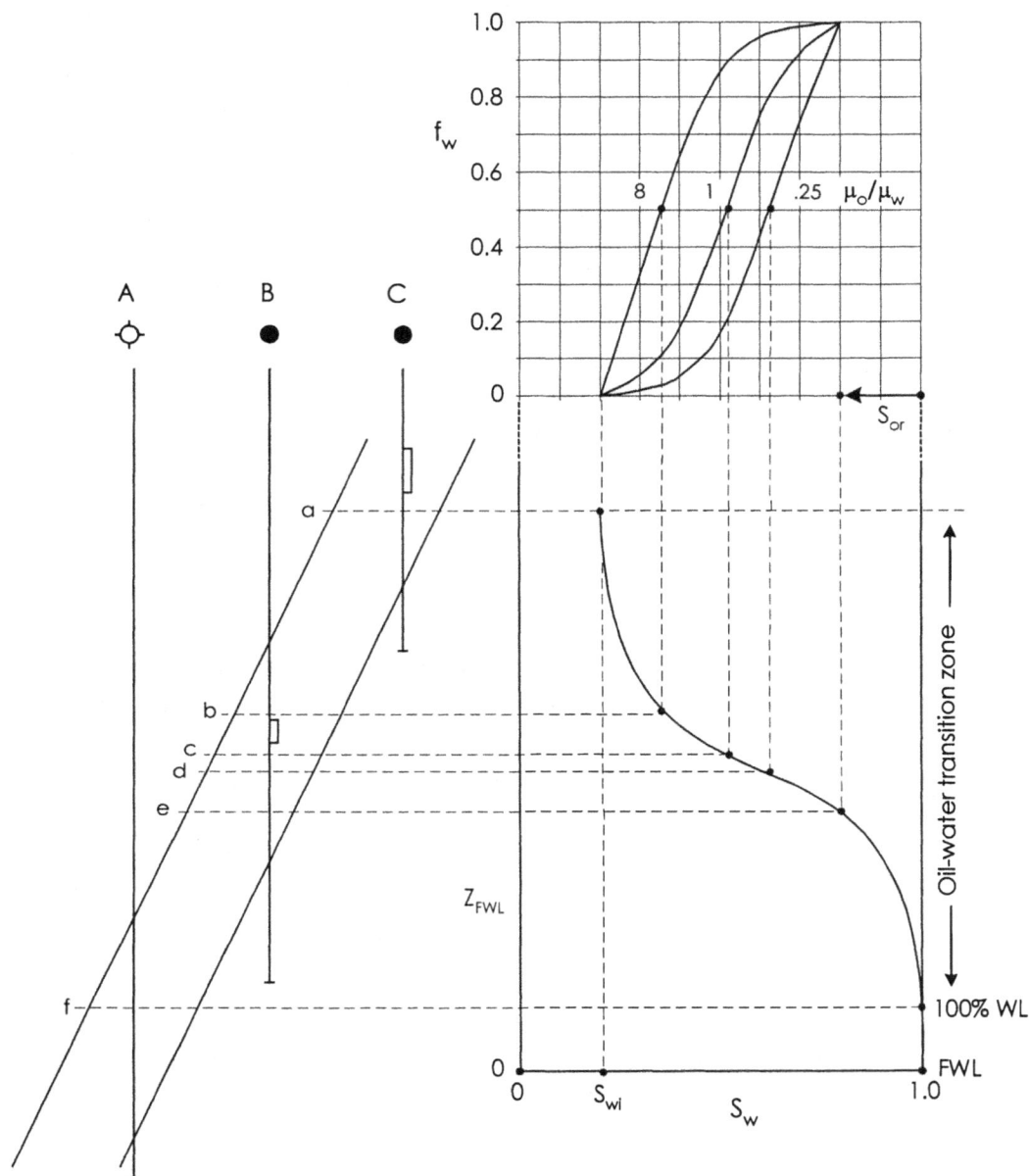

Fig. 3.1—Typical fractional flow curves and capillary pressure curve illustrating a method to estimate the vertical position of the 50% water level for various oil/water viscosity ratios.

- The "50% water level," defined as the depth below which a well completion interval will produce in excess of 50% water,[33] may be observed by testing[34] or calculated

---

[33] Depending on circumstances, the capillary properties of reservoir rocks may exhibit significant spatial variation. Thus, the emphasis here is on wells, rather than the reservoir "in-toto."

from capillary pressure, PVT, and water/oil relative permeability data, as done for Fig. 3.1. As illustrated, for favorable water/oil mobility ratios the 50% water level generally is observed at log water saturations of about 50% or greater, depending on rock quality; for unfavorable water/oil mobility ratios the 50% water level generally is observed at log water saturations less than 50%, depending on rock quality and actual water/oil mobility ratio.

- The "100% water level," defined[35] as the depth below which a well completion interval will produce 100% water, may be observed by testing or may be calculated, as above. As illustrated by Fig. 3.1, the 100% water level generally is observed at log water saturations approaching $(1 - S_{or})$, depending on rock quality.

- The "free water level" (FWL), defined as the datum of zero capillary pressure, below which there generally will be no "shows" in the cores,[36] generally cannot be observed on resistivity logs. In a common reservoir, however, the FWL is the zero datum for vertical adjustment of all capillary pressure curves. As discussed in Sec. H.4.2, in a common reservoir, the FWL is the point of intersection of pressure/depth traverses of the oil and water columns.

In hydrodynamic traps, the planar surface corresponding to the FWL typically is tilted, as discussed by Hubbert [1954]. In low-porosity heterogeneous rocks, the 50% and the 100% water levels typically will not be planar but will be controlled by the local capillary properties of the reservoir rock/fluid system, as illustrated by **Fig. 3.2.**

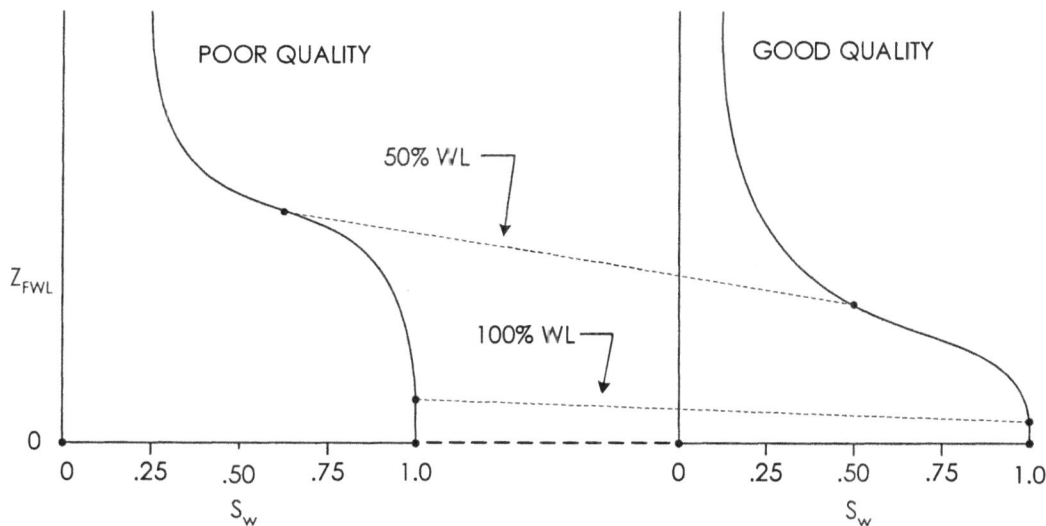

Fig. 3.2—Influence of rock quality on vertical position of 100% and 50% water levels.

In stratigraphic traps, where there may be an areal gradation in rock properties, the initial water level may appear to be tilted, which might erroneously be attributed to hydrodynamic effects.

---

[34] In this context, the term "testing" includes drillstem tests (DST), repeat formation tests (RFT) and production tests.

[35] As shown by Fig. 3.1, the "100% water level" also is defined as the elevation (above the free water level) at which the drainage capillary pressure curve departs from 100% water saturation.

[36] The presence of "shows" in cores below an apparent FWL usually is interpreted to mean there has been upward movement of the oil column subsequent to initial migration of oil into the reservoir.

For reservoirs producing with water influx, it may be appropriate to map the current, or producing, water level, for example, to estimate volumetric sweep efficiency. Such water levels may be determined by:

- openhole logs in wells penetrating the reservoir after production commenced,
- through-casing logs in wells penetrating the reservoir but completed in deeper zones, and
- flowmeters in wells completed in the reservoir.

Caution should be exercised in placing too much reliance on flowmeter data because wellbore coning may cause producing water levels to be higher than in interwell areas. Also, producing water levels, in general, may not be planar, being influenced by water/oil mobility ratio and permeability heterogeneities.

3.6.3 Estimating Net Pay

Estimation of net pay is one of the most important steps in volumetric mapping. Unfortunately, estimation of net pay is one of the most subjective aspects of such mapping. It is influenced by: (a) the amount and quality of log, core, and test data, (b) the nature of the rock/fluid system, and (c) the anticipated drive/recovery mechanism, among other factors. In the following discussion, the term "net pay" refers to "true vertical net pay"—i.e., logged net pay thickness corrected for borehole inclination.[37]

Snyder [1971] observed that net pay in a given reservoir might be determined for one of several different purposes. The procedure to determine net pay and the results might be different for each purpose, as the following examples illustrate:

- When a well log initially is being evaluated to select an interval for completion, net pay generally will include only those intervals judged likely to contribute to well inflow at commercial rates—i.e., the log analyst will tend to be conservative in estimating net pay.

- When a reservoir is being evaluated to determine total hydrocarbons in place (for example, as an independent check on material balance calculations), net pay usually will include all hydrocarbon-bearing intervals likely to contribute to the "energy balance." Net pay determined for this purpose generally will be more than that estimated during initial completion.

- If a waterflood is being considered, net pay should include only those intervals considered "floodable." This criterion implies interwell continuity. Also, the methods used to define "floodable" intervals are subjective and may exclude intervals that will contribute to recovery by imbibition. Net pay determined for this purpose may be less than that determined for either of the first two purposes.

- If a reservoir is to be unitized and net pay is part of the unitization formula, the determination of net pay may be subject to arbitrary rules to insure "uniformity," which may result in determining net pay unrelated to any of the considerations noted above; see, for example, Sec. 7.3.

To facilitate visualization of pertinent data, composite logs should be prepared for all wells in the reservoir under study. Regional and individual variations are to be expected. In *general,* however, the following steps are involved:

---

[37] If "Wharton's method" is used for volumetric mapping, distinction must be made between "net pay" and "net sand."

1. Determine "gross interval" by establishing the top and bottom of the zone of interest. In sand-shale sequences, the inflection point of the SP[38] curve or the midpoint of the gamma-ray deflection[39] usually are considered to be zone boundaries. In carbonates, it may involve using one or more of the porosity curves to establish a minimum porosity or "porosity cutoff,"[40] or it may involve using a combination of logs to determine lithologic top and bottom.

2. Exclude nonpay intervals within the gross interval, considering a maximum shale cutoff, a minimum porosity cutoff, a maximum water saturation, exclusion of intervals where there is a specified degree of reversal in either the SP or the gamma-ray curve, or some combination of the foregoing.

PS-CIM [1994] reports "generally accepted (minimum) cutoffs" as follows:

Permeability*
    Medium to high gravity oil: 1 md
    Wet gas: 0.5 md**
Porosity#
    Carbonates: 2 to 4%
    Sandstones: 7 to 10%
    Heavy oil (sandstones)   26 to 28%
Water saturation   50 to 60%

* Unstressed, horizontal permeability from measurements on cores.
** As low as 0.1 md for dry gas under special circumstances.
\# Unstressed, from cores.

For other regions, the Schlumberger[41] Well Evaluation Conference publications contain typical cutoff data.

Despite its antiquity, a Microlog™ can be very useful in estimating net pay. A common rule of thumb is that zones with less than about 1 md permeability will not exhibit Microlog™ separation. Also, depending on hole condition, a caliper log may be useful in identifying vertical intervals where the presence of filter cake may indicate permeable intervals.

In another commonly used log analysis procedure, called the ratio method, the water saturation calculated in the uninvaded zone, S_w, is compared to the water saturation calculated in the invaded zone, S_{xo}. If the ratio S_w/S_{xo} is less than about 0.7 in sandstones (less than about 0.6 in carbonates), the presence of movable hydrocarbons may be inferred, and the zone could be considered net pay.[42]

In reality, the tacit objective of initially estimating net pay is to determine which intervals in each well zone can be expected to contribute to fluid flow into the wellbore under the anticipated method of well and/or reservoir operation. As noted by Cobb and Marek [1998], the criteria for determining net pay under primary drive should include fluid mobility, k/μ, and pressure differential. Permeability data, however, are seldom available on all wells in an

---

[38] SP = spontaneous (or self) potential.
[39] Deflection from the shale line—from the maximum reading through the section of interest—to clean sand—to the minimum reading.
[40] Choice of minimum porosity, or porosity "cutoff," typically is quite subjective and is related to the minimum permeability considered productive in the area, which depends on petrophysics, reservoir fluid, and drive mechanism.
[41] This reference is not intended to be limiting and does not imply endorsement.
[42] For additional details of this procedure, see Asquith and Gibson [1982].

accumulation. Typically permeability is estimated based on: (a) correlations between porosity and permeability measured in cores and/or (b) empirical correlations.

In some cases (for example, shaly, laminated sand reservoirs[43] and many carbonate reservoirs), it may be difficult, at best, to determine net pay with an acceptable degree of confidence.[44] A net pay isopach made under this type of uncertainty may be subject to frequent revision as interpretive procedures are revised or as new data become available. In this situation, it may be desirable to map gross pay[45] and apply an average net-to-gross ratio to account for nonproductive rock.[46] The net-to-gross factor may be revised as warranted, thereby avoiding the need to remap the entire reservoir. If there are significant areal variations in the net-to-gross ratio, such a procedure would not be appropriate.

Irrespective of the methodology used to determine net pay, the porosity and water saturation values used in the volumetric equations—Eqs. 3.1 and 3.3—should be consistent with the net pay cut-off values. This point is discussed further in Sec. 9.9.2.

Havlena [1966], George and Stiles [1978], Sneider *et al.* [1977], Wilhite [1986], PS-CIM [1994] and Cobb and Marek [1998] provided additional discussions regarding net pay.

3.6.4 Porosity/Net Pay/Saturation Mapping (Isovols)

In reservoirs where there are significant spatial variations in porosity and/or water saturation, it may be preferable to map these variations, rather than attempting to apply average porosity or average water saturation over the entire reservoir.[47] Depending on reservoir stratigraphy, isovols may be needed for each stratigraphic unit. An isovol map is similar to an isopachous map, except that the contours represent the product

$$(POROSITY) \times (NET\ PAY) \times (HYDROCARBON\ SATURATION).$$

Planimetered volume is hydrocarbon pore volume (HCPV). Amyx *et al.* [1960] and Tearpock and Bischke [1991], among others, discuss various methods to determine the volume from isopachous maps.

3.6.5 Reservoir Limits

In general, two types of reservoir limits may be identified:

- those that externally bound a fluid accumulation and define the zero isopach contour for either oil or free gas, here defined as external limits, and
- those that are internal to a fluid accumulation and may affect the flow of fluids in the reservoir, here defined as internal limits.

External Limits

External reservoir limits, as described above, include:

- gas/oil contacts,
- oil/water contacts,

---

[43] For examples of log response in this type of formation, which is called low resistivity, low contrast (LRLC), the reader is referred to publications of the Houston and New Orleans Geological Socs. [1993] and the Rocky Mountain Assn. of Geologists [1996].
[44] In such cases, a probabilistic approach may be warranted, as discussed in Sec 9.10.2.
[45] Gross pay is defined here as a hydrocarbon-bearing, true vertical, stratigraphic interval mapped as a common reservoir, which may contain intervals of both commercially productive and noncommercially productive rock.
[46] As discussed in Sec. 9.10.2, it may be appropriate to treat the net-to-gross ratio as a stochastic variable, a procedure followed by some companies in making probabilistic estimates of reserves.
[47] Averaging procedures were discussed in Sec 3.4.2.

- gas/water contacts,
- bounding—i.e., sealing—faults,
- porosity or permeability pinchouts, and
- unconformities and/or other truncations—e.g., diapiric salt or shale.

In many cases, especially early in the drilling and production stages of reservoir life, it may be difficult to determine the spatial position of these limits with tolerable uncertainty.

Depending on circumstances, 3D seismic surveys may provide insights to help reduce the uncertainty in mapping these limits. In this regard, new techniques have been developed to process seismic data to map discontinuities like faults and stratigraphic features [Bahororich and Farmer 1995].

In cases where a well has penetrated a gas/oil contact, log and core data may not be sufficient to locate the fluid/fluid boundary. RFT tests may be precluded because of poor hole conditions. Such conditions are not uncommon in Nigeria, for example, and gas chromatography has been successful in helping to determine reservoir fluid type and fluid/fluid boundaries in that area [Baskin *et al.* 1995].

In *structurally controlled* reservoirs, for example, wells may not have penetrated an oil/water contact, and the subsurface information may be limited to "lowest known oil" (LKO) and "highest known water" (HKW). Procedures to map and classify accumulations with such uncertainties vary, depending on objectives. Industry practice and recommended procedures to estimate the structural position of fluid/fluid contacts with different degrees of confidence are summarized in Appendix G.

In *stratigraphically controlled* reservoirs, the porosity or permeability limit may be only approximately delineated by dry holes around the periphery of the accumulation. The undrilled area between commercial wells and dry holes define a "band of uncertainty" around the accumulation that may be one or more locations wide. Undeveloped reserves—proved, probable, or possible—might be assigned to locations in this area, depending on:

- the quality of reservoir rock observed in the commercial wells around the periphery,
- the depositional environment of the reservoir rock,
- the distance to the nearest dry hole,
- the apparent rate of thinning of reservoir rock, and/or
- the minimum thickness of reservoir rock needed to support a commercial well.

Internal Limits[48]

In addition to external limits, as discussed above, there may be internal limits, which may have a significant influence on recovery efficiency. Discussed briefly are two such limits: internal faulting and depositional/diagenetic discontinuities.

Internal faults may be partial or complete barriers to fluid flow along all or part of their length. Under pressure gradients imposed by production, such barriers may respond differently than they would under static conditions.[49]

Usually, internal faults exhibit smaller vertical displacements than bounding faults. In many cases, such faulting may not be detectable with existing well control or even seismic

---

[48] The term "compartmentalization" has become part of reservoir engineering lexicon to describe such limits.

[49] There has been considerable dialog within the geologic community regarding the criteria to distinguish between sealing and nonsealing faults. Unfortunately, publications are sparse because much of the work is considered proprietary.

data. Such faults may not become apparent until after a period of sustained production, as was the case in the Cormorant field in the North Sea [Stiles and McKee 1991] and other fields in this area.

Depositional discontinuities internal to a reservoir may have an adverse influence on recovery efficiency. In the U.S. Gulf Coast, for example, Hartman and Paynter [1979] reported on composite reservoirs producing by strong waterdrive where high-permeability channel sands flooded out in preference to low-permeability fringe sands. Such occurrences in comparable geologic and reservoir settings elsewhere have been reported by others.

Several methods have been used to detect internal limits, including: (a) RFT depth/pressure traverses, (b) differences in hydrocarbon isotopes, (c) $^{87}Sr/^{86}Sr$ isotope ratios [Mearns and McBride 1999], and (d) differences in saturation pressure.

3.6.6 Capillary Transition Zones

The relationship between reservoir capillary pressure and vertical distance, or height, above the FWL is

$$Z_{FWL} = P_c / (\rho_w - \rho_h) , \dots\dots\dots\dots\dots\dots\dots\dots\dots\dots\dots\dots\dots\dots\dots\dots\dots\dots (3.9a)$$

where Z_{FWL} = height above FWL, ft,
$\quad\quad\ P_c$ = capillary pressure, psi,
$\quad\quad\ \rho_w$ = density of reservoir brine, psi/ft,* and
$\quad\quad\ \rho_h$ = density of reservoir hydrocarbon, oil or gas, psi/ft.*

* Units also may be expressed as pounds per cubic foot, in which case the constant 144 must be introduced for dimensional consistency, as in Eq. G.2. If units are grams per cubic centimeter, the constant is 0.433.

As observed in Sec. 3.2.1, if a reservoir includes a significant oil/water transition zone, it might be appropriate to calculate oil initially in place separately for (a) the volume above the transition zone and (b) the volume within the transition zone. Depending on circumstances, it might be appropriate to calculate oil initially in place in the transition zone by integrating S_w vs. Z_{FWL}.

Not all the oil in the transition zone may be considered potential reserves, depending on anticipated drive mechanism. Consideration must be given to both the relative permeability characteristics and the drive mechanism before estimating reserves attributable to the transition zone. Cronquist [1968] reported an estimate of waterflood reserves from an oil/water transition zone in a light oil reservoir (39°API). The transition zone occupied 18% of the initial reservoir rock volume, but contained only 10% of the (initially estimated) ultimate recovery.

Long oil/water transition zones usually are associated with heavy oil, low-permeability rocks, and/or brackish formation water. Short zones usually are associated with medium- to high-gravity oil, high-permeability rocks, and/or highly saline formation water. Gas/oil and gas/water transition zones usually are negligibly small.

In heavy-oil reservoirs, a significant portion of the oil initially in place may be in the oil/water transition zone. On a low relief structure, the entire reservoir may be in the oil/water transition zone, and it might not be possible to make a water-free completion, even at the crest of the structure. For example, Arps [1964] reported a structure in the Tensleep

(Pennsylvanian) sandstone in Wyoming that contained 19°API oil where there was insufficient vertical closure for a commercial completion.

In Sec. 3.4.1, it was noted that the best capillary pressure data to use for calculations of water saturation vs. height are from rock samples that have approximately mean porosity and median permeability. The vertical distribution of reservoir rock above the transition zone level may be determined using a vertical distribution curve, as discussed in Sec. 3.9.1.

Caution should be exercised in using laboratory capillary pressure data to calculate the relation between water saturation and vertical distance above the FWL in a subsurface reservoir. Laboratory data typically are determined using either a mercury/air system or an air/brine system under ambient conditions. In either case, the conversion from laboratory capillary pressure, P_{cL}, to reservoir capillary pressure, P_{cR}, is of the form

$$P_{cR} = P_{cL}(\sigma_R \cos\theta_R)/(\sigma_L \cos\theta_L) , \dots\dots\dots\dots\dots\dots\dots\dots\dots\dots\dots (3.9b)$$

where σ_L = interfacial tension between laboratory fluids, usually mercury/air or air/brine, dynes/cm,

θ_L = fluid/fluid/solid contact angle at laboratory conditions, degrees,

σ_R = interfacial tension between reservoir fluids, either oil/water or gas/water, dynes/cm, and

θ_R = fluid/fluid/solid contact angle at reservoir conditions, degrees.

Values for σ and θ typically used in Eq. 3.9b are tabulated below.

	Interfacial Tension, σ (dynes/cm)	Contact Angle, θ (degrees)
Laboratory		
Mercury/air	470 to 480	140
Air/brine*	70 to 75	0
Reservoir		
Oil/brine*	30	30
Gas/brine*	50	0

* The term "brine" is used here to denote an aqueous fluid that may (or may not be) chemically "equivalent" to interstitial water in the reservoir pore space.

The interfacial tensions and contact angles for laboratory systems typically used to measure capillary pressure are known with a reasonable degree of accuracy. However, these properties under reservoir conditions are difficult to measure directly and usually are estimated. Thus, conversion from laboratory to reservoir capillary pressure using Eq. 3.9b may be subject to significant error, the degree of which cannot be determined. Accordingly, there may be significant (but indeterminate) error in calculations of reservoir water saturation vs. vertical distance above the FWL.

An additional complication to using capillary pressure curves for reservoir characterization is the likelihood the capillary properties of friable rocks may be different under reservoir stress than they are under ambient conditions. For example, Ajufo [1993] has reported that stressed cores exhibited higher irreducible water saturation than unstressed cores.

3.7 Equations To Estimate Primary Recovery Efficiency

3.7.1 Fluid-Type/Drive Mechanism Matrix

Fig. 3.3 shows analytical equations and correlations to estimate primary recovery efficiency, which are organized by reservoir fluid type and drive mechanism. The principal horizontal axis across the top of Fig. 3.3 shows the two principal hydrocarbon fluid phases in the reservoir, oil and gas. The secondary horizontal axes show, from left to right, the spectrum of fluid types from black oil to nonretrograde gas, with typical API gravities and GOR's and/or CGR's. As noted, nonretrograde gases, where $T_f > T_c$, include wet gas and dry gas. The vertical axis along the left side represents pressure, decreasing downward from initially undersaturated conditions to saturation pressure to abandonment pressure. Primary drive mechanisms for each of the two fluid phases, oil and gas, are shown in vertical lettering on this axis. The pressure-fluid region of potential two-phase flow (oil plus free gas) is bounded by the heavy dashed lines, which identify the region inside the phase envelopes for black and volatile oils shown in **Fig. 3.4a**. The pressure-fluid region of retrograde condensation is shown by the medium dotted lines, which identify the region inside the phase envelope for retrograde gas shown in **Fig. 3.4b**.

Reservoir Fluid Types

For the purposes of this work, hydrocarbon reservoir fluids are subdivided into black oils, volatile oils, retrograde gases, and nonretrograde gases.[50] As observed by Novosad [1996], "classification is based on the position of a fluid two-phase envelope relative to reservoir and surface conditions of pressure and temperature." Fig. 3.4a shows typical phase envelopes[51] for each of these fluids, which may be described as follows:[52]

- *"Black" oil*: oils with initial solution GOR's up to about 1,750 scf/STB; stock-tank gravity ranges from about 8 to 40°API.[53]

- *Volatile oil*: transitional between black oil and retrograde gas; reservoir temperature near, but less than, critical temperature; initial solution GOR ranges from about 1,750 to 3,200 scf/STB; stock-tank gravity ranges from about 40 to 60°API.

- *Retrograde gas*: as shown by Fig. 3.4b, reservoir temperature, T_f, between the critical temperature, T_c, and the cricondentherm,[54] T_{ct}; subject to retrograde condensation (and possible loss) of liquids in the reservoir if reservoir pressure decreases below the dewpoint pressure, p_d.

- *Nonretrograde gas*: as shown by Fig. 3.4b, reservoir temperature greater than the cricondentherm; not subject to retrograde condensation as the reservoir pressure decreases, because the path of isothermal expansion (a-b) does not enter the two-phase envelope; some engineers subdivide nonretrograde gases into wet gases and dry gases, with the demarcation at about 10 STBC/MMscf.

---

[50] McCain [1990] and others recognize five fluid types, subdividing nonretrograde gas into wet gas and dry gas. For material balance analysis, discussed in Sec. 4.4.1. both wet and dry gas may be treated as constant composition reservoir fluids.

[51] The author is indebted to Maria Barrufet, Texas A&M U., for providing these phase envelopes for typical reservoir fluids.

[52] For detailed discussions of reservoir fluid types and phase envelopes, please refer to Standing [1952], Amyx *et al.* [1960], Craft and Hawkins [1959], Moses [1986], McCain [1990], McCain and Bridges [1994], Novosad [1996], and Barrufet [1998].

[53] The 8°API lower limit is consistent with the definition for crude oil in Sec. 1.1.2; 15°API, shown on Fig. 3.3, is the lower limit of the gravity range for several empirical correlations for recovery efficiency.

[54] The cricondentherm is defined as the maximum temperature for a given reservoir fluid at which two phases can coexist.

FLUID -->	OIL ($T_f < T_c$)		GAS ($T_f > T_c$)	
TYPE -->	BLACK	VOLATILE	RETROGRADE ($T_f < T_{ct}$)	NON RETROGRADE ($T_f > T_{ct}$)
				WET DRY
OIL °API -->	15 30 40	50 60	40 20	10
GOR/CGR-->	100	2000/500		

DRIVE MECHANISM (GAS / OIL)

$p > p_s$

WATER DRIVE — complete ($W_e = F_{pR}$)

$$[E_{Ro}]_{xp} = C_e(p_i - p_b)B_{oi}/B_{ob}$$

$$[E_{Rsg}]_{xp} = [E_{Ro}]_{xp}$$

$$[E_{Rg}]_{wd} = 1 - R_{pz}(1 - E_V E_D)$$

$$[E_{Rc}]_{wd} = [E_{Rg}]_{wd}$$

$$[E_{Ro}]_{wd} = 0.55[\phi(1-S_w)/B_{oi}]^{0.0422}[k\mu_{wi}/\mu_{oi}]^{0.077}[S_w]^{-0.1903}[p_i/p_a]^{-0.2159}$$

$p = p_s$

SOLUTION / DISSOLVED — partial ($W_e < F_{pR}$)

$$[E_{Rg}]_{pd} = 1 - p_a Z_i/p_i Z_a$$

$p_a \ll p_s$ ($p \sim 0.1 p_i$)

$$[E_{Ro}]_{sg} = 0.42[\phi(1-S_w)/B_{ob}]^{0.1611}[k/\mu_{ob}]^{0.0979}[S_w]^{-0.3722}[p_b/p_a]^{-0.1741}$$

$$\ln(C_{pvl}) = 2.8\ln(°API) + \ln(p_i) - 0.65\ln(R_t) - 20$$

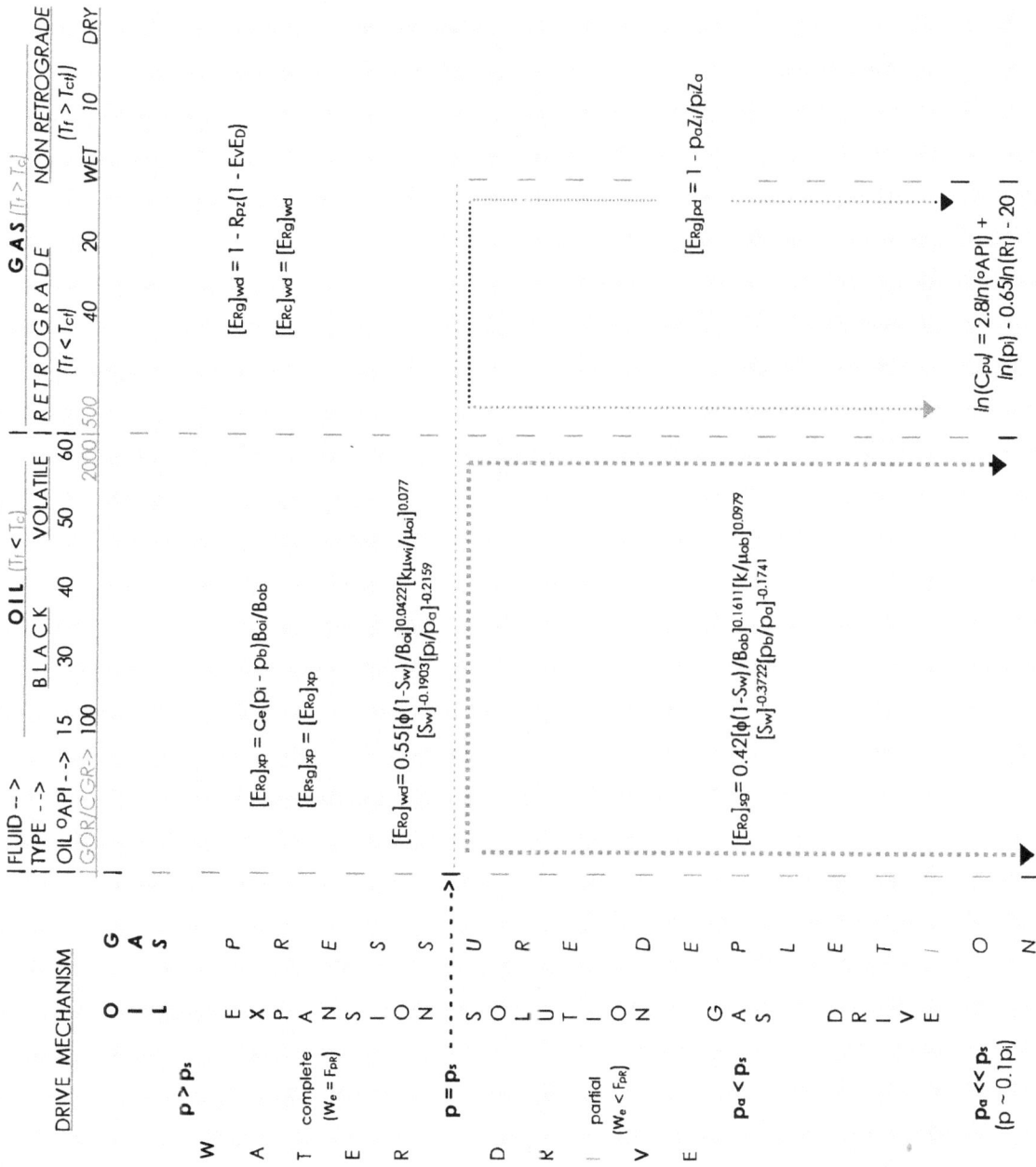

Fig. 3.3—Formulas to estimate primary recovery efficiency for various reservoir fluid types and drive mechanisms.

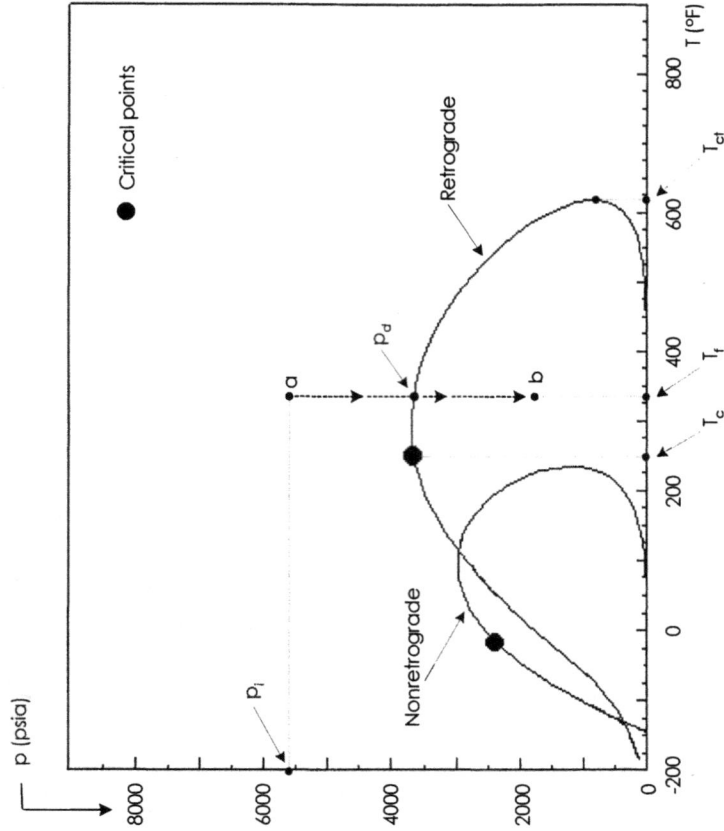

Fig. 3.4b—Phase envelopes illustrating retrograde and nonretrograde gases [courtesy Barrufet].

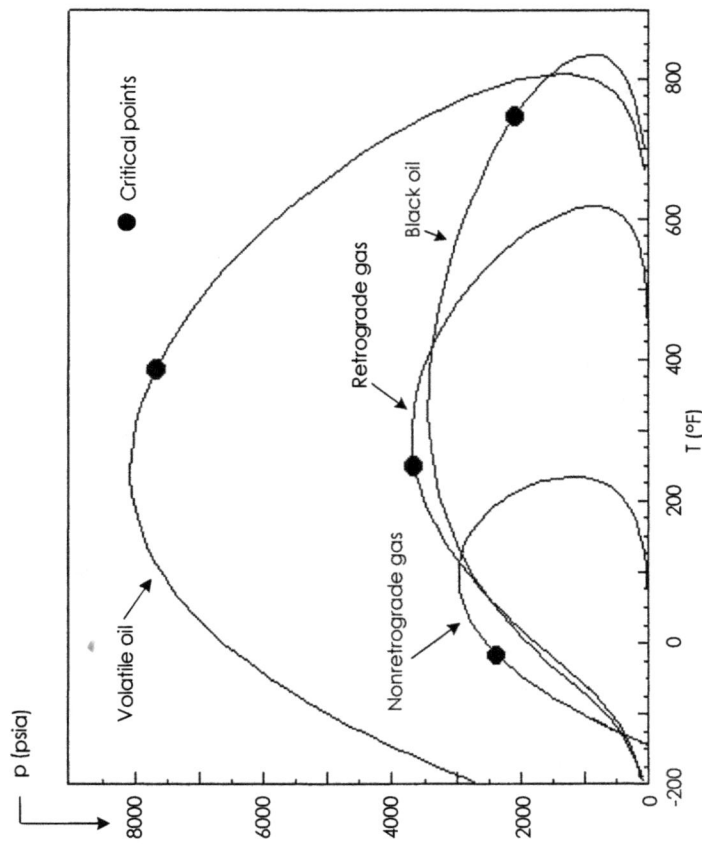

Fig. 3.4a—Phase envelopes for four typical hydrocarbon reservoir fluids [courtesy Barrufet].

Composition ranges for each of the four reservoir fluids are shown on **Fig. 3.5**, in which composition is expressed in terms of three "pseudocomponents:" (a) heptanes plus, (b) intermediates plus carbon dioxide and hydrogen sulfide, and (c) methane and nitrogen. [Cronquist 1979]. Compositions of the four fluids represented by the phase diagrams in Fig. 3.4a are indicated by triangles. Solid triangles show black oil (BO) and volatile oil (VO); open triangles, retrograde gas (RG) and wet gas (WG).

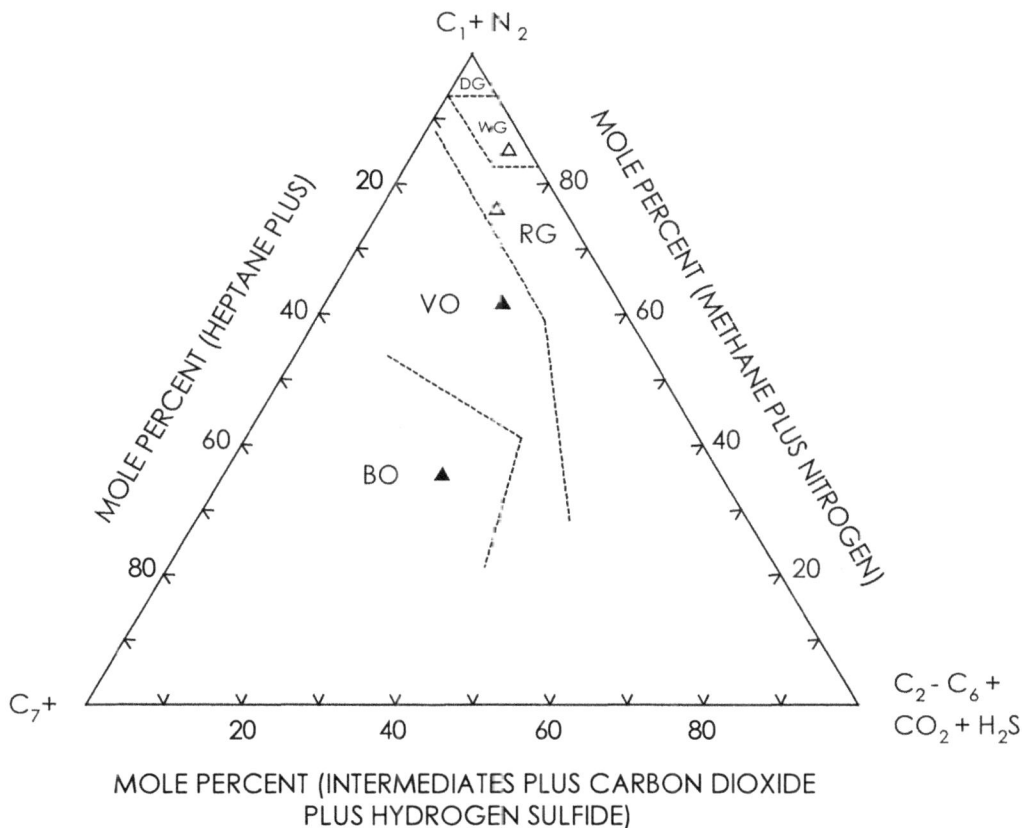

Fig. 3.5—Triangular diagram illustrating pseudocomponents of typical reservoir fluids.

The 15°API lower limit for black oil shown on Fig. 3.3 is the lower limit of the gravity range for several empirical correlations for recovery efficiency; an 8°API lower limit for black oil is consistent with the definition for crude oil in Sec. 1.1.2.

Reservoir Drive Mechanisms

In Fig. 3.3, reservoir mechanisms are defined with reference to fluid type, reservoir abandonment pressure, p_a, and the initial saturation pressure, p_s, of the reservoir fluid. Two broad types of scenarios may be identified:

a) reservoirs from which production during all or a significant part of life occurs at or above the saturation pressure (outside the regions identified by the heavy dashed line and the medium dotted line in Fig 3.3), and

b) reservoirs from which most of the production occurs below the saturation pressure
    (inside the regions identified by the heavy dashed line and the medium dotted line
    in Fig 3.3).

Scenario (a) would be expected for oil or gas reservoirs subject to strong waterdrive or oil
columns with large, segregated gas caps. Scenario (b) would be expected for oil or gas
reservoirs initially at saturation pressure that produce by pressure depletion or partial
waterdrive—waterdrive insufficient to maintain reservoir pressure above the saturation
pressure.

From another perspective, the two scenarios might be considered to represent sequential
pressure-production stages in a reservoir with an undersaturated fluid at initial conditions that
declined to significantly below the saturation pressure before being abandoned. Thus, the first
stage, Scenario (a), represents production by expansion of undersaturated fluid—i.e.,
production outside the two-phase envelope. The second stage, Scenario (b), represents
production inside the two-phase envelope.

Scenario (a) is further described by the boldface notation in the left column, which
indicates $p > p_s$ and $p = p_s$. Under these conditions, the primary drive mechanism is identified
in Fig. 3.3 as "complete" waterdrive.

Scenario (b) is further described by the notation in the left column, $p = p_s$, $p_a < p_s$, and
$p_a \ll p_s$. Under these conditions, the primary drive mechanism is identified as "partial"
waterdrive, in general, or "solution gas drive" for oil reservoirs or "pressure depletion" for gas
reservoirs.

Reservoir drive mechanisms typically are defined with reference to the presence (or
absence) of an active aquifer and the presence (or absence) of a significant gas cap. The
equations and correlations in Fig. 3.3 are for oil and/or gas accumulations in which either, or
both, of the initial fluid columns can be produced under conditions of essentially complete
gravity segregation. The influence of a significant initial gas cap on oil recovery for both
unsegregated and segregated flow is discussed in Sec. 3.7.2, Pressure Depletion—Initial Gas
Cap.

3.7.2 Black Oil Reservoirs

Methods are presented to estimate recovery of oil and gas from pressure depletion reservoirs
($W_e = 0$)—with and without an initial gas cap—and from reservoirs with water influx
($W_e > 0$).

Pressure Depletion—No Initial Gas Cap

Two stages may be recognized in the production of oil reservoirs by pressure depletion:

*   the *undersaturated stage*—i.e., the period during which the reservoir pressure
    decreases from initial pressure to bubblepoint pressure, and
*   the *saturated stage*—i.e., the period during which the reservoir pressure decreases
    from the bubble point pressure to abandonment pressure.

Undersaturated Stage

Hawkins [1955] has shown that, in general, recovery efficiency of oil attributable to rock and
fluid expansion caused by a pressure decrease from initial reservoir pressure, p_i, to the bubble
point pressure, p_b, may be calculated as

$$(E_{Ro})_{xp} = c_e(p_i - p_b)B_{oi}/B_{ob} , \dots\dots\dots\dots\dots\dots\dots\dots\dots\dots\dots\dots\dots\dots\dots(3.10)$$

where $(E_{Ro})_{xp}$ = recovery efficiency of oil attributable to expansion of the rock/fluid system from p_i to p_b

B_{oi} = oil formation volume factor at initial pressure, RB/STB,

B_{ob} = oil formation volume factor at bubblepoint pressure, RB/STB, and

c_e = effective compressibility of the rock/fluid system, defined as

$$(S_o c_o + S_w c_w + c_p)/S_o, \dotfill (3.11)$$

where S_o = oil saturation, fraction,

c_o = oil compressibility, 1/psi,

S_w = water saturation, fraction,

c_w = water compressibility, 1/psi, and

c_p = pore compressibility, 1/psi.

Eq. 3.11 tacitly assumes that the compressibility terms are constant over the pressure range of interest. In certain cases—friable sands, geopressured reservoirs, and highly undersaturated light oils—this is not a valid assumption. Pore compressibility is discussed in Sec. 6.4.5.

If PVT data are not available for the subject or a similar reservoir oil, c_o may be estimated using methods developed by Trube [1957] or by Vasquez and Beggs [1980]. Water compressibility may be estimated using data published by Dodson and Standing [1945] or by Osif [1988]. Regression equations to estimate such data are discussed in Appendix E.

As discussed in Sec. 6.5.5, there are no general correlations published to estimate pore compressibility, and this parameter should be determined in the laboratory. Typically, in commercial laboratories, compressibility is measured under hydrostatic—i.e., uniform triaxial—stress. Most authors, however, indicate that uniaxial stress more closely represents reservoir conditions [Geertsma 1957, van der Knaap 1959, Teeuw 1971, Andersen 1988, Yale et al. 1993].

There have been a number of methods proposed to adjust laboratory (hydrostatic) compressibility data to reflect reservoir (uniaxial) stress [Andersen 1988, Teeuw 1971]. The method proposed by Yale et al. [1993], however, appears to be the most direct

$$(c_p)_U = K_3(c_p)_H, \dotfill (3.12)$$

where $(c_p)_U$ = pore volume compressibility under reservoir (uniaxial) stress,

$(c_p)_H$ = pore volume compressibility under laboratory (hydrostatic) stress, and

where K_3 is a constant, depending on rock type,[55] as follows:

Rock type	K_3
Consolidated sandstone	0.45
Friable sandstone	0.60
Unconsolidated sand	0.75
Carbonates	0.55

---

[55] A major commercial laboratory reportedly uses K_3 = 0.61, irrespective of rock type, unless requested otherwise by clients.

If laboratory data are not available, pore compressibility may be estimated (with caution) as discussed in Sec 6.5.5.

For naturally fractured reservoirs, Reiss [1980] noted that total effective compressibility can be expressed as

$$c_e = \frac{c_o + \phi_{ma} S_{wma} c_w + \phi_{ma} c_{pma} + \phi_f c_{pf}}{\phi_{ma}(1 - S_{wma}) + \phi_f} , \quad\dots\dots\dots\dots\dots\dots\dots\dots\dots (3.13)$$

where: ϕ_{ma} = matrix porosity, fraction,
             ϕ_f = fracture porosity, fraction,
             c_{pma} = matrix pore compressibility, 1/psi,
             c_{pf} = fracture pore compressibility, 1/psi,
             S_{wma} = matrix water saturation, fraction,

and other terms are as previously defined.

While Eq. 3.13 is theoretically correct, independent determination of matrix porosity and fracture porosity and their respective compressibilities may be difficult. If fracture porosity is significantly less than matrix porosity, fracture compressibility may be ignored [Reiss 1980].

In some cases, pore and water compressibilities are negligibly small, and c_e may be equated to c_o, where

$$c_o = (B_{ob} - B_{oi})/B_{oi}(p_i - p_b) , \quad\dots\dots\dots\dots\dots\dots\dots\dots\dots\dots\dots\dots\dots\dots (3.14)$$

where c_o is the *integrated* value of oil compressibility between p_i and p_b. In this case, Eq. 3.10 becomes

$$(E_{Ro})_{xp} = (B_{ob} - B_{oi})/B_{ob} \quad\dots\dots\dots\dots\dots\dots\dots\dots\dots\dots\dots\dots\dots\dots\dots (3.15)$$

There is no free gas evolved in the reservoir during this stage of pressure depletion—i.e., between p_i and p_b. Thus, the recovery efficiency of solution gas should be the same as that for oil. This is indicated on Fig. 3.3 with the notation

$$[E_{Rsg}]_{xp} = [E_{Ro}]_{xp}$$

Saturated Stage

After reservoir pressure decreases below the bubblepoint pressure and the critical gas saturation is exceeded, the recovery efficiency of oil and gas will be governed by:

- oil and gas viscosity,
- initial solution GOR,
- the gas/oil relative permeability characteristics of the reservoir rock,
- the gravity/viscous force ratio,
- the ratio of horizontal to vertical permeability, and
- the method of well completion[56] and production policy.

---

[56] In the context of this discussion, a "horizontal well" is considered a "method of well completion."

There are no simple analytical equations available to estimate recovery efficiency of oil and solution gas attributable to solution gas drive. However, several investigators, for example, Arps and Roberts [1955], have published the results of step-wise calculations to estimate *oil* recovery efficiency by this mechanism in which they used the Muskat [1945] form of the differential material balance equation.

Arps and Roberts [1955] used PVT correlations to calculate the recovery factors for 12 different reservoir oils producing by solution gas drive. They used six different gas/oil relative permeability curves, three to represent sandstones and three to represent carbonates. The results are shown in **Table 3.1**.[57] The gas/oil relative permeability curves used in these calculations are identified as maximum, average, and minimum in **Fig. 3.6a** and **3.6b**.[58] These designations correspond to the column headings in Table 3.1. Other boundary conditions included the following:

- Recovery efficiencies were calculated from bubblepoint pressure to an abandonment pressure that was 10% of the bubblepoint pressure.[59]

- The methodology assumes complete dispersion of the liberated solution gas—no gravity segregation—with relative production of gas and oil at each stage controlled by relative mobilities of gas and oil determined by the k_g/k_o curve and fluid viscosities.

- Irreducible water saturation was assumed to be 25% in the sandstones and 15% in the carbonates.

- Recovery efficiencies in parentheses were estimated from calculations discontinued before abandonment pressure.

In addition, the computational procedure used by the authors does not account for additional stock-tank liquids recovered from rich solution gas produced from volatile-oil reservoirs, as would be expected for 50°API oil with a solution GOR of 2,000 scf/STB. Please refer to Sec. 4.1.2 for a brief discussion of the theoretical limitations of black oil material balance calculations.

Using a calculation procedure similar to that used by Arps and Roberts [1955], Wahl *et al.* [1958] developed a set of nine nomographs to estimate recovery efficiency of oil attributable to solution gas drive. They used k_g/k_o relations similar to those used by Arps and Roberts [1955]. The nine nomographs[60] cover combinations of three different k_g/k_o curves and three different residual oil viscosities that range from 0.5 to 10 cp. The results were presented as graphical functions of bubblepoint pressure, p_b, formation volume factor, B_{ob}, and bubblepoint gas/oil ratio, R_{sb}. **Fig. 3.7** is an example of one of these graphs, which is based on a k_g/k_o curve somewhat more favorable than the average curve shown on Fig 3.6a.

---

[57] In their paper, Arps and Roberts expressed the recovery factors as "stock tank barrels per acre foot per percent porosity" (STB/AF/%). These factors were converted here to fractional recovery $(E_{Ro})_{pd}$, which is somewhat simpler to use. Note that $(E_{Ro})_{pd}[77.58(1 - S_w)/B_{ob}] = (STB/AF/\%)$

[58] The individual curves, from which the average curves were estimated, are numbered to identify the data source, to which the interested reader is referred.

[59] For initially undersaturated reservoirs, recovery efficiency from Table 3.1 should be added to recovery efficiency calculated using Eq. 3.10 or 3.15. In this case, Table 3.1 recovery efficiency would be applied to OIP multiplied by the ratio B_o/B_{ob}.

[60] Discussion of this material might seem dated; however, such figures can be useful to make a quick estimate and serve as a reality check on more sophisticated methods.

TABLE 3.1—RECOVERY EFFICIENCY OF OIL FROM SOLUTION-GAS-DRIVE RESERVOIRS FOR PRODUCTION FROM BUBBLEPOINT PRESSURE TO ABANDONMENT PRESSURE EQUAL TO 10% OF BUBBLEPOINT PRESSURE[a,b]

Solution GOR, R_{sb} (RB/STB)	Stock-Tank Gravity (°API)	Sandstones			Carbonates		
		Max[c]	Avg[d]	Min[e]	Max[f]	Avg[g]	Min[h]
60	15	0.128	0.086	0.026	0.280	0.040	0.006
	30	0.213	0.152	0.087	0.328	0.099	0.029
	50	0.342	0.248	0.169	0.390	0.186	0.080
200	15	0.133	0.088	0.033	0.275	0.045	0.009
	30	0.222	0.152	0.084	0.323	0.098	0.026
	50	0.374	0.264	0.176	0.398	0.193	0.074
600	15	0.180	0.113	0.060	0.266	0.069	0.019
	30	0.243	0.151	0.084	0.300	0.096	(0.025)[i]
	50	0.356	0.230	0.138	0.361	0.151	(0.043)
1,000	15	NA[j]	NA	NA	NA	NA	NA
	30	0.344	0.212	0.126	0.326	0.132	(0.040)
	50	0.337	0.202	0.116	0.318	0.120	(0.031)
2,000	15	NA	NA	NA	NA	NA	NA
	30	NA	NA	NA	NA	NA	NA
	50	0.407	0.248	0.156	0.328	(0.145)	(0.050)

a) After Arps and Roberts [1955]. Please see Sec. 3.7.2, *Undersaturated Stage.*
b) Use 0.75 as a *preliminary* estimate of recovery efficiency of solution gas for reservoir pressure depleted to about 10% of bubblepoint pressure.
c) Maximum recovery efficiency usually associated with unconsolidated offriable, well sorted sands, with porosity greater than about 25%.
d) Average recovery efficiency usually associated with moderately cemented sandstores, with porosity between about 15 and 25%.
e) Minimum recovery efficiency usually associated with highly cemented sandstones, with porosity less than about 15%.
f) Maximum recovery efficiency calculated using average k_g/k_o data from 26 samples of Permian dolomite from west Texas. Because average primary recovery efficiency from these reservoirs typically is less than tabulated here, these results should be used with caution.
g) Average recovery efficiency usually associated with intergranular, slightly vugular limestones and dolomites.
h) Minimum recovery efficiency usually associated with highly vugular or fractured carbonates or with fractured cherts.
i) Values in parentheses are estimates made from calculations discontinued before reaching abandonment pressure.
j) An unlikely combination of solution GOR and stock-tank gravity; therefore, calculations not made.

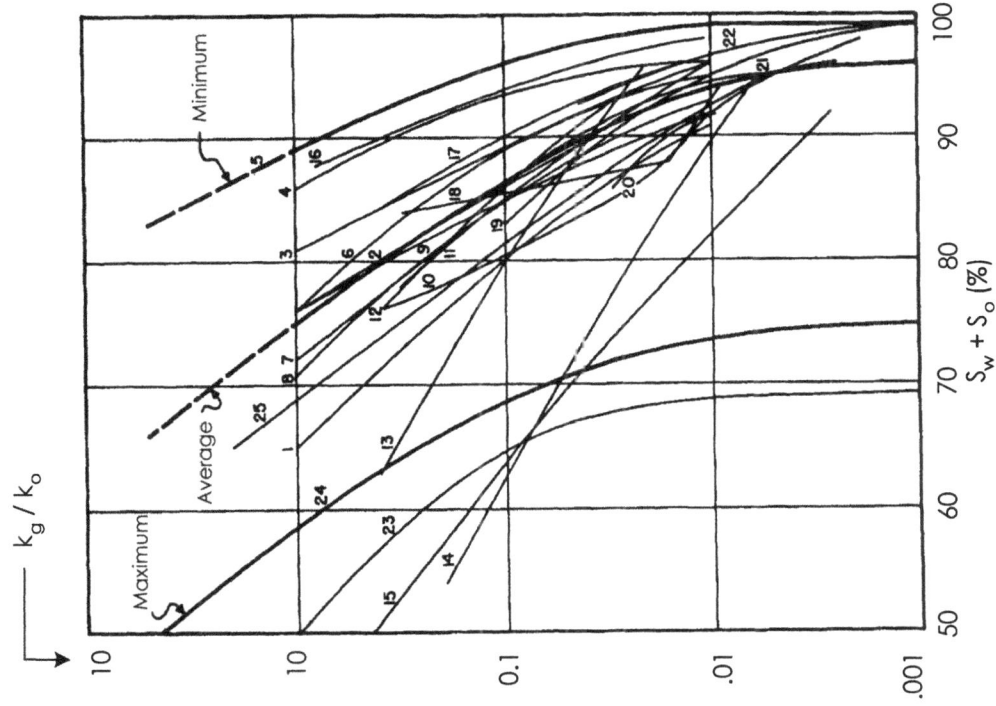

Fig. 3.6b—Gas/oil relative permeability, k_g/k_o, vs. total liquid saturation, $S_w + S_o$, for 23 limestones, dolomites, and cherts, showing curves used to calculate maximum, average, and minimum oil recovery by solution gas drive [after Arps and Roberts 1955].

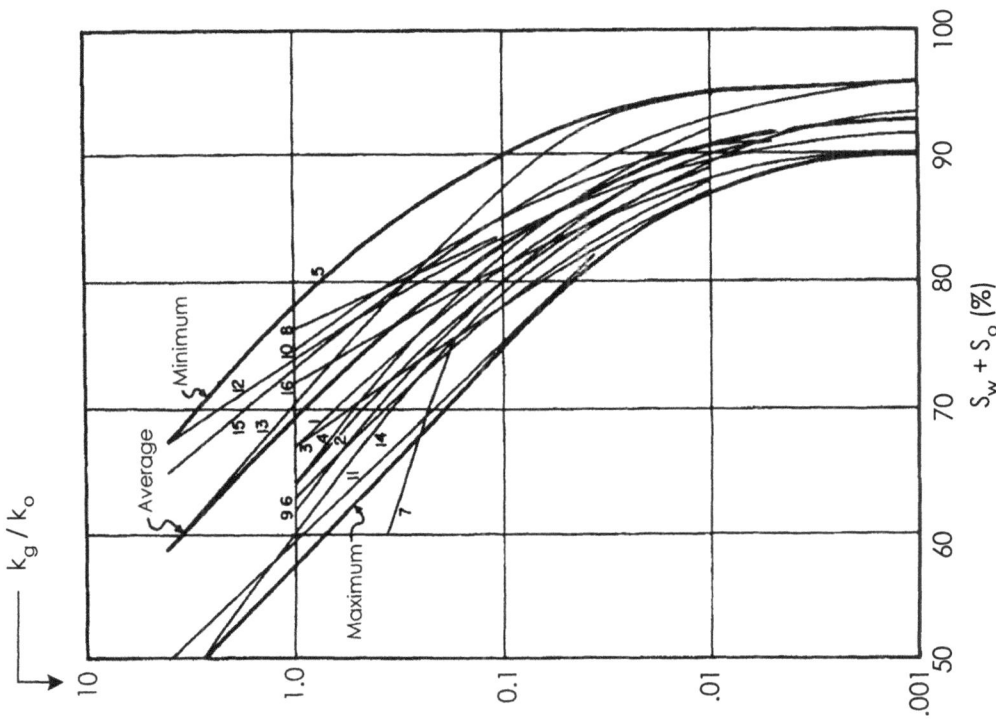

Fig. 3.6a—Gas/oil relative permeability, k_g/k_o, vs. total liquid saturation, $S_w + S_o$, for 16 sands and sandstones, showing curves used to calculate maximum, average, and minimum oil recovery by solution gas drive [after Arps and Roberts 1955].

The assumptions in the Wahl *et al.* [1958] work are the same as those in the Arps and Roberts [1955] work, except for the following:

- Recovery efficiency was calculated assuming abandonment at atmospheric pressure; the Arps and Roberts [1955] calculations assumed abandonment at 10% of the bubblepoint pressure.

- Critical gas saturation, S_{gc}, was fixed at 5%; in the Arps and Roberts [1955] work, S_{gc} was determined by the k_g/k_o curve.

- Irreducible water saturation was varied from 10 to 50%; in the Arps and Roberts [1955] work, it was fixed at 25% for the sandstones and 15% in the carbonates.

Use of either of these studies requires knowledge of the gas/oil relative permeability characteristics of the reservoir in question. If laboratory k_g/k_o data are not available, reference may be made to correlations developed by Corey [1954, 1994] and Felsenthal [1959]. Wyllie [1962], Rose [1987], and Honarpour *et al.* [1994] provided comprehensive discussions on relative permeability.

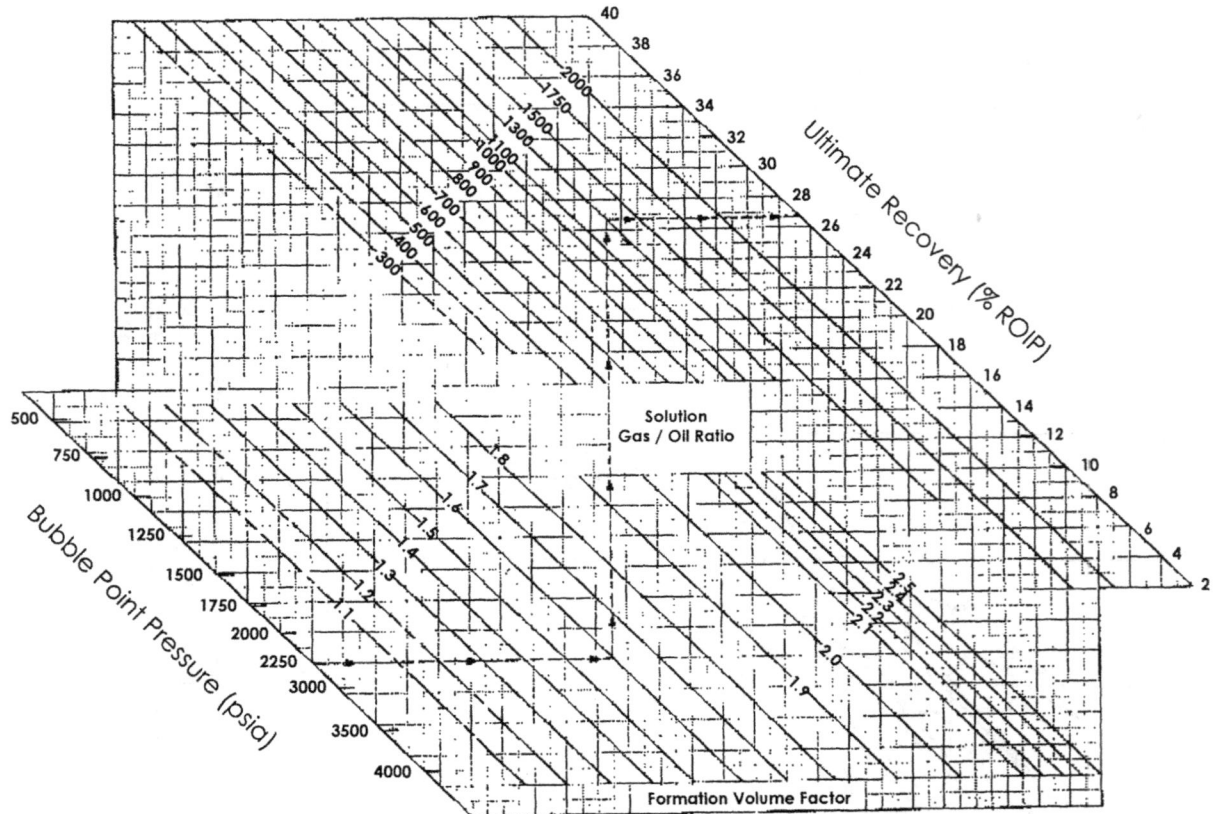

Fig. 3.7—Example chart for preliminary estimate of recovery efficiency of oil from solution-gas-drive oil reservoir [after Wahl *et al.* 1958].

API 1967 Correlation

In 1967 the American Petroleum Inst. (API) published the results of a regression analysis of oil recovery efficiency from 80 reservoirs in the U.S. that had produced by solution gas drive with no initial gas cap [Arps *et al.* 1967].[61] The correlation is

$$(E_{Ro})_{sg} \approx 0.41815[\phi(1 - S_w)/B_{ob}]^{0.1611}[k/\mu_{ob}]^{0.0979}[S_w]^{0.3722}[p_b/p_a]^{0.1741} , \dots\dots\dots\dots(3.16)$$

where $(E_{Ro})_{sg}$ = recovery efficiency of oil attributable to solution gas drive, fraction,
        ϕ = porosity, fraction,
        S_w = water saturation, fraction,
        B_{ob} = formation volume factor at the bubblepoint, RB/STB,
        k = arithmetic average absolute permeability, darcies,
        μ_{ob} = oil viscosity at the bubblepoint, cp,
        p_b = bubblepoint pressure, psi, and
        p_a = abandonment pressure, psi.

The regression coefficient is 0.932. The standard error of the estimate is 22%—i.e., there is a 68% probability that the correct value is within 22% of the estimated value. The data for this equation came from 67 sandstone reservoirs and 13 carbonate reservoirs. The ranges of rock and fluid properties for the 80 reservoirs are shown below.

Parameter	Sandstones		Carbonates[62]	
	Minimum	Maximum	Minimum	Maximum
K, darcies	0.006	0.940	0.001	0.252
ϕ, fraction	0.115	0.299	0.042	0.200
S_w, fraction	0.150	0.500	0.163	0.350
γ_{API}, °API	20.	49.	32.	50.
R_{sb}, scf/STB	60.	1,680.	302.	1,867.
p_b, psi	639.	4,403.	1,280.	3,578.
$(E_{Ro})_{sg}$, fraction	0.095	0.460	0.155	0.207

This work was updated in 1984, and the section titled API 1984 Study should be reviewed before using Eq. 3.16.

Solution Gas

There is little historical data available on the recovery efficiency of solution gas from solution-gas-drive oil reservoirs. In part, this is due to the historically low economic value of this product compared to that for crude oil. Also, with the increasing application of pressure-maintenance techniques, fewer oil reservoirs are produced to depletion by solution gas drive. Nevertheless, prudent asset management mandates a reasonable estimate of this resource. This is especially so when an evaluation is required to compare economics of operation by solution gas drive vs. operation under waterflood.[63]

---

[61] As it falls more logically in the discussion, this material is included here, rather than in Sec. 2.3, Statistical Methods. Please refer to Sec. 9.2.4 for a discussion of the implications of using the results of regression analyses to estimate reserves.

[62] The term "carbonates" refers to limestones, dolomites, dolomitic limestones, cherts, etc.

[63] Some years ago, a volatile-oil accumulation developed competitively in the U.S. Rocky Mountains was produced wastefully (almost to depletion) because of conflicting interests between two groups of operators. One group had made significant investments in a gasoline plant and pressed for pressure depletion. The other group, with no such investments, pressed for pressure maintenance.

In an oil reservoir produced at pressures below the bubblepoint pressure, so that solution gas is evolved in the reservoir and becomes mobile, the producing GOR will be determined by: (a) the gas/oil mobility ratio, (b) the degree of gravity segregation, and (c) the position of impermeable streaks relative to well completion intervals, among other factors. In the absence of significant gravity segregation, about 75% of the gas initially in solution can reasonably be expected to be produced if the reservoir is produced to an abandonment pressure approaching atmospheric.

Roberts and Ellis [1962], using calculation procedures similar to those used by Arps and Roberts [1955], developed a set of graphs of gas/oil ratio history and bottomhole pressure vs. cumulative oil recovery. **Fig. 3.8** is an example of one of these graphs.[64]

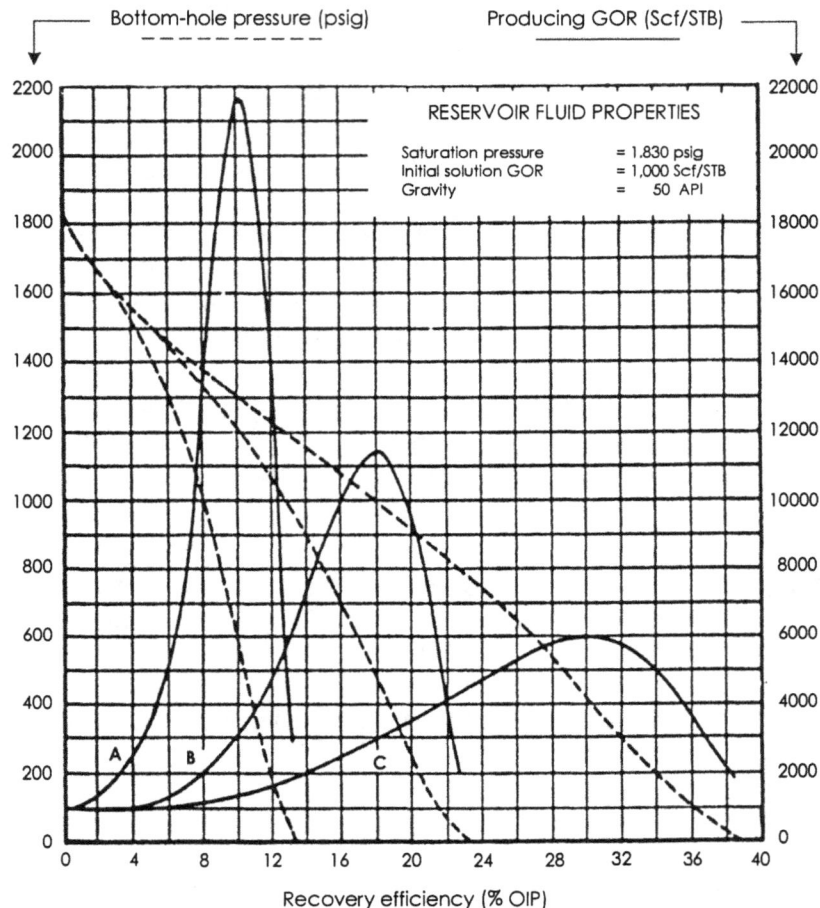

Fig. 3.8—Producing gas/oil ratio and bottomhole pressure vs. oil recovery efficiency for high GOR oil, calculated from minimum, average, and maximum k_g/k_o curves of Fig. 3.6a [after Roberts and Ellis 1962].

Note that the computations assume $p_i = p_b$. This work is subject to the same limitations discussed previously. The graphs can be used to make a reasonable *initial* estimate of GOR performance, pending development of individual well or reservoir performance or history matching by computer simulation. A reasonable preliminary estimate of ultimate recovery,

[64] The reader is referred to the original work for a complete set of these graphs.

pending obtaining definitive performance data, would be 75% of the solution gas initially in place, as previously noted.

Pressure Depletion—Initial Gas Cap

As observed by Muskat [1945 and 1949], the recovery efficiency of oil by pressure depletion of an oil reservoir with an initial gas cap will be governed mainly by the size of the initial gas cap and the degree of gravity segregation. It is noted here that other factors also should be considered, including: (a) ratio of horizontal to vertical permeability, (b) well spacing, and (c) type of completion.[65] There are no simple analytical equations or correlations available to estimate oil recovery efficiency in this type of reservoir.

No Gravity Segregation

The case of no gravity segregation, the least favorable case, was studied analytically by Muskat [1945] for an oil reservoir with a bubblepoint solution GOR of 534 scf/STB and a bubblepoint oil viscosity of about 1 cp. Tabulated below as an example is Muskat's calculated oil recovery efficiency by pressure depletion from initial pressure to a reservoir abandonment pressure of 100 psi for various initial gas cap sizes.

Initial Gas Cap Size (m)	Oil Recovery Efficiency (% OIP)
0	21.2
0.2	25.3
0.4	28.1
0.7	30.1
1.0	33.1

Gravity Segregation

In nature, there always is some degree of gravity segregation of free gas and oil during production of oil by solution gas drive. The relative importance of gravity segregation may be expressed by a gravity number, N_G, of the form [Smith 1953]

$$N_G = k_o (\rho_o - \rho_g)\sin\alpha/\mu_o , \dots\dots\dots\dots\dots\dots\dots\dots\dots\dots\dots\dots\dots\dots\dots\dots\dots(3.17)$$

where k_o = effective oil permeability, md,

μ_o = oil viscosity, cp,

ρ = density of reservoir oil or gas, g/cm^3, and

α = dip angle, degrees.

For guidance in making a *preliminary* estimate of the relative importance of gravity segregation, **Table 3.2,** compiled from published studies of actual reservoirs, may be useful. As may be observed for this data set, there is no correlation between gravity number and oil recovery efficiency.

Gas was injected into some of these reservoirs. The term m_{inj} indicates the volume injected as a multiple of initial gas cap volume. In the presence of gravity segregation, gas cap injection reasonably can be expected to have the same effect as a large initial gas cap.

---

[65] In this context, horizontal drainholes are considered here to be a type of completion.

From the data in this table, it may be observed that, in an oil reservoir with a gravity number greater than about 10, effective gravity segregation is a reasonable expectation [Smith 1953].

TABLE 3.2—RECOVERY EFFICIENCY OF OIL FROM GAS-CAP DRIVE AND/OR GRAVITY DRAINAGE RESERVOIRS

Field (Pool), Location	Gravity Number	Rec. Eff. (% OIP)
Elk Basin (Tensleep),[a] Wyoming, U.S.	15[b]	61[c]
Midway Sunset (Lakeview), California, U.S.	41[d]	62[e]
Mile Six, Peru	47[f]	62[g]
Oklahoma City, Oklahoma, U.S.	30[b]	37[h]
Shuler (Jones), Arkansas, U.S.	340[b]	60[i]

a) Reportedly, reservoir is oil-wet.
b) Estimated from published data.
c) $m = 0$; $G_i = 110$ Bscf [Stewart et al. 1955, 1962, 1975].
d) Smith [1953].
e) $m = 0$; $m_{inj} = 0$ [Sims and Frailing 1950, Frailing 1962].
f) Craft and Hawkins [1959].
g) $m = 0$; $m_{inj} > 8.7$ [Anders 1953].
h) Arps [1962].
i) $m = 0.027$; $m_{inj} = 14.8$. 1962 estimated ultimate recovery (70%) includes effects of water injection [Tarner 1944, Tarner et al. 1951, 1962, Tarner and Curzon 1975].

Richardson and Blackwell [1971] published guidelines for simple computational procedures to estimate performance of oil reservoirs subject to gas cap encroachment.

Waterdrive

Waterdrive may vary from "complete" to "partial." If the volume of water encroaching into the initial hydrocarbon-bearing pore space maintains average static reservoir pressure above the bubblepoint over reservoir life ($p_a \geq p_s$), waterdrive is considered here to be "complete." If the volume of water is insufficient to maintain reservoir pressure above the bubblepoint, and reservoir pressure continues to decline over life ($p_a < p_s$), waterdrive is considered to be "partial." In Fig. 3.3, the area above the notation **p = p$_s$** is noted as "complete"; the area below, as "partial."

API 1967 Correlation

In 1967, the API published the results of a regression analysis of oil recovery efficiency from 70 waterdrive reservoirs in the U.S. [Arps et al. 1967]

$$(E_{Ro})_{wd} \approx 0.54898[\phi(1 - S_w)/B_{oi}]^{0.0422}[k\mu_{wi}/\mu_{oi}]^{0.077}[S_w]^{-0.1903}[p_i/p_a]^{-0.2159}, \ldots\ldots\ldots (3.18)$$

where $(E_{Ro})_{wd}$ = oil recovery efficiency attributable to waterdrive, fraction,
    μ_{wi} = initial water viscosity, cp,
    μ_{oi} = initial oil viscosity, cp,
    p_i = initial reservoir pressure, psi,
and other notation is as previously defined.

The regression coefficient is 0.958. The standard error of the estimate is 18%—i.e., there is a 68% probability that the correct value is within 18% of the estimated value. The data for this regression came from 70 sandstone reservoirs with the following ranges of rock and fluid properties:

Parameter	Minimum	Maximum
K, darcies	0.011	4.000
ϕ, fraction	0.111	0.350
S_w, fraction	0.052	0.470
γ_{API}, °API	15.	50.
μ_{oi}, cp	0.2	500.
$(E_{Ro})_{wd}$, fraction	0.278	0.867

The validity of this correlation is discussed further in Sec. 3.7.2 *API 1984 Study*.

Bottomwater Drive

There is no discussion in the Arps *et al.* [1967] work regarding the relative efficiency of bottomwater drive compared to edgewater drive. The nature of the water encroachment is an important factor in estimating reserves, and several classic studies of bottomwater drive are summarized in **Appendix I.**

Viscous Oils

From the table accompanying the discussion of Eq. 3.18, it may be noted that 500 cp is the upper range of oil viscosity included in this correlation. Above this viscosity range, the confidence level of the regression decreases substantially.

Additional insight regarding recovery efficiency from waterdrive oil reservoirs with viscous oil is provided by Van Meurs and van der Poel [1958]. Based on model experiments, they developed an equation of the form

$$(E_{Ro})_{wd} = S_{wi} + S_{om}[R_{\mu ow} - f_{wa}^2(R_{\mu ow} - 1)]/[R_{\mu ow} - f_{wa}(R_{\mu ow} - 1)]^2 , \dots\dots\dots\dots\dots (3.19)$$

where
$(E_{Ro})_{wd}$	= recovery efficiency of OIP,
S_{wi}	= irreducible water saturation, fraction,
S_{om}	= movable oil saturation $(1 - S_{or} - S_{wi})$,
f_{wa}	= economic limit water cut, fraction, and
$R_{\mu ow}$	= oil/water *viscosity* ratio.

This work is applicable to reasonably homogeneous, edgewater drive reservoirs, where the displacement is not gravity stable, but is dominated by "viscous fingering."[66]

Reportedly, results from Eq. 3.19 are in good agreement with laboratory-scale model experiments and the performance of a strong waterdrive reservoir producing 250 cp oil.

Eq. 3.19 was modified for this work to solve for volumetric sweep efficiency, E_V, by setting S_{wi} equal to 0 and S_{om} equal to 1. Solutions to the modified equation for a range of water cuts and oil/water viscosity ratios are shown in **Fig. 3.9.** It may be noted that for oil/water viscosity ratios in the range about 10 to 500, volumetric sweep efficiency is strongly

---

[66] As discussed by Lake *et al.* [1992], the existence of the viscous fingering phenomenon observed in (homogeneous) laboratory sand packs is not likely to occur in heterogeneous rocks observed in nature. Eq. 3.15 does, however account for variations in viscosity ratio and, with adjustment, may be used to compare otherwise analogous reservoirs.

influenced by the water cut, f_w. For example, for a reservoir with an oil/water viscosity ratio of 100, volumetric sweep efficiency could—theoretically—be increased from 0.30 to approximately 0.75 if the economic limit water cut could be increased from 0.95 to 0.99.

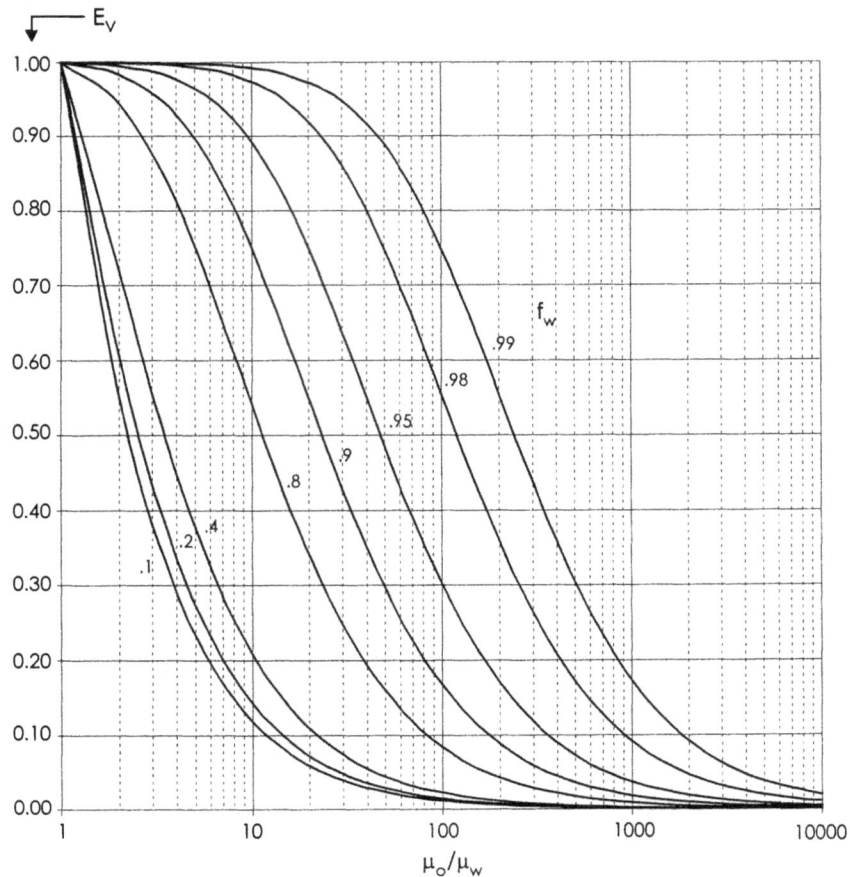

Fig. 3.9—Volumetric sweep efficiency, E_V, vs. oil/water viscosity ratio, μ_o/μ_w, for various producing water cuts, f_w [after van Meurs and van der Poel 1958].

In addition, Timmerman [1982] published graphs that may be used to make preliminary estimates of the recovery efficiency from viscous oil reservoirs producing by edgewater drive. These graphs are based on experiments with scaled models and cover water/oil mobility ratios, M_{wo}, from 1 to approximately 500.

API 1984 Study

As noted previously, in an update to the 1967 work the API [1984] studied an expanded database that included 376 solution-gas-drive and 244 natural waterdrive oil reservoirs in six states in the U.S. In reviewing the 1967 work, the API [1984] observed that the reservoir data sets were weighted in accordance with the quality of the data. A reservoir considered to have good data was counted three times; one with poor data, one time. When the reservoirs were not weighted for data quality in the course of the 1984 review, the statistical reliability of the correlations decreased significantly. Regarding the 1984 study, it was concluded that "no statistically valid correlation has been found between oil recovery and definable reservoir parameters."[67]   The API cautioned "against continued use of the (1967) correlations...to predict recovery or recovery efficiency for any one reservoir."

Regarding the API's "caution," in the writer's experience, the API 1967 correlations, both Eqs. 3.16 and 3.18, provide estimates that, in many cases, are in *reasonable* agreement with observed recovery efficiency.   Because these equations are empirical correlations of large amounts of data, it should not be expected they would provide acceptable estimates in every case.   If usage is tempered with good judgement and cross-checked with other data, there is no reason to reject—in toto—the 1967 API work.   Additionally, if analogy methods (discussed in Chap. 2) are being used to estimate reserves, the dimensionless groups in the API correlation may be used to adjust recovery efficiencies between prototype and model.

In further discussion of the 1984 work, it was noted that "calculated average recoveries in a single geologic trend are significant."   API [1984] data for 533 U.S. reservoirs on oil initially in place and primary recovery, both solution gas drive and natural waterdrive, are shown in **Table 3.3**.

TABLE 3.3—OIL INITIALLY IN PLACE AND PRIMARY RECOVERY FOR 533 U.S. RESERVOIRS [API 1984]

Production Mechanism	Lithology	State	No. of Res.	OIP (STB/NAF) Range	OIP (STB/NAF) Avg.	Recovery (STB/NAF) Range	Recovery (STB/NAF) Avg.	Avg. RE
Solution Gas Drive	Sandstone	California	51	328 – 2146	1148	55 – 644	273	0.24
		Louisiana	17	360 – 1540	1167	110 – 613	315	0.27
		Oklahoma	41	386 – 1030	604	40 – 227	117	0.19
		NW Texas	24	NR[2]	578	NR	86	0.15
		SE Texas	38	NR	827	NR	205	0.25
		W Virginia	13	546 – 882	678	100 – 250	143	0.21
		Wyoming	23	386 – 1135	692	82 – 361	176	0.25
	Carbonate	All	100	59 – 1067	526	5 – 190	93	0.18
Natural Water Drive	Sandstone	California	9	508 – 1770	1020	169 – 568	367	0.36
		Louisiana	42	535 – 1960	1203	219 – 1470	722	0.60
		Texas	101	547 – 1801	1167	125 – 1094	633	0.54
		Wyoming	27	NR	860	NR	313	0.36
	Carbonate	All	47	85 – 705	265	29 – 370	117	0.44

1. API [1984]; please see Sec.3.7.2, API 1984 Study.
2. Not reported.

3.7.3 Volatile-Oil Reservoirs

As noted in Sec. 3.7.1, volatile oils are intermediate in composition between black oils and retrograde gases. Typically, they produce with initial solution gas/oil ratios in the range 1,750 to 3,500 scf/STB; gravity of stock-tank oil usually is greater than about 40°API. The differential formation volume factor at the bubblepoint usually will be greater than about 2.0 RB/STB.

---

[67] It is suggested here that the apparent absence of a "statistically valid correlation" may be due in part to the API's failure to include well spacing and/or heterogeneity terms in the correlation.

If a strong waterdrive is expected ($p_a \geq p_b$), the API [1967] correlation for waterdrive recovery efficiency may be used to estimate reserves. This type of accumulation, however, generally is associated with deep, high-temperature reservoirs. Under these conditions, reservoir rocks are subject to substantial diagenetic alteration, which usually results in reduced permeability. Strong waterdrives are not common in low-permeability reservoirs. Thus, it is more likely that these reservoirs will produce chiefly by pressure depletion, with significant water encroachment being unlikely.

Most of the reservoirs used in the API 1967 correlation for depletion-drive recovery efficiency contained black oils. From the range of fluid properties shown in the previous section, however, it may be noted that the correlation includes oils that could be classified as volatile. Thus, with caution, the API 1967 correlation could be used to estimate depletion-drive recovery efficiency of volatile oils. It is recommended that the flash liberation, rather than the differential liberation, oil formation volume factor be used to determine B_{ob} for use in this equation.

As discussed in Sec. 3.7.4, Eaton and Jacoby [1965] developed a correlation to estimate the recovery of condensates from retrograde gas systems produced by pressure depletion. Two of the reservoir fluids used in the correlation were volatile oils. This is not surprising because near the critical point, the thermodynamic behavior of volatile oils and retrograde gas is almost indistinguishable. Thus, *with caution,* this correlation may be used to estimate the recovery efficiency of liquids from *near-critical* volatile oil reservoirs produced by pressure depletion.

In comparing the API 1967 and Eaton and Jacoby [1965] correlations, Eqs. 3.16 and 3.20, respectively, it may be noted the parameters used in the two correlations are completely different. Thus, use of these two correlations could provide independent estimates of condensate recovery. For this reason, consideration might be given to using both the API 1967 correlation and the Eaton a Jacoby [1965] correlation to estimate recovery and taking the average of the two.

A third source of information on the recovery from volatile-oil reservoirs, albeit analogy data, has been provided by Cronquist [1979]. Data published elsewhere were summarized for seven volatile oil reservoirs in southeastern U.S. As shown in **Table 3.4**, actual (or estimated) recovery efficiency attributed to pressure depletion of these reservoirs ranged from 17 to 32% of OIP. These recoveries are consistent with estimates that might have been made using Table 3.1 for 50°API oil with an initial GOR of 2,000 scf/STB.

3.7.4 Retrograde-Gas Reservoirs

Pressure Depletion

As with solution-gas-drive oil reservoirs, there are no simple analytical expressions available to calculate recovery efficiency of gas and condensate from retrograde-gas reservoirs producing by pressure depletion. Given a PVT analysis, the usual procedure is to make a series of flash calculations to determine recovery of gas and condensate to an assumed abandonment pressure, as discussed by Jacoby *et al.* [1959] and others. Frequently, however, a PVT analysis is not available. Eaton and Jacoby [1965] published the following correlation, which may be used to estimate recovery of *condensate* from a depletion-drive retrograde-gas reservoir as reservoir pressure declines from the dewpoint pressure to abandonment pressure, assumed by the authors to be 500 psi.[68]

---

[68] If $p_i > p_d$ —i.e., if the accumulation is initially undersaturated—then as the reservoir pressure declines from p_i to p_d, the

$$\ln(C_{pa})_{pd} \approx 2.7958\ln(\gamma_{API}) + \ln(p_i) - 0.65314\ln(R_T) - 20.243 \,, \dots\dots\dots\dots\dots\dots (3.20)$$

where $(C_{pa})_{pd}$ = ultimate recovery of condensate at 500 psi abandonment pressure, bbl/bbl hydrocarbon pore space,[69]

γ_{API} = stock-tank gravity ($^{\circ}$API) of condensate,

p_i = initial reservoir pressure, psi, and

R_T = total GOR from entire separator train, scf/STB.

The regression coefficient is 0.814; the standard error is 14.1%.

TABLE 3.4—RECOVERY EFFICIENCIES FOR VOLATILE OIL RESERVOIRS IN THE U.S. [after Cronquist 1979]

Field/ Reservoir	Depth (ft)	Lithology	Permeability (md)	Spacing (acre)	Recovery Efficiency With Injection (fraction OIP) Primary	Method	Est. RE	Remarks
Blackjack Creek/ Smackover	15,700	Dolomite	110	320	0.19*	Waterflood	0.39	100 MMSTB OIP; 10.3% H₂S; peripheral injection
Jay-LEC/ Smackover	15,400	Dolomite	35	160	0.17*	Waterflood	0.48	728 MMSTB OIP; 9% H₂S; staggered line drive
Pickton/ Rodessa	7,900	Oolitic limestone	250	40	0.19*	Gas injection	0.61#	Crestal gas injection at 500 psi below bubblepoint pressure; # RE of OIP plus NGL; RE of OIP = 0.52
Raleigh/ Hosston	12,600	Sandstone	40	40	0.32*	Gas injection	0.66#	Crestal gas injection at 2,000 psi above bubblepoint pressure # RE of propanes plus
Shoats Creek/ Cockfield	8,950	Sand	35	140	<0.20*	Gas injection	0.41	Peripheral gas injection at 1000 psi above bubblepoint pressure
North Louisiana/ Smackover	10,000	Oolitic limestone	175	160	0.22	**		Est. RE of OIP extrap from 700 psi BHP R_p = 8.8 Mscf/STB
Un-named/ "A"	8,200	Sand	750	310	0.25	**		Est. RE of OIP extrap from 1500 psi BHP R_p = 8.7 Mscf/STB

* Estimated; reservoir produced by indicated injection method.
**Produced by pressure depletion.

This correlation was developed using 27 different reservoir-fluid systems with the following ranges of properties:

Parameter	Minimum	Maximum
γ_{API}, $^{\circ}$API	45	65
p_i, psi	4,000	12,000
R_T, scf/STB)	2,500	60,000
T_f, $^{\circ}$F	160	290

In this work, calculated *gas* recovery efficiency (to an abandonment pressure of 500 psi) ranged from 88 to 97%, with an average of 92.6%. In reservoirs with initial pressures in the range indicated, 500 psi abandonment pressure may be too optimistic. Accordingly, it is

producing CGR should equal the initial CGR, and the condensate production from this stage of pressure depletion should be added to that estimated from the correlation.
[69] This unit, STB/RB, may be converted to STBC/MMscf gas initially in place by multiplying by B_{gi} with units RB/MMscf.

suggested that, for an *initial volumetric estimate,* gas recovery efficiency be estimated at 85%, unless there are factors that would warrant a different recovery efficiency, as discussed in the following paragraphs.

While the Eaton and Jacoby [1965] work may account for condensate losses due to retrograde condensation, it does not account for the flow mechanics in retrograde-gas reservoirs, as discussed below.

These simple guidelines may be adequate for retrograde-gas reservoirs in good-quality rocks. In poor-quality rocks, however, estimating reserves from such reservoirs is (still) one of the major problems in reservoir engineering. When such reservoirs are produced by pressure depletion at reservoir pressures less than the dewpoint pressure, retrograde liquids will condense in the pore space. Such condensation typically results in a condensate buildup of no more than a few percent of initial pore space in the interwell areas of the reservoir. Around wellbores, however, such condensation may build up to saturation levels that may substantially reduce the effective permeability to gas. Depending on circumstances, well productivity may be severely reduced, possibly resulting in premature abandonment.

Reliable prediction of the depletion performance of such reservoirs, especially those with "rich gases" (high initial CGR) in low-permeability rocks, has been a major problem for the last 20 years. With advances in simulation technology, some of the problems are being overcome, but there are no simple guidelines in the open literature.

Water Influx/Pressure Maintenance

The above correlation provides a useful method to estimate recovery of condensate from retrograde-gas reservoirs produced by pressure depletion—a volumetric reservoir with no water influx. If, however, there is a partial waterdrive, or if pressure maintenance (by water injection) is being considered, it may be necessary to estimate recovery of condensate from such reservoirs at an intermediate stage of pressure depletion. Two correlations have been published of condensate/gas ratio vs. reservoir pressure, Joiner and Long [1978] and Garb 1978].

Joiner and Long

Joiner and Long's [1978] correlation is of the form

$$\ln(R_T)_{0.5d} \approx 31.49 - 0.000217 p_d/2 - 92.03(C_{4+})/T_f + 110.8(C_{5+})/T_f$$

$$+ 0.0215 T_f + 6.833 \gamma_{gR} - 26.98/(R_T)_d - 6.632\ln(T_f), \ldots\ldots (3.21)$$

where $(R_T)_{0.5d}$ = total producing GOR[70] (all stages of separation) at 50% of the dewpoint pressure, Mscf/STBC,

p_d = dewpoint pressure, psi,
C_{4+} = butanes plus in dewpoint fluid, mole %,[71]
C_{5+} = pentanes plus in dewpoint fluid, mole %,
T_f = reservoir temperature, °F,
γ_{gR} = specific gravity of reservoir gas (air = 1),
$(R_T)_d$ = total producing GOR at the dewpoint pressure, Mscf/STBC,

and where the butanes plus, C_{4+}, and pentanes plus, C_{5+}, fractions can be estimated as follows:

---

[70] Reported as GOR, not CGR, by the authors!
[71] Units in these equations are mole %, not mole fraction.

$$C_{4+} \approx 6.547 + 25.52\gamma_{gR} + 30.38/(R_T)_d + 0.02633(R_T)_d - 30.3\rho_c - 0.00417T_f \dots \dots (3.22a)$$

and

$$C_{5+} \approx (-)8.53 + 7.83\gamma_{gR} + 56.26/(R_T)_d + 0.0109(R_T)_d + 0.07286(°API) - 0.00424T_f , \dots (3.22b)$$

where ρ_c = density of condensate, g/cm^3, and where the specific gravity of reservoir gas, γ_{gR}, was approximated by

$$\gamma_{gR} = \gamma_{gs} + 17.3/[(R_T)_d(°API)^{0.47}] , \dots \dots (3.23)$$

where γ_{gs} = gravity of separator gas (air = 1.0).

Joiner and Long [1978] observed that, between the dewpoint pressure and approximately 30% of the dewpoint pressure, the trend of R_T vs. p/p_d may be approximated by

$$\log(R_T) \approx 2(1 - p/p_d)\log[(R_T)_{0.5d}/(R_T)_d] + \log(R_T)_d \dots \dots (3.24)$$

To use these correlations: (1) use Eq 3.21 to calculate $(R_T)_{0.5d}$, (2) use Eq 3.24 to calculate R_T between the dewpoint pressure and 30% of the dewpoint pressure, and (3) assume R_T is constant at pressures less than 30% of the dewpoint pressure.

The Joiner and Long [1978] correlations were based on data from 10 retrograde-gas systems with properties as follows:

Parameter	Minimum	Maximum
p_d, psi	2,440	6,907
T_f, °F	140	269
R_{cd}, STBC/MMscf	15	141
γ_{API}, °API	55	61

In comparing predicted vs. actual GOR for the 10 fluids, Joiner and Long [1978] reported, at 50% of the dewpoint pressure, errors ranging from (+)40 to (-)23%, averaging (+)17%.

Garb

Garb's [1978] correlation is of the form

$$R_c \approx R_{cd}[1 + nD(1 - p/p_i)]^{-1/n} , \dots \dots (3.25)$$

where R_{cd} is the CGR at the dewpoint pressure and n and D are defined as

if $R_{cd} \leq 15$ STBC/MMscf: $n \approx 24.5 - 1.37R_{cd} , \dots \dots (3.26a)$

if $R_{cd} > 15$ STBC/MMscf: $n \approx 3.91[1 + 2.71(R_{cd} - 15)]^{-0.129} , \dots \dots (3.26b)$

if $R_{cd} \leq 150$ STBC/MMscf: $D \approx 4.05 + 0.0099(R_{cd})^{1.48} , \dots \dots (3.27a)$

and

if $R_{cd} > 150$ STBC/MMScf: $D \approx 24.3\ln(R_{cd}) - 101$ . ... (3.27b)

Reportedly, Garb's [1978] correlation was based on four typical hydrocarbon compositions; details were not published.

Each of these two correlations was used here to calculate producing CGR vs. reservoir pressure for Joiner and Long's Cases 1 and 10. The initial CGR for Case 1 is 141 STBC/MMscf; for Case 2, 15 STBC/MMscf. The results are plotted as dimensionless CGR, R_c/R_{cd}, vs. dimensionless pressure, p/p_d, on **Fig. 3.10**.

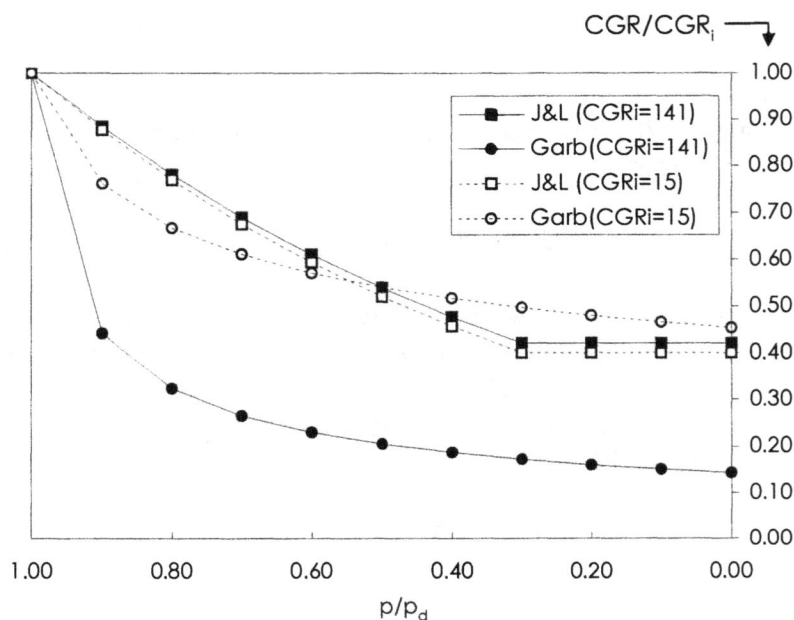

CGR/CGR$_i$

	J&L (CGRi=141)
	Garb(CGRi=141)
	J&L (CGRi=15)
	Garb(CGRi=15)

p/p$_d$

Fig. 3.10—Dimensionless CGR vs. dimensionless pressure calculated using Garb [1978] and Joiner and Long [1978] correlations.

From examination of this figure, it is apparent that, for the cases examined, the two correlations yield significantly different results. The Joiner and Long [1978] correlation, plotted as open and solid squares, indicates very little difference in dimensionless CGR vs. dimensionless pressure, despite the wide range in initial CGR. The Garb [1978] correlation, plotted as open and solid circles, shows substantial difference.[72]

Given the disparity between these two published correlations, it seems apparent that for any reasonable size retrograde-gas accumulation, a PVT analysis is essential.

3.7.5 Nonretrograde-Gas Reservoirs

For nonretrograde-gas reservoirs, two reservoir drive mechanisms are discussed: (a) no water influx and (b) water influx.

Pressure Depletion

If there is no, or negligibly small, water influx and the compressibility of the interstitial water and pore volume are negligibly small, ultimate gas recovery efficiency may be estimated as

---

[72] Considering that both correlations were based on limited data, the disparity is not surprising.

$$(E_{Rg})_{pd} = 1 - p_a Z_i / p_i Z_a , \dots\dots\dots\dots\dots\dots\dots\dots\dots\dots\dots\dots\dots\dots\dots\dots(3.28)$$

where $(E_{Rg})_{pd}$ = recovery efficiency of gas attributable to pressure depletion, fraction,

p_i = initial reservoir (or well) pressure, psia,

p_a = abandonment reservoir (or well) pressure, psia,[73]

Z_i = gas deviation factor[74] at initial conditions, dimensionless, and

Z_a = gas deviation factor at abandonment conditions, dimensionless.

Because reservoir temperature in these reservoirs is greater than the cricondentherm, there is no retrograde loss of condensate as reservoir pressure decreases. Thus, the recovery efficiency of condensate should, in *theory*, be the same as the recovery efficiency of gas. This is indicated in the nonretrograde area of Fig. 3.3 with the notation [E$_{Rc}$]$_{pd}$ = [E$_{Rg}$]$_{pd}$.

In *practice*, however, as flowing wellhead pressure declines, some types of lease separation equipment (for example, low-temperature separators) become less effective. Under these conditions, unless the lease separation equipment is modified, the producing CGR may decrease gradually over life.[75]

Lamont [1963], in a study of 37 pressure-depletion gas reservoirs along the U.S. Gulf Coast, reported volume-weighted average recovery efficiency of gas to be 84.1%; reservoir average was 82.9%. It is noted that these reservoirs generally are moderately consolidated sand, with permeabilities of several hundred millidarcies. Although not reported, well spacing probably was about 160 acres. Lamont's [1963] study is consistent with results published by Stoian and Telford [1966]. In their study of 158 pressure depleted, or nearly pressure depleted, gas reservoirs in Canada average recovery efficiency of gas was estimated to be about 85%.

Hale [1981], Harlan [1966], and Stewart [1970], in separate papers on gas wells in *low-permeability* reservoirs in the central U.S., reported data that indicate significantly lower recovery efficiency of gas as the transmissibility, k_{gh}, decreases below about 100 md-ft. Recovery efficiencies calculated from data in these papers are plotted on **Fig. 3.11.** Additional comments regarding low-permeability gas reservoirs are in Sec. 6.8.

Water Influx

The recovery efficiency of gas from a gas reservoir subject to partial waterdrive can be expressed analytically with an equation of the form: [Agarwal *et al.* 1965, Cronquist 1980]

$$(E_{Rg})_{wd} = (1 - R_{pZ}) + (R_{pZ} E_V E_D) , \dots\dots\dots\dots\dots\dots\dots\dots\dots\dots\dots (3.29)$$

where $(E_{Rg})_{wd}$ = gas recovery efficiency, fraction,

R_{pZ} = ratio of abandonment p/Z to initial p/Z, or $p_a Z_i / p_i Z_a$,

E_V = fraction of the initially gas bearing volume swept by the aquifer—
i.e., volumetric sweep efficiency—and

---

[73] Before definitive well performance data are obtained and in the absence of experience with analogous reservoirs, abandonment pressure typically will be assumed to be approximately 10% of initial pressure. However, in low-permeability reservoirs, a higher value may be warranted.

[74] To avoid confusion with the term "gas compressibility,' the term "gas deviation factor" or "Z-factor" is used in this work rather than the historically used terms "super compressibility" or "compressibility factor."

[75] While the natural gas liquids not recovered on the lease might be recovered by a gas plant, consideration should be given to possible different ownership between lease and plant in assigning total net reserves (lease plus plant) to a specific interest.

E_D = microscopic displacement efficiency, defined as

$$(1 - S_{wi} - S_{gr})/(1 - S_{wi}) , \dots\dots\dots\dots\dots\dots\dots\dots\dots\dots\dots\dots\dots\dots\dots (3.30)$$

where S_{gr} is the residual gas saturation in the flooded portion of the reservoir. In the absence of laboratory k_w/k_g data, it may be estimated as: [Katz *et al.* 1966][76]

$$S_{gr} \approx 0.62 - 1.3\,\phi . \dots\dots\dots\dots\dots\dots\dots\dots\dots\dots\dots\dots\dots\dots\dots\dots\dots\dots\dots (3.31)$$

In the development of Eq. 3.29, it is assumed that gas is trapped in the flooded portion of the reservoir and is abandoned at the same average pressure as the gas in the unflooded portion of the reservoir at abandonment, p_a. Depending on reservoir dynamics, this may not be a valid assumption. The downdip areas of the reservoir may be flooded out at pressures higher or lower than reservoir abandonment pressure, even in moderate permeability reservoirs [Cronquist 1980]. In low-permeability reservoirs, this probably is not a valid assumption. Thus, in cases where the floodout pressure might be significantly different from the average reservoir abandonment pressure, it might be desirable to use a gas reservoir simulator to estimate reserves.

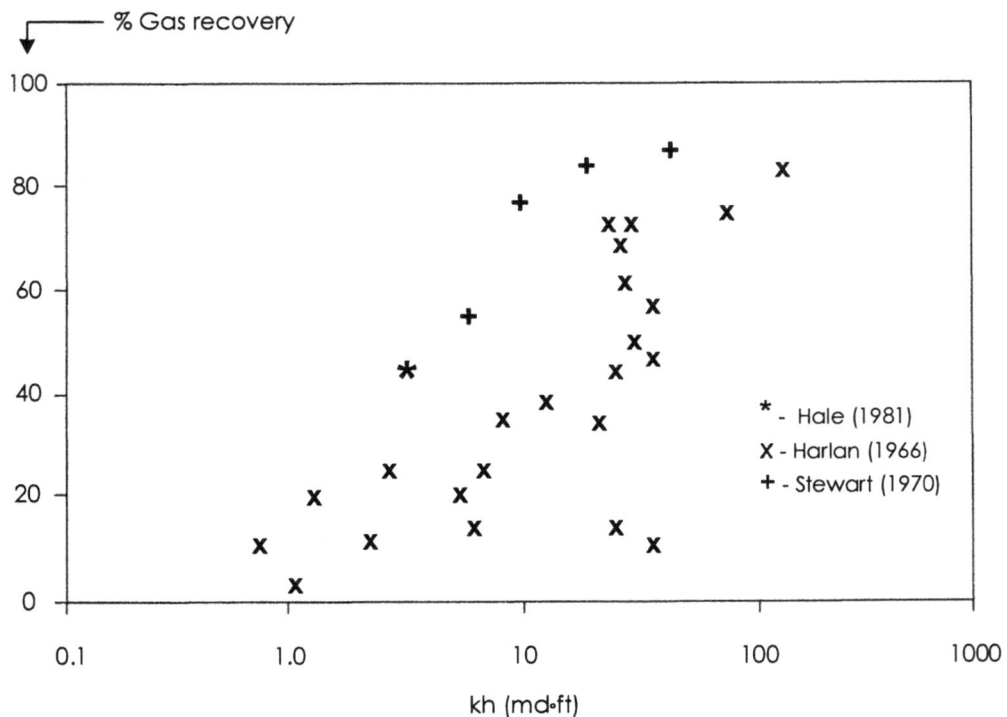

Fig. 3.11—Percent recovery of gas by pressure depletion vs. gas transmissibility, *kh*, [after Harlan 1966, Stewart 1970, and Hale 1981].

From examination of Eq. 3.29, it may be noted that the term $(1 - R_{pz})$ is the portion of the recovery efficiency attributable to pressure depletion. (It is equal to the term $(E_{Rg})_{dd}$, previously defined.) The term $(R_{pz}E_V E_D)$ is the portion of the recovery efficiency attributable

[76] De Leebeeck [1987] reported data from the North Sea Frigg field, from which residual gas in the flooded portion of the reservoir is approximated here by: $S_{gr} \approx 0.82 - 1.85\phi$.

to water influx. If there is complete pressure maintenance by the aquifer (such that $p_a = p_i$), the term $(1 - R_{pz})$ becomes zero, and the recovery efficiency is the product of the volumetric sweep, E_V, and microscopic displacement, E_D, efficiencies.

From "what if" calculations using Eqs. 3.30 and 3.31, it is apparent that the displacement efficiency of reservoir gas by water might approach 80% in very highly porous sands with low irreducible water saturation. Thus, under highly favorable conditions recovery efficiency of gas by strong waterdrive might approach 75 to 80% of GIP. Pressure depletion of such reservoirs, in contrast, could reasonably be expected to exceed 75 to 80% of GIP.

Recovery efficiency of gas from partial waterdrive gas reservoirs is rate sensitive [Agarwal *et al.* 1965, Cronquist 1980]. It has been advocated that such reservoirs should be produced at high rates to minimize abandonment pressure, thereby maximizing recovery efficiency of GIP [Agarwal *et al.* 1965]. However, depending on the nature of permeability distribution in a reservoir, high production rates may cause irregular water encroachment, which could result in less ultimate recovery than if the reservoir was produced at a lower rate [Cronquist 1980]. Please refer to Sec. 6.6 for additional discussion of this point.

The most difficult parameter to estimate in Eq. 3.29 is E_V. Even with considerable experience in a given area, in the absence of well performance data for the reservoir in question it is unlikely an accurate estimate can be made of the ultimate volumetric sweep efficiency. In moderate- to high-permeability sand reservoirs, typical of those encountered in clastic Tertiary basins, an initial volumetric estimate of 60% recovery efficiency of gas would be reasonable for reservoirs in which waterdrive was anticipated.[77]

In a study of a nearly depleted, major gas field in the U.S. Gulf Coast, actual recovery efficiencies of gas from partial waterdrive gas reservoirs were observed to be in the range 40 to 75%. Variations between individual reservoirs were attributed to differences in aquifer strength, reservoir heterogeneity, and operating practices [Cronquist 1984]. In the Frigg field, a major gas field in the North Sea that produced with a partial waterdrive, gas recovery is estimated to be 76% of GIP; volumetric sweep efficiency by water "should approach 97-98%."[78]

3.8 Reservoir Heterogeneity

Reservoir heterogeneity may be defined as the nonuniform spatial distribution of the physical and chemical properties of a reservoir. During the last 20 years or so, there has been increasing recognition that reservoir heterogeneity has a major influence on the recovery of oil and gas from subsurface reservoirs. Several developments have contributed to a growing knowledge of reservoir heterogeneity:

- increasing scrutiny and infill drilling of major waterfloods in the U.S., especially of the carbonate waterfloods in the Permian Basin of west Texas/New Mexico,

- development of better downhole logging devices, both open and cased hole, and of better interpretation techniques,

- advances in geophysical technology and interpretation techniques, including seismic stratigraphy,

- rapid advances in computer technology, which have enhanced capabilities for geophysical processing, log analysis, and reservoir simulation,

---

[77] Provided drainage point(s) are strategically located.
[78] Personal communication, Ole Bengt Hegreberg, 2000.

- development and application of improved recovery processes, which have spurred research on reservoir characterization, and
- use of multidiscipline teams for coordinated geophysical, geologic, and engineering studies of developing fields.

In preceding sections, little mention was made of the influence of reservoir heterogeneities on recovery efficiency of oil and/or gas. In scale, heterogeneities range from microscopic to megascopic and may be either favorable or unfavorable to recovery processes. Two examples —thin, low-permeability layers and natural fractures—are discussed briefly.

Shale stringers or thin, low-permeability layers near the bottom of a hydrocarbon column underlain by water may retard bottomwater coning [Wellings 1975]. Shale stringers or low-permeability layers near a gas/oil contact may retard gas coning. Richardson *et al.* [1978] discuss other examples and provide guidelines for calculating the effect of discontinuous shales on oil recovery.

Natural fractures may be either favorable or unfavorable, depending on reservoir drive mechanism. Without natural fractures, many low-permeability oil and gas reservoirs would be noncommercial; examples include the Spraberry (a siltstone) and the Austin Chalk trends in Texas. Natural fractures in the Asmari limestone (Iran) contribute to enormous well productivity. Natural fractures, however, cause major problems in waterflooding; the same natural fractures that contributed to primary recovery from the Spraberry trend[79] were detrimental to waterflooding this reservoir.

Although examples can be cited where reservoir heterogeneities are favorable to oil and gas recovery, in most cases they tend to be unfavorable. Heterogeneities cause two general types of problems:

- they make it difficult to *characterize* and *monitor performance* of wells and reservoirs, and
- they tend to cause nonuniform recovery of oil and gas.

Characterization of a reservoir includes determining the spatial distribution of porosity, oil and gas saturation, oil and gas properties, permeability, relative permeability, net and gross thickness, fractures, dip, rock texture and mineralogy, and other physical/chemical properties relevant to commercial recovery of oil and gas.

Seldom is it possible to characterize an oil or gas reservoir adequately from "static" data alone. All too often, one encounters the comment "the field (is) substantially more complex than initially interpreted." Adequate characterization almost always requires ongoing monitoring and analysis of "dynamic" data—e.g., individual-well bottomhole pressure, well performance, and changes in wellbore fluid saturations. In many cases, 3D seismic surveys can help resolve mapping anomalies. Additional discussion of performance monitoring is provided in Chap. 8.

Monitoring reservoir performance involves observation of spatial and temporal changes in pressure, fluid saturation, and fluid phase as the reservoir is produced—i.e., the periodic procurement of "dynamic" reservoir data. In addition, monitoring reservoir performance may involve periodic material balance or computer simulation studies. Reservoir characterization and monitoring methodology and the degree of reliability of the techniques used are strongly influenced by the nature and degree of reservoir heterogeneity. For detailed discussions of monitoring techniques in various reservoir settings, see Denny and Heusser-Maskell [1984],

---

[79] This accumulation is discussed in Sec. 7.5.

Harpole [1979], Haugen et al. [1988], LaChance and Winston [1987], Poston et al. [1983], Talash [1988], and Mills [1993].

Heterogeneities in sandstone reservoirs tend to be controlled primarily by the original environment of deposition [Le Blanc 1977]. Heterogeneities in carbonate reservoirs tend to be controlled by both original environment of deposition and post depositional diagenesis [Jardine et al. 1977].

3.8.1 Sandstone Reservoirs

Heterogeneities of concern in sandstone reservoirs include:

- different depositional units in the same reservoir,
- lateral discontinuities and multiple reservoirs in apparently "blanket" sands and
- shale layers of indeterminate areal extent.

The impact of these phenomena on reservoir characterization and performance are discussed briefly in the ensuing paragraphs.

Both irreducible water and residual hydrocarbon saturations are strongly influenced by rock texture, which is controlled by depositional environment [Sneider et al. 1977]. Fine-grained sediments, usually characteristic of low-energy depositional environments, tend to have high irreducible water and high residual hydrocarbon saturations. Coarse-grained sediments, usually characteristic of high-energy environments, tend to have low irreducible water saturation and low residual hydrocarbon saturation. In addition, fine-grained sediments tend to have lower-permeability than coarse-grained sediments. In waterdrive reservoirs, high-permeability zones tend to be swept by water before low-permeability zones. In solution-gas-drive reservoirs, high-permeability zones tend to pressure deplete before low-permeability zones.

The presence of lateral discontinuities and multiple reservoirs in apparently "blanket" sands has been discussed by Knutsen et al. [1971], Seal and Gilreath [1975], Hartman and Paynter [1979], McCubbin [1982] and Weimer et al. [1982]. Typically, a deltaic environment includes many subenvironments. Although sands deposited in these subenvironments often are interconnected, in many cases they are separated by thin, impermeable shale breaks that are not always detectable by wireline logs. If these separate sand bodies are not included in well-completion intervals, significant volumes of oil and gas may be left undrained.

Shale layers within a reservoir may have a major influence on fluid flow, depending on reservoir drive mechanism and spatial distribution of the shale breaks [Begg et al. 1989, Hazeu et al. 1988, Richardson and Blackwell 1971].

3.8.2 Carbonate Reservoirs

In addition to problems similar to those discussed for sandstone reservoirs, heterogeneities of concern in carbonate reservoirs include:

- lateral discontinuities in pay zones and
- very erratic, frequently vugular and often fracture-controlled, porosity.

The shelf carbonates in the U.S. Permian Basin typically contain numerous thin pay zones over a gross interval of several hundred feet [Ghauri et al. 1974, Stiles 1976]. Many of these pay zones are not laterally continuous between 40-acre spaced wells. Recovery efficiency from this type of rock sequence has been shown to be dependent on well spacing [Driscoll 1974, Barber et al. 1983, Barbe and Schnoebelen 1987, Wu et al. 1988, Godec and Tyler

1989]. It may be expected that recovery efficiency in shelf carbonates in other areas of the world will exhibit similar spacing dependency.

The original depositional fabric of most carbonates is quite heterogeneous because of the wide variety of material comprising these sediments [Jardine *et al.* 1977, Horsfield 1958]. In addition, carbonates are highly susceptible to post-depositional diagenesis, which often leads to more complex pore networks [Jardine *et al.* 1977]. Highly vugular sections frequently are difficult to core and evaluate with logs.

This section is not intended to be a complete treatment of difficulties in exploitation and reserve estimation caused by reservoir heterogeneities. The subject warrants an entire monograph. The reader is referred to specific field case histories [McCaleb and Willingham 1967, Wayhan and McCaleb 1969, and Lelek 1983] and to the bibliography provided by Le Blanc [1977] for more comprehensive discussions.

3.9 Allocation of Reserves to Wells and Leases [80]

Previous sections have dealt with reservoirs and their reserves "in-toto." However, the well-completions through which reserves are produced may not all have the same ownership.[81] *Net* reserves attributable to a specific interest may differ from well to well.

Even if the ownership is the same for all wells in a reservoir, good reservoir management mandates allocating reserves estimated for a reservoir to the wells expected to produce those reserves. This is especially important in the case of fields with multiple productive horizons, where reserves may be assigned to zones behind pipe.

Efficient exploitation of a multihorizon field requires efficient and timely utilization of available wellbores, especially in offshore fields, where utilization of well "slots" must be optimized. Monitoring of individual well performance over reservoir life facilitates detection of deviation from initial reserve estimates and timely implementation of appropriate remedial action.

In developing a plan to optimize utilization of wellbores, the following points should be considered:

- Reserves allocated to a well should be producible within the expected mechanical life of the well and the associated production facilities. A reserve life of about 20 years would be a reasonable initial design target.

- Consideration should be given to a multiple completion or a twin well if the estimated reserve life of a well greatly exceeds that of others in the field.

- Reserves should be producible within the term of the concession (or other) agreement, if one is in force.[82]

The initial allocation of reserves to wells usually will involve the initial volumetric estimate of reserves for the reservoir. Typically, each well will be assigned a pro-rata share of reserves based on the well's net pay and estimated drainage area. Later in reservoir life, however, individual well reserves probably will have to be modified to reflect individual well performance. Significant differences between the initial volumetric estimate of ultimate recovery and that determined from well performance should be investigated.

---

[80] Procedures discussed in this section may be moot if reserves have been estimated using a computer simulation model with a grid that includes all anticipated completions in the reservoir under study.

[81] For example, in the U.S. Gulf Coast, a geologic setting in which wells may penetrate multiple reservoirs, the ownership in each reservoir may differ, depending on the configuration of the overlying leases with respect to the mapped areas of the reservoirs.

[82] While this might be an appropriate criterion for a concession operator, the views of the host government might differ.

For a given reservoir, in general, reserves may allocated to:

- wells producing from the reservoir—i.e., proved, producing (PvPd) reserves,
- wells producing from deeper reservoirs where the reservoir in question is behind pipe—i.e., proved or probable, behind pipe (PvBP or PbBP) reserves, and
- wells yet to be drilled—i.e., proved, probable, or possible, not drilled (PvND, PbND, or PsND) reserves.

Procedures for allocating an initial volumetric estimate to individual wells or locations in a reservoir include vertical distribution curves, potential tests, and porosity/net pay/saturation maps. Each of these procedures is discussed briefly in the following sections.

3.9.1 Vertical Distribution Curve

In light-oil reservoirs[83] (or nonassociated gas reservoirs), if there is substantial vertical relief and waterdrive is anticipated, a vertical distribution curve provides a reasonable method to make an initial allocation of reserves to individual wells. This procedure may not be satisfactory in heavy-oil reservoirs where water under-running may occur [Dietz 1953]. After wells have started to produce water, it usually will be necessary to revise the initial allocation for all the wells in the reservoir, as reserve estimates shift from a volumetric to a performance basis.

A vertical distribution curve, such as that shown in **Fig. 3.12** for an oil reservoir, is prepared as follows:

1) From a structure map on the *top* of porosity, determine the area (acres) enclosed by the oil/water contact and the areas enclosed by progressively shallower structural contours.

2) Repeat Step (1) for a structure map on the *base* of porosity.

3) Plot contour depth (y axis) vs. area on top of porosity and vs. area on the base of porosity (x axis).

4) Post current and anticipated wells on the vertical distribution curve by their position on the top of porosity; also post the perforated interval for current wells.

5) Subdivide the reservoir into interwell *volumes,* identified as "A," "B," "C_1," "C_2," "D," and "E" on Fig. 3.12; note these are *gross* volumes.

6) Allocate initial reserves estimated for the entire reservoir to the interwell volumes in the proportion the interwell volumes bear to the whole.

7) Allocate interwell reserves to individual wells in accordance with the *drainage plan* for the reservoir.

For the purpose of this discussion, it is assumed that: (a) the oil is slightly undersaturated at initial conditions and (b) the aquifer will maintain reservoir pressure above the bubblepoint. Thus, the (initial) drainage plan considers Well 4 to be contingent on the strength of the aquifer and would not be drilled if a secondary gas cap were to develop.

---

[83] In this context, light oil is characterized here as having a stock-tank gravity greater than about 25°API, and would be expected to have reservoir mobility greater than that of the encroaching water. Heavy oil is characterized here as having a stock-tank gravity less than about 25°API and would be expected to have a reservoir mobility less than that of encroaching water so that under-running would be expected [Dietz 1953].

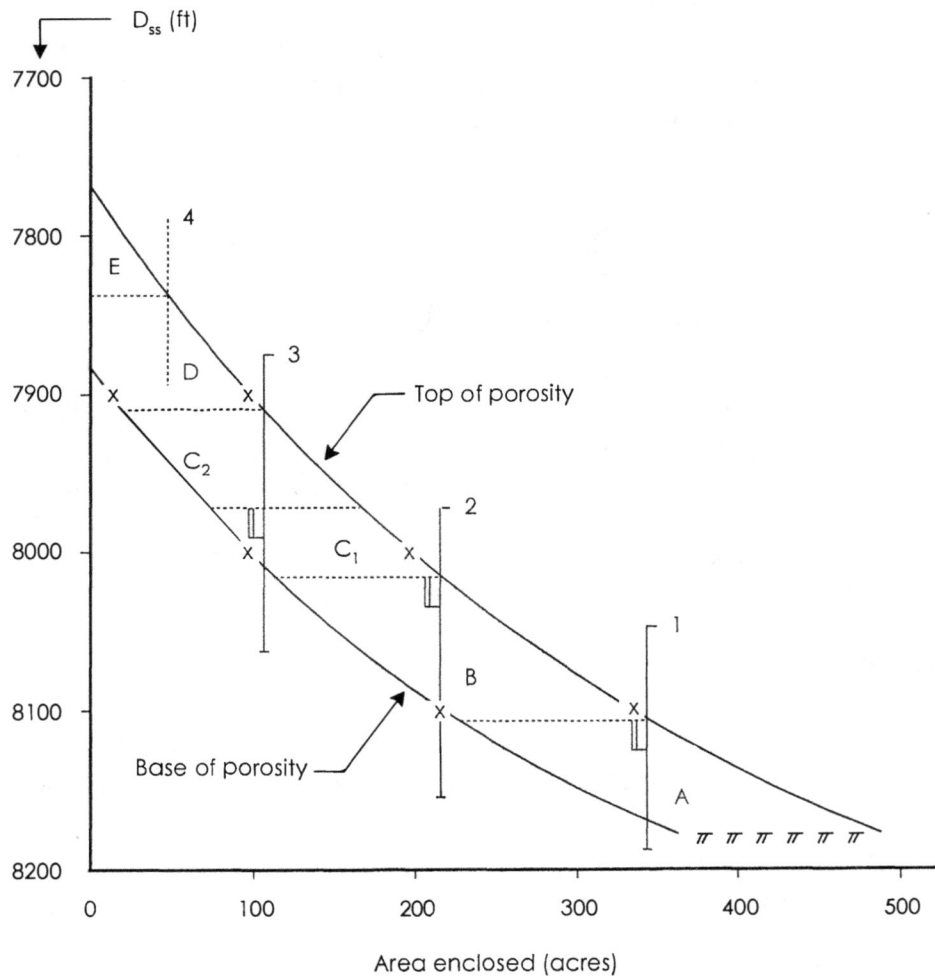

Fig. 3.12—Example of distribution of reservoir volume vs. depth, Illustrating method to allocate reserves to wells.

In the example shown by Fig. 3.12, the *drainage plan* for this reservoir is to drill and complete Well 4 after Well 1 waters out. Also, Well 3 will be plugged back to the top of the formation after the perforations in the base of sand in this well water out. Thus, *for this drainage plan,* the initial allocation of interwell reserves to individual wells would be as follows:

Well 1: A/3
Well 2: A/3 + B/3
Well 3: A/3 + B/3 + $(C_1 + C_2)/2$
           Well 4: B/3 + $(C_1 + C_2)/2$ + D

For Well 3, the reserves $C_2/2$ would be classified as proved, behind pipe (PvBP); reserves for Well 4 would be classified as probable, not drilled (PbND). Reserves in volume "E" would be classified either Pb or Ps, depending on expectations about drive mechanism, and assigned to Well 4. All other reserves would be classified as proved, producing (PvPd).

This method is a logical way to make an *initial* allocation of reserves to individual wells in a multiwell waterdrive reservoir. The procedure assumes: (a) uniform spatial distribution of the ratio of net to gross pay and (b) uniform vertical encroachment of water. Reasonably uniform vertical encroachment of water may be expected when the displacement is controlled by gravity forces. Examples include the Coldwater (oil) field in Michigan [Criss and McCormick 1954] and the Turtle Bayou "U" Sand (gas) reservoir [Cronquist 1984]. However, if the displacement is controlled by viscous forces, which may occur in low-permeability or heterogeneous reservoirs, the displacement may not be uniform. The Turtle Bayou "Z" Sand (gas) reservoir [Cronquist 1984] and the South Pass Block 27 "M6" Sand (oil) reservoir [Hartman and Paynter 1979] are examples of nonuniform water encroachment.

A procedure analogous to that discussed would be followed for oil reservoirs producing from an expanding gas cap, if effective gravity segregation were anticipated. If, for example, volumes "D" and "E" on Fig. 3.12 represented an initial gas cap, the allocation of oil reserves to wells would be as follows:

Well 3: $C_2/3$
Well 2: $C_2/3 + C_1/2$
Well 1: $C_2/3 + C_1/2 + B$

One contingency in such a plan would be whether to complete Well 1 deeper, which would depend on aquifer strength. If well performance and/or material balance (or other studies) indicated insignificant influx, a recompletion to the base of sand in Well 1 might be warranted.

3.9.2 Well Potential

In reservoirs producing by pressure depletion, well potential or average flow rate provides a reasonable *initial* method to allocate reserves. Matthews *et al.* [1954] have shown that in a bounded reservoir, once semisteady state conditions have been established, each well *should* drain a volume proportional to its production rate. This theoretical work was corroborated by Stewart [1966] for gas wells in Wilcox (Eocene) sandstones in south Texas. Similar behavior has been noted elsewhere.

3.9.3 Porosity/Net Pay/Saturation

In the absence of, or to supplement, well test information, volumetric parameters may be used in the initial allocation of reserves to individual wells. The product,

(POROSITY) x (NET PAY) x (HYDROCARBON SATURATION),

here called the PNH product, includes data that should be available for each well after development is complete. Depending on well spacing, the timing of production initiation, and production policy, each well reasonably can be initially expected to drain a volume proportional to its PNH product.

3.9.4 Other Considerations

Other factors to consider in allocating reserves to wells include commerciality of undrilled locations, possible prior drainage of behind-pipe reserves, and recharge from deeper zones.

Commerciality of Undrilled Areas

Caution should be exercised in classifying the entire mapped volume of hydrocarbons as "reserves." For example, a net hydrocarbon isopach of a partially developed reservoir may include areas that are too thin to support drilling commercial wells. If these areas are not expected to be drained by existing (or planned) wells, the hydrocarbons in those areas cannot be considered reserves.

Additional complications are posed by areally extensive accumulations offshore that must be developed by fixed platforms: (a) the minimum acceptable reserve size to justify a well or platform is substantial and (b) the extremities of some accumulations may not be reachable, even with highly deviated or extended-reach wells. The industry's ability to drill extended-reach horizontal wells, however, has increased substantially in the last few years. In late 1997, for example, 2.5- to 3-mile offsets were more or less routine, with the record being 5 miles. Thus, the question of unreachable accumulations may be more one of economics than technology.

Possible Drainage of Behind-Pipe Reserves

Frequently, reserves expected to be produced from a zone behind pipe in a well or wells will be produced instead from adjacent wells completed in the same reservoir. This may happen, for example, because of unanticipated drainage patterns caused by reservoir heterogeneities. In *multiple zone* fields, behind-pipe reserves for a zone may be produced by wells completed in overlying or underlying zones because of faulty primary cementation of production casing.

When behind-pipe reserves are included in the reserve base, the performance of all completions in a field should be monitored to ensure those reserves are not being drained by other wells in the field. Regarding producing reserves in such a field, *significant* deviation between an initial volumetric reserve estimate for a well and actual recovery (either high or low) should be investigated. While there are many reasons for such deviation, one reason for a well's producing significantly more reserves than initially expected might be that the well is draining behind pipe reserves anticipated to be produced by other wells in the field.

Recharge From Deeper Zones

Although it seems to be quite unusual, there is a possibility that some fields are being recharged from deeper zones—i.e., zones at depths significantly below the field productive limits proved by drilling. Reportedly [Browne 1995], estimated (remaining) reserves in the Eugene Island Block 330 field, offshore Louisiana, have decreased much less than anticipated with production. It was noted that the "chemical signature" of some of the oil currently being produced from this field suggested sourcing from a deeper stratigraphic section. The concept of reserve replenishment of producing fields by deep sourcing is intriguing and ought to be retained in one's set of multiple working hypotheses.[84]

---

[84] Since completion of this section, an additional paper on the Eugene Island Block 330 field was published that (in the author's view) apparently confirms the recharge hypothesis [Losh 1998].

CHAPTER 4

MATERIAL BALANCE METHODS

4.1 Expanded Material Balance Equation

4.1.1 Principles

The so-called Schilthuis [1936] material balance equation is one of the fundamental relations in reservoir engineering. In "expanded" form—i.e., including water influx—it may be stated

PRODUCTION OF OIL & GAS (PLUS WATER)	=	EXPANSION OF OIL & FREE GAS INITIALLY IN PLACE	+	WATER INFLUX.

Assuming an initial gas cap and, at this time, ignoring compressibility of pore volume and interstitial water, this relation may be expressed in SPE [1993] notation as:

$$\underline{\text{PRODUCTION}} \quad = \quad \underline{\text{EXPANSION}} \quad + \quad \underline{\text{INFLUX}}$$

$$\underline{\text{GAS CAP}} + \underline{\text{OIL COLUMN}} + \underline{\text{AQUIFER}} = \underline{\text{GAS CAP}} + \underline{\text{OIL COLUMN}} + \underline{\text{AQUIFER}}$$

$$\underline{\text{TOTAL}} - \underline{\text{SOLUTION}}$$
$$\underline{\text{GAS}} \quad \underline{\text{GAS}}$$

$$N_p R_p B_g - N_p R_{si} B_g + N_p B_t + W_p B_w = N_i m B_{ti}(B_g - B_{gi})/B_{gi} + N_i(B_t - B_{ti}) + W_e, \quad ......(4.1a)$$

and gathering terms,

$$N_p[B_t + B_g(R_p - R_{si})] + W_p B_w = N_i[(B_t - B_{ti}) + m B_{ti}(B_g - B_{gi})/B_{gi}] + W_e ....(4.1b)$$

Units on both sides of the equal sign are reservoir barrels. Also, units for B_g in this and subsequent equations in this section are RB/scf, in contrast with units for this term in Eq. 3.3a and 3.3b, where they are res cf/scf.

Eq. 4.1b is consistent with the formulation of Schilthuis [1936], who ignored compressibility of reservoir pore space and interstitial water.

Inclusion of compressibility terms should be considered for material balance calculations involving oil reservoirs above the bubblepoint and for all geopressured reservoirs.[1]

Depending on the magnitude of pore and water compressibility compared to overall system compressibility, it may be desirable to include compressibility terms for oil material balance calculations both above and below the bubblepoint pressure. If compressibility terms were included for both the initial gas cap and the oil column, Eq. 4.1b would be written[2]

$$N_p[B_t + B_g(R_p - R_{si})] + W_p B_w = N_i[(B_t - B_{ti}) +$$

$$(1 + m)(c_p + c_w S_w)B_{ti}(p_i - p)/S_o + mB_{ti}(B_g - B_{gi})/B_{gi}] + W_e, \dots\dots\dots (4.1c)$$

where the pore compressibility, c_p, is the *integrated* value between p_i and p, where p is the pressure at which Eq. 4.1c is evaluated.[3]

The formulation of Eq. 4.1c differs from that of Havlena and Odeh [1963], who ignored compressibility effects in the initial gas cap. In high-pressure accumulations in friable rocks, it may be necessary to consider compressibility effects in initially gas-bearing zones, as has been the case for the North Sea Ekofisk field [Johnson and Rhett 1986].

In Eq. 4.1b and 4.1c, it is assumed that the reservoir can be treated as a "tank"—i.e., that a reservoir average value of reservoir pressure and the corresponding PVT terms may be used at each timestep. There are three unknowns in each of these two equations:

- stock-tank barrels of oil initially in place, N_i,
- size of the initial gas cap as a fraction of the initial oil zone volume, m, and
- cumulative water influx, W_e.

Depending on how compressibility is treated, either Eq. 4.1b or 4.1c is solved at the end of successive time periods, typically quarterly,[4] using the cumulative production data at the end of each period. PVT properties are evaluated at the average static reservoir pressure at the end of the period.

Theoretically, with a sufficiently long pressure/production history and repetitive solutions of one or the other of these equations, it should be possible to solve for all three unknowns. In practice, this is rarely possible, mainly because of errors in measuring produced fluids and problems in interpreting and averaging bottomhole pressures. If the presence of both a significant initial gas cap and/or water influx is a possibility, efforts should be made to determine initial gas cap size, m, by volumetric mapping.

4.1.2 Limitations

Limitations to reliable application of the material balance equation are both theoretical and practical.

[1] Geopressured reservoirs are discussed in Sec. 6.5.
[2] This formulation tacitly assumes that S_w in the gas cap is the same as that in the oil column.
[3] Compressibility terms—c_p and c_w—were defined in Sec. 3.7.2. As discussed there, if dealing with consolidated rocks, one may prefer to (a) treat pore compressibility as a constant (independent of reservoir stress) or (b) ignore it.
[4] The frequency of material balance calculations typically depends on the frequency of reservoir pressure measurements. In a new accumulation of reasonable size, it is good practice to monitor reservoir pressure quarterly or semiannually until the accumulation has been developed, after which the frequency of measurements may be reduced to an annual basis.

Theoretical Limitations

Theoretical limitations are imposed by the following assumptions, which are necessary for a tractable methodology:

- Oil and free gas in the reservoir are assumed to be in thermodynamic equilibrium. Wieland and Kennedy [1957] reported about 20 psi supersaturation in experiments conducted using East Texas and Slaughter field cores. Comparable observations have been made by others, especially in low-permeability rocks.

- PVT data, obtained from differential liberation, are assumed to replicate the liberation process in the field. As discussed by Dodson et al. [1953] and others, both flash and differential liberation of gas may occur at various times and places between the reservoir and the stock tank, with the differences in PVT properties between the two processes increasing with more volatile oils.

- Free gas in the reservoir is assumed to have the same composition as free gas on the surface, differing only in volume, as expressed by the gas formation volume factor. With progressively more volatile oils, free gas in the reservoir will contain progressively more liquids in the vapor phase, which are recovered as stock-tank liquids but are not accounted for by the differential liberation process.[5]

Practical Limitations

Practical limitations are imposed by data requirements and reservoir conditions. Data required for reliable application of the material balance equation include:

- PVT analyses of fluid samples *representative of initial reservoir conditions,*[6]

- accurate static bottomhole pressure history of key wells in the reservoir,

- accurate monthly production data for oil, gas, and water, and

- depending on circumstances, pore volume compressibility vs. net overburden pressure, as discussed in Sec. 3.7.2.

The accuracy requirements for pressure and production data usually exceed the routine needs for many field operations. Reservoir conditions that may limit the reliability of a material balance estimate include:

- Strong waterdrive and/or a large initial gas cap may maintain reservoir pressure at nearly initial pressure. Under these conditions, the material balance equation generally will not yield stable solutions because the small pressure decrements in the reservoir frequently are of the same magnitude as errors in the measurements. Because there is little fluid expansion, small errors in the pressure measurements cause disproportionately large errors in the fluid expansion terms.

- Areally extensive reservoirs may have different areas at different stages of development and production. This may lead to wide variations in gas saturation and reservoir pressure that cannot readily be averaged.

- Areally extensive reservoirs with low values of kh/μ may make it difficult to determine the static bottomhole pressure reliably and often cause large areal variations in pressure that are difficult to average.

---

[5] Walsh [1995] proposed a method that may overcome this limitation.

[6] This requirement is not met for an oil reservoir with a gas cap that has not been sampled. If material balance methods must be used to determine m, an estimate must be made of the gravity of the gas cap gas, thereby introducing possible error in the calculations.

- Very heterogeneous reservoirs with zones of high permeability interbedded with zones of low permeability, or highly fractured reservoirs,[7] conditions under which the low-permeability zones, or the matrix blocks, usually pressure deplete slower than the high-permeability zones, or the fractures, and it is practically impossible to determine volumetrically weighted average reservoir pressure.

- Reservoirs with significant spatial variations in initial fluid properties, especially those with variations in initial GOR and bubblepoint pressure, may preclude representing the accumulation with "average" fluid properties.

As discussed in Sec. 4.3.1, *some* of these limitations may be overcome by using a multidimensional computer simulator rather than a zero-dimensional material balance, or tank, model.

Irrespective of the material balance method used (tank model or multidimensional simulator), it is good practice to plot all static bottomhole pressure data vs. time on the same graph for all the wells suspected of being in a common reservoir. Such a plot frequently will provide valuable insight regarding the degree of communication between wells. It may help to identify wells that are in separate reservoirs, which might be contrary to the existing geologic interpretation.

4.2 Havlena-Odeh Methodology

Havlena and Odeh [1963] have shown that Eq. 4.1c may be written as an equation of a straight line by making the following simplifications:

- Define F_{pR} as the reservoir volume of cumulative oil, gas, and water production

$$F_{pR} = N_p[B_t + B_g(R_p - R_{si})] + W_p B_w . \dotfill (4.2)$$

- Define E_o as the expansion of a unit volume of oil and dissolved (solution) gas initially in place

$$E_o = (B_o - B_{oi}) + (R_{si} - R_s)B_g = (B_t - B_{ti}) . \dotfill (4.3)$$

- Define E_c as the "expansion" due to pore and water compression effects, which term is identified elsewhere as the "compression term"[8]

$$E_c = (1 + m)(c_p + c_w S_w)B_{ti}(p_i - p)/S_o . \dotfill (4.4)$$

- Define E_g as the expansion of the initial gas cap, if one is present

$$E_g = (B_g - B_{gi}). \dotfill (4.5)$$

Substituting these new terms, Eq. 4.1c becomes

$$F_{pR} = N_i[E_o + (1 + m)E_c + mB_{ti}E_g/B_{gi}] + W_e . \dotfill (4.6)$$

---

[7] Please refer to the discussion of the Palm Valley (Gas) field in Sec. 7.4 for a historical summary of reserve estimates in a highly fractured reservoir rock with a low-permeability matrix.

[8] The formulation of this term, which accounts for expansion of pore volume and interstitial water in the gas cap in addition to that in the oil column, is taken from Dake [1978]. Havlena and Odeh [1963] neglected a compressibility term for reservoirs with an initial gas cap.

Depending on pressure/production history and available subsurface data, it may not be apparent in early stages of the analysis if a reservoir being studied had an initial gas cap or if there is water influx, or both. Eq. 4.6 may be used in various forms to estimate the amount of oil and gas initially in place and the probable reservoir drive mechanism. The possible scenarios are discussed in Sec. 4.2.1 for oil reservoirs and in Sec. 4.2.3 for gas reservoirs. To estimate (remaining) reserves, one of the predictive forms of the material balance equation, or other method, must be used, as discussed in Sec. 4.3.

4.2.1 Oil Reservoirs

The Havlena-Odeh [1963] methodology is examined for oil reservoirs for each of four possible scenarios:

- no water influx and no initial gas cap,
- no water influx and initial gas cap of unknown size,
- water influx and no initial gas cap, and
- water influx and initial gas cap of unknown size.

No Water Influx/No Initial Gas Cap

If there is no initial gas cap—i.e., if $m = 0$—and if there is no water influx, $W_p B_w$ is dropped from Eq. 4.2. After dropping W_e, Eq. 4.6 reduces to

$$F_{pR} = N_i (E_o + E_c) . \dotfill (4.7)$$

A plot of F_{pR} vs. $(E_o + E_c)$ should be a straight line,[9] passing through the origin, as shown by **Fig. 4.1**.

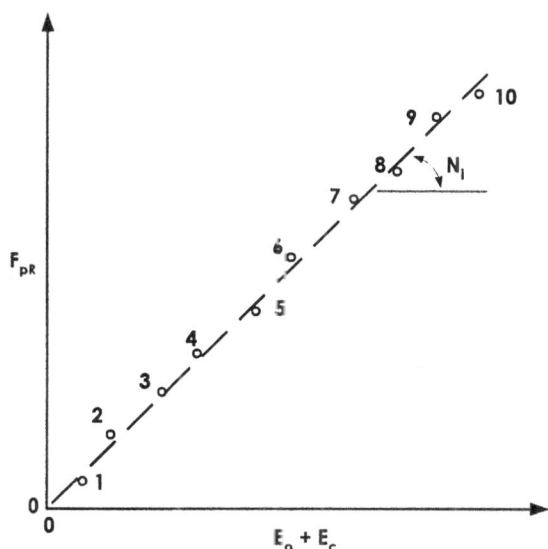

Fig. 4.1—Reservoir volume of cumulative production, F_{pR}, vs. expansion of oil and initially dissolved gas, E_o, and compression, E_c, illustrating method to estimate OIP, N_i, for case of no water influx and no initial gas cap [after Havlena and Odeh 1963].

---

[9] For consolidated rocks in normally pressured reservoirs, the compressibility term, E_c, may be ignored. This simplification would lead to a plot of F_{pR} vs E_o.

The forward sequence of points represents successive solutions of the fluid production and expansion terms over the life of the reservoir. As indicated by Havlena and Odeh [1963], "the origin is a *must* point; thus, one has a fixed point to guide the straight line plot." The first few points may be erratic and may be ignored in constructing the straight line. As indicated by Eq. 4.7, the slope of the line is $F_{pR}/(E_o + E_c)$, which has units RB/(RB/STB), which equals N_i —STB oil initially in place.

No Water Influx—Initial Gas Cap of Unknown Size

Considering the reservoir evaluated with Eq. 4.7, if there had been an initial gas cap or if there has been water influx, or both, a plot of F_{pR} vs. E_o would have curved upward. If the reservoir oil had been saturated at initial conditions (even though free gas may not have been encountered in drilling or testing) an initial gas cap might reasonably be suspected. If there is no water influx, Eq. 4.6 becomes

$$F_{pR} = N_i[E_o + (1 + m)E_c + mB_{ti}E_g/B_{gi}] , \quad\dots\dots\dots\dots\dots\dots\dots\dots\dots\dots\dots (4.8a)$$

and a plot of F_{pR} vs. $[E_o + (1 + m)E_c + mB_{ti}E_g/B_{gi}]$ for the correct m should be a straight line passing through the origin. As indicated by Eq. 4.8a, the slope will be N_i. As shown by **Fig. 4.2**, if the (estimated) value for m is too small, the plot will curve upward; if it is too large, downward. If there is significant water influx, the plot will curve upward, regardless of the m value used.

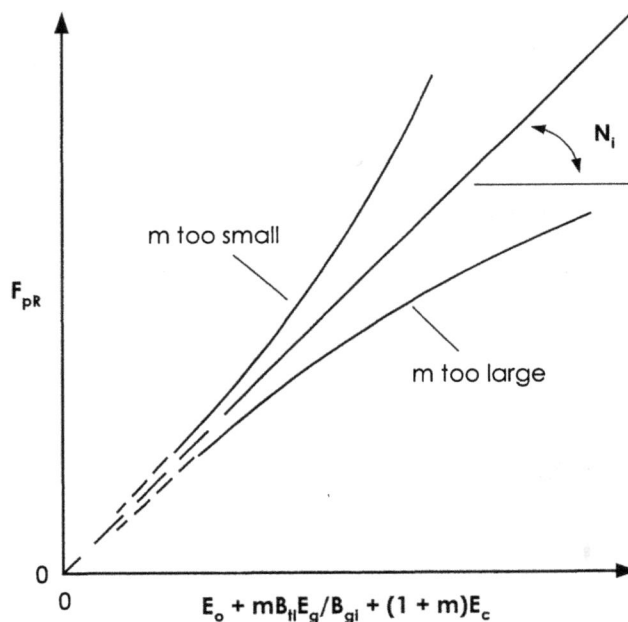

Fig. 4.2—Reservoir volume of cumulative production, F_{pR}, vs. expansion of oil, initially dissolved gas, and initial gas cap, $E_o + mB_{ti}E_g/B_{gi}$, plus compressibility term, $(1+m)E_c$, illustrating method to estimate OIP, N_i, and m for case of no water influx and unknown gas cap size [after Havlena and Odeh 1963].

Dividing both sides of Eq. 4.8a by $[E_o + (1 + m)E_c]$ leads to

$$F_{pR}/[E_o + (1 + m)E_c] = N_i + [mN_iB_{ti}E_g/B_{gi}]/[E_o + (1 + m)E_c] . \quad\dotfill(4.8b)$$

A plot of $F_{pR}/[E_o + (1 + m)E_c]$ vs. $E_g/[E_o + (1 + m)E_c]$ for the correct m should be a straight line with slope equal to mN_iB_{ti}/B_{gi}, which equals G, and a y intercept equal to N_i. Havlena and Odeh [1963] observed that the Eq. 4.8a plot is more powerful because it must pass through the origin, but that both plots should be used in every case.

Water Influx—No Initial Gas Cap

If the oil initially in place had been undersaturated, but there was reason to suspect water influx, W_pB_w should be retained in Eq. 4.2. Eq. 4.6 can be rearranged and written as follows:

$$F_{pR}/[E_o + E_c] = N_i + C[\Sigma\Delta pQ_D(\Delta t_D,r_D)]/[E_o + E_c] , \quad\dotfill(4.9)$$

where the water influx term, W_e, has been written in terms of the water influx constant, C, and the van Everdingen and Hurst [1949] solution to the diffusivity equation for unsteady-state dimensionless water influx, $\Sigma\Delta pQ_D(\Delta t_D,r_D)$.[10] Use of the van Everdingen-Hurst formulation is not mandatory; other methods to compute W_e may be used, as discussed in Sec. 4.2.2.

As shown by **Fig. 4.3**, a plot of $F_{pR}/(E_o + E_c)$ vs. $\Sigma\Delta pQ_D(\Delta t_D,r_D)/(E_o + E_c)$ should be a straight line for the correct water influx parameters (t/t_D and r_e/r_i). N_i is the y intercept, and C is the slope. If the water influx term is too small, the plot will curve upward; if it is too large, downward.

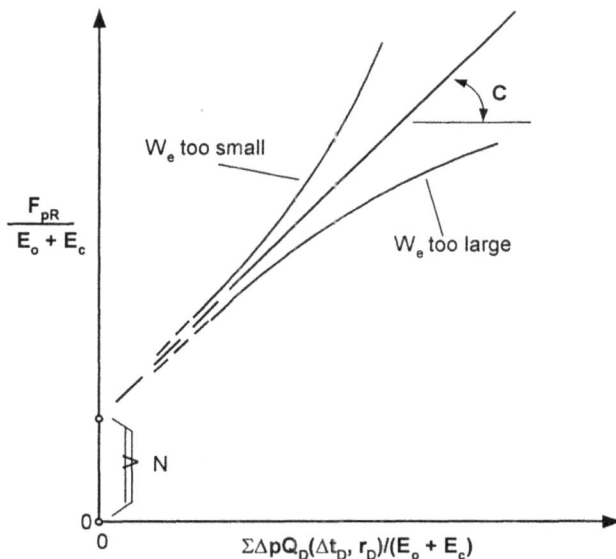

Fig. 4.3—Reservoir volume of cumulative production divided by oil expansion and rock-fluid compression term, $E_o + E_c$, vs. estimated cumulative water influx divided by oil expansion and rock/fluid compression term, $E_o + E_c$, illustrating method to estimate OIP, N_i, and water influx constant, C, for case of water influx and no initial gas cap [after Havlena and Odeh 1963].

---

[10] Depending on the strength of the aquifer, compressibility effects may be negligibly small, and the compressibility term, E_c, may be dropped from Eq. 4.9.

Regarding limited aquifers, Havlena and Odeh [1963] noted that the early (unsteady-state) period may be successfully fitted by assuming an infinite aquifer, during which time the calculated points will plot in a forward sequence. However, if the aquifer boundary is reached during the history-match period, the onset of semisteady-state water influx may result in the calculated points plotting in a reverse sequence. This point also has been discussed by Dake [1978].

Stewart [1980] discussed an application of this methodology to the Piper field (North Sea). Reportedly, he was able to obtain an acceptable straight-line fit after only 6 months of production history. The material balance estimate, 1.4 billion STB, was within 10% of the volumetric estimate.[11]

In the presence of uncertainties in reservoir pressure, the Havlena-Odeh methodology may not be sensitive enough to define a linear trend from which oil initially in place can be determined. In this event, a method proposed by McEwen [1962] may work better.

Water Influx—Initial Gas Cap of Unknown Size

If there is an aquifer and an initial gas cap, Eq. 4.6 may be written as

$$\frac{F_{pR}}{E_o + (1 + m)E_c + mB_{ti}E_g/B_{gi}} = N_i + C\frac{\Sigma\Delta p Q_D(\Delta t_D, r_D)}{E_o + (1 + m)E_c + mB_{ti}E_g/B_{gi}}. \quad\dots\dots\dots\dots (4.10)$$

If the gas cap and the aquifer have been characterized correctly (correct m, t/t_D, and r_e/r_i), then a plot of the left term in Eq. 4.10 vs. the fractional part of the second right term should be linear, with the y intercept being N_i and the slope being C.

Instead of plotting only the fractional part of the second right side term of Eq. 4.10 on the x axis, Dake [1978] proposed plotting the entire second right term. He noted that, if the gas cap and the aquifer have been characterized correctly, the plot should be linear and have a slope of 45 degrees. The slope requirement of this type of plot places an additional constraint on the interpretation and should be considered as an additional, albeit confirming, step in the analysis.

4.2.2 Water Influx Term

Estimation of the water influx term, W_e, when required to develop a linear Havlena-Odeh [1963] plot, typically is the source of the most difficulty and error in the application of material balance methods. Although not mandatory, as previously noted, the van Everdingen-Hurst [1949] solutions to the diffusivity equation historically have been the most widely used analytical methods to calculate W_e. These solutions have been published in both graphical and tabular form, with arguments being dimensionless time, t_D, and dimensionless radius, r_D. Historically used table-lookup procedures for calculation of water influx, however, may be avoided by use of numerical approximations published by Klins et al. [1988]. Dake [1978, 1994] provided comprehensive discussions of the problems and procedures for calculation of water influx, and no attempt is made to summarize them here.

4.2.3 Gas Reservoirs

Two cases of the Havlena-Odeh [1963] methodology are examined for nonassociated gas reservoirs:

---

[11] It is noted here, however, that the subsequent 4 years of pressure history gave greater confidence in the estimate of OIP than appears warranted by the claimed fit after only 6 months.

- water influx and
- no water influx.

Water Influx

For a nonassociated gas reservoir, or one with a negligibly small oil rim, F_{pR} (from Eq. 4.2) may be redefined as

$$F_{pR} = G_p B_g + W_p B_w \,. \quad\text{...}(4.11)$$

Ignoring the compression term, Eq. 4.6 then may be written as

$$F_{pR} = G_{Fi} E_g + C[\Sigma \Delta p Q_D(\Delta t_D, r_D)] \,. \quad\text{...}(4.12)$$

Dividing by E_g leads to

$$F_{pR}/E_g = G_{Fi} + C[\Sigma \Delta p Q_D(\Delta t_D, r_D)]/E_g \,. \quad\text{...}(4.13)$$

As shown by **Fig. 4.4**, if the aquifer is characterized correctly (correct t/t_D and r_e/r_i), a plot of F_{pR}/E_g vs. $[\Sigma \Delta p Q_D(\Delta t_D, r_D)]/E_g$ should be a straight line with the y intercept equal to G_{Fi}, gas initially in place, and slope equal to C.

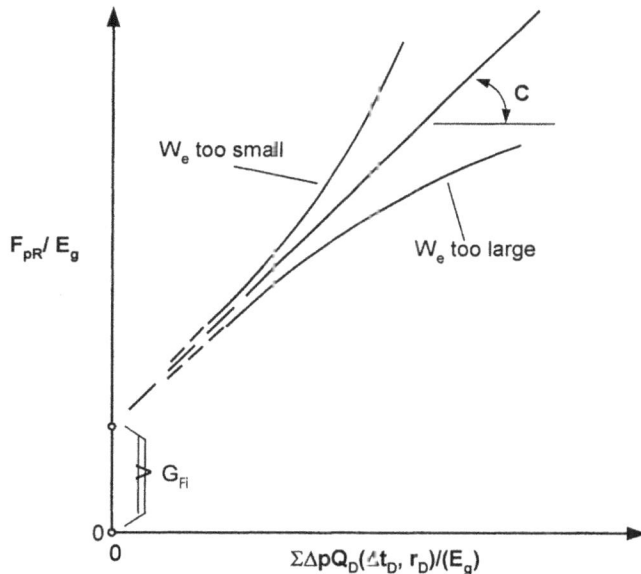

Fig. 4.4—Reservoir volume of cumulative production divided by expansion of initial gas, F_{pR}/E_g, vs. estimated cumulative water influx divided by expansion of initial gas, $\Sigma \Delta p Q_D(\Delta t_D)/E_g$, illustrating method to estimate GIP, G_{Fi}, and water influx constant, C, for case of gas reservoir with water influx [after Havlena and Odeh 1963].

No Water Influx

In this case, a plot of G_pB_g vs. E_g should be a straight line through the origin with the slope equal to G_{Fi}. Usually, however, it is easier to use another form of the material balance equation—i.e., a plot of p/Z vs. G_p, as discussed in Sec. 4.4.[12]

4.2.4 Guidelines for Application[13]

From a review of the foregoing, it should be apparent that, in general, the Havlena-Odeh [1963] methodology involves a series of trial-and-error calculations and plots that involve various combinations of the following terms:

E_o	- oil expansion term, a function of reservoir pressure, fn (p),
E_g	- gas expansion term, fn (p),
E_c	- compressibility term, fn (p),
F_{pR}	- cumulative fluid production term, a function of time and pressure, fn (t,p),
B_{ti}/B_{gi}	- ratio of initial formation volume factors, constant,
m	- ratio of initial gas cap volume to oil column volume, assumed or estimated from subsurface data and/or performance data; and
W_e	- cumulative water encroachment, fn (C,t,r_e,p), calculated analytically.

The above terms may be calculated for the reservoir being analyzed using a spreadsheet. A suggested layout is shown in **Table 4.1**. In support of the material balance calculations, separate columns outside the displayed worksheet area have been set up to curve fit the PVT data to average the historical pressure data. Using a spreadsheet for the computations facilitates: (a) trial-and-error combinations of the above terms, (b) making trial-and-error plots, and (c) linear regression analysis of the trial plots. The following procedure is suggested for an oil reservoir:

1. If the oil was undersaturated at initial conditions, and there is no evidence (from production or pressure data) of water influx, plot F_{pR} vs. $(E_o + E_c)$ as discussed in connection with Eq. 4.7. Then: (a) if the plot is linear, the slope is equal to N_i, or (b) if the plot curves upward, there apparently is water influx, despite the evidence, and it will be necessary to calculate $\Sigma\Delta pQ_D$ $(\Delta t_D, r_D)/(E_o + E_c)$ for various estimated values of t/t_D and plot $F_{pR}/(E_o + E_c)$ vs. $\Sigma\Delta pQ_D(\Delta t_D, r_D)/(E_o + E_c)$, as discussed in connection with Eq. 4.9. The "best" value of t/t_D must be determined by trial and error. (For the first trial, it is recommended that r_D be assumed to be infinite, unless there is evidence to the contrary from subsurface data.)[14]

2. If the oil was saturated at initial conditions, but there is no evidence (from pressure or production data) of an initial gas cap or water influx, assume for the first trial that m equals 0, and make a plot like in Step 1. Then: (a) if the plot is linear, the assumption that $m = 0$ apparently is valid, and the slope is equal to N_i, as in Step 1(a), or (b) if the plot curves upward, there may have been an initial gas cap, water influx, or both. The simplest procedure is next to assume the largest m value

---

[12] However, as discussed in Sec. 4.4.5, in the presence of water influx, plots of p/Z vs. G_p frequently have been misinterpreted. Accordingly, Dake [1994] strongly urged use of the Havlena-Odeh approach in such scenarios.

[13] The Havlena-Odeh [1963] methodology involves making a "best fit" through data points calculated with linearized equations. The implications of this procedure to classifying oil and/or gas initially in place and reserves are discussed in Sec. 9.2.4.

[14] If, during the history-match period, aquifer influx becomes semisteady state, then, providing the aquifer has otherwise been properly characterized before the semisteady-state period, the plot should be linear, then flatten with the onset of semisteady-state influx.

TABLE 4.1—SPREADSHEET ILLUSTRATING METHOD OF HAVLENA AND ODEH [1963]

PRESSURE & PRODUCTION DATA					PVT DATA		HAVLENA-ODEH PRODUCTION & EXPANSION TERMS							TERMS TO ESTIMATE N_i AND m FOR NO WATER INFLUX		TERMS TO ESTIMATE N_i FOR WATER INFLUX	
t	G_p	W_p	p	B_t	B_g	F_{pR}	E_o	E_c	E_g	E_o+E_c	$F_{pR}/(E_o+E_c)$	$m_{max}B_{ti}E_g/B_{gi}$	$E_o+(1+m_{max})E_c +m_{max}B_{ti}E_g/B_{gi}$	$E_o+(1+m_{new})E_c + m_{new}B_{ti}E_g/B_{gi}$	$\Sigma\Delta pQ_D(\Delta t_D,r_D)$	$\Sigma\Delta pQ_D(\Delta t_D,r_D)/(E_o+E_c)$
N_p																
0	0	0	p_i	B_{ti}	B_{gi}	0	0	0	0	0	0				0	0
0	0															

supportable by subsurface mapping and production performance (identified here as m_{max}), calculate a new spreadsheet column equal to $[E_o + (1 + m_{max})E_c + m_{max}B_{ti}E_g/B_{gi}]$,[15] and plot F_{pR} vs. this new column, as discussed in connection with Eq. 4.8a.

- If the plot is linear, the assumed value of m is correct and the slope is equal to N_i. This should be verified by computing and plotting two new spreadsheet columns, $F_{pR}/[E_o + (1 + m_{max})E_c]$ and $E_g/[E_o + (1 + m_{max})E_c]$ as discussed in connection with Eq. 4.8b.
- If the plot curves downward, assume a value of m equal to half m_{max} (identified here as m_{new}) calculate a new column equal to $[E_o + (1 + m_{new})E_c + m_{new}B_{ti}E_g/B_{gi}]$, and plot F_{pR} vs. this new column, continuing this procedure until a linear plot is obtained. When a linear plot is obtained, follow the procedure outlined in the previous step to verify this interpretation.
- If the plot curves upward, there apparently is water influx, and it will be necessary to estimate W_e for various assumed values of t/t_D, as in Step 1(b).

3. If there is evidence of an initial gas cap from subsurface or well performance data but no evidence of water encroachment, follow the procedure outlined in Step 2, but make an estimate of m from volumetric mapping for the first trial plot of F_{pR} vs. $[E_o + (1 + m)E_c + mB_{ti}E_g/B_{gi}]$, as discussed in connection with Eq. 4.8a. Then: (a) if the plot curves upward, the estimated m may be too small or there may be water influx, and (b) the simplest procedure is first to follow Step 2(b); if the plot curves downward, then it will be necessary to follow the procedure discussed in connection with Eq. 4.10.

4.3 Prediction Methods

The Havlena-Odeh [1963] methodology facilitates estimation of oil and gas initially in place and an indication of the probable drive mechanisms. To estimate (remaining) reserves, however, this methodology must be supplemented by one of the predictive forms of the material balance equation or another method to estimate reserves. Historically, two predictive forms of the material balance equation have been used; one is attributed to Tarner [1944], the other to Muskat [1945]. (Results of this type of calculation were discussed in Sec. 3.7.2.)

These predictive methods, however, are limited to solution-gas-drive reservoirs (no gas cap, no water influx, and no gravity segregation) and generally have been replaced by computer simulation.

4.3.1 Computer Simulation

As noted in Sec. 4.1.2, there are several scenarios that preclude reliable use of a tank model, or zero-dimensional material balance, to estimate oil and gas initially in place. Provided necessary and sufficient data are available, multidimensional computer simulation should be considered for the following situations:

- Areally extensive reservoirs with low values of, or large areal variations in, kh/μ
  These conditions often lead to large areal variations in bottomhole pressure that are difficult to average for use in a tank model.

---

[15] If this analysis procedure is followed, the gravity of the gas cap gas must be determined to calculate E_g. In the absence of any other data, use the gravity of gas liberated in the first pressure decrement below the bubblepoint, assuming there is a PVT analysis for the oil where these data typically are reported.

- Areally extensive reservoirs at different stages of development and/or production where there are large areal variations in reservoir pressure, which may range from initial pressure to below the bubblepoint pressure. These conditions are impossible to handle with a tank model.

- Very heterogeneous reservoirs with large spatial variations in permeability due to any combination of natural fractures,[16] interbedding, variations in depositional environment, or diagenesis.

- Oil reservoirs with substantial vertical relief, especially those with large gas caps or significant vertical variations in initial oil properties.

Since the mid 1970's, computer simulation has become an almost indispensable aid for the development planning of major oil and gas fields. This methodology is especially applicable where there is considerable monetary risk and uncertainty regarding optimal exploitation. As discussed by Kingston and Niko [1975] and by Tollas and McKinney [1991], extensive computer simulation studies were used to assist in the selection of the operating scheme for the Brent field (North Sea). Similar applications have been reported for other North Sea fields, such as the Statfjord field [McMichael 1978], the Forties field [Hillier *et al.* 1978], the Fulmar field [Valenti and Buckles 1986], and the Heidrun field [Till *et al.* 1986].

In addition, apparently successful application of this technology has been reported for a waterflood in an extremely heterogeneous carbonate reservoir in the U.S. Permian Basin, which had been discovered and developed more than 30 years before [Harpole 1980].

However, despite the widespread use of computer simulation, the reader is cautioned that *these methods typically oversimplify the representation of the spatial distribution of reservoir properties.* Results of computer simulation typically are sensitive to changes in reservoir characterization. The degree of sensitivity may render moot an estimate of ultimate recovery based on this method. Any such estimates should be checked against simple models, as discussed by Richardson and Blackwell [1971], and against actual performance in analogous reservoirs.[17] As observed by Saleri and Toronyi [1988] "determination of ultimate recovery through simulation is an unattainable goal, particularly for heterogeneous reservoirs that are coarsely gridded."

Jacks [1990] noted that "an estimate of ultimate recovery from any single simulation is subject to considerable uncertainty, especially if it is developed early in the life of a reservoir."

There may be some validity to claims that computer simulation provides the only acceptable answer to analysis of complex reservoirs. However, there are many examples where forecasts of future performance have been significantly in error, despite the users' best efforts at reservoir characterization and history matching. Thus, continuous monitoring of performance is an essential element of reserve estimation by computer simulation—indeed, by any method!

---

[16] There seems to be a consensus that naturally fractured reservoirs (so-called dual-porosity systems) cannot be analyzed reliably without dual-porosity computer simulation. However, in view of the complexities of such simulation models, reliable results cannot be expected unless there has been a reasonable period of production, preferably below the bubblepoint, that can be history matched. Problems with analyzing this type of reservoir are discussed in Sec. 6.4. Case histories of naturally fractured reservoirs are presented in Secs. 7.4 and 7.5.

[17] Similar guidelines have been provided by PS-CIM [1994]: "if sufficient amounts of good geological and performance data are available to allow for a reasonable history match, and if the estimator is using an appropriate simulation model that has been used successfully in reservoirs similar to the one being studied, projections of recovery under primary mechanisms...might be considered proved."

4.3.2 Uncertainties

As discussed in Sec. 1.1.2, all methods to estimate reserves are subject to varying degrees of uncertainty, with the nature and degree of uncertainty depending on, among other factors, the estimation method.

Material balance methods, which include multidimensional computer simulation, utilize reservoir engineering equations to calculate both oil and/or gas initially in place and future production (reserves). The reliability of these calculations depends on the accuracy with which the mathematical model and the available data simulate the reservoir under study.

Aside from economics, reservoir heterogeneity usually has the most influence on recovery efficiency. Reservoir drive mechanism also is important, but by the time there are sufficient performance data for a material balance estimate of oil and gas initially in place, the drive mechanism usually has been determined with reasonable confidence. The efficiency of that mechanism, however, may be strongly influenced by reservoir heterogeneity.

Reservoir heterogeneity is manifested in material balance calculations in several ways:

- Reservoir heterogeneity affects the relative permeability terms, which control the relative production rate of the driving and the driven fluids and the abandonment saturations of oil and gas.

- Reservoir heterogeneity—e.g., permeability stratification or natural fractures—makes it difficult to interpret bottomhole pressure surveys. Errors in bottomhole pressure (both measurement and interpretation) are among the major contributors to errors in material balance calculations.

- In multidimensional simulation, reservoir heterogeneity is manifested in the permeability level of each gridblock, which controls the cumulative calculated recovery from the block at abandonment.

In addition, uncertainties may be caused by the following:

- History may be insufficient for a reliable material balance. For example, in the case of a waterdrive reservoir with a limited aquifer, if the pressure sink has not reached aquifer boundaries during the history-match period, reserve calculations may be based mistakenly on infinite aquifer behavior.

- Actual relative permeability relations may be different from those used. This may occur if, for example, calculations of future production are made in an edgewater-drive reservoir where the k_w/k_o relations had been matched in only a few downdip wells.

4.4 Volumetric Gas Reservoirs

For gas reservoirs where there is no—or insignificant—aquifer influx and pore volume compressibility is small, the volume of the initial gas-bearing pore space typically is *assumed* to remain constant over life.[18] This is the so-called *volumetric* gas reservoir, for which the material balance equation may be stated

GAS PRODUCTION = EXPANSION OF FREE GAS INITIALLY IN PLACE

or

---

[18] Strictly speaking, there is no such situation in nature, as all aquifers expand and all porous systems compress in response to fluid withdrawal and consequent pressure reduction. When these phenomena are *negligibly small* compared to gas compressibility, Eq. 4.14 is a convenient approximation.

$$G_p B_g = G_{Fi}(B_g - B_{gi}) \,. \dots\dots\dots\dots\dots\dots\dots\dots\dots\dots\dots\dots\dots\dots\dots (4.14)$$

B_g may be expressed in terms of the real gas law

$$B_g = p_{sc} Z T_f / T_{sc} p \,. \dots\dots\dots\dots\dots\dots\dots\dots\dots\dots\dots\dots\dots\dots (4.15)$$

Using the real gas law for B_g and B_{gi}, Eq. 4.14 may be written

$$G_p p_{sc} Z T_f / T_{sc} p = G_{Fi}[p_{sc} Z T_f / T_{sc} p - p_{sc} Z_i T_f / T_{sc} p_i] \,. \dots\dots\dots\dots\dots (4.16)$$

Solving for p/Z gives

$$p/Z = p_i/Z_i - G_p p_i / G_{Fi} Z_i \,, \dots\dots\dots\dots\dots\dots\dots\dots\dots\dots\dots\dots (4.17)$$

which is of the form

$$y = b - mx \,. \dots\dots\dots\dots\dots\dots\dots\dots\dots\dots\dots\dots\dots\dots\dots\dots\dots (4.18)$$

Thus, for a *volumetric* gas reservoir, a plot of p/Z vs. G_p will be linear. Extrapolating a best fit through the data points results in a y intercept (at $G_p = 0$) equal to p_i/Z_i and the x intercept (at $p/Z = 0$) equal to G_{Fi}, as shown by **Fig. 4.5.** Some engineers make such a fit visually; others prefer linear regression analysis.

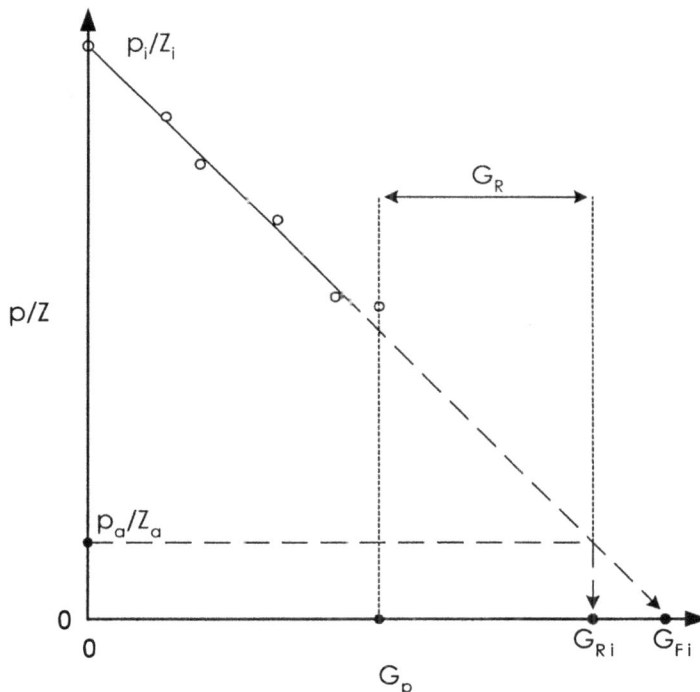

Fig. 4.5—Reservoir pressure divided by gas deviation factor, *p/Z*, vs. cumulative gas production, *G_p*, illustrating method to estimate free gas initially in place, *G_{Fi}* and initial reserves *G_{Ri}* for volumetric gas reservoir.

Such a best-fit interpretation is considered here to be deterministic. The reader is referred to Sec. 9.7.3 for a discussion of potential sources of uncertainty and a probabilistic procedure to analyze data from volumetric gas reservoirs.

The term G_p is total cumulative *wellhead* gas, not just first-stage separator gas, as usually is reported, and not just "sales gas," which is reported in some areas. Methods to estimate total wellhead gas from first-stage separator gas are discussed in the section following and in Appendix E.

With reference to Fig. 4.5, ultimate recovery (initial reserves) may be estimated as the value of G_p when p/Z declines to p_a/Z_a, where p_a is estimated reservoir pressure at abandonment (usually the economic limit).

Depending on the amount of pressure and production data available at the time of the estimate, some engineers may prefer to estimate the gas recovery efficiency based on arbitrary rules of thumb rather than using the procedure discussed in the previous paragraphs. With reference to Eq. 3.28, it is noted that to achieve gas recovery efficiency in excess of 90%, a common rule of thumb, the average reservoir abandonment pressure must be reduced to (on the order of) 5% of initial reservoir pressure. Reservoir abandonment pressures this low generally require multistage compression. Considering the relatively high cost to install and operate such gas-compression equipment, such a reduction may not be commercially feasible. Thus, engineers are cautioned regarding estimates of average reservoir abandonment pressure, or recovery efficiency, based on arbitrary rules of thumb, especially for newly discovered reservoirs in remote areas where there is little operating experience.

Two general types of volumetric gas reservoirs are discussed: (1) nonretrograde and (2) retrograde.

4.4.1 Nonretrograde Gases

As noted in Sec. 3.7.1, nonretrograde gas reservoirs are distinguished from retrograde gas reservoirs by the position of the two-phase envelope with respect to the temperature of the reservoir. As shown by Fig. 3.4b, in nonretrograde gas reservoirs, the cricondentherm is less than reservoir temperature. The path of isothermal expansion *of the gas remaining in the reservoir* does not enter the two-phase region as the reservoir is pressure depleted. Accordingly, (a) there is no retrograde condensation and (b) the composition of the reservoir gas remains unchanged during production. Thus the initial composition may be used to determine the Z-factor for material balance calculations over the life of the reservoir. In the absence of a PVT analysis, reservoirs may be treated as "nonretrograde" for application of Eq. 4.17[19] if the initial producing condensate/gas ratio is less than about 65 STBC/MMscf gas [McCain and Piper 1994].

Regarding the Z-factor, two situations regarding data availability are discussed: (a) PVT data or composition of separator fluids available and (b) only well test data available.

PVT Data or Composition of Separator Fluids Available

A PVT report usually will include a table of Z-factors vs. reservoir pressure. For reservoir computations one may interpolate between the tabular values in the laboratory report to determine Z-factors at reservoir pressures of interest for Eq. 4.17. A better procedure, however, is to fit the data with a regression equation, using, for example, the statistics module in spreadsheet programs like EXCEL™ or LOTUS 1-2-3.™ [20]

---

[19] Estimation of condensate reserves, however, should follow the procedures discussed in Sec. 3.7.4.
[20] Reference to these products does not constitute endorsement.

Before using these data, however, it is good practice cross check the laboratory Z-factors against an empirical correlation. The Standing-Katz [1942] correlation charts, Figs. E-7 and E-8 (Appendix E), are recommended for this purpose. Note that Fig. E-7 is a composite of two figures. The upper part of the figure covers pseudoreduced pressures from 0 to 8; the lower part, from 7 to 15. Fig. E-8 covers pseudoreduced pressures between 15 and 30, which typically are encountered in geopressured reservoirs.[21]

Pseudoreduced pressure, p_{pr}, and pseudoreduced temperature, T_{pr}, which are the arguments for Fig. E-7 and Fig. E-8, are defined as

$$p_{pr} = p/p_{pc} \dotfill (4.19)$$

and

$$T_{pr} = T_f/T_{pc} . \dotfill (4.20)$$

The pseudocritical properties of a reservoir gas, which are not normally reported in a standard PVT analysis, are a function of gas composition and may be calculated as outlined in Appendix E.

If there are no PVT data, but composition analyses are available for the first-stage separator fluids—both gas and condensate—the composition of the reservoir gas may be calculated as shown in Appendix E.

Only Well Test Data Available

If only well test data are available, the Z-factors used in Eq. 4.17 may be estimated from empirical correlations. Generally, this is a three step process:

1) Estimate the gravity of the *reservoir gas* from well test data.[22]

2) Estimate the pseudocritical properties of the reservoir gas from empirical correlations.

3) Estimate the Z-factor from the Standing-Katz [1942] correlations discussed in the previous section.

Steps 1 and 2 are outlined briefly in the ensuing paragraphs.

Standing [1952, 1977] has shown that the gravity of reservoir gas, γ_{gR}, may be calculated from well test data as[23]

$$\gamma_{gR} = \frac{\gamma_{gS}[240 - 2.22(\gamma_{API})][(28.97\gamma_{gS})(131.5 + (\gamma_{API}) + 18.76R_{ci}]}{28.98\gamma_{gS}[31{,}560 - 52.2(\gamma_{API}) - 2.22(\gamma_{API})^2 + 18.76R_{ci}]}, \dotfill (4.21)$$

where γ_{gS} = gravity of separator gas (air = 1),
  γ_{API} = gravity of condensate, °API, and
  R_{ci} = initial condensate/gas ratio, STBC/MMscf.

---

[21] The performance of geopressured reservoirs is discussed in Sec. 6.5

[22] For wells producing with CGR's less than about 10 STBC/MMscf, the gravity of the separator gas may be assumed to be equal to the gravity of the reservoir gas.

[23] The fluid property correlations attributed to Standing, which are discussed here and elsewhere in this work, are in the eighth printing of Standing [1952, 1977], which s available from SPE. As discussed by Standing in that work, reservoir gas gravities calculated from Eq. 4.21 are essentially the same as values calculated using a correlation published by Eilerts [1957].

Fig. E-1 (Appendix E) is a graph of Eq 4.21. In applying this equation, the terms γ_{gS} and R_{ci} should account for *all* the surface gas, not just that produced through the first-stage separator, as usually is done. Depending on the CGR and on the volatility of the condensate, ignoring gas liberated between the first-stage separator and the stock tank could introduce substantial error in subsequent calculations. Gold *et al.* [1989] provided a correlation to estimate reservoir gas gravity from test data for two- or three-stage flash separation, as discussed in Appendix E.

The pseudocritical properties of reservoir gas may be estimated using correlations by Standing [1952] and Sutton [1985] provided in Appendix E. Although the Standing correlation has a long history of usage, the Sutton correlation may be more accurate, especially for high-molecular-weight gases.

Estimate of Wellhead (Reservoir) Gas Equivalent

The term G_p in Eq. 4.17 is cumulative *wellhead* gas production, which is the reservoir "vapor equivalent" of total gas, including all stages of separation and stock tank vapors plus condensate plus water of vaporization.

As observed previously, gas production from other than the first stage of separation is rarely reported. Thus, depending on circumstances, it may be appropriate to estimate cumulative reservoir gas production using the procedures discussed in Sec. E.1.6.

4.4.2 Retrograde Gases

As noted in Sec. 3.7.1, retrograde gas reservoirs are distinguished from nonretrograde gas reservoirs by the position of the two-phase envelope with respect to the temperature of the reservoir, as shown by Fig. 3.4b. In retrograde gas reservoirs, the reservoir temperature, T_f, is between the pseudocritical temperature, T_c, and the cricondentherm, T_{ct}. Thus, the path of isothermal expansion *of the gas remaining in the reservoir* (Path a-b in Fig. 3.4b) enters the two-phase region if the reservoir is pressure depleted below the dewpoint pressure, resulting in retrograde condensation. The *composition of the fluid system remaining in the reservoir will change during production below the dewpoint pressure.* In this case, the initial composition should not be used to determine Z-factors for material balance calculations.

If there is a PVT analysis available, the laboratory-determined two-phase Z-factor should be used at reservoir pressures below the dewpoint pressure.

If there is no PVT analysis available, an empirical correlation by Rayes *et al.* [1992], provided in Sec. E.2.5, may be used to estimate the two-phase Z-factor.

4.4.3 Sour Gases

For natural gases containing more then about 7 mol % carbon dioxide and hydrogen sulfide in total, corrections must be made to the pseudocritical properties before using the "industry standard" Z-factor chart discussed above. The method of Wichert and Aziz [1972] is summarized in Sec. E.2.8.

4.4.4 Moderate- to High-Permeability Reservoirs

In moderate- to high-permeability gas reservoirs, there generally is good pressure communication between wells. In such cases, p/Z vs. G_p plots should be maintained for each *reservoir* being analyzed, using reservoir average static bottomhole pressure[24] and aggregate reservoir production.

---

[24] Dietz [1965], Matthews *et al.* [1954], and Earlougher [1977] have provided guidelines for estimating average bottomhole

It has been shown that, under semisteady-state conditions, the drainage volume of each well in a multiwell reservoir is proportional to its production rate [Matthews *et al.* 1954]. Thus, in *volumetric* gas reservoirs where there is good interwell pressure communication, p/Z vs. G_p plots for individual *wells* may deviate from linearity and lead to erroneous conclusions about reservoir drive mechanism and well drainage volume, as discussed in subsequent paragraphs.

Under semisteady-state conditions in a gas *reservoir,* the p/Z vs. G_p plot for each well should indicate the gas volume being drained by that *well.* If, after a period of semisteady-state production, the production rate of a well is increased relative to the production rate of other wells in the same reservoir, the drainage volume of that well will increase relative to the drainage volume of other wells in the reservoir.

Under these conditions the p/Z vs. G_p plot of the well will deviate from the previously established trend, indicating a larger drainage volume, proportional to the new rate [Stewart 1966]. Under the new semisteady-state conditions, the individual p/Z vs. G_p plots of one or more of the other wells in the reservoir also will deviate from previously established trends, indicating proportionally smaller drainage volumes.

Initial reserves for a volumetric *reservoir* may be estimated by extrapolating the p/Z vs. G_p plot to p_a/Z_a, where p_a is the estimated *average* reservoir abandonment pressure at the economic limit. In some cases, it may be necessary to estimate reserves for *individual wells* in a large reservoir, in which case p/Z vs. G_p plots would be needed for individual wells. As a check on reserves estimated using this procedure, it may be desirable to use the backpressure test curves for each well to estimate the abandonment bottomhole pressure for the well at the economic limit.

4.4.5 Water Influx

If water influx is a reasonable expectation,[25] caution should be exercised in using p/Z vs. G_p plots to estimate gas initially in place and reserves. Bruns *et al.* [1965] have shown that an *apparently* linear plot of p/Z vs. G_p does not necessarily mean that there is no water influx. Their work showed that it is theoretically possible for plots of p/Z vs. G_p to vary from concave upward to concave downward, depending on the size and permeability of the aquifer. *Apparently* linear plots could be generated by limited aquifers; these plots exhibited small inflections at the origin, which often are ignored or attributed to measurement error.

Chierici *et al.* [1967] corroborated the Bruns *et al.* [1965] work with examples of several gas reservoirs in the Po Valley, Italy. These reservoirs were being invaded by water, but exhibited apparently linear plots of p/Z vs. G_p, as revealed in subsequent discussion [Hurst 1967].

In cases where water influx is suspected, subtle departures from linearity may be detected by repeated linear regression analyses of progressively more p/Z vs. G_p points. The procedure involves a regression analysis for the first four or five p/Z vs. G_p data points, then a series of regressions, including progressively more data points with each successive regression. If the data points reflect a volumetric reservoir, then the regression coefficient for each successive regression should approach 1.0. If, on the other hand, if the data points gradually deviate

---

pressure in volumetric reservoirs.

[25] For example, water influx is a reasonable expectation in sandstone reservoirs in unconsolidated sequences of shale and sand, irrespective of the apparent size of the contiguous aquifer. Examples include gas reservoirs in such sequences in Southeast Asia.

from linearity, reflecting possible water influx, then the regression coefficient for each successive regression should become progressively less than 1.0.[26]

To further test the water influx hypothesis, it might be appropriate to plot the "residuals" (calculated from the regression) vs. cumulative production—i.e., $[(p/Z)_{FIT} - (p/Z)_{ACT}]$ vs. G_p. In the case of water influx, such a plot should exhibit a trend of points that is initially negative, then positive, then negative, depending on the quality of the basic data and the trend of the fitted plot.

Gruy and Crichton [1950] advocated plotting "cumulative pressure decrement," actually $(p_i/Z_i - p/Z)$, vs. cumulative gas production on log-log paper to determine whether there is influx into, or drainage from, a gas well drainage volume. Influx should result in downward curvature; drainage, upward curvature. While the procedure is theoretically sound, log-log plots are quite insensitive, and variation from linearity may be obscured by data errors.

4.4.6 Low-Permeability Reservoirs

In low-permeability gas reservoirs developed on wide spacing, where there usually is no apparent pressure communication between wells, it is advisable to maintain a p/Z vs. G_p plot for each well. It is good practice to use the backpressure test curve for each well to estimate the abandonment bottomhole pressure at the economic limit flow rate.

Estimates of *gas initially in place* (GIP) for each well (not reserves) should be made using volumetric methods, assuming a provisional drainage area equal to the well spacing. These estimates of GIP should be compared with estimates of GIP made by extrapolating the p/Z vs. G_p plot to zero pressure. In many cases, it will be observed that the volumetric estimate of GIP is substantially larger than the p/Z vs. G_p estimate. In such cases, infill drilling may be warranted to increase recovery efficiency. If infill drilling can be justified, some portion of the difference between GIP estimated volumetrically and that estimated from the p/Z vs. G_p plot might be considered *undeveloped* reserves, pending successful infill drilling.

In many cases, the transmissibility of the reservoir may be too low to determine bottomhole pressure after a reasonable shut-in period, in which case reliable p/Z vs. G_p plots cannot be made. Possible solutions to this problem are discussed in Sec. 6.8.

4.5 Reconciliation of Material Balance and Volumetric Estimates

The volumetric and material balance methods are independent ways to estimate oil and/or gas initially in place (O/GIP). The basic assumptions for each method, however, are different. Thus, calculations of O/GIP for the same reservoir using each method might result in significantly different results.

The volumetric method accounts for net pay observed in wells and historically has been based on the assumption that the pay is continuous between wells.[27] The material balance method accounts for production and the resulting pressure decline and is based on the assumption that the pressure decline is caused by withdrawals from a common fluid system.

Depending on the depositional environment of the reservoir rock and on well spacing, pay may not be continuous between wells [Ghauri *et al.* 1974, Sneider *et al.* 1977, George and Stiles 1978]. Also, intervals counted as pay for volumetric calculations, but not perforated for production, may be in reservoirs separate from intervals that have been perforated. The

---

[26] Any such statistical analysis tacitly assumes data scatter to be attributable only to random errors. Thus, the data should always be reviewed carefully to eliminate spurious measurements before doing such analysis.

[27] Since the middle 1990's there has been increasing utilization of geostatistical methods to map interwell areas. In this context, the historical assumption of pay continuity must be considered only one interpretation of the possible spatial distribution of pay in heterogeneous reservoirs.

unperforated intervals may not contribute to the pressure response and may not be accounted for by material balance calculations. Because of these and other factors, there may be substantial differences between the results of volumetric and material balance estimates of O/GIP.

Differences of this type are commonly reported in complex carbonate reservoirs [Stiles 1976]. If such differences are observed and cannot be reconciled by careful review of the data and computational procedures, then a detailed geologic and engineering review may be warranted. Such a review might lead to the drilling of additional wells or the perforating and/or stimulation of additional intervals for production.

CHAPTER 5

PERFORMANCE/DECLINE TREND ANALYSIS

"I know of no way of judging the future but by the past."

Patrick Henry 1775

5.1 Overview

After a well, or aggregate of wells,[1] has been producing long enough for the producing characteristics to develop clearly defined trends, it may be feasible to extrapolate these trends to the economic limit to estimate (remaining) reserves. (The term "economic limit" is discussed in Sec. 5.1.1.)

Frequently, two stages, or periods, may be recognized in the performance of a well:

1. the period before there is a continuing decline in the production rate of the "principal product,"[2] but during which there are trends in "performance indicators" like gas/oil ratio (GOR), water/oil ratio (WOR),[3] etc., which may be extrapolated to economic limit *conditions*, (discussed in Sec. 5.1.1) and

2. the period during which the production rate of the principal product exhibits a continuously declining trend, which can be extrapolated to the economic limit *production rate.*

The term "performance/decline trend analysis" (P/DTA) is introduced here to describe these two stages and the methodologies for analysis. There are several cautions, however, regarding these methods, which are discussed in Secs. 5.1.2 and 5.1.3.

5.1.1 Economic Limit

Definition and Application

As implied by the foregoing discussion, the general term "economic limit" may refer to:

(a) a producing characteristic that typically is used locally to establish the practical limit of production—e.g., a maximum WOR or maximum GOR—which is identified here as economic limit *conditions*, or

---

[1] Engineers historically have used "trend analysis" to estimate reserves for single wells and for aggregates of wells (i.e., for multiwell leases, reservoir units, or fields.) As noted in Sec. 5.1.3, however, caution should be exercised if trend analysis is used to estimate reserves for aggregates of wells.

[2] The term "principal product," previously introduced, refers to oil for oil wells or gas for gas wells. Performance/decline trend analysis (P/DTA) of "associated products" (i.e., solution and/or free gas from oil wells and condensate from gas wells) are discussed in Sec. 5.3.6.

[3] Purvis [1985] called these "production-performance graphs."

(b) a minimum rate of production at which income from such production is insufficient to pay the cost of continued operation, which is identified as the economic limit *production rate*, or simply, the economic limit.

An economic limit may refer to:

- a single well,

- an aggregate of wells—e.g., a lease or unit, or other economic aggregation or financial grouping—or,

- a production facility for an aggregate of wells with processing equipment that must be operated as a unit—e.g., offshore production platforms or fluid injection projects.

The economic limit generally is defined as the production *rate* at which the net revenue to the operator's working interest that is attributable to production from a well, aggregate, or facility equals the "out-of-pocket" cost to operate the well, aggregate, or facility.[4] State and federal income tax and corporate overhead usually are excluded[5] in determining out-of-pocket costs. Net revenue is gross revenue less: (a) production and ad valorem taxes, (b) royalties, (c) transportation, and (d) treating expense, if any. Out-of-pocket costs, sometimes called direct operating costs, are costs that would be saved if the well or facility were shut in, such as costs for power and materials. Out-of-pocket costs would include labor only if shutting in the well or facility would save that cost.

For a reserve estimate for a producing entity, good practice involves:

- forward projection, on a monthly (or annual) basis, of an observed trend in production using procedures discussed in Sec 5.3,

- calculation of anticipated future monthly (or annual) net revenue attributable to estimated future production,

- calculation of monthly (or annual) operating costs attributable to the producing entity, and

- termination of projected future production during month (or year) when calculated net revenue is equal to calculated operating costs.

Typically, such calculations are done using in-house or commercially available computer software.

Annual reserve estimates made for publicly owned companies in the U.S. are subject to regulations of the U.S. Securities and Exchange Commission (U.S. SEC). Historically, that agency's regulations mandate use of wellhead prices at the end of the fiscal year. Other than those provided by contract, escalation of prices for anticipated future economic conditions is not allowed. For additional discussion of reserve estimates prepared using U.S. SEC guidelines, the reader is referred to *Guidelines for Application of Petroleum Reserve Definitions* [SPEE 1998].

Depending on the purpose of the estimate, corporate guidelines, and/or the agency for which the estimate is being provided, it might be appropriate to escalate future wellhead prices for oil and gas and future costs for operations and capital expenses. Depending on circumstances, such changes in anticipated future economic conditions might have a

---

[4] While the definition of economic limit may seem reasonably straightforward, the application is subject to interpretation, especially regarding out of pocket costs. For further discussion of this topic, the reader is referred to to *Guidelines for Application of Petroleum Reserve Definitions* [1998].

[5] Depending on circumstances, the economic limit may refer to total costs, including corporate and other overhead.

significant influence on reserves. For additional discussion of this point, the reader is referred to *Guidelines for Application of Petroleum Reserve Definitions* [SPEE 1998].

Economic Limit Conditions

Economic limit conditions, which are usually established locally, may include, for example, a maximum WOR, a maximum GOR, or a minimum flowing tubing pressure (FTP). Such limits typically are used to estimate reserves for a well by extrapolation of an observed trend in one of these performance indicators vs. cumulative oil (for oil wells) or vs. cumulative gas (for gas wells).

Economic limit conditions also may be established for an aggregate of wells producing to a common facility. Such conditions may include, for example, a maximum gas production rate or a maximum water production rate, which maxima might be constrained by available processing equipment.

In the absence of a declining trend in the production rate of the principal product, such limiting conditions should be used with caution. For example, one might extrapolate an observed trend of WOR vs. cumulative oil in a producing well to a locally established WOR limit of 20:1. If, however, the well is incapable of making more then 25 B/D total liquids, even on pump, extrapolation of the WOR trend to 20:1 would imply an economic limit production rate of 1.2 BOPD, which might not be realistic. The converse is also true. If the well is capable of being pumped at 5,000 bbl total liquids produced/D, an extrapolation of the WOR trend to 20:1 would result in a economic limit production rate of 238 BOPD, which might be unrealistically high.

Depending on circumstances, substantial variation in economic limit WOR might be observed in a single field. For example, in a study of the Pembina Cardium oil pool in Alberta (Canada), Purvis and Bober [1979] reported WOR at a calculated economic limit production rate of 4 BOPD ranged from 0.7 to 61 bbl water/bbl oil.

Caution should be exercised in extrapolating WOR trends for production data aggregated to leases or units being waterflooded. The trend of WOR vs. cumulative oil in a waterflood project might depend on field operations, and long-term extrapolation of an observed short-term trend could lead to substantial error. These points are discussed in Sec. 5.1.3.

5.1.2 Reservoir Performance or Wellbore/Mechanical Problem?

A performance/decline trend in a well might in some instances be attributable to a wellbore and/or mechanical problem rather than to the performance of the well and/or reservoir. Extrapolation of such trends might lead to estimates of future production that do not reflect production that might actually be recovered if the wellbore and/or mechanical problem is remedied.

For example, in oil reservoirs where reservoir pressure is being maintained by natural waterdrive or water injection, individual wells usually will be produced at a stable flowing rate. After water breakthrough in such wells, if the total liquid production rate can be maintained by pumping as the WOR increases, the oil production rate typically will decline continuously. To be meaningful reserve indicators, the declining trends in oil production should reflect an increase in the WOR in such wells, not a decrease in the production rate of total liquids. If there is a continuing decrease in the rate of total liquid production from such wells, gradual wellbore plugging or pump wear should be suspected.

In a gas reservoir producing by pressure depletion, for example, it would be reasonable to expect a trend of declining shut-in tubing pressures vs. cumulative gas for individual wells. If the wells are being produced at approximately constant rates, it also would be reasonable to expect a trend of declining FTP's vs. cumulative gas. If the wells were being produced against a constant backpressure imposed by sales line or compressor intake pressure, then it would be reasonable to expect a trend of declining production rate vs. cumulative gas. In contrast, if reservoir pressure is being maintained by strong waterdrive, declining trends in wellhead pressure or production rate might be indicative of wellbore plugging or buildup of liquids in the wellbore.

5.1.3 Lease/Pool Trends vs. Individual Well Trends

Historical production data on individual wells might not always be available, and reserve estimates might have to be made using production data aggregated to leases or to reservoir units that contain more than one well. In such aggregates, individual wells might be in different stages of decline and might have different GOR's and/or different WOR's. Remedial operations, infill drilling, and modifications to production equipment might have influenced the production history and may be expected to influence future trends.

Before attempting to estimate reserves from production trends from multiwell aggregates, the engineer should become familiar with past, current, and anticipated future operations. Routine lease operations that affect well performance may be reported only informally, if at all. Frequently, pumpers or lease foremen maintain records of well tests, pressures, and equipment changes in so-called "daily gauge reports" that are not generally available elsewhere. Depending on circumstances, the engineer responsible for reserve estimates should consider an on-site visit with field personnel to determine the nature of field operations and the possible influence of those operations on production rate.

Brons [1963] showed that, for multiwell aggregates, the decline characteristics of the high-rate wells dominate the aggregate decline. Thus, for such aggregates where the high-rate wells are declining more rapidly than the low-rate wells, as might be observed in aggregates producing by pressure depletion, extrapolation of the aggregate production trend might lead to a pessimistic estimate of reserves. Conversely, where the high-rate wells are declining less rapidly than the low-rate wells, as might be observed in aggregates producing by waterdrive, extrapolation of the aggregate production trend might lead to an optimistic estimate of reserves. It may be necessary to use well test data to construct individual-well decline curves to help interpret aggregate production trends where there are large variations in production rates between wells on the lease or unit.

An additional complication may occur in the analysis of an aggregate of wells producing from a common reservoir. If all the wells in the reservoir have reached semisteady-state conditions—i.e., if interwell interference boundaries have been established—shutting in one or more such wells might result in expansion of the drainage volumes of surrounding wells. Such expansion typically will result in slight increases in reserves for the affected wells, a phenomenon that might not be apparent in an aggregate decline.

In areas where development drilling is ongoing, a meaningful trend might be observed from a plot of barrels of oil (or cubic feet of gas) per month per active producing well.

Another difficult aspect of interpreting multiwell aggregates of production is determination of the number of producing wells in future years and the minimum number of wells at the economic limit. For most operations not requiring centralized facilities, a single well usually will be the minimum operating unit.[6] On the other hand, for operations requiring

---

[6] Provided other conditions are similar, direct operating costs for a single remote well usually will be substantially more than for

centralized facilities (for example, an offshore platform, or a waterflood unit), the minimum number of wells at the economic limit might be a significant proportion of the whole.

From studies of the production performance of aggregates of wells in oil fields in Alberta (Canada) by Purvis [1990], the following may be observed:

- At any time during the productive life of such aggregates, the statistical distribution of well potentials is approximately log normal.

- During the period when the number of producing wells is more or less constant, the slopes[7] of such distributions, determined periodically, tend to be more or less parallel.

- During the period when the number of producing wells is decreasing due to abandonment, the slopes of such distributions, determined periodically, tend to decrease progressively as progressively more wells are abandoned.

- The decline rate for the statistical median well is the significant decline for such aggregates.

These observations are consistent with observations elsewhere in this work—i.e., for wells producing from a comparable geologic/reservoir setting—the statistical distributions of initial potential and ultimate recovery tend to be log normal.[8]

In estimating reserves for large aggregates of producing wells in fully developed reservoirs or fields, engineers typically assume the number of producing wells will decrease more or less uniformly over remaining life as wells reach the economic limit. Such a procedure is not consistent with Purvis' [1990] observations. In the absence of actual abandonment history, it might be more appropriate to decrease the number of producing wells so the remaining lives have a log-normal distribution.

For additional details regarding analysis of production/decline trends in wells and of the performance of aggregates of wells, the reader is referred to Purvis and Bober [1979], Purvis [1985, 1987, 1994], and Purvis and Dick [1991].

5.1.4 Curtailed Wells or Leases

Production of oil and/or gas may be curtailed for many reasons, including: (a) capacity limitations on pipelines or plants, (b) market restrictions, (c) inability to handle all produced water or gas, (d) limitations on treating facilities, (e) contract arrangements, (f) regulatory limits, and (g) governmental interference.

For example, if wellhead gas is being processed through a gasoline plant, total gas production may be limited by plant capacity. Production from low-pressure gas wells may have to be compressed before being sold; such production might be constrained by available compressor capacity. Production from high-WOR oil wells may have to be limited because of limited capacity to separate, treat, and dispose of produced water. Thus, before attempting to analyze historical production trends, especially on multiwell aggregates, the engineer should determine: (a) whether there has been curtailment of production during the period being analyzed and (b) whether there are facilities constraints that might cause curtailment in the future.

Regarding possible future curtailments, the engineer should ensure that extrapolations of historical trends to estimate reserves do not generate forecasts of future fluid production

---

a single well being operated in a large, multiwell aggregate.
[7] The "slope" of a log-normal distribution is analogous to the Dykstra-Parsons [1950] coefficient.
[8] In this context, the reader is referred to Secs. 2.3.2 and 9.8.2.

inconsistent with existing or anticipated processing capacity. For example, a forecast of future oil and gas production from a waterdrive reservoir involves the tacit assumption that, after breakthrough, water production will gradually increase until wells reach the economic limit. Depending on circumstances, such forecasts might not be realistic because of physical limitations to handle the produced water.

5.1.5 Performance Estimates Compared to Volumetric Estimates

For each well or reservoir, ultimate recovery estimated using P/DTA should be compared to ultimate recovery estimated using volumetric methods.[9] The following possibilities should be considered:

1. If the ultimate recovery estimated using volumetric methods is significantly greater than that estimated using performance methods, (a) the performance data might be representative of only the transient period (when the pressure sink around wells is still expanding) and not of semisteady-state conditions (when the drainage area has stabilized), which usually is more representative of well performance; (b) the volumetric parameters[10] might not be representative of the well or reservoir; (c) wells may need to be stimulated or equipped with higher capacity equipment; or (d) infill wells may be needed.

2. If the ultimate recovery estimated using volumetric methods is significantly less then that estimated using performance methods, (a) the volumetric parameters might not be representative of the well or reservoir, or (b) the mapped area or volume might be too small and additional development may be warranted.[11]

3. Pending determination of the reason for the difference in ultimate recovery estimated using the two methods, consider assigning proved reserves only to those volumes demonstrated to be recoverable from performance methods.

As discussed in Sec. 3.8, reservoir heterogeneities might have an adverse effect on the recovery of oil and gas. A significant difference between ultimate recovery estimated using volumetric methods and that estimated using performance methods might be an indication of inefficient drainage caused by reservoir heterogeneities.

5.2 Performance Trends

One or more of the "performance indicators" of a well and/or reservoir might exhibit a trend before the production rate of the principal product begins to decline. Depending on reservoir type and drive mechanism, these performance indicators[12] include:

- WOR,
- water/gas ratio (WGR),
- GOR,
- condensate/gas ratio (CGR),
- bottomhole pressure (BHP),

---

[9] Analogy methods also should be used, if warranted by available data. Such methods, however, are not included in the context of the discussion following.

[10] In this context, "volumetric parameters" include porosity, water saturation, formation volume factor, net pay, well drainage area, and/or recovery efficiency.

[11] Reportedly, the Fairway field, in the U.S., was discovered because of the lead provided by production from an edge well that produced substantially in excess of a reasonable volumetric estimate!

[12] In general, a prudent operator will be alert to changes of any kind, as they might be significant, however subtle, indications of changes in reservoir, well, and/or equipment performance.

- FTP, and
- shut-in tubing pressure (SITP).

5.2.1 Oil Reservoirs

Typical performance indicators for solution-gas-drive and waterdrive (or waterflood) oil reservoirs are discussed below.

Solution Gas Drive

Arps [1956], in discussing the "cumulative gas-cumulative oil method," observed that, in solution-gas-drive oil reservoirs, a plot of cumulative gas production, G_p, vs. cumulative oil production, N_p, might develop a trend that could be extrapolated to estimate ultimate oil recovery. In a fully developed reservoir, the solution gas initially in place, G_{Si}, can be estimated with reasonable confidence by volumetric methods. Extrapolation of the trend of the logarithm of cumulative gas production, G_p, to the logarithm of solution gas initially in place, G_{Si}, should provide a reasonable estimate of maximum ultimate oil recovery, N_{pa}, as shown on **Fig. 5.1.** Arps' [1956] example showed a log-log plot, but a semilog plot might work better.

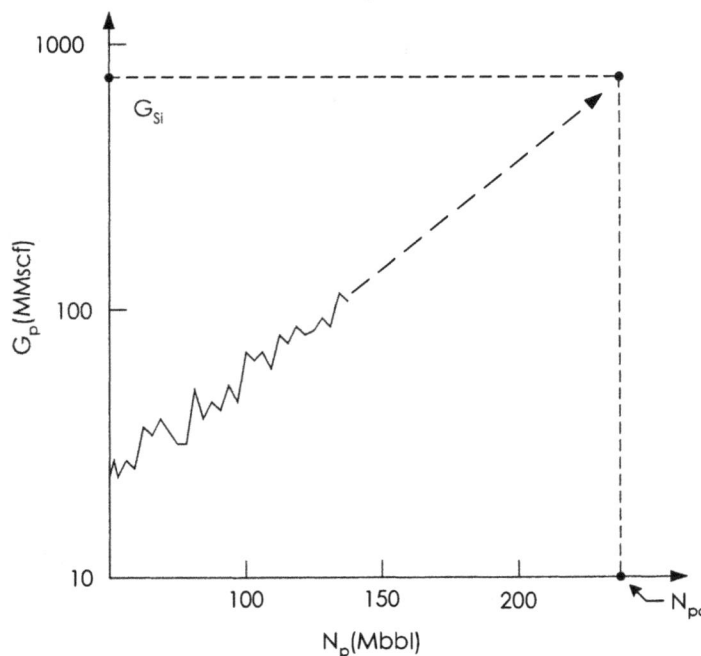

Fig. 5.1—Cumulative gas production, G_p, vs. cumulative oil production, N_p, illustrating a method to estimate ultimate oil recovery, N_{pa}, from a solution-gas-drive oil reservoir [after Arps 1956].

This extrapolation is based on the assumption that the abandonment reservoir pressure can be reduced to very nearly atmospheric pressure by pumping. Under these conditions, gas remaining in the reservoir will be a very small fraction of the gas in the reservoir at initial conditions. Thus, extrapolation to the logarithm of G_{Si}, rather than to a value slightly less than G_{Si}, should introduce negligible error.

This method might not be reliable if there is significant gravity segregation. In such reservoirs, the producing GOR in the downstructure wells might actually decrease with time, while the upstructure wells "gas out" as the gas cap expands into their drainage volume. Such performance has been reported for the Lower Lagunillas reservoir in Venezuela, for example, where permeabilities are in the multidarcy range.

In such a scenario, a procedure analogous to that shown by Fig. 5.1 might be used for the updip wells. Such wells typically exhibit a semilog trend of GOR vs. cumulative oil. An economic limit GOR might be established by reservoir voidage considerations and/or local operating or regulatory practices.

Waterdrive

Several performance methods have been used to estimate oil reserves from individual wells in waterdrive (or waterflood) reservoirs. Two of the most widely used procedures involve plotting either (a) fractional oil flow, f_o, or (b) WOR vs. cumulative oil, N_p. These and other methods are discussed briefly in the following paragraphs.

Plotting f_o vs. N_p

In waterdrive (or waterflood) reservoirs, after breakthrough of water in individual wells, plots of $\log(f_o)$ vs. N_p for each well might exhibit linear trends, as illustrated by **Fig. 5.2.**[13]

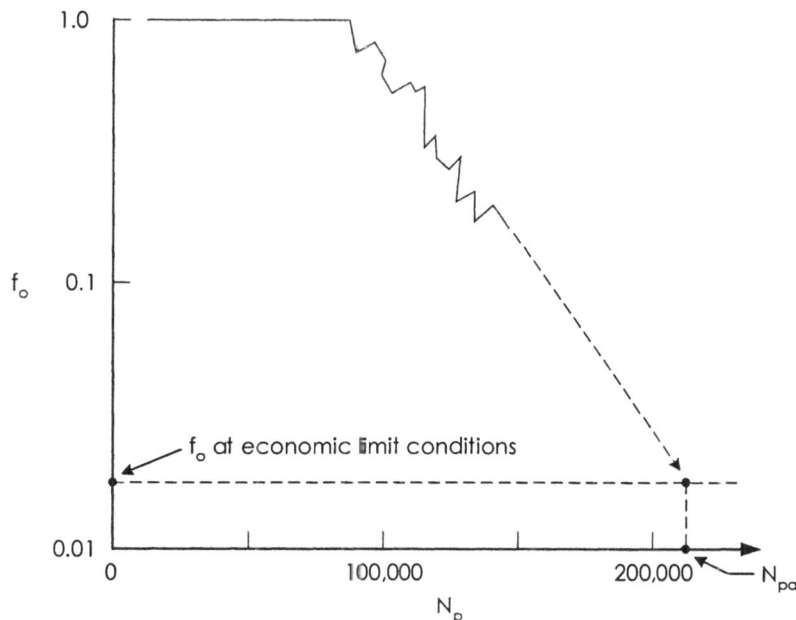

Fig. 5.2—Fraction oil in total fluids, f_o, vs. cumulative oil production, N_p, for typical well in a waterdrive (or waterflood) oil reservoir, illustrating a method to estimate ultimate oil recovery, N_{pa}.

Usually, the trend may be extrapolated to f_o at economic limit conditions to estimate reserves. The fractional flow of oil at the economic limit will be governed by the total liquids capacity (oil plus water) of the well and the completion equipment used to produce the well.

---

[13] In Arps' [1956] presentation of this method, reference was made to a plot of *reservoir* data rather than *individual-well* data. If individual well data are not available, such an approach might be the only feasible solution. However, this approach might lead to significant error unless all wells in the reservoir are producing significant water and there are no plugbacks available.

Thus, plots of total liquid production per day vs. time like those shown by **Fig 5.3** should be prepared for every well in which this method is used to estimate oil reserves.

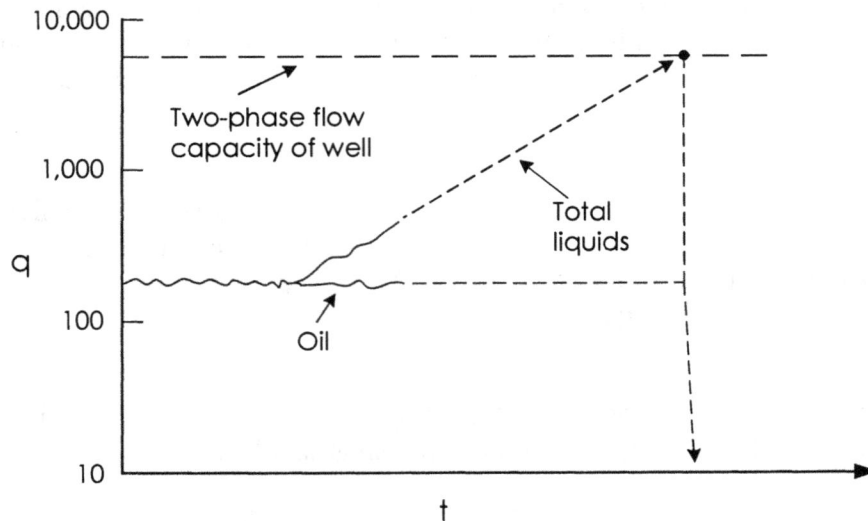

Fig. 5.3—Production rate, q, vs. time, t, for oil well exhibiting water encroachment, illustrating a method to estimate reserves, corroborating methodology illustrated by Fig. 5.2.

Brons [1963] noted that, in many cases, a linear trend of f_o vs. N_p might turn down as f_o approaches small values. The downturn reportedly occurs at smaller values of f_o for heavy (viscous) oils than for light oils, as illustrated by **Fig. 5.4.**

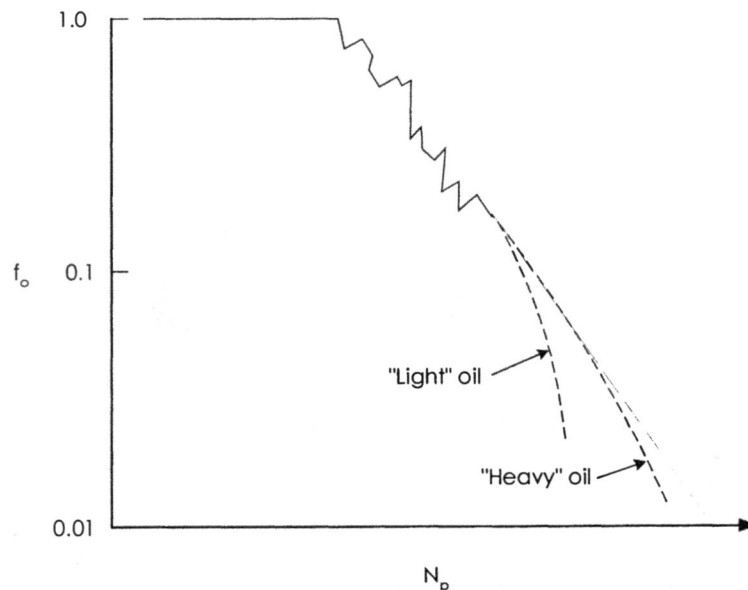

Fig. 5.4—Fraction oil in total fluids, f_o, vs. cumulative oil production, N_p, for "light" ($M_{wo} < 1$) and "heavy" ($M_{wo} > 1$) oils illustrating influence of water/oil mobility ratio on performance [after Brons 1963].

In some areas (offshore Nigeria, for example), it has been observed that a Cartesian plot of f_o vs. N_p, rather than a semilog plot, provides a more reliable basis for extrapolation.[14]

In some cases, the completion interval may extend across a shale break, as shown schematically by **Fig. 5.5a.**

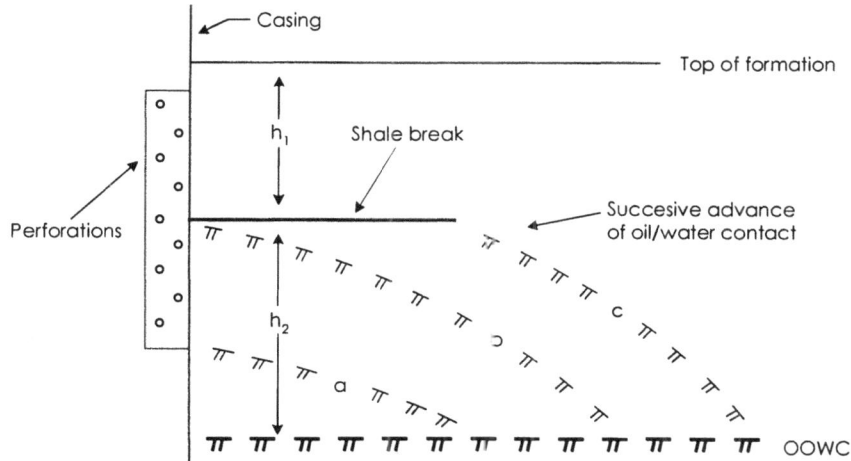

Fig. 5.5a—Typical water encroachment behavior for well in bottomwater-drive oil reservoir with shale break in completion interval.

In bottomwater-drive reservoirs with such completions, discontinuities have, in many cases, been observed in the trend of f_o vs. N_p, as shown by **Fig. 5.5b.**

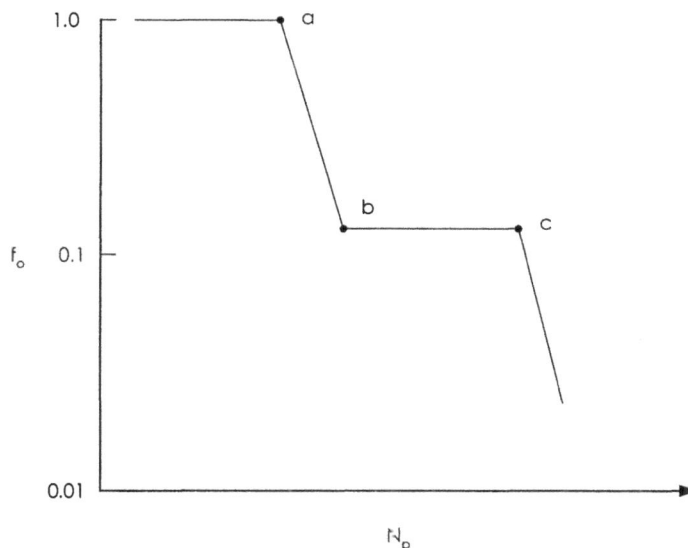

Fig. 5.5b—Fraction of oil in total fluids, f_o, vs. cumulative oil production, N_p, for well completed with shale break in completion interval, Fig. 5.5a, illustrating influence of typical reservoir heterogeneity on well performance.

---

[14] A linear trend of f_o vs. N_p for a well would suggest a uniform water level rise within the well's drainage area. Such a phenomenon would not be unexpected for offshore Nigeria oil wells given the generally favorable water/oil mobility ratios and high permeabilities that characterize many of these reservoirs. Similar performance trends might reasonably be expected in comparable reservoirs elsewhere.

After water breakthrough (Point "a" on both figures), the trend of f_o vs. N_p will decrease as the perforations below the shale break water out (Point "b" on both figures). The trend will stabilize until the encroaching water rises above the shale break (Point "c" in both figures). After water breakthrough into the perforations above the shale break, the trend will again decline. The rate of decline of f_o after the stabilization period might be greater or less than the rate of decline before the stabilization period, as discussed below. During Period b-c, the fractional flow of oil should be approximately

$$f_o = \frac{k_o h_{n1}}{\mu_o(k_o h_{n1}/\mu_o + k_w h_{n2}/\mu_w)},\dots\dots\dots\dots\dots\dots\dots\dots\dots\dots\dots\dots\dots\dots\dots\dots\dots (5.1a)$$

where k_o/μ_o = mobility of oil at S_{wi},
            h_{n1} = net thickness of upper zone,
           k_w/μ_w = mobility of water at S_{or}, and
            h_{n2} = net thickness of the lower zone.

If such behavior is observed—i.e., Trend a-b-c—and if well-completion geometry can be verified by Eq. 5.1, then, as a first approximation,

$$N_{pu} \geq N_{pc}(1 + h_1/h_2),\dots\dots\dots\dots\dots\dots\dots\dots\dots\dots\dots\dots\dots\dots\dots\dots\dots (5.1b)$$

where N_{pc} = cumulative oil production at Point c, and
           N_{pu} = ultimate oil production.

If $h_1 > h_2$, the rate of decline in f_o after the stabilization period usually will be less than the rate of decline before the stabilization period, and vice versa.

In a scenario like that shown by Figs. 5.5a and 5.5b, extrapolation of Trend a-b could lead to significant underestimation of reserves. Thus, faced with such a possibility, the engineer should check reserves from such an extrapolation against a reasonable volumetric estimate. In general, reserves estimated from any such extrapolation should be checked against those estimated from volumetric mapping and/or compared with reserves from analogous wells.

For wells produced at an approximately constant oil rate after breakthrough of water, the logarithm of the rate of total liquid production might increase linearly with time, as illustrated by Fig. 5.3. This behavior has been observed for high-capacity oil wells in waterdrive reservoirs being produced by gas lift. Oil reserves may be estimated by extrapolating the total liquids produced rate to the maximum two-phase flow capacity of the well. Plots like Figs. 5.2 and 5.3 should be prepared for each well producing significant water. In using such plots to estimate reserves, note that the semilog trend of f_o vs. N_p might be expected to turn down sharply when total liquids produced rate reaches the maximum capacity of the well.

Plotting WOR vs. N_p

Trends of WOR vs. N_p also may be used to estimate reserves from waterdrive, or waterflood, wells. Typically, semilog plots of such data tend to become linear at WOR's greater than about 1.0. Purvis [1985] discussed the use of semilog plots of (WOR + 1) and total fluids vs. time, which help define trends in oil rate. It is noted here that plots of (WOR + 1) vs. N_p, which tend to be linear at WOR's less than 1.0, might help to define trends at low values of WOR.

Other Methods

Ershagi and Omoregie [1978] and Ershagi and Abdassah [1984] observed that $\log(k_{rw}/k_{ro})$ is an approximately linear function of S_w over a significant range. They noted that a plot of

$$[1/f_w - \ln(1/f_w - 1)] \text{ vs. } N_p$$

should be linear and could be used to estimate reserves for a waterdrive reservoir.[15] These authors noted that, due to the inflection in the water/oil fractional flow curve at $f_w = 0.5$, such plots should be used only for water cuts greater than 0.5. Also, it was noted that the proposed method is more precise than the traditional semilog plot of WOR vs. N_p. Additional discussion of the method, together with details of analyses in two reservoirs in Egypt, was provided by Macary and Al Hamid [1999].

Startzman and Wu [1984] observed that this procedure should be compared with the conventional semilog plot of WOR vs. N_p, which should yield better results at low water cuts than the Ershagi-Omoregie-Abdassah method.

In continuing discussion of the Ershagi-Omoregie-Abdassah method, Liu [1989] proposed plotting

$$[\ln(N_p/W_p) - N_p/W_p] \text{ vs. } N_p.$$

At high water cuts, however, the plots accompanying the Liu [1989] paper exhibit departure from the early linear trend, a point noted by Ershagi [1989] in subsequent analysis and discussion of the Liu [1989] method.

Warren[16] proposed plotting cumulative fractional flow vs. cumulative oil production,

$$N_p/(N_p + W_p) \text{ vs. } N_p.$$

Reportedly, reserves estimated with this procedure were in reasonable agreement with computer simulation studies of a major field in the North Sea.

In summary, it seems apparent that developing the correct performance plot to estimate reserves in waterdrive reservoirs is highly empirical. Many of the published procedures are based on the observation that a significant part of a semilog plot of (k_{rw}/k_{ro}) vs. S_w is linear. The floodout performance of wells and reservoirs, however, is governed by not only relative permeability, but also rock heterogeneity and gravity/viscous forces, among other factors. Thus, different types of plots should be considered, with the best plot being determined by experience in analogous scenarios. It should be kept in mind, however, that the best plot at low water cuts might not be the best plot at high water cuts.

In fractured reservoirs, the WOR trend might be influenced by the highly heterogeneous nature of the reservoir rock and might exhibit sharp variations as a function of cumulative production and/or production rate. For example, sharp variations in the WOR trend have been observed in wells producing from the fractured, vugular Chappel limestone (north central

---

[15] As discussed previously, caution should be exercised in using such procedures for aggregates of wells. Unless all wells are producing with significant WOR's, such plots might result in erroneous estimates.

[16] Warren, J.E.: "A Simple Model for Predicting the Performance of a Complex Field," informal presentation at 1998 SPEE luncheon, Houston, 7 January.

Texas), which produces by strong waterdrive. Also, El Banbi [1974] reported that the water cut in the Alamein field (Egypt), a fractured carbonate producing with a strong bottomwater drive, was sensitive to production rate. Reportedly, individual wells were choked back as water cut increased, which resulted in sharp decreases in water cut. Such behavior might also be observed in naturally fractured sandstones if imbibition of water from fractures into matrix blocks is a significant production mechanism.

Timmerman [1982] provided additional discussion of the analysis of performance trends in waterflood reservoirs that might provide additional insight regarding performance trends in natural waterdrive reservoirs.[17]

5.2.2 Gas Reservoirs

Performance indicators are discussed briefly for pressure-depletion and waterdrive reservoirs.

Pressure Depletion

In nonretrograde gas reservoirs with moderate to high permeability, periodic SITP measurements on individual wells can be used to estimate shut-in bottomhole pressure (SIBHP), provided such wells are not producing significant liquids. SITP measurements may be used to estimate SIBHP, from which p/Z vs. G_p plots[18] may be prepared to estimate reserves, as discussed in Sec. 4.4.1. In cases where it is not practical to obtain periodic SITP data, consideration might be given to using a procedure proposed by Lewis [1985], which involves using adjusted performance data from individual well backpressure curves.

Over the life of such reservoirs, a plot of SITP vs. G_p for individual wells should be slightly concave up. However, late in life, the trend should be approximately linear. Under constant terminal rate conditions, a plot of FTP should be approximately parallel to the SITP plot. After it is adjusted downward to approximate the economic limit flow rate, the FTP plot may be extrapolated to sales line or compressor intake pressure to estimate reserves.

As discussed in Sec. 4.4.4, if backpressure test data representative of well performance approaching economic limit conditions are available, these data also may be used to estimate BHP at the economic limit flow rate.

Waterdrive

In gas reservoirs subject to water influx, after water breakthrough in individual wells, semilog plots of producing WGR vs. cumulative gas produced from each well usually will be linear.[19] Typically in such wells, the time of abandonment will be governed by the two-phase flow characteristics of the tubing string, and wells usually will become incapable of sustained flow before they reach the economic limit gas flow rate. In this case, reserves may be estimated by extrapolating the trend of FTP vs. cumulative gas to the sales line pressure. The reliability of such estimates may be improved by incorporating a plot of FTP vs. WGR in the analysis.

In some areas, however, flow from such wells may be maintained with multistage compression. Well abandonment usually will be at the economic limit flow rate, which will be a function of compressor operating cost and the cost to handle produced water, among other factors. In this case, each well might have a different abandonment WGR, as illustrated by

---

[17] Depending on well/reservoir geometry, the nature of aquifer encroachment into a reservoir might not be the same as the frontal advance observed in waterfloods. Techniques used to analyze the performance of wells in pattern waterfloods might not be appropriate to analyze the performance of wells in reservoirs with natural waterdrive.

[18] Such plots are available from commercial vendors of historical production data; e.g., PI/Dwights.

[19] Some operators report "%BS&W," the percent water in total liquids production (water plus condensate), for gas wells producing water. This unit is meaningless as a performance indicator. The proper criterion is the ratio of water to the principal product; i.e., the WGR.

Fig. 5.6. Reserves for each well may be estimated by extrapolating a plot of WGR vs. G_p for the well to the economic limit WGR for the facility. A similar procedure to determine the economic limit may be appropriate for waterdrive oil wells, as discussed in Sec. 5.2.1.

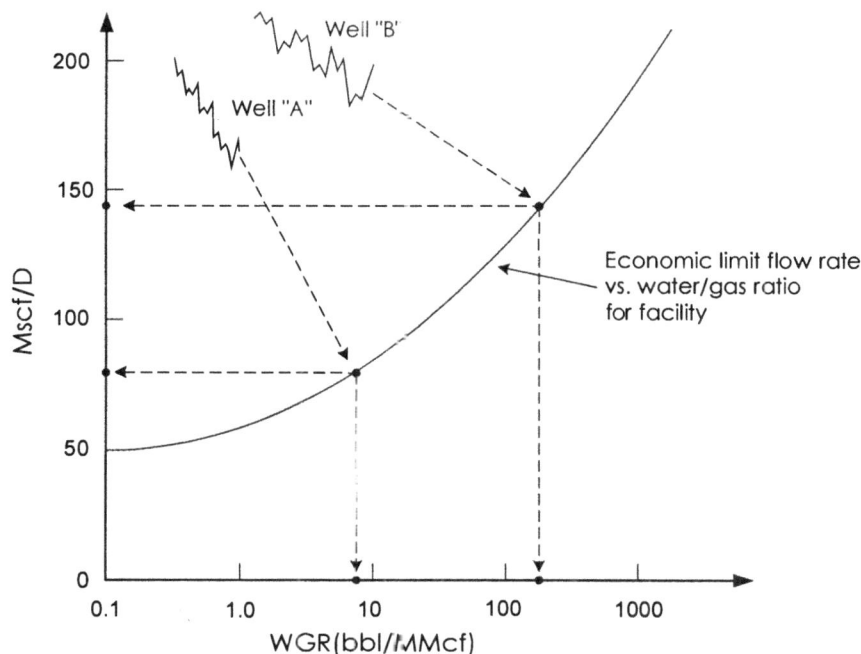

Fig. 5.6—Gas production rate, MScf/D, vs. water/gas ratio, WGR, for different gas wells exhibiting water encroachment, illustrating a method to estimate an economic limit WGR and flow rate for wells producing to a common production facility.

5.3 Production Decline Equations

5.3.1 Background

Lewis and Beal [1918] and Cutler [1924] were among the earliest to publish on the analysis of declining trends in production in oil wells. Historically, three types of equations have been used: hyperbolic, harmonic, and exponential [Arps 1945]. As discussed in the following sections, the harmonic and exponential equations are special cases of the hyperbolic equation.

Historically, the Arps [1945] equations have been considered to be empirical fits[20] to observed trends of production rate vs. time (or vs. cumulative production). As discussed by numerous authors, such equations should in theory be used only to fit the constant terminal pressure period of production—i.e., the semisteady-state period during which flowing BHP is maintained at a constant value. In practice, however, this operating condition is seldom maintained over long periods of time, which contributes to uncertainties in reserve estimation.

5.3.2 Hyperbolic Declines

In the following sections, the development follows that of Arps [1945]. Notation, however, has been modified to reflect SPE [1993] standard, except as noted otherwise.

---

[20] Such fits typically are made using in-house or commercially available computer software. Such software typically provides the user with a number of options, including: (a) regression analysis of all the data points within a user-specified time period, (b) regression analysis of specified data points within a user-specified time period, and (c) a visual "fit," based on the engineer's professional judgment. Application of such software is discussed further in Sec. 5.3.5.

Mathematics

Arps [1945] proposed fitting observed trends of production rate vs. time with a hyperbolic equation of the form

$$a = (-) \frac{dq/dt}{q} = Cq^b , \dots\dots\dots\dots\dots\dots\dots\dots\dots\dots\dots\dots\dots\dots\dots\dots (5.2)$$

where a is the nominal (or instantaneous) decline rate and is defined as the negative slope of the natural logarithm of the production rate vs. time, or

$$a = (-) \, d(\ln q)/dt , \dots\dots\dots\dots\dots\dots\dots\dots\dots\dots\dots\dots\dots\dots\dots (5.3)$$

where C is a constant that is defined under "initial" conditions, identified by the subscript 1, as

$$C = a_1/q_1{}^b , \dots\dots\dots\dots\dots\dots\dots\dots\dots\dots\dots\dots\dots\dots\dots\dots\dots (5.4)$$

and where b is the hyperbolic decline exponent.[21] Note that in Eqs. 5.2 and 5.4, $0 < b \le 1$.[22] However, in subsequent equations in this section (after integration of Eq. 5.2), $0 < b < 1$.

After integration, Eq. 5.2 becomes

$$q_2 = q_1(1 + ba_1t)^{-1/b} , \dots\dots\dots\dots\dots\dots\dots\dots\dots\dots\dots\dots\dots (5.5)$$

which is the rate/time equation for hyperbolic declines. Note that t, which should be written as Δt, is the incremental time $(t_2 - t_1)$ required for the production rate to decline from q_1 to q_2.

After a second integration with respect to time, Eq. 5.5 becomes

$$Q_{12} = q_1{}^b[q_1{}^{(1-b)} - q_2{}^{(1-b)}]/[a_1(1 - b)] , \dots\dots\dots\dots\dots\dots\dots\dots (5.6a)$$

which is the rate/cumulative equation for hyperbolic declines. Q is a general term to denote cumulative production, which may be either gas or oil. Note that in Eq. 5.6a the term Q_{12} denotes the incremental cumulative production, ΔG_p or ΔN_p, as the production rate declines from q_1 to q_2.

Eq. 5.5 may be solved to determine the incremental time[23] for the production rate to decline from q_2 to the economic limit rate, q_{el}

$$t_{el} = [(q_2/q_{el})^b - 1]/a_1b , \dots\dots\dots\dots\dots\dots\dots\dots\dots\dots\dots\dots\dots (5.7)$$

where t_{el} is abbreviated notation for incremental time, $(t_{el} - t_2)$. Incremental production (reserves) produced during this time increment may be calculated by using a variation of Eq. 5.6a:

---

[21] This notation is not consistent with SPE [1993] standard, which defines h as the hyperbolic decline exponent. (Note that $b = 1/h$.) This notation is, however, consistent with the original usage by Arps [1945] and with usage by Gentry and McCray [1978] and other authors. Note that b is equivalent to n, which was used in the *Petroleum Engineering Handbook* [SPE 1987].

[22] This limit is merely for definition of the three types of equations introduced by Arps [1945]. As discussed elsewhere, values of b greater than 1.0 might be required to fit observed data to such equations.

[23] In this and subsequent equations in this section the abbreviated notation "t," which is consistent with petroleum engineering literature, will be used, rather than the explicit notation, $(t_2 - t_1)$. Also, the subscripts "1" and "2" will be used to denote the curve-fitted period; the subscripts "2" and "el," the period for which reserves are estimated.

$$Q_{2el} = q_2{}^b[q_2{}^{(1-b)} - q_{el}{}^{(1-b)}]/[a_2(1-b)] \dots (5.6b)$$

Q_{2el} is the incremental production as the production rate declines from q_2 to q_{el}.
Fig. 5.7 illustrates the notation introduced here and used in the remainder of this section.

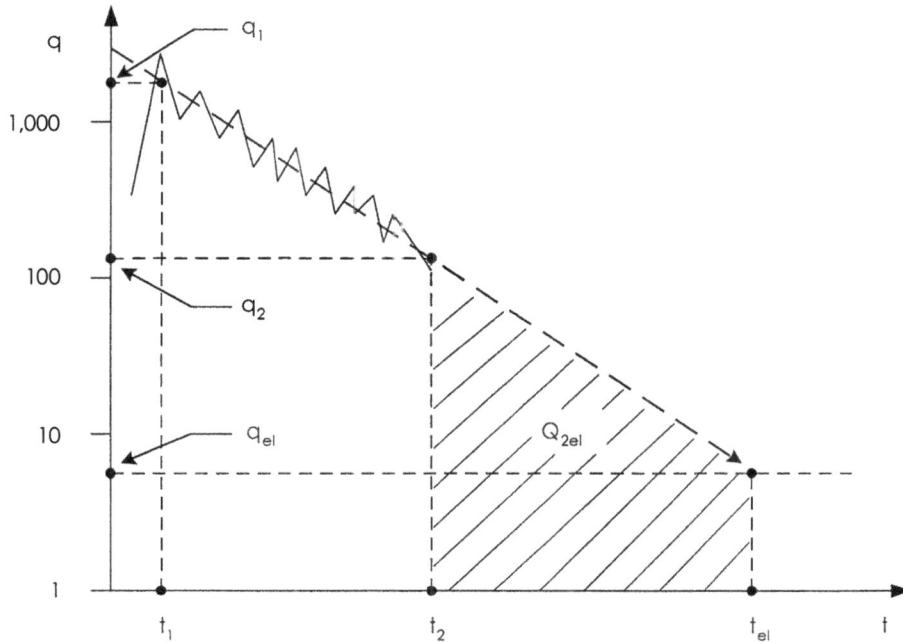

Fig. 5.7—Decline curve nomenclature.

Nominal vs. Effective Decline Rate

The nominal decline rate, a, defined in Eq. 5.3, frequently is confused with the effective decline rate, D, which is defined as

$$D = (q_1 - q_2)/q_1 \dots (5.8)$$

The effective decline rate is a stepwise, rather than a continuous, function and is dimensionless.[24] Usually, D is calculated over either a 1-month or 1-year time period.[25] For hyperbolic declines, the initial effective decline is

$$D_i = 1 - (1 + ba_i)^{-1/b}, \dots (5.9)$$

which is related to the initial nominal decline as

$$a_i = [(1 - D_i)^{-b} - 1]/b \dots (5.10)$$

---

[24] The reciprocal of the effective decline rate, called the "loss ratio," was introduced by engineers in the late 1920's to help analyze historical production data.
[25] Use caution! For these calculations to be consistent with others, a year must be considered to have 12 months, each with an average of 30.4 days.

Applications

Matthews and Lefkovits [1956], reporting on theoretical studies of oil production from dipping, homogeneous reservoirs producing by gravity drainage, concluded:

- Wells with free gas above oil—i.e., wells penetrating the secondary gas cap— should exhibit a hyperbolic decline with $b = 0.5$.

- Wells downdip from the secondary gas cap should exhibit an exponential decline— i.e., $b = 0$.

In subsequent investigations of actual reservoirs producing by gravity drainage, Lefkovits and Matthews [1958] observed that wells exhibited hyperbolic declines with b ranging from less than 0.1 to 2.5. The higher b factors were attributed to reservoirs with zones of different permeability or to wells with zones having different skin effects.

Russell *et al.* [1966] observed that in theory gas wells producing early in life under constant terminal pressure conditions and cylindrical flow should exhibit hyperbolic production declines with b equal to approximately 0.5.

Stewart [1970] observed that, during the transient period, gas wells completed in the very heterogeneous, low-permeability Cretaceous sandstones (approximately 0.5 md) in the Green River basin of the western U.S. exhibited hyperbolic declines.

Wong and Ambastha [1995] studied the production decline trends in 78 oil pools in Alberta that were being waterflooded. They reported that "the majority...follow a hyperbolic decline with...b values less than 0.5." They noted that there was no functional relationship between b and the rock/fluid properties of the pools studied.

Arps [1945] observed that "most decline curves seem to be characterized by b values between 0 and 1, with the majority between 0 and 0.4." Since that time, however, b values greater than 1 have been reported in many reservoirs. These include reservoirs with low permeability, or low-permeability reservoirs that are naturally or hydraulically fractured [Bailey 1982; Long and Davis 1988].

In addition, Gentry and McCray [1978] demonstrated mathematically that b should be expected to be larger than 1 in very heterogeneous reservoirs.

Fetkovich [1980] observed that "rate data existing only in the transient period...would require values of 'b' much greater than 1 to fit the data." This point is discussed in Sec. 5.3.8.

It is noted that, although b values greater than 1 might be required to fit the initial period of a well's production, extreme caution should be exercised in using such b values to extrapolate short-term trends to estimate reserves. Faced with such initial b values, most engineers arbitrarily adjust projected decline rates to a minimum constant rate consistent with experience in the area.

5.3.3 Exponential Declines

Because of the relative ease of use, extrapolation of the exponential, or constant percentage, decline trend is one of the most widely used performance methods to estimate reserves. As noted in Sec. 5.3.2 Applications, however, this method may lead to significant underestimation of reserves if applied during the transient flow period. Transient flow commonly is observed in the early life of wells producing from fractured or low-permeability oil and/or gas reservoirs and from (shallow) low-pressure gas reservoirs. Examples include the Cotton Valley formation in east Texas and the Green River formation in the Piceance basin of Colorado.

Mathematics

The exponential curve is a special case of the hyperbolic decline curve—i.e., where $b = 0$. In this case, Eq. 5.2 becomes

$$a = (-) \frac{dq/dt}{q} = C \dots \dots \dots \dots \dots \dots (5.11)$$

After integration, this becomes

$$q_2 = q_1 e^{-at} , \dots \dots \dots \dots \dots \dots (5.12)$$

which is the rate/time equation for exponential declines. As discussed in connection with Eq. 5.5, t is abbreviated notation for the incremental time for the production rate to decline from q_1 to q_2.

From inspection of Eq. 5.12, it should be apparent that for production declining exponentially, a plot of $\ln(q)$ vs. t can be fitted with a straight line with a slope equal to $(-)a$.

After a second integration, Eq. 5.12 becomes

$$Q_{12} = (q_1 - q_2)/a , \dots \dots \dots \dots \dots (5.13a)$$

which is the rate/cumulative equation for exponential declines.[26]

From inspection of Eq. 5.13a, it should be apparent that for production declining exponentially, a Cartesian plot of q vs. Q can be fitted with a straight line with a slope equal to $(-)a$. When semilog plots of q vs. t are very erratic and subject to considerable uncertainty, it is good practice to also plot q vs. Q to confirm the more traditional semilog, rate/time plot and to use this plot to estimate a.

Eq. 5.12 may be solved to determine the (incremental) time for the production rate to decline from q_2 to the economic limit rate, q_{el}

$$t_{el} = \ln(q_2/q_{el})/a \dots \dots \dots \dots \dots (5.14)$$

Reserves produced during this time increment may be calculated with a variation of Eq. 5.13a

$$Q_{2el} = (q_2 - q_{el})/a \dots \dots \dots \dots \dots (5.13b)$$

The effective decline rate, D, for exponential declines is

$$D = 1 - e^{-a} \dots \dots \dots \dots \dots (5.15)$$

The effective decline rate is related to the nominal decline rate, a, previously defined, as follows

$$a = (-)\ln(1 - D) \dots \dots \dots \dots \dots (5.16)$$

---

[26] Caution! Units for q and a must be consistent; B/D and 1/D, or bbl/month and 1/month, or bbl/yr and 1/yr.

Applications

Transient flow in fractured or very heterogeneous, stratified reservoirs usually is characterized initially by a relatively steep decline in production rate vs. time. These early decline trends may appear to be exponential; extrapolation of these trends to estimate reserves might result in significant underestimates.[27] There is a high degree of uncertainty in interpreting production trends in this type of reservoir. Thus, reserve estimates made from decline curve analysis should always be checked against: (a) reserve estimates based on volumetric methods and/or (b) the performance of analogous wells.

Later in life, such wells might exhibit exponential declines at much lower decline rates. For example, Brons [1963] and Stewart [1970] observed that the gas flow rates from gas wells producing late in life at constant terminal pressure conditions generally exhibit an exponential decline. Blasingame and Lee [1986] presented a method to determine well drainage area and shape factor for homogeneous and naturally or hydraulically fractured reservoirs that are producing with constant BHP.[28]

5.3.4 Harmonic Declines

The harmonic decline is a special case of the hyperbolic decline, where the decline rate is proportional to the production rate—i.e., where $b = 1$.

Mathematics

When $b = 1$, Eq. 5.2 becomes

$$a = (-) \frac{dq/dt}{q} = Cq . \quad\quad\quad\quad\quad\quad\quad\quad\quad\quad\quad\quad\quad\quad\quad (5.17)$$

The term C is defined under initial conditions as

$$C = a_1/q_1 . \quad\quad\quad\quad\quad\quad\quad\quad\quad\quad\quad\quad\quad\quad\quad\quad (5.18)$$

After integration, Eq. 5.17 becomes

$$q_2 = \frac{q_1}{1 + a_1 t} , \quad\quad\quad\quad\quad\quad\quad\quad\quad\quad\quad\quad\quad\quad (5.19)$$

which is the rate/time equation for harmonic declines. After a second integration, Eq. 5.19 becomes

$$Q_{12} = q_1 \ln(q_1/q_2)/a_1 , \quad\quad\quad\quad\quad\quad\quad\quad\quad\quad\quad\quad (5.20a)$$

which is the rate/cumulative equation for harmonic declines.

From examination of Eq. 5.20a it may be observed that for wells declining harmonically, a plot of $\ln(q)$ vs. Q can be fitted with a straight line.

Eq. 5.19 may be solved to determine the (incremental) time for the production rate to decline from q_1 to the EL rate, q_{el}

---

[27] Oil wells producing at constant pressure by fluid expansion above the bubblepoint pressure also tend to exhibit exponential declines [Brons 1963].

[28] As noted by the authors, "These methods are applicable only to systems of small and constant compressibility and single phase flow. If water influx, solution gas evolution, or multiphase flow is evident, these methods should not be used."

$$t_{el} = (q_1/q_{el} - 1)/a_1 . \quad \text{..(5.21)}$$

Reserves produced as the production rate declines from q_2 to q_{el} may be calculated with a variation of Eq. 5.20a

$$Q_{2el} = q_2\ln(q_2/q_{el})/a_1 . \quad \text{..(5.20b)}$$

The initial effective decline rate for harmonic declines, D_i, is

$$D_i = a_i/(1 + a_i) . \quad \text{..(5.22)}$$

The initial nominal decline rate, a_i, is related to the initial effective decline rate as follows:

$$a_i = D_i/(1 - D_i). \quad \text{...(5.23)}$$

Applications

With the exception of waterdrive (or waterflood) reservoirs, harmonic declines are rarely observed. For example, Purvis and Bober [1979], in their study of the Pembina Cardium oil pool in Canada, a mature waterflood project, reported that, "Harmonic decline was observed for many of the study areas and for the pool as a whole."

5.3.5 Methodology

To estimate reserves, given a declining trend in the production rate of the principal product[29] of a well, the following general procedure is recommended:

1. Examine the historical data, rejecting both anomalously low values and/or anomalously high values, for example: (a) production during months when wells were temporarily shut in for remedial work or temporarily off production due to mechanical failure or power outage, etc., (b) production directly after a well has been shut in or curtailed, which might be anomalously high,[30] or (c) production after fracturing, because some operators may report "frac-oil" as production.

2. Regress the remaining data with the rate/time equation[31] that results in the "best" fit.[32] Use caution with hyperbolic or harmonic fits, because the calculated future decline rate may become unrealistically low and result in calculation of reserves that exceed a reasonable volumetric estimate. It is good practice to set a minimum decline rate consistent with experience in the area and to project production using an exponential curve—i.e., a constant percentage decline—after this minimum decline rate has been reached.

3. Calculate the economic limit producing rate, q_{el}, for the well or facility being evaluated.

---

[29] Caution should be exercised in the analysis of gas production. In some areas, "sales" or "residue" gas is reported, rather than total wellhead gas, and adjustments may be needed if reserves of total wellhead gas are required.

[30] For example, use caution in analyzing low-permeability gas wells that have not been produced at capacity. For such wells curtailed 50% of the time, Hale [1986] observed transient production "spikes" of up to 150% of the stabilized rate.

[31] For wells with frequent periods of downtime, resulting in an erratic trend of rate vs. time, it may be preferable to analyze a rate/cumulative trend.

[32] As noted in Sec. 5.3.1, such fits typically are computed using in-house or commercially available software. Caution should be exercised in accepting the results of such analyses without ascertaining if the results represent good engineering. A computed fit with a regression coefficient approaching 1.0 may represent good math, but poor engineering.

4. Calculate reserves to be produced as the production rate declines from q_2, the rate at the effective date of the estimate, to q_{el}, the economic limit.[33]

Historically, plots of monthly production, together with appropriate notes on field operations, were maintained on semilog (rate vs. time) graph paper for wells and/or leases of interest. The widespread application of computer technology has facilitated storing appropriate numerical data, which may be retrieved, aggregated, and plotted as required. Unfortunately, however, the vital notes on field operations typically are not included in many such databases. Thus, as noted in Sec. 5.1.3, a review of such operations with field personnel is an important aspect of P/DTA.

It is good practice to check reserve estimates made from extrapolations of historical rate/time trends against estimates made using historical trends of one or more performance indicators, as discussed in Sec. 5.2, whenever such data are available. Depending on circumstances, however, the two types of estimates may differ substantially. If there is no obvious resolution of the difference, it might be advisable to consider the smaller of the two estimates to be "proved." The larger of the two estimates might be classified as "proved plus probable," or "proved plus probable plus possible," depending on circumstances.

Even if reserve estimates from P/DTA seem reasonable, it also is good practice to calculate ultimate recovery and compare the results with volumetric estimates and/or the performance of analogous wells in the area and/or of wells in analogous geologic/reservoir settings.

In areas where the completion intervals of wells extend across different lithologies, complex performance behavior may preclude performance analysis with the simple mathematical models discussed in Secs. 5.3.2 through 5.3.4. For example, in the Four Corners area of the southwestern U.S., Paradox formation reservoirs typically consist of localized high-permeability, low-porosity carbonate algal mounds that have been deposited on an areally extensive low-permeability, high-porosity carbonate shelf. Wells completed across these two lithologies typically exhibit a relatively short period of high productivity, attributed to depletion of the high-permeability algal mound, followed by a relatively long period of significantly lower productivity, attributed to slow depletion of the low-permeability shelf.

In many areas, special methods have been developed to estimate reserves from analysis of production trends using other than plots of rate vs. time. For example, Gurley [1963] discussed a method to analyze Clinton Sand gas wells (Ohio) by plotting "percent of best month" capacity vs. time. Stright and Gordon [1983] developed special procedures to analyze the performance of fractured low-permeability gas wells in the Piceance basin (USA).

For additional details, please refer to Thompson and Wright [1984] and Laustsen [1996].

5.3.6 Associated Hydrocarbons

As noted in Sec. 1.2, a complete reserve estimate should include all the commercial production attributable to the entity being analyzed. Such production includes: (a) the principal products—i.e., oil for oil wells plus gas from gas wells—and (b) associated hydrocarbons—i.e., solution and/or free gas from oil wells, condensate from gas wells, and plant products.[34] Some writers have used the term "secondary streams" in lieu of the term "associated hydrocarbons."

---

[33] Typically, the calculations in Steps 3 and 4 are performed using in-house or commercially available computer software.

[34] For procedures to estimate reserves of plant products, the reader is referred to PS-CIM [1994].

Solution and/or Free Gas From Oil Wells

The historical data available to a reserve estimator for a producing entity typically includes monthly gas and oil (and sometimes monthly water). Given a reasonably well-defined declining trend of oil, the historical data may be extrapolated to the economic limit to determine oil reserves. Usually, gas reserves attributed to the same set of wells are determined using the same procedure. While this may be accepted practice, it may lead to a forecast of gas that is not consistent with reservoir mechanics. For example, for a well, or aggregate of wells, producing oil and only solution gas by strong waterdrive, the projected GOR would reasonably be expected to be constant. Minor perturbations in the historical gas production data, however, might lead to a projection of gas production that implied an increasing (or decreasing) GOR, which is inconsistent with reservoir mechanics. Thus, in analyzing historical data for oil wells it is good practice to: (a) determine if the historical GOR trend is consistent with reservoir mechanics, and (b) determine gas reserves by extrapolation of a GOR that is consistent with reservoir mechanics.

Condensate From Gas Wells

Comments analogous to those for oil wells in the preceding paragraph are appropriate for gas wells. For example, the historical production trends for gas and condensate for a gas well might reflect production while reservoir pressure is greater than the dewpoint pressure—i.e., production of undersaturated reservoir gas. Such a condition might be detected by a plot of CGR vs. cumulative gas. A declining trend in gas production might indicate that continued (future) production will reduce reservoir pressure to less than the dewpoint pressure. Given such a scenario, estimated future production of condensate must be based on the PVT properties of the gas, rather than extrapolation of production trends observed during production of undersaturated reservoir gas. For additional discussion regarding the behavior of retrograde gases, the reader is referred to Sec. 4.4.2.

5.3.7 Effects of Workovers/Artificial Lift

Two scenarios are discussed briefly: (a) stimulation of an existing completion interval and (b) installation of artificial lift in oil wells or compression for gas wells.

Stimulation of an Existing Completion

Stimulation of an existing completion might be successful in increasing the well's capability to produce fluids by: (a) reducing the "skin," (b) increasing the effective wellbore radius, (c) opening additional intervals for production, or (d) any combination of the foregoing. If such results facilitate sustained production at a significantly higher average rate than before stimulation, an increase in reserves might reasonably be expected. However, in a competitive drainage situation, if the offset wells in the same reservoir also are stimulated, such apparent increase might be short-lived. The net result might be only acceleration of production, rather than an increase in reserves.[35] Good practice mandates monitoring stimulation results for a reasonable period before booking additional reserves. In competitive reservoirs, monitoring offset operators might also be good practice.

---

[35] For example, Skov [personal communication, 2000] observed that in the Kansas part of the Hugoton (gas) field, gas allowables were based on calculated absolute open flow potential (CAOFP). Huge fracturing jobs increased CAOFP significantly, but after all operators had fractured their wells, gas allowables were increased for all operators, resulting in essentially the same competitive position as initially. The expenditure of significant monies resulted in production acceleration, but not significant reserve increases.

If, on the other hand, such stimulation results in opening intervals not previously contributing to flow, or if such wells are not in competitive reservoirs, the reserve increase might be real, rather than apparent.

Installation of Artificial Lift

Installation of artificial lift for oil wells or compression for gas wells typically results in reduction of wellhead pressure with the consequent reduction in flowing BHP. The net effect might be to increase the drainage volume of the well, which might increase reserves,[36] depending on the competitive situation, as discussed in the preceding paragraph. Another effect might be to reduce the BHP at abandonment, which also tends to increase reserves.

Offsetting these factors, however, is the higher operating cost associated with artificial lift, which typically results in a higher economic limit production rate than would be the case without artificial lift. Each case must be evaluated on its own merit, considering prior experience in the area with such installations.

5.3.8 Type Curve Analysis

Until the late 1950's, production decline trend analysis was considered as merely empirical fits to observed data. Since that time, however, it has become apparent that there is a mathematical relationship between well mechanics and production-decline trends. In this context, for example, the reader is referred to the discussion of gravity drainage reservoirs in Sec. 5.3.2 Applications.

In 1973, Fetkovich [1980][37] demonstrated the mathematical relationship between the cylindrical flow diffusivity equation and well production performance, thereby establishing a sound theoretical basis for decline trend analysis. Since that time there have been numerous papers[38] discussing application of the so-called "Fetkovich method."

Theory

The Fetkovich [1980] methodology, called "advanced decline curve analysis," is similar to type curve analysis used to analyze bottomhole pressure transient data.

In his development of the theory, Fetkovich [1980][39] demonstrated that, as might be expected with well pressure performance, well production performance may be characterized by two periods:

1. the transient period, when the well drainage boundary is expanding, and
2. the semisteady-state or depletion period, when the bottomhole pressure is declining more or less uniformly in the well drainage volume.

Units for the x and y axes were defined by Fetkovich [1980] as dimensionless time, t_D, and dimensionless flow rate, q_D, respectively. The equations for the transient flow period are

$$t_D = [0.00634kt/\phi\mu ctr_w^2]/\tfrac{1}{2}[[(r_e/r_w)^2 - 1][\ln(r_e/r_w) - \tfrac{1}{2}]] \quad \dots\dots\dots\dots\dots (5.24)$$

and

---

[36] Such reserve increases have been reported following the installation of high-volume centrifugal pumps for oil wells producing from waterdrive reservoirs in southeast Sumatra [Ventura and Temansja 1991]. Reserve increases have also been reported after the installation of high-volume pumping equipment in the Talco field, Texas, which produces heavy oil from a strong waterdrive reservoir [Rose 1982].

[37] First presented at the 1973 SPE Annual Technical Conference and Exhibition, Fetkovich's paper was not published until 1980.

[38] The reader is referred to, for example, Gentry and McCray [1978], Fetkovich *et al.* [1985, 1987], Fetkovich [1997], and Mannon and Porter [1989].

[39] The Fetkovich [1980] model is based on slightly compressible, single-phase, cylindrical flow from a *homogeneous* reservoir.

$$q_D = 141.3q\mu B[\ln(r_e/r_w) - \frac{1}{2}]/[kh(p_i - p_{wf})] \quad\text{...(5.25)}$$

Equations for the depletion period are

$$t_D = a_i t \text{ , ...(5.26)}$$

and

$$q_D = q/q_i = \exp(-at_D) \text{ , for } b = 0 \text{ ....(5.27a)}$$

and

$$q_D = [1 - ba_i t_D]^{-1/E} \text{ , for } 0 < b \le 1 \text{ ....(5.27b)}$$

Eqs. 5.27a and 5.27b are functionally the same as Eqs. 5.12 and 5.5, which were introduced Secs. 5.3.3. and Sec. 5.3.2, respectively.

Fig. 5.8 shows the Fetkovich type curves, which are a composite of two sets of curves, one set for the transient (unsteady-state) period, the other set for the depletion (semisteady-state) period.

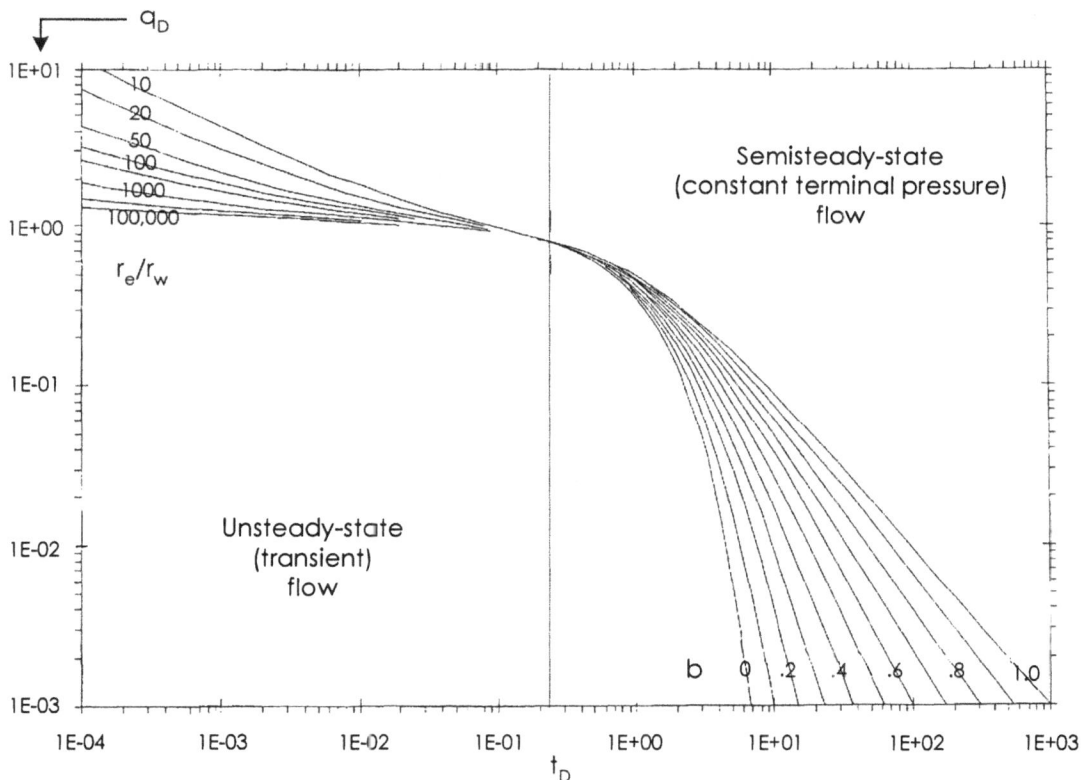

Fig. 5.8—Dimensionless time/rate behavior during unsteady-state and semisteady-state periods [after Fetkovich 1980].

As shown on Fig 5.8, the transition from transient to semisteady-state behavior occurs at dimensionless time between about 0.2 and 0.3. Transient pressure and/or production data observed prior to that time are insufficient to determine depletion behavior after that time. Thus, until a well reaches semisteady-state conditions, decline curve analysis (alone) cannot be used to calculate reserves.

In concept, the method involves generating a log-log plot of production rate vs. time and fitting this plot to the appropriate type curve. Since the method was introduced, there have been several computer programs developed to perform these functions that are commercially available.

Examples of the application of this method have been discussed by Fetkovich et al. [1985, 1987] and Fetkovich [1997]. Mannon and Porter [1989], in discussing application of the MIDA[40] computer program to implement Fetkovich's [1980] method, noted that the ability to discern transient and depletion conditions is one of the strengths of this approach.

An additional strength of the Fetkovich [1980] methodology is the capability to "back-calculate" reservoir parameters of interest, which provides a "realty check" on the fit.

Using a technique similar to that of Fetkovich [1980], DaPrat et al. [1981] presented type curves that may be used to analyze production trends from wells in dual-porosity (naturally fractured) reservoirs for both finite and infinite systems. Their analysis was based on the Warren and Root [1963] model for fractured reservoirs. They observed that "the initial decline in production rate often is not representative of the final state of depletion...type curve matching based only on the initial decline can lead to erroneous values for...(the drainage volume)...if the system is considered homogeneous." In addition, they noted that both dimensionless production and dimensionless time are controlled by the dimensionless fracture storage, S_{fD}, and by the ratio of matrix to fracture permeability, R_{kmf}. Thus, a different type curve would be required for different values of these parameters. Accordingly, they recommended that S_{fD} and R_{kmf} be estimated using transient pressure analysis so that the correct type curve may be used to estimate reserves.

Applications

Many engineers consider type curve analysis to be a procedure distinctly different from curve fitting.

The procedure is especially useful to analyze the performance of individual wells. For example, Fetkovich [1997] demonstrated that advanced decline curve analysis could be used to identify wells in west Texas in which differential depletion was occurring because of layered, no-cross-flow production. Once such wells were identified, the low-permeability layers were fracture stimulated to improve drainage efficiency.

Fetkovich's [1980] "advanced decline curve analysis" is a powerful analytical procedure that has been used successfully to analyze well behavior in a variety of scenarios [Fetkovich et al. 1985, 1987, 1988, Fetkovich 1997, and Mannon and Porter 1989]. However, for routine decline trend analysis, most engineers use the classic Arps [1945] hyperbolic equation to fit observed trends and set a minimum annual decline rate consistent with experience in the area to ensure that computed (future) decline rates do not become unrealistically small.

Fig 5.9 is an example of this procedure for a well completed in the Spraberry field, west Texas.[41] Production from early 1993 to mid-1996 declined rapidly, apparently reflecting

---

[40] Reference to this program does not imply endorsement.
[41] This field is discussed in Sec. 7.5.

transient flow. This period was fitted with a *b* factor of 1.8. Based on experience in the area, however, the minimum annual (computed) decline rate was set at 10% per year. Ultimate recovery was calculated to be 25,000 bbl, with a total well life of 15 years. Had a minimum (computed) annual decline rate not been set, the calculated ultimate recovery would have been 37,000 bbl, with a total well life of 30 years. This well life is not consistent with analogous wells in the area, and the latter interpretation was rejected.

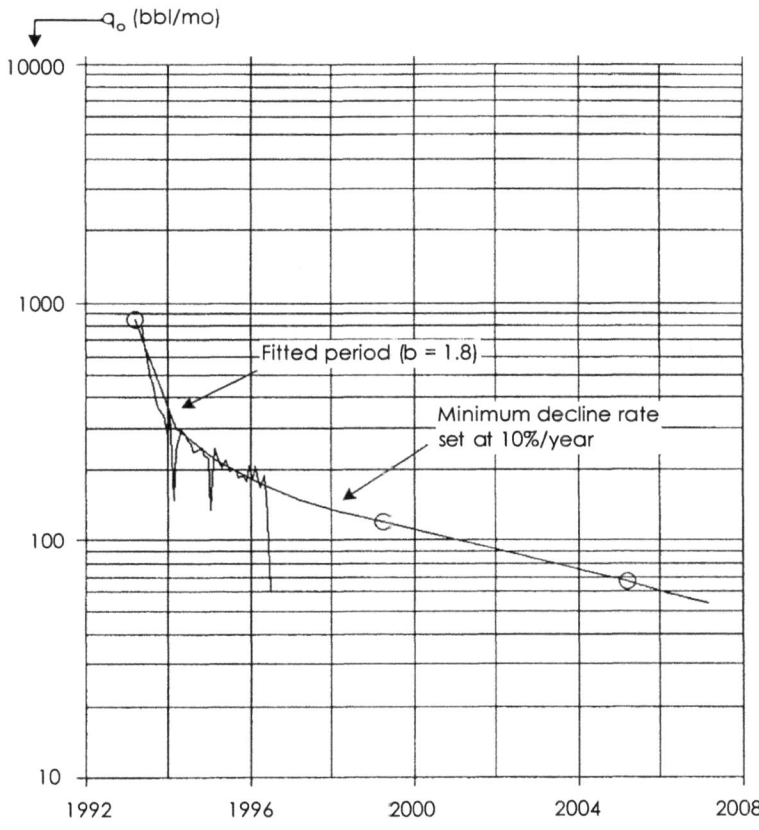

Fig. 5.9—Method to fit transient period and set minimum decline rate for projection of future production.

CHAPTER 6

SPECIAL PROBLEMS IN RESERVE ESTIMATION AND CLASSIFICATION

6.1 Inconsistent Reserve Definitions and Classifications

Currently, there is not a common set of reserve definitions in worldwide use. As may be noted from a review of Sec. 1.3, there seem to be as many definitions as there are countries and organizations. Some definitions rely on deterministic calculations to estimate and classify reserves; others, on probabilistic calculations. This lack of standardization has contributed to misunderstandings between companies and countries and to miscommunication between geologists and engineers, industry and the media, and the media and the general public.

Deterministic calculations, which typically result in "best" or "most likely" estimates of proved, probable, and possible reserves, do not quantify the risk in such estimates. Probabilistic calculations, in contrast, result in a distribution of reserve estimates, with the risk of recovery and classification of such estimates being determined by the cumulative frequency, as discussed in Sec. 9.3.

In mature areas where geologic conditions are well known and there is a broad base of experience regarding well and reservoir performance—e.g., the U.S. Permian Basin— deterministic calculations have been and continue to be used with wide acceptance. On the other hand, in frontier areas where geologic conditions are not well known and there is no experience regarding well and reservoir performance—e.g., the south China Sea— probabilistic calculations might be more appropriate.

Under conditions of even moderate uncertainty, reserves estimated and classified using deterministic calculations may differ from reserves estimated and classified using probabilistic calculations [Cronquist 1991]. Thus, caution should be exercised in comparing results of the two procedures. These points are discussed further in Sec. 9.6.

6.2 Remote/Frontier Areas and Harsh Operating Conditions

Harsh operating conditions are defined here as physical environmental conditions that require extraordinary provisions to ensure the safety and comfort of personnel and integrity of equipment. The Beaufort Sea, off the northern coast of the U.S. and Canada, and the Norwegian Sea might be considered such areas.

Remote/frontier areas are defined here as areas that are so isolated from existing oilfield infrastructure that provision of supplies and support services represents major cost and logistical problems. The Timor Sea, off the northwestern coast of Australia, and the south China Sea, off the coast of Vietnam, might be considered such areas.

Reserve estimation and classification in remote and/or frontier areas is especially difficult because: (a) there usually is a high degree of geologic uncertainty and (b) the economics of developing, producing, processing, and transporting oil and/or gas to market are highly uncertain.

Harsh operating conditions impose extraordinary costs that are difficult to quantify during the assessment phase. Under such conditions, the minimum size of potentially economic accumulations may be so large as to preclude development of all but giant fields. For example, during their exploration in the Canadian Beaufort Sea during the late 1970's, Dome Petroleum indicated minimum reserves for a commercial field to be 400 million bbl of oil.

Remote areas initially require very large discoveries to justify development. For example, regarding Mobil's exploration operations during the early 1990's, offshore Malaysia, Wood Mackenzie estimated 300 million bbl of oil would be required for commerciality [*Oil & Gas J.* 1993].

6.3 Heavy and Extra Heavy Crude Oil

6.3.1 Background

Discovered world resources of heavy and extra heavy crude oil, defined in Sec. 1.2.2, are estimated to be approximately 4,600 billion bbl, two thirds of which are in Canada and Venezuela [Briggs *et al.* 1988]. Bitumen and tar sands are excluded from this estimate.

Published data on recovery efficiency from this resource by primary drive mechanisms are sparse. Estimated primary recovery efficiency from the Lloydminster area of western Canada, where stock-tank gravities range from 13 to 17°API, ranges from 3% to 8% OIP [Adams 1982]. Primary recovery efficiency vs. API gravity for "heavy crude oil pools" in Alberta [Energy Resources Conservation Board[1] 1990] are plotted on **Fig. 6.1**. From these data, it may be inferred that, for crude oils with gravities less than about 10°API, primary recovery is nil.

Estimated primary recovery efficiency from the Orinoco area of Venezuela, where stock-tank gravities range from 8 to 13°API, ranges from 8% to 12% OIP [Martinez 1987].

Incremental recovery attributable to waterflooding in the Lloydminster area, where crude oil gravity in project areas averages about 16°API, typically does not exceed 1% to 2% OIP [Adams 1982]. Comparable data are not available on the Orinoco area. Thermal recovery methods—e.g., steam injection—probably will be required to develop significant reserves from this resource. Significant reserves also may be developed by "cold" production from horizontally completed wells. Such efforts were in the early stages during the late 1990's, and it probably will be several years before definitive performance data are available.

For 12 mature steamflood projects, Matthews [1983] noted estimated ultimate recovery efficiencies were in the range from 39% to 68% OIP.[2] These projects ranged in depth from 535 to 2,600 ft and had reservoir oil viscosities ranging from 85 to 6,400 cp. Well spacing for these projects generally is on the order of 1 to 2 acres/well, which is required for efficient steamdrive operations. Projects were in California, The Netherlands, and Venezuela. Other data for these projects are in Appendix F.

---

[1] In 1995 the Alberta Energy Resource Conservation Board was merged with other agencies and became part of the Alberta Energy and Utilities Board.

[2] Due to continuing improvements in technology and reductions in cost, even higher recovery efficiencies are anticipated from several mature areas in California.

Meyer and Mitchell[3] estimated worldwide ultimate recovery from heavy and extra heavy crude oils to be approximately 476 billion bbl, which is 10% of the Briggs *et al.* [1988] estimate of the discovered resource initially in place. Fiorillo [1987] estimates ultimate recovery from the Orinoco oil belt to be approximately 245 billion bbl, which is approximately 20% OIP.

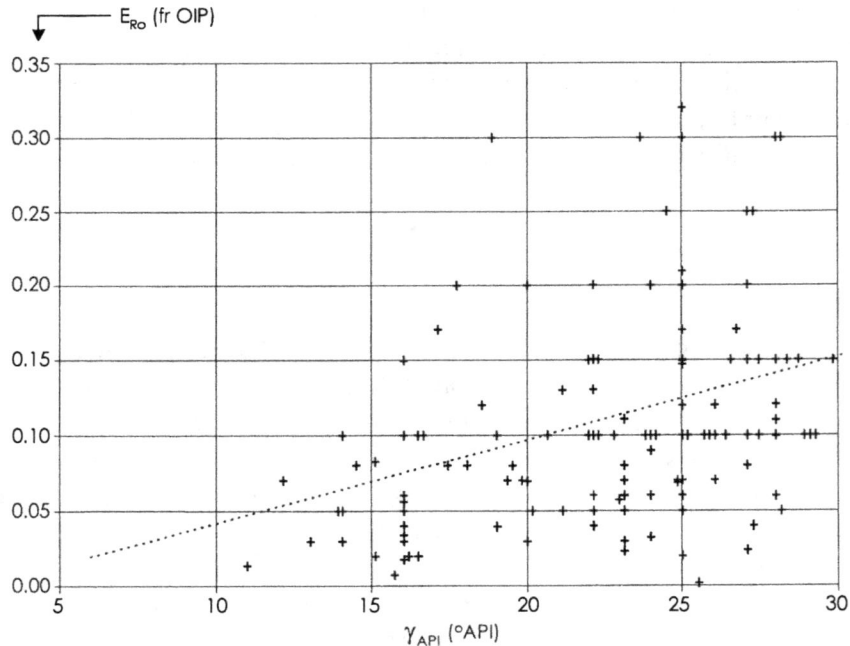

Fig 6.1—Estimated primary recovery efficiency, E_{Ro}, vs. API gravity, γ_{API}, for heavy oil pools in Alberta, Canada.

6.3.2 Estimating Reserves

Difficulties in estimating reserves from these resources may be attributed to: (a) problems in characterizing the accumulations, (b) uncertainties in quantifying the response to steam injection, and (c) generally marginal economics.

Problems in characterizing these accumulations include the following:

• They usually are associated with relatively shallow reservoirs,[4] which typically contain brackish interstitial water, which causes problems in evaluating resistivity logs.

• Most of the resource is contained in friable sandstones, causing problems in retrieval and analysis of cores.

---

[3] Meyer, R.F. and Mitchell, R.W.: "A Perspective on Heavy and Extra Heavy Oil, Natural Bitumen, and Shale Oil," oral presentation at the 12th World Petroleum Congresses, Houston [1987].
[4] Exceptions to this observation include the Rospo Mare field (located in the Adriatic Sea), which produces 12°API oil from approximately 4,600 ft subsea, the Vega field (offshore Sicily), which produces 16 °API oil from approximately 8,500 ft subsea, and the Gela field (onshore Sicily) which produces 10 °API oil from approximately 10,800 ft subsea.

- Analysis of pressure transient tests is hampered by the long times required to reach semisteady state in the viscous, low-mobility oil.

- Because of the small density contrast between the reservoir oil and the interstitial water, significant fractions of these accumulations may be in oil/water transition zones.

Uncertainties in quantifying the response to steam injection may be attributed to: (a) significant spatial variations in the viscosity of the oil, which are exhibited in many such accumulations, and (b) the highly variable response to steam injection, which makes it difficult to develop a reliable prediction model.

Marginal economics are caused by:

- extra costs for complying with local air pollution requirements,

- high energy needs to generate steam,

- additional costs to handle these bituminous, sulfurous, and asphaltic crudes, and

- the frequent presence of heavy metals in the oil, which causes problems in refining, thereby depressing wellhead price for the crude oil.

6.4 Fractured/Vugular Reservoirs

6.4.1 Background

Fractured reservoirs have been observed in most producing areas of the world. Occurrences have been reported in igneous/metamorphic rocks, in sandstones, in carbonates, and in shales. Reiss [1980] and others have identified two broad categories of fractured reservoirs: (a) those with a porous matrix and (b) those with a nonporous matrix. In the former category, which is the most common, most of the hydrocarbons are stored in the matrix porosity; the fractures serve as the principal flow conduits. Such reservoirs typically are identified as "dual-porosity" systems. Examples include many of the Iranian fields, Ekofisk (North Sea), Palm Valley (Australia), and Spraberry (Texas).[5] In the latter category, which are less common, are reservoirs in fractured igneous and/or metamorphic rocks, fractured shales, and fractured cherts. Such reservoirs frequently are associated with basement rocks [P'an 1982]. Examples include the Bach Ho field (offshore Vietnam), the Augila field (Libya), the Edison field (California), the Big Sandy gas field (Kentucky), and the Santa Maria basin fields (California).

When they occur in carbonates, fractures tend to facilitate extensive leaching and diagenesis, which may result in the development of vugular, sometimes karstic, porosity. Examples include the Albion-Scipio trend (Michigan) and the Rospo Mare field (offshore Italy).[6]

6.4.2 Reservoir Engineering Aspects

This type of reservoir poses formidable difficulties in estimating reserves. These difficulties are attributable to the heterogeneity of the reservoir rock, which causes substantial uncertainties in (a) estimates of oil and/or gas initially in place and (b) recovery efficiency. Because of uncertainties in determining the flow characteristics of dual-porosity systems, estimates of reserves using volumetric methods are subject to substantial uncertainty. When feasible, such estimates should be compared with observed recovery in analogous reservoirs.

---

[5] The Palm Valley and Spraberry fields are discussed in Secs. 7.4 and 7.5, respectively.
[6] These two fields are discussed in *AAPG Treatise on Petroleum Geology*, Stratigraphic Traps I and II, respectively.

In dual-porosity reservoirs producing by pressure depletion, the early performance of wells typically is characterized by relatively rapid decline in production rate, which usually is caused by transient pressure behavior. Not until wells have passed through the transient pressure period and settled into semisteady-state conditions can reserves be estimated with any degree of confidence using decline curve analysis.[7] In general, problems are caused by:

- Boreholes frequently are severely washed out, making log interpretation difficult or impossible.

- Core recovery frequently is fragmental, at best.[8]

- Even in good-quality boreholes, detection of fractures and measurement of fracture porosity using logging devices is highly empirical.

- In accumulations where severe lost circulation has occurred, operators typically stop drilling at the top of the reservoir section;[9] this practice, while necessary for safe operations, precludes characterization of the objective section.

- Well performance frequently will be strongly influenced by proximity to major fractures and by completion technique.

- Although transient pressure analysis may provide useful data, application of modern interpretation techniques mandates using highly accurate quartz pressure transducers.

- The accuracy of type curve matching is dependent on the accuracy of the mathematical model used for the type curves. An invalid model cannot yield a valid interpretation. Even if the model is valid, analysis of results may not provide unique answers.

- In pressure-depletion reservoirs, the rate/time performance of wells typically is hyperbolic.[10] While the behavior of an average well might be used to estimate reserves, wide variation in performance between wells should be expected.

Aguilera [1999] provided guidelines for estimating recovery efficiency in fractured reservoirs, classifying fractured reservoirs by pore type and "storage ratio"—i.e., the relative amount of storage in matrix vs. fractures. The following material is adapted from his Tables 2 and 3:

	RECOVERY EFFICIENCY (%)					
	OIL RESERVOIRS STORAGE RATIO			GAS RESERVOIRS STORAGE RATIO		
DRIVE MECHANISM	A^1	B^2	C^3	A^1	B^2	C^3
Solution gas (*Depletion*)	10 to 20	20 to 30	30 to 35	70 to 80	80 to 90	>90
Solution gas plus gas-cap expansion plus water influx	35 to 45	45 to 55	55 to 65			
Strong waterdrive				15 to 25	25 to 35	35 to 45

1. Storage mostly in matrix, with small amount in fractures.
2. Storage about 50/50 in matrix and fractures.
3. All storage in fractures.

---

[7] This aspect of decline curve analysis is discussed in Sec. 5.3.8.
[8] Even if good recovery is achieved, cores from highly vugular carbonates—e.g., reef facies—may be severely invaded by drilling mud solids, which may be difficult to remove and may require special techniques to measure porosity and other petrophysical properties [Sprunt 1989]. For additional information regarding cores from fractured reservoirs, the reader is referred to Kulander *et al.* [1990].
[9] For example, this is common practice in exploiting the Chappel limestone in north central Texas.
[10] Hyperbolic production declines are discussed in Sec. 5.3.2.

Although not so noted in Aguilera's paper, it is presumed that "oil" means light oil—i.e., stock-tank gravities greater than about 25°API.

6.5 Geopressured[11] Reservoirs

6.5.1 Geologic Overview

The term "geopressure," introduced in the late 1950's by Charles Stuart of Shell Oil Co., refers to reservoir fluid pressure that significantly exceeds hydrostatic pressure and may approach overburden pressure. Hydrostatic pressure, defined in Sec. 3.5.3, Initial Reservoir Pressure, is a function of the local fluid gradient and varies from about 0.4 to 0.5 psi/ft depth (Appendix H). Overburden pressure usually is estimated to be about 1.0 psi/ft depth.

Abnormally high reservoir pressures have been observed in many areas of the world. Such pressures have been attributed to several causes, including tectonic stress [Berry 1973], undercompaction [Dickinson 1953], geochemical and/or diagenetic processes [Parker 1973, Jansa and Urrea 1990], and PVT effects in sealed, subsiding gas reservoirs [Stuart and Kozik 1977].

Geopressures in sand/shale sequences in tertiary basins—e.g., in the U.S. Gulf Coast—generally are attributed to undercompaction of thick sequences of marine shales. Geopressured reservoirs in this type of depositional sequence tend to be geologically complex and exhibit producing mechanisms that are not well understood. Both of these factors cause considerable uncertainty in reserve estimates at all stages of reservoir maturity. Geologic complexity contributes to uncertainty in estimates of oil and gas initially in place based on volumetric mapping. Poorly understood producing mechanisms contribute to uncertainty in estimates of reserves based on pressure/production performance.

Geopressured reservoirs in tertiary sand/shale sequences frequently are associated with substantial faulting and complex stratigraphy, which may make correlation, structural interpretation, and volumetric mapping subject to considerable uncertainty.

The resistivity of interstitial water in geopressured sections may approach that of fresh water, which may suppress the spontaneous potential log. Under these conditions, it may be difficult to estimate net pay unless a gamma ray log also has been run. In addition, the relatively fresh waters frequently encountered in geopressured sections complicate interpretation of resistivity logs, especially in shaley sands. Cases have been reported where reserves were "booked" based on high resistivity observed in porous sands that later investigation proved were freshwater-bearing.

6.5.2 Reservoir Overview

In Secs. 6.5.3 and 6.5.4, the application of material balance methods to estimate reserves in geopressured reservoirs is discussed. As noted in Sec. 4.5, however, there may be a substantial difference between the volumetric estimate and the material balance estimate of oil and gas initially in place in a given reservoir. Given the geologic complexity of the geopressure environment, such differences are quite likely. In view of the high costs usually associated with drilling, completing, and operating geopressured reservoirs, it is especially important to reconcile such differences to ensure cost-effective exploitation of this resource.

Frequently, wells completed in geopressured reservoirs produce substantial amounts of water, even though the wells may be considerable distances from any obvious water source.

---

[11] The term "overpressured" also is used in the same context.

If mechanical problems can be ruled out, the water may be attributable to expulsion of water from "proto-shale"[12] zones interbedded with the productive sands.

6.5.3 Geopressured Gas Reservoirs[13]

Background

As discussed in Sec. 4.4, if gas production is attributed to gas expansion only, a plot of p/Z vs. G_p should be a straight line. Because geologists considered them to be closed accumulations, during the early years of exploitation, it was assumed that geopressured gas reservoirs would produce by pressure depletion and exhibit linear plots of p/Z vs. G_p. While this was observed to be true in many cases [Wallace 1969, Prasad and Rogers 1987], it soon became apparent it is not universally true.

It has been observed that p/Z vs. G_p plots for many geopressured reservoirs initially appear to be linear but exhibit downward curvature as reservoir pressure approaches hydrostatic pressure. Extrapolation of the initial part of such a plot may result in an estimate of gas initially in place approximately twice that estimated using volumetric methods. The anomalously low initial slope of the p/Z vs. G_p plot has been attributed to several factors, including pore volume compression [Hammerlindl 1971], expansion of interstitial water, and partial waterdrive. The downward curvature of the p/Z vs. G_p plot has been attributed to other factors, including depletion of a limited "proto-shale water" aquifer[14] [Duggan 1972] and "rock collapse."[15] [Harville and Hawkins 1969].

Production mechanisms in geopressured gas reservoirs have been observed to vary from strong waterdrive to pressure depletion. In general, the producing mechanisms in geopressured gas reservoirs may include:

- gas expansion,
- compressibility of the reservoir pore volume,
- expansion of the interstitial water,
- water influx due to water expansion from a contiguous aquifer,
- water influx due to dewatering of interbedded proto-shale, and
- evolution of natural gas dissolved in interstitial and aquifer water.[16]

Any one, or all, of these mechanisms may be active in a geopressured gas reservoir.

There is disagreement regarding the relative importance of these mechanisms, especially compressibility of reservoir pore volume [Bernard 1987] and water influx from interbedded proto-shale[17] [Wallace 1969, Bourgoyne 1990, Duggan 1972, Chierici *et al.* 1978]. As a

---

[12] The American Geological Inst. [1972] defines shale as an "indurated (hardened)...sedimentary rock formed by the consolidation...of clay." Because geopressures in tertiary basins generally are attributed to undercompaction, the term "proto-shale" is considered more appropriate and is adopted here to make this distinction.

[13] After this section was written, SPE published a book by Poston and Berg [1997] on overpressured gas reservoirs, to which the reader is referred for additional discussion.

[14] As discussed in Sec. 4.4.5, Bruns *et al.* [1965] demonstrated that downward curving plots of p/Z vs. G_p can be generated by partial waterdrive gas reservoirs, with the actual shape of the plot depending on the size and the permeability of the aquifer.

[15] It is noted here that "rock collapse"—i.e., nonelastic response to increases in net overburden pressure—would result in reduction of initial reservoir pore space, which would result in flattening of the p/Z vs. G_p trend, not downward curvature.

[16] Such a mechanism might seem improbable. However, at reservoir pressures typically encountered in geopressured accumulations, on the order of 10 to 15 scf/STB water can be dissolved in formation water, depending on salinity and actual pressure and temperature. During the late 1970's and early 1980's this resource was the subject of a U.S. federally funded program to determine commerciality.

[17] Fertl and Timko [1972] reported a case of an "overpressured section in the Louisiana Gulf Coast area," where a replacement well was drilled approximately 50 ft away from a well that had produced for about 22 years. The log in the new well measured increased resistivity in the shale section immediately above and below the producing sand, which was attributed to "decrease in shale porosity because some shale water...moved into adjacent...sands."

result, several different empirical methods have been developed to adjust the p/Z vs. G_p plot. As there is no generally accepted methodology, a brief discussion of work to date follows.

Methods Proposed to Date

Both Hammerlindl [1971] and Ramagost and Farshad [1981] developed methods to account for pore volume compressibility and expansion of interstitial water. Ramagost and Farshad [1981] proposed the following equation to adjust the p/Z vs. G_p plot:

$$p/Z\,[1 - \Delta p(c_w S_w + c_p)/(1 - S_w)] = p_i/Z_i - G_p p_i/G_{Fi} Z_i \,. \quad\text{................................ (6.1)}$$

This equation differs from Eq. 4.17 by inclusion of a "p/Z adjustment factor," which is the left-side term in square brackets.

Hammerlindl [1971] proposed adjusting the "apparent gas in place" (AGIP)—that estimated by extrapolation of the initial part of the p/Z vs. G_p plot—by multiplying AGIP by the ratio, (gas compressibility)/(effective compressibility).

Both methods assume pore volume compressibility remains constant over the life of the reservoir being evaluated. In addition, there is no provision in these methods to account for possible water encroachment.

The assumption of constant pore volume compressibility is contrary to work by numerous investigators, which is discussed in Sec. 6.5.5. The notation in Eq. 6.1 is consistent with that of the original authors. As discussed in Sec. 6.5.5, compressibility is a function of net overburden pressure.[18] Thus, the term c_p in Eq. 6.1 should be the integrated value of c_p over the pressure range p_i to p.

Bernard [1987] noted several discrepancies in publications discussing drive mechanisms in apparently volumetric, geopressured reservoirs. He proposed use of a "catch-all" approximation term, C, to account for the effects of rock and water compressibility, a small steady-state-acting aquifer, and proto-shale water influx. Incorporating Bernard's proposed C in Eq. 4.17 would yield

$$p/Z\,[1 - C\Delta p] = p_i/Z_i + G_p p_i/G_{Fi} Z_i \,. \quad\text{..(6.2)}$$

Using performance data from volumetric, geopressured gas reservoirs in the U.S. Gulf Coast in which gas initially in place (GIP) was known, Bernard [1987] calculated the C factor for 13 different reservoirs. Reportedly, C ranged from 1.5 to 46.5 msip[19] and did not correlate with either depth or initial pressure gradient. The C factor was used to calculate a correction to the initial slope of the p/Z vs. G_p plot, so extrapolation of the early part of the plot would intersect the known GIP on the x axis. Bernard's [1987] correction factor ranged from about 0.96 to about 0.45. The factor appeared to correlate reasonably well with the inverse of AGIP, which was determined by extrapolation of the early part of the p/Z vs. G_p plot for each reservoir examined. The correlation, however, is based on only 13 geopressured gas reservoirs in the U.S. Gulf Coast and may not be valid generally.

---

[18] Fatt [1958] defined net overburden pressure as external pressure—0.85 x internal fluid pressure; elsewhere, it is defined as (EXTERNAL PRESSURE) – (INTERNAL FLUID PRESSURE). Irrespective of the definition, if reservoir fluid pressure is reduced by production, net overburden pressure will increase. Thus, if reservoir fluid pressure decreases over reservoir life, pore volume compressibility will decrease, not remain constant.

[19] A microsip (msip) is equal to psi/10^6.

Fetkovich *et al.* [1991][20] proposed an equation of the form

$$p/Z \, [1 - c_e(p)(\Delta p)] = p_i/Z_i + G_p p_i/G_{Fi} Z_i \, , \quad\quad\quad\quad\quad\quad\quad\quad (6.3)$$

where $c_e(p)$ was defined as "pressure-dependent cumulative effective compressibility," which included cumulative pore volume compressibility, $c_p(p)$, cumulative total water compressibility, $c_{tw}(p)$, and the total pore and water volumes in pressure communication with the reservoir, as

$$c_e(p) = \{S_w \, c_{tw}(p) + c_p(p) + M[c_{tw}(p) + c_p(p)]\}/(1 - S_w) \, , \quad\quad\quad\quad (6.4)$$

where M was defined as the ratio of the pore volume of interbedded shale and the nonpay volume, V_{pNP}, plus the ratio of limited aquifer volume, V_A, to the pore volume of reservoir net pay, V_{pR}, as[21]

$$M = (V_{pNP} + V_A)/V_{pR} \, . \quad\quad\quad\quad\quad\quad\quad\quad\quad\quad\quad\quad\quad\quad (6.5)$$

Cumulative pore volume compressibility was defined as

$$c_p(p) = 1/V_{pi}[(V_{pi} - V_p)/(p_i - p)] \, , \quad\quad\quad\quad\quad\quad\quad\quad\quad\quad (6.6)$$

thereby recognizing the pressure dependence of this term, consistent with previously reported laboratory data.

Cumulative total water compressibility was defined as

$$c_{tw}(p) = [B_{tw}(p) - B_{tw}(p_i)]/[B_{tw}(p_i)(p_i - p)] \, . \quad\quad\quad\quad\quad\quad\quad (6.7)$$

In contrast to the Fetkovich *et al.* [1991] method, Yale *et al.* [1993][22] proposed accounting for variable formation compressibility with a term identified as a "pore volume formation factor," defined as

$$B_f = V_p/V_{psc} \, , \quad\quad\quad\quad\quad\quad\quad\quad\quad\quad\quad\quad\quad\quad\quad\quad (6.8)$$

where V_p and V_{psc} were defined as reservoir pore volume at reservoir and standard conditions, respectively. For non-associated gas reservoirs, Yale *et al.* [1993] developed a material balance equation of the form

$$G_{Fi} = \frac{G_p B_{gf} + (W_p - W_e) B_{wf}}{B_{gfi}\{[(B_{gf}/B_{gfi}) - 1] + [(S_w + R_{pa})/(1 - S_w)][B_{wf}/B_{wfi} - 1]\}} \, , \quad\quad (6.9)$$

where $B_{gf} = B_g/B_f$,
$\quad\quad\;\; B_{wf} = B_w/B_f$, and
$\quad\quad\;\; R_{pa}$ = pore volume ratio (potential aquifer/gas zone).

---

[20] The Fetkovich [1991] paper has not been peer-reviewed and published in an SPE journal. Papers in SPE conference proceedings are "subject to correction by the author(s)," so caution should be exercised in using this material.

[21] These special terms do not appear elsewhere in petroleum engineering literature and, accordingly, are not included in Notation.

[22] The Yale *et al.* [1993] paper has not been peer approved and published in an SPE journal. Papers in SPE conference proceedings are "subject to correction by the author(s)," so caution should be exercised in using this material.

The two, square-bracketed, B terms in the denominator of Eq. 6.9 represent the expansion of unit volumes of gas and interstitial water, including its dissolved gas, and the contraction of the associated pore space.

R_{pa} accounts for bottomwater and/or a small contiguous aquifer.

Proposed Method

Methods outlined in previous sections either treat compressibility as a constant or require more information than generally may be available. As an alternative, a method is proposed here that parallels that of Havlena and Odeh [1963]. Thus, Eq. 4.1c may be written for a gas reservoir as

$$G_p B_g + W_p B_w = G_{Fi}(B_g - B_{gi}) + G_{Fi} B_{gi} \Delta p(c_p + c_w S_w)/(1 - S_w) + W_e . \qquad (6.10)$$

Similar to the development in Sec. 4.2, define

$$F_{pR} = G_p B_g + W_p B_w \qquad (6.11)$$

and

$$E_g = B_g - B_{gi} . \qquad (6.12)$$

Substituting Eqs. 6.11 and 6.12 into Eq. 6.10 leads to

$$F_{pR} = G_{Fi} E_g + G_{Fi} B_{gi} \Delta p(c_p + c_w S_w)/(1 - S_w) + W_e$$

$$= G_{Fi}[E_g + B_{gi} \Delta p(c_p + c_w S_w)/(1 - S_w)] + W_e . \qquad (6.13)$$

Substitute the van Everdingen-Hurst [1949] water influx term for W_e, and divide by the gas expansion and rock/fluid compression term in square brackets

$$\frac{F_{pR}}{E_g + B_{gi} \Delta p(c_p + c_w S_w)/(1-S_w)} = G_{Fi} + C \frac{\Sigma \Delta p Q_D(\Delta t_D, r_D)}{E_g + B_{gi} \Delta p(c_p + c_w S_w)/(1-S_w)} . \qquad (6.14)$$

If the water influx term[23] and the rock/fluid expansion/compression terms are estimated correctly, a plot of the left-side term vs. the fractional part of the second right-side term of Eq. 6.14 will be a straight line. The y intercept should be equal to G_{Fi}, the free gas initially in place. The slope of the line should equal C, the water influx constant.[24]

The reader is cautioned that the c_p term in these equations is $c_p(p)$ integrated over a change in net overburden pressure that corresponds to the value $(p_i - p)$. This point is discussed in Sec. 6.5.5.

The water influx term probably will be the most difficult term to evaluate, as water influx in a given reservoir may be attributable to expansion from a contiguous aquifer, to dewatering of interbedded proto-shale, or to both. Conditions favoring proto-shale water influx include:

---

[23] Eq. 6.14 reflects the generalized (cylindrical) form of the van Everdingen-Hurst [1949] equation. In many cases, a linear influx model might be more appropriate.

[24] This procedure is analogous to that discussed in Sec. 4.2.3.

(a) considerable interbedding of proto-shale with the gas-bearing sand, (b) a small contiguous aquifer, and (c) a high initial fluid pressure gradient. Opposite conditions would favor aquifer influx. Depending on the size and shape of the contiguous aquifer, W_e may be calculated using a limited linear aquifer model or a limited cylindrical aquifer model. If proto-shale dewatering is suspected, a limited linear aquifer model may be more appropriate.

Depending on the degree of pressure support from water influx, the procedures discussed for partial waterdrive gas reservoirs in Sec. 3.6.5 also may be used to estimate recovery efficiency from these reservoirs.

Retrograde Gas Accumulations

Geopressured gas reservoirs may exhibit retrograde behavior. When reservoir pressure decreases below the dewpoint pressure, liquids will condense in the reservoir pore space. This will result in a change in the PVT properties of the reservoir hydrocarbon fluid from a single phase to a two-phase system.

If a plot of p/Z vs. G_p is used to estimate gas reserves from retrograde gas reservoirs, the two-phase Z-factor should be used for reservoir pressures below the dewpoint pressure. If expanded material balance calculations are made, the two-phase B_g should be used below the dewpoint pressure. Methods to analyze retrograde gas reservoirs and to estimate two-phase Z-factors are discussed in Sec. 4.4.2 and in Sec. J.1.5, respectively.

Condensation of liquids in the reservoir pore space around the wellbore may cause development of a severe "skin," which may lead to a rapid decline in well productivity.[25] In such cases, it may be extremely difficult to estimate average reservoir abandonment pressure. Any such reserve estimates should be corroborated by decline trend analysis, Sec. 5.3.

6.5.4 Geopressured Oil Reservoirs

Oil reservoirs are encountered less frequently than gas reservoirs in the geopressured section and rarely are discussed in the literature. Comments similar to those for geopressured gas reservoirs are appropriate regarding drive mechanism in geopressured oil reservoirs. With reference to the development of Eq. 4.1c, the expanded material balance may be written

$$N_p[B_t + B_g(R_p - R_{si})] + W_pB_w = N_i[(B_t - B_{ti}) + B_{ti}\Delta p(c_p + c_wS_w)/(1 - S_w)] + W_e . \quad .....(6.15)$$

Most geopressured oil accumulations are undersaturated; thus, the term for gas cap expansion is excluded from Eq. 6.15. As done previously, define

$$F_{pR} = N_p[B_t + B_g(R_p - R_{si})] + W_pB_w . \quad ....(6.16)$$

and

$$E_o = B_t - B_{ti} . \quad ....(6.17)$$

Eq. 6.15 may be restated as

$$F_{pR} = N_i[E_o + B_{ti}\Delta p(c_p + c_wS_w)/(1 - S_w)] + W_e . \quad ....(6.18)$$

---

[25] If reservoir pressure decreases significantly during production, the reservoir rock may be compressed due to the increase in net overburden pressure, as discussed previously. Compression of the reservoir rock may cause a substantial decrease in permeability, which also may be manifested by a severe "skin."

Substitute the van Everdingen-Hurst [1949] water influx term for W_e, and divide by the expansion/compression term in square brackets:

$$\frac{F_{pR}}{E_o + B_{ti}\Delta p(c_p + c_w S_w)/(1 - S_w)} = N_i + C\frac{\Sigma\Delta p Q_D(\Delta t_D, r_D)}{E_o + B_{ti}\Delta p(c_p + c_w S_w)/(1 - S_w)} \quad\ldots\ldots\ldots\ldots(6.19)$$

If the water influx term and the rock/fluid expansion/compression terms are estimated correctly, a plot of the left-side term vs. the fractional part of the second right-side term of Eq. 6.19 will be a straight line. The y intercept should equal N_i (OIP.) The slope should equal C, the water influx constant. Comments in the previous section regarding the water influx term for geopressured gas reservoirs are appropriate for geopressured oil reservoirs discussed here. Evaluation of the c_p term should follow the procedure outlined in connection with Eq. 6.14, as discussed in Sec. 6.5.5.

6.5.5 Pore Volume Compressibility

Background

There have been numerous studies published on the influence of reservoir pressure on pore volume compressibility [Geertsma 1957, Fatt 1958, van der Knaap 1959, Dobrynin 1962, Chierici *et al.* 1967, Von Gonten and Choudhary 1969, Teeuw 1971, Newman 1973, Jones 1988, and Yale *et al.* 1993]. From these studies, it seems apparent that:

- The pore volume compressibility of porous rocks is dependent on the stress conditions in the reservoir.
- Compressibility decreases as the stress increases.
- Compressibility decreases as rocks become more consolidated.
- Compressibility may increase as temperature increases.

There does not appear to be a correlation between compressibility and rock properties that is generally valid across a broad spectrum of lithologies and pressures. Hall's [1953] correlation between compressibility and porosity (still widely cited) covers only a narrow range of stress conditions and apparently reflects only data from well-consolidated rocks. Fatt [1958] reported "pore volume compressibilities...could not be correlated with porosity."

It is intuitive that sandstone texture—e.g., sorting coefficient—might be an important factor. However, its influence on pore volume compressibility has not been quantified in work published to date. For example, Newman [1973] observed that "pore volume compressibilities for a given porosity can vary widely according to rock type...the data are too widely scattered for correlations to be reliable."

Reportedly, some geopressured sandstones have compressibilities substantially lower than those indicated by Hammerlindl's [1971] correlation—i.e., compressibilities approaching those usually associated with consolidated rock [Bernard 1987]. These data, however, apparently were measured on rock samples taken from geopressured aquifers, rather than from hydrocarbon reservoirs. In the high temperatures usually associated with geopressured environments, sandstones undergo rapid diagenesis that can result in a geologically young rock becoming tightly cemented. Such diagenesis is more likely to occur in aquifers, where the interstitial water is mobile, than in hydrocarbon reservoirs, where the interstitial water is immobile. These tightly cemented sandstones can be expected to be less compressible than relatively uncemented sands.

In view of the foregoing observations, it is recommended that compressibility measurements be made on samples taken from the hydrocarbon-bearing zone and not on samples from the aquifer. Extreme caution should be exercised in using compressibility data from rocks that appear to be similar to the zone of interest or rocks that have comparable porosity and permeability.

Previous Publications

In the absence of laboratory data, industry publications may provide guidance regarding the relation between reservoir stress and pore volume compressibility. Several of the papers cited in the previous section are discussed briefly in the ensuing paragraphs.

From limited data reported by Fatt [1958] on eight sandstones with porosity in the range 10 to 15%, it may be observed that, for poorly sorted sandstones, the pore volume compressibility varied from a maximum of approximately 35 msip (at a net overburden pressure of 1,000 psi) to a minimum of approximately 5 msip (at a net overburden pressure of 12,000 psi). For well-sorted sandstones, pore volume compressibility varied from approximately 10 to 2 msip over a comparable range of net overburden pressures. Fatt [1958] suggested the data could be fitted with a relation of the form

$$c_p \approx A - B\ln(p_{obn}) \,, \qquad\qquad\qquad\qquad\qquad\qquad\qquad\qquad\qquad (6.20)$$

where "A" and "B" are constants for each sample.

For *poorly sorted* sandstones the relation between pore volume compressibility and net overburden pressure for Fatt's [1958] data is estimated here to be approximately

$$c_p \approx 100 - 10\ln(p_{obn}) \,, \qquad\qquad\qquad\qquad\qquad\qquad\qquad\qquad\qquad (6.21)$$

where c_p units are microsip ($1/10^6$ psi).

Van der Knaap [1959] reported pore volume compressibility data for a variety of porous media and observed an approximately linear log-log relation between pore volume compressibility and net overburden pressure.

Data reported by Dobrynin [1962] appear to be consistent with those reported by Fatt [1958]. Dobrynin reported that "between a certain minimum (net overburden) pressure and a certain maximum (net overburden) pressure, the relation between pore (volume) compressibility and logarithm of pressure can be approximated by a straight line," or

$$c_p \approx c_{px}\ln[(p_{obn})_x/p_{obn})]/\ln[(p_{obn})_x/(p_{obn})_n] \,, \qquad\qquad\qquad\qquad (6.22)$$

where c_p = pore volume compressibility, 1/psi,
   c_{px} = pore volume compressibility at minimum net overburden pressure, 1/psi,
   $(p_{obn})_x$ = maximum net overburden pressure, psi, and
   $(p_{obn})_n$ = minimum net overburden pressure, psi.

Eq. 6.22 is consistent with one suggested by Fatt [1958]—i.e., Eq. 6.20. Reportedly, the experimental procedures used by Dobrynin [1962] were "practically the same as those described by Fatt [1958]."

Newman's [1973] compressibility data were reported at a common pressure of 75% of initial overburden pressure. For unstressed porosity in the range 20 to 30%, compressibilities for unconsolidated and friable sandstones ranged from approximately 5 to 80 msip. Reportedly[26] [Craft *et al.* 1991], Newman developed the following correlations

Sandstone: $c_p \approx 97.32(10)^{-6}/(1 + 55.8721\phi)^{1.42859}$ ....(6.23)

and

Limestone: $c_p \approx 0.853531/(1 + 2.47664(10)^6\phi)^{0.9299}$ . ...(6.24)

In discussing his work, Newman [1973] observed that "pore-volume compressibility values shown in this study are, in most cases, pressure dependent," and "pore-volume compressibilities for a given porosity can vary widely according to rock type."

Jones [1988] reported fitting pore volume compressibilities with an equation of the form

$$V_p \approx V_{p0}\exp\{a_v[\exp(-p_{Hn}/p_D) - 1]\}/(1 + Cp_{Hn}) , \quad\quad\quad\quad\quad\quad\quad\quad (6.25)$$

where p_{Hn} = net hydrostatic pressure (external pressure minus internal pressure), psi,
$\quad\quad V_p$ = pore volume of sample at p_{Hn},
$\quad\quad V_{p0}$ = pore volume of sample at $p_{Hn} = 0$,
$\quad\quad p_D$ = decay constant, psi, and
$\quad\quad a_v, C$ = constants, dependent on rock properties.

Relations between rock properties and the constants in Eq. 6.25 were not reported.

Yale *et al.* [1993] reported fitting formation compressibility[27] data for a large number of sands and sandstones with an equation of the form

$$c_p \approx A[K_1 p_{ob} - K_2 p_i + K_3(p_i - p) - B]^C + D , \quad\quad\quad\quad\quad\quad\quad\quad (6.26)$$

where p_{ob} = overburden pressure, psi,
$\quad\quad$ A,B,C,D = constants, dependent on rock properties, and
$\quad\quad$ K_1, K_2, K_3 = constants, dependent on rock properties.[28]

The Yale *et al.* [1993] constants for sandstones and carbonates are as follows:

	Sandstones			Carbonates
	Unconsolidated	Friable	Consolidated	
A	(-)2.805E-5	1.054E-4	2.399E-5	NR*
B	300	500	300	NR
C	0.1395	(-)2.2250	0.0623	NR
D	1.183E-4	(-)1.103E-5	4.308E-5	NR
K_1	0.95	0.90	0.85	0.85
K_2	0.95	0.90	0.80	0.85
K_3	0.75	0.60	0.45	0.55

* Not reported

---

[26] The numerical correlations reported by Craft *et al.* [1991] do not appear in Newman's 1973 *JPT* paper. They do, however, appear to reflect the graphical correlations identified as "consolidated sandstones" and "limestones" in Newman's Fig. 7.
[27] In their work, Yale *et al.* [1993] distinguished between pore volume compressibility as measured in the laboratory and formation compressibility as observed in the field.
[28] K_3 was introduced in Eq. 3.12, in Sec. 3.7.2.

Proposed Methodology To Estimate Compressibility

For reservoirs in which pore volume compressibility is expected to be a significant production mechanism, laboratory compressibility data should be obtained. It is recognized, however, there are many situations where such data are not available.

Under these circumstances, the following values may be used to estimate c_{px}—i.e., maximum pore volume compressibility—for substitution* in Eq 6.22:

Rock Texture	C_{px} (msip)
Loosely consolidated to friable, poorly sorted, with 20 to 35% intergranular detritus, for example, a U.S. Gulf Coast "graywacke"	50
Well sorted, well rounded grains with up to about 15 % intergranular detritus, for example, a "quartzite"	10

* At $(p_{obn})_n = 200$ psi and with $(p_{obn})_x = 20,000$ psi

Curves generated using this procedure appear to be comparable to those reported by Yale *et al.* [1993] for unconsolidated and consolidated sands, respectively. Thus, the Yale *et al.* [1993] correlation, Eq. 6.26, might also be used for this same purpose.

Adjustment of Laboratory Compressibility Data

As discussed by van der Knaap [1959], Teeuw [1971], and others, during pressure reduction of reservoir fluids, the resulting stresses on reservoir rocks differ from the stresses imposed on core samples during hydrostatic testing in the laboratory. In the subsurface, as reservoir fluid pressure is reduced by production, the reservoir rock is compacted by the weight of the overburden, which results in uniaxial reduction of the bulk volume of the rock with a consequent reduction pore volume. Although this process can be replicated in the laboratory, such tests require special equipment, not used by most commercial laboratories. Most laboratory compressibility data are measured using hydrostatic stress. An equation relating reservoir and laboratory stress, called the "effective stress equation," has been developed by numerous investigators

$$p_n = K_1 p_{ob} - K_2 p_i + K_3(p_i - p) , \quad\quad\quad\quad\quad\quad\quad\quad\quad\quad\quad\quad\quad\quad\quad (6.27)$$

where p_n = laboratory net (hydrostatic) pressure (confining pressure less pore pressure), psi,

p_{ob} = overburden pressure, generally 1 psi/ft of depth,

K_1, K_2, K_3 = constants determined by rock type, defined in connection with Eq. 6.26.

Eq. 6.27 was used by Yale *et al.* [1993] to develop Eq. 3.12, which relates pore volume compressibility measured under laboratory (hydrostatic) stress to pore volume compressibility under reservoir (uniaxial) stress.

6.6 Partial Waterdrive (Rate-Sensitive) Gas Reservoirs

In 1965, Agarwal *et al.* demonstrated theoretically that, in an isotropic gas reservoir subject to edgewater encroachment, "gas recovery from certain...reservoirs may be very sensitive to gas production rate." Their calculations suggested that low-permeability reservoirs were more sensitive to gas production rate than high-permeability reservoirs. They concluded by stating: "If practical, the field should be produced at as high a rate as possible, and field curtailment should be avoided. This may result in a significant increase in gas reserves by lowering the abandonment pressure." As discussed in the next paragraph, the author recommends against the adoption of such a production policy.

The Agarwal *et al.* [1965] work was based on a mathematically isotropic computer model. Such a model is not representative of subsurface reservoirs, which typically exhibit considerable permeability heterogeneity. Depending on the nature of the heterogeneity, well spacing, and reservoir geometry, a high-rate production policy could lead to inefficient sweep efficiency by the encroaching aquifer, resulting in a reduction in overall recovery efficiency. This point is discussed further in subsequent sections.

As discussed in Sec. 3.7.5, the recovery efficiency of gas from partial waterdrive gas reservoirs may be estimated with Eq. 3.18

$$(E_{Rg})_{wd} = (1 - R_{pZ}) + (R_{pZ}E_V E_D) , \dots\dots\dots\dots\dots\dots\dots\dots\dots\dots\dots\dots\dots\dots\dots\dots\dots\dots(3.18)$$

where E_V is the volumetric sweep efficiency and R_{pZ} is the ratio of abandonment to initial p/Z. From examination of Eq. 3.18, it is apparent that these two terms are of comparable significance in the determination of gas recovery efficiency from partial waterdrive reservoirs.

Since the Agarwal *et al.* publication, it has become increasingly apparent that seemingly minor variations in permeability can have a major influence on volumetric sweep efficiency in reservoirs subject to water influx [Hartmann and Paynter 1979, Cronquist 1984]. Thus, *theoretical* calculations for recovery efficiency of gas of the type used by Agarwal and Ramey [1965] should not be used to establish production policy or to estimate gas reserves from this type of reservoir.

6.7 Thin Oil Columns Between Free Gas and Water

Thin oil columns overlain by free gas and underlain by water pose difficult problems in well spacing and completion method, production policy, and reserve estimation. In this context, "thin" is a relative term. Whether an oil column is considered "thin" depends on costs to drill and produce the accumulation.[29] For example, in the Bream field (Australia Bass Strait, 230 ft water depth), 44 ft was considered "thin" [Sulaiman and Bretherton 1989]. In the Troll field (offshore Norway, 980 ft water depth), 79 ft was considered "thin" [Seines *et al.* 1994]. A pragmatic approach was taken by Irrgang [1994], who defined thin oil columns as those that "will cone both water and gas when produced at commercial rates."

The overall recovery efficiency of oil from such accumulations can be influenced by: (a) well spacing and completion method and (b) gas-cap management policy. Exploitation economics, however, tend to be controlled by individual-well recoveries and production capacities, rather than by average recovery efficiency from the reservoir. Thus, the focus in exploitation planning is the economics of individual wells—i.e., the cost to drill, complete, and operate vs. the oil rate/time profile. Ultimate oil recovery from individual wells tends to be controlled by: (a) the gross thickness of the oil column and (b) the ratio of horizontal to vertical permeability in the well's drainage area. There may be significant variation in this ratio over the areal extent of the reservoir, depending on the depositional environment of the

---

[29] The LeDuc D-3 reservoir (Alberta, Canada), discussed in Sec. 7.2, is another example of a thin oil column.

reservoir rock. From experience with such reservoirs, it is observed that this parameter is typically underestimated, resulting in underestimates of oil recovery.

Based on limited data from conventional well completions in several such fields in Australia, Irrgang [1994] developed the following relation

$$N_{paw} \approx \text{fn}[\phi(1 - S_w - S_{or})k_H h_t{}^{2.5} R_{ng}{}^{1.5}/\mu_o B_o] , \quad \dots\dots\dots\dots\dots\dots\dots\dots\dots\dots\dots\dots\dots\dots\dots \text{(6.28)}$$

where the term in square brackets was identified as a "correlating parameter," and
where N_{paw} = ultimate oil recovery per well,
              h_t = gross oil column thickness,
              k_H = horizontal permeability, and
             R_{ng} = net-to-gross (pay) ratio.

Other notation is SPE [1993] standard. Details regarding estimation of k_H were not provided; it is assumed here that the median permeability would be appropriate. Also, in a private communication Irrgang [1994] observed "a higher power may be appropriate for permeability—possibly even 2." As discussed in Sec. 3.7.2 and Appendix I, the ratio of vertical to horizontal permeability, k_V/k_H, influences volumetric sweep efficiency in bottomwater-drive reservoirs. The absence of this term in Irrgang's [1994] correlation is puzzling, but may be due to measurement difficulties.

Depending on the water/oil mobility ratio and the ratio of horizontal to vertical permeability, oil wells completed in this type of accumulation may exhibit coning of the overlying gas and/or coning of the underlying water early in life. These phenomena may result in rapidly increasing GOR or WOR and relatively short economic life. Thus, efficient exploitation of this type of accumulation depends on the degree to which premature coning of gas and/or water can be avoided by appropriate completion methods and production policy.[30] In one of the earliest published analyses of the problem, Van Lookeren [1965] advocated perforating below the initial oil/water contact to minimize gas coning. The simple isotropic model used in his analysis, however, essentially negates the practical application of this approach. In the last few years, horizontal drainholes[31] have been used to exploit these accumulations [Irrgang 1994, Seines *et al.* 1994, Sognesand 1997]. This technology is still evolving, and readers should consider with caution apparent successes in analogy reservoirs.

Determination of optimum well spacing and estimation of oil reserves in such reservoirs is subject to substantial uncertainty, at least until a reasonably well defined performance trend has been established for each well. Before such performance trends are established, however, reserves typically are estimated using volumetric mapping combined with analogy or analytical methods. In this context, computer simulation can be extremely useful in establishing sensitivity of recovery efficiency to various assumed scenarios, thereby assisting in determining optimum well spacing and commerciality.

Regarding potential analogs, Irrgang [1994] provided the following data for such reservoirs in Australia:

---

[30] Zonal isolation is essential to efficient exploitation of these accumulations, and consideration might be given to routinely "block squeezing" before initiating production.
[31] The term horizontal "drainhole" is here considered a more appropriate term than horizontal "well," used by most authors.

Lithology	Field			
	Bream[a] Sandstone	South Pepper[a] Sandstone	Chookoo[b] Sandstone	Taylor[b] Sandstone
Porosity, fraction	0.21	0.21	0.17	0.08 to 0.16
Net-to gross ratio	0.65	0.96	0.9	0.4 to 1
Permeability, md	2,900	80 to 4,800	500	100 to 2,000
k_v/k_H	0.01	0.01 to 0.2	Not rep	Not rep
Oil viscosity, cp	0.24	0.48	0.21	.26
Oil column, m	13.5	9.7	3.7	0 to 11.3
Gas column, m	105[d]	18	19[d]	33
Well spacing, m[c]	300 to 800	300 to 800	600 to 1,200	700 to 1,500
Oil recovery, 10^6 bbl/well	2.0	1.5	0.26	0.09

a. Offshore.
b. Onshore.
c. Author's estimate.
d. Injected produced gas into gas cap.

Potential analogs in carbonate reservoirs are discussed in Sec. 7.2.

Regarding production policy, several correlations have been developed to estimate critical rates to produce such wells, below which, in theory, coning should not occur [Muskat and Wyckoff 1935, Chaney et al. 1956, Chierici et al. 1964, Bournazel and Jeanson 1971, Richardson and Blackwell 1971, Kuo and DesBrisay 1983, Kuo 1989, and Hoyland et al. 1989].

From a critical review of this literature, which spans an interval of more than 50 years, it is apparent the industry has yet to develop a general treatment of coning that includes the influence of gas cap, aquifer influx, and other relevant parameters. Some authors, for example, have investigated the problem of coning in the presence of an inactive aquifer, analogous to the classic coning problem first discussed by Muskat and Wyckoff [1935]. Others, however, have investigated the problem in the presence of an active aquifer. It seems apparent that, in the presence of an active aquifer, the critical rate to avoid water coning would be less that the critical rate in the presence of an inactive aquifer, other factors being the same.

In addition to aquifer strength, another critical parameter needed to apply the correlations cited above is the ratio of horizontal to vertical permeability over each well's drainage area. Laboratory measurements of vertical and horizontal permeability of small core samples are not adequate to estimate this parameter. In theory, this parameter can be determined by vertical interference testing or vertical pulse testing, as discussed by Earlougher [1977]. However, the test procedure involves two sets of perforations separated by a packer, an expense many operators might be reluctant to incur. Another possible approach is computer simulation of wells exhibiting coning to establish the ratio of horizontal to vertical permeability that results in an acceptable match to observed behavior. Depending on the depositional environment of the reservoir rock and the degree of lateral heterogeneity, however, it is unlikely that results from a few such wells would be applicable to all wells in the reservoir. As a practical matter, it might be easier to test wells at gradually increasing rates to determine a maximum rate at which each well can be produced without coning.

Regarding the ratio of horizontal to vertical permeability, Haldorsen and Lake [1984] developed a statistical method to estimate this parameter in reservoirs containing thin, discontinuous shale zones, which they identified as stochastic shales.

In the presence of a strong aquifer and a gas cap, the combination of water encroachment and gas-cap coning may result in displacement of part of the oil column into the gas cap. Depending on the size of the initial gas cap and the degree to which gas-cap voidage occurs, significant volumes of oil may be lost. In some cases, this loss may be avoided by injection of the produced free gas into the gas cap to maintain constant gas-cap volume.

Assuming acceptable operating practices, recovery efficiency of oil from wells completed in this type of reservoir will depend on:

- well spacing and completion method,
- the ratio of horizontal to vertical permeability,
- the size of the initial gas cap, if any,
- the thickness of the initial oil column,
- the strength of the aquifer, if any, and
- whether or not produced gas is injected into the gas cap to maintain the size constant.

6.8 Very Low-Permeability Gas Reservoirs—"Tight Gas"

As defined by the U.S. Federal Energy Regulatory Commission, low-permeability gas reservoirs are those with an average in-situ permeability of 0.1 md or less.[32] Others have placed the upper limit at 1 md. Such accumulations in the U.S. contain substantial resources. Estimates of these resources and ultimate recovery vary widely and depend chiefly on assumptions of wellhead gas price. For example, based on an assumed wellhead price of $10/Mscf, ultimate recovery from this resource in the U.S. could range from 400 to 550 Tscf [Baker 1981]. Others, using more optimistic assumptions, have placed ultimate recovery as high as 2,600 Tscf [Baker 1981].

Finley [1984], the American Assn. of Petroleum Geologists [1986], and Dutton *et al.* [1993] have provided descriptions of the geologic and engineering characteristics of the major low-permeability gas reservoirs in the U.S. Similar accumulations have been reported elsewhere—e.g., eastern Europe.

With the increasing activity in developing and producing these reservoirs came the realization that methods to estimate gas reserves in moderate- to high-permeability reservoirs are not reliable in very low permeability reservoirs. The problems can be attributed to: (a) the geologic setting in which these reservoirs occur and (b) the completion methods required to make them commercial.

In general, the geologic setting of these reservoirs is characterized by:

- a high degree of permeability heterogeneity,
- lateral discontinuities in apparently blanket sands,
- stratigraphic, rather than structural, traps, and
- complex mineralogy, frequently with high-grain-density minerals randomly dispersed throughout the section, and water-sensitive clays.

These attributes make it very difficult to determine porosity and interstitial water saturation by conventional log and core analysis [Brown *et al.* 1981, Kukal *et al.* 1983, Spencer 1985, Witherbee *et al.* 1983, Lewis and Perrin 1992] These problems, plus the high degree of lateral discontinuity, lead to substantial uncertainties in volumetric estimates of gas

---

[32] Defining and determining the "average," however, is no trivial matter. See, for example, Rollins *et al.* [1992] and Jensen [1992], whose discussions reveal critically different views regarding sampling and analysis of reservoir rocks.

initially in place. In many cases, it is not possible to distinguish between commercial and noncommercial intervals from log analysis alone. Drillstem tests rarely provide any useful information because formations often are damaged during drilling. Massive hydraulic fracturing usually is required to obtain commercial flow rates. However, despite more than 35 years of experience with fracturing technology, the industry is unable to design a treatment and predict the results with a high degree of confidence relying solely on analytical methods [Kazemi 1982, Veatch 1983]. Reliance in most areas typically is placed on analytical models coupled with analogy.

In the U.S. Gulf Coast, such accumulations frequently occur in geopressured sections.[33] [Robinson et al. 1986, Lewis and Perrin 1992]. In this environment, understanding the influence of reservoir stress on rock properties is an important aspect of differentiating between productive and nonproductive formations.

It generally is recognized that, over life, the gas production rate from wells completed in these reservoirs typically exhibits a hyperbolic decline [Stewart 1970, Brown et al. 1981]. As discussed in Sec. 5.3.1, apparent b values generally are greater than 1.

In addition to the so-called conventional decline curve models discussed in Sec. 5.3, empirical log-log rate/time models[34] may provide useful short-term information for such wells—i.e., before the onset of significant pressure depletion. Hale [1986] reported the following equations had been used to estimate flow rates and near-term reserves for damaged fracture flow for more than 2,500 relatively new wells in the U.S. Rocky Mountains:

Flow rate

$$q_g = At^n , \qquad\qquad\qquad\qquad\qquad\qquad\qquad\qquad\qquad (6.29)$$

and

$$q_g = A/(B + t^{0.5}) . \qquad\qquad\qquad\qquad\qquad\qquad\qquad (6.30)$$

Near term reserves:

$$G_p = A^{n+1}/(n + 1) \qquad\qquad\qquad\qquad\qquad\qquad\qquad (6.31)$$

$$G_p = 2ABln[B/(B + t^{0.5})] + 2At^{0.5} , \qquad\qquad\qquad\qquad (6.32)$$

where   A, B = constants,
     n = decline exponent (not the same as b, discussed in Sec. 5.3)
and other terms are as defined in Notation.

In presenting these equations, Hale [1986] noted that for an undamaged fractured well, initial values of n should equal about (-)0.5. Because of damage, however, initial n values as small as (-)0.15 had been observed in the wells studied; the average was (-)0.34. With the onset of depletion, Hale [1986] noted, n will increase to values of (-)1.0 or more.

Eqs. 6.29 and 6.31 may be used to account for the effects of a damaged fracture using the field-observed value of n for each such well. Depending on circumstances, however, Eqs. 6.30 and 6.32 may provide a better fit to the observed data.

---

[33] Geopressured accumulations are discussed in Sec. 6.5.
[34] These models are identified here as "empirical" to distinguish them from Fetkovich's theoretical rate/time models discussed in Sec. 5.3.8.

In an earlier work, Hale [1981] demonstrated that elliptical flow equations provide a good approximation to well behavior in these reservoirs, and the reader is referred to this paper for a complete discussion of that work.

Recovery efficiency of gas from some of these reservoirs was plotted vs. $k_g h$ in Fig. 3.11, which was discussed briefly in Sec. 3.7.5.

Because of the high degree of permeability heterogeneity, drainage areas of individual wells vary widely. In the Green River basin (U.S.), for example, effective drainage areas reportedly have ranged from 100 to 640 acres [Cipolla and Kyte 1992]. Depending on economics, such situations may offer opportunities for significant increase in reserves by infill drilling.

6.9 Pitfalls in Reserve Estimation[35]

There have been significant advances in the industry's ability to characterize subsurface accumulations, to monitor well and reservoir performance, and to forecast future behavior. Despite these advances, however, reserve estimation and classification tends to be very subjective. Significant differences between estimates by independent engineers for the same entity are not uncommon. Some of these differences may be due to honest differences of opinion by equally competent engineers. Other differences, however, may be due to one or more pitfalls[36] into which the unwary engineer has fallen.

The objective of this section is to identify various pitfalls that may trap the unwary engineer and cause significant errors in reserve estimation. The material is organized by stage of maturity (Secs. 6.9.1 through 6.9.3) and by fluid type (Secs. 6.9.4 and 6.9.5). Miscellaneous pitfalls are collected in Sec. 6.9.6.

6.9.1 Analogy Stage

Reserves estimated using analogy methods often are classified as "proved" when "proved plus probable" might be a more appropriate classification. Typically, analogy estimates are based on "best estimates" of performance or "best estimates" of key reservoir parameters. Such estimates usually are based on arithmetic averages. As noted in Secs. 2.3 and 3.4.1, however, many reservoir parameters, including ultimate recovery, tend to exhibit log-normal distributions. As discussed in Sec. 9.2, the arithmetic average of such distributions tends to be biased on the high side, usually significantly greater than the mode (most likely value). The median of such distributions, however, reflects an outcome that has a 50% chance of being exceeded. Depending on the degree of skew of a distribution, the arithmetic average may be approximately equal to, or significantly greater than, the median. To estimate reserves using analogy methods, median values of key parameters should be used, not arithmetic averages.[37]

6.9.2 Volumetric Stage

Porosity

As noted in Sec. 6.5.5, most commercial laboratories measure pore volume compressibility under hydrostatic stress. Such stress conditions, however, do not reflect those in subsurface reservoirs. While the distinction may not be important in consolidated rocks, it can be

---

[35] This section was inspired by "Pitfalls in Oil & Gas Evaluations," a talk given by Harry Gaston Jr. (Ryder Scott Co., Houston) at the 1992 Annual Convention of the Soc. of Petroleum Evaluation Engineers. Some of the material presented here was taken from that talk and is so identified. Other material has been compiled from the author's experience.

[36] Pitfall: "a trap or danger for the unwary."

[37] This comment tacitly assumes that "proved" reserves are defined by the tenth percentile, which corresponds to an expectation of 90%.

significant in unconsolidated rocks. Failure to make appropriate corrections to laboratory compressibility data is a common pitfall.

Net Pay

Use of net pay determined with "mechanical ground rules," rather than net pay determined with geologic and engineering insight, is a common pitfall in reserve estimation. Mechanical ground rules typically are established as part of a unitization formula to ensure uniformity. While such rules may be an unavoidable aspect of unitization agreements, they seldom result in a reliable determination of net pay. Such an instance is discussed in Sec. 7.3.

Volumetric Mapping

In the construction of net pay isopachs, care should be taken in extrapolation beyond areas of subsurface control. In the construction of net porous interval maps, all available data should be used, even that from nearby dry holes.

For example, Gaston[38] discussed a case where a net gas isopach showed a 70 ft contour. This interpretation was based on extrapolation updip from well data within the hydrocarbon accumulation. However, the maximum net sand observed in any of the wells in the vicinity, including dry holes, was only 45 ft.

Reservoirs with sands deposited in complex deltaic-marine sequences typically exhibit significant spatial variations in sand thickness and sand quality. Examples may be found throughout the world, including the U.S. and Nigerian Gulf Coasts, Indonesia, and the North Sea.

Care should be taken in projecting faults downdip from existing well control. A common pitfall is to assume faults have a constant dip. In areas of growth faulting, faults typically "flatten" with depth. As shown by **Fig. 6.2**, structure maps made on horizons at depths where there is no fault control may show an erroneously large productive area updip from existing well control on the mapped horizon.

6.9.3 Performance Stage

Computer Simulation

No knowledgeable reservoir engineer will deny that computer simulation is a powerful tool for the analysis and prediction of well and/or reservoir performance. However, the same engineer, if prudent, will check carefully such simulation to ensure that the results are reasonable.[39] What is "reasonable?" Perhaps brief discussion of a classic "pitfall" will make the point.

Based on computer simulation studies of a gravity-stable miscible flood in a pinnacle reef in Canada (the Golden Spike field), ultimate oil recovery initially was estimated to be 95% OIP. After several years of operation and extensive infill drilling, however, it became apparent that there were permeability barriers to vertical flow that were unsuspected at the time of the initial study. Consequently, the operator revised estimated ultimate recovery to 67% OIP [Stalkup 1983].

Could this outcome reasonably have been anticipated? More to the point, is 95% recovery efficiency a reasonable expectation?

---

[38] Op. Cit.

[39] Carlson [1997] has provided an excellent summary of procedures whereby one may check the reasonableness of a simulation study.

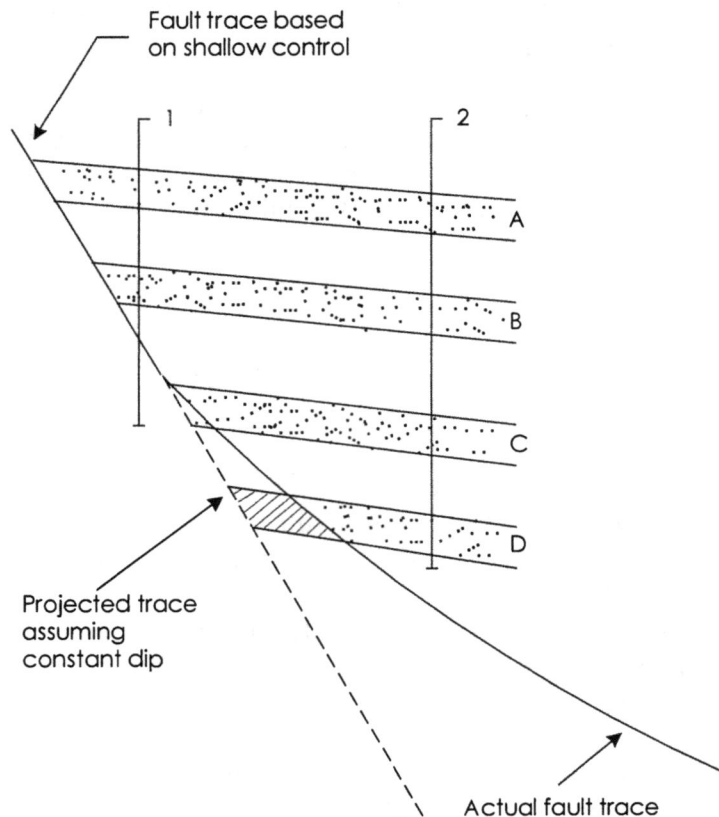

Fig. 6.2—Potential volumetric mapping error (Zone D) caused by erroneous projection of fault.

Material Balance

Material balance estimates of OIP early in the life of a waterdrive reservoir typically assume an infinite aquifer in the absence of data to the contrary. Usually, such an assumption will yield (initially) acceptable results. However, with continued production, the reservoir pressure "sink" will advance progressively farther into the aquifer. Depending on the relative volume of the aquifer compared to that of the oil reservoir, the pressure disturbance may reach a boundary before the reservoir reaches the economic limit. Once a boundary is reached, reservoir pressure behavior will change from transient to semisteady state. Under these conditions, the (initially assumed) infinite aquifer model will result in calculated water influx greater than actual; calculated reservoir pressure will be greater than actual. Thus, it is important to continue monitoring bottomhole pressue (BHP) behavior and periodically update material balance estimates. Failure to do so is a common pitfall.

Many years ago, such a pitfall became apparent regarding the behavior of a major oilfield in the south Louisiana Miocene trend. The major reservoir in this field, Field "A," produced for many years with (apparent) infinite aquifer behavior. After some 20 years of production, static BHP had decreased only slightly below initial BHP. As a cost saving measure, it was decided to discontinue routine BHP surveillance. Such a practice was normal for the area for mature reservoirs because most of wells exhibited high WOR's and gas lift was required to maintain production. Routinely shutting in such wells to run BHP gauges was not viewed favorably by cost (and safety) conscious production superintendents with good reason.

As WOR's in this reservoir continued to increase, production engineers attempted to continue increasing total fluid (oil plus water) production to maintain oil production rates. However, such increases became progressively more difficult to achieve. Thus, it gradually became apparent there was a "problem." Rates of total fluid production became difficult to maintain, much less increase. In attempting to determine the causes, BHP gauges were run in key wells. It soon became apparent that static BHP had decreased substantially during the years when routine BHP surveillance had been discontinued. The reduced reservoir pressure had resulted in lower standing fluid levels, resulting in less efficient gas lift.

About the same time this condition was observed in Field "A," the operator purchased a small field, Field "B," in the same major fault block and about 20 miles away from Field "A," that produced from the same Miocene sand. The purchase had been motivated by the observation that the former operator's production operations had not been efficient, which had resulted in inefficient recovery of the oil. In the course of appraising Field "B," it soon became apparent that the BHP in the objective reservoir was subnormal, and to a degree not commensurate with the low cumulative production from the field. Oddly enough, the BHP observed in the objective sands in Field "B" was in good agreement with that observed in the same sands in Field "A." The conclusion was inescapable; sustained high-volume production of total fluids from Field "A" had resulted in sufficient cumulative aquifer voidage to affect reservoir pressure in Field "B."

Performance/Decline Trend Analysis

There are many pitfalls for the unwary who use performance/decline trend analysis to estimate reserves. Two pitfalls are discussed here. The first involves analysis of production decline trends to estimate reserves for wells still in the transient stage. The second involves extrapolation of performance indicators to unrealistic economic limit conditions.

Transient flow from initial well completions is characterized by an expanding drainage radius and (initially) steep, hyperbolic declines[40] in production rate. Estimates of reserves during this stage of production that are based solely on decline trend analysis usually will be too low. The problem is especially severe in shallow, low-permeability gas reservoirs, where wells may exhibit transient flow over a substantial part of their productive life [Russell *et al.* 1966].

As discussed in Sec. 5.2.1, semilog plots of f_o vs. N_p are commonly used to estimate reserves for oil wells in waterdrive reservoirs. Such plots frequently are used before there is a clearly defined decline in oil production rate, which may be a pitfall for the unwary engineer. Plots of f_o vs. N_p may be extrapolated to a local average "f_o cutoff," which may be too low for wells with low productivity. For example, in many areas the average "f_o cutoff" is 0.01. If, however, a well is capable of producing only 250 B/D (oil plus water) and the economic limit is 5 B/D, then the "f_o cutoff" for this well should be 0.02, not 0.01.

6.9.4 Gas Reservoirs

Initial Reservoir Pressure

Accurate measurement of initial formation pressure is vitally important in all significant gas reservoirs, especially those that are geopressured. While accurate determination of initial

---

[40] Hyperbolic declines are discussed in Sec. 5.3.2.

pressure may not be considered too important for initial volumetric mapping of GIP, it is very important for analysis of reservoir drive mechanism.

As discussed in Sec. 7.6, initial reservoir pressure in the Turtle Bayou "AA" Sand was not measured. It was estimated. This pitfall contributed (initially) to erroneous conclusions regarding drive mechanism and (later) to erroneous estimates of GIP.

Spacing Units and Drainage Areas

In most jurisdictions in North America, gas wells are assigned statutory spacing units. In new gas reservoirs in some areas (for example, Alberta, Canada and western U.S.), wells typically are assigned spacing units of 640 acres [259 ha]. Not infrequently, an unwary engineer will (initially) assume that well drainage area equals the area of the spacing unit. This pitfall became apparent to a major U.S. operator when a "look-back" analysis involving plots of p/Z vs. G_p was made of the performance of 44 wells in the Anadarko basin (Oklahoma). From the analysis, it was concluded that in most cases the apparent drainage volumes were significantly less than the initially assigned 640-acre drainage area.

Shrinkage and Lease Usage

In many fields producing natural gas for commercial sale, the gas must be processed prior to sale by removal of noxious compounds—e.g., hydrogen sulfide—and by removal of noncombustible compounds—e.g., nitrogen and carbon dioxide. Also, depending on the sales contract, hydrocarbon vapors and entrained liquids may have to be removed. This processing results in sales gas[41] volumes being less than wellhead gas[42] volumes, and this shrinkage must be accounted for in estimating salable gas—i.e., reserves.

Also, some of the produced, processed gas may be used to operate field equipment—e.g., to operate compressors, pumps and/or dehydration equipment.

Depending on circumstances, volumes attributed to processing and field usage may account for a significant fraction of the total wellhead gas produced. Depending on the manner in which gas production is measured and reported, it may be difficult to accurately account for these volumes.[43] Failure to account for such volumes, however, may lead to significant errors in reserve estimation.

Regarding shrinkage of wellhead gas attributed to removal of (only) condensate, Gaston[44] provided the following rules of thumb:

CGR (STBC/MMscf)	Adjustment (residue/raw)
50	0.95
100	0.91
150	0.88
200	0.84

Geopressured Reservoirs

A common pitfall is extrapolating a plot of p/Z vs. G_p to estimate GIP and reserves, without checking the reasonableness of such an estimate by using volumetric mapping and/or

---

[41] Called "residue gas" in some areas.

[42] Called "raw gas" in some areas.

[43] Well tests, typically taken at the first-stage separator, may not provide a good measure of the producing condensate/gas ratio (CGR), especially for wells producing rich gas. Historical monthly volumes of sales gas and lease condensate production might be a better measure of producing CGR than well tests. Care should be taken, however, to discount abnormal fluctuations that might be caused by operational problems in the field.

[44] Op. Cit.

analogous wells in the area.[45] As discussed in Sec. 6.5, geopressured gas reservoirs may exhibit a variety of production mechanisms, ranging from pressure depletion to strong waterdrive. In many cases, the geologic setting may not provide any clues as to whether waterdrive is a reasonable assumption.

The author is familiar with a single-well, geopressured gas accumulation in south Louisiana where offsets on four sides did not encounter any sand in the objective horizon. The well, however, flowed gas at high rates for several years with negligible pressure drop, then watered out normally. When shut in after reaching the economic limit, the well had a full column of saltwater with several thousand psi shut-in tubing pressure. Inferred BHP was very close to initial. Where did the water come from? Certainly not from an aquifer because there was none! It seems apparent that this well produced with strong (proto) shale water influx.

6.9.5 Oil Reservoirs

Initial Formation Volume Factor

Two types of pitfalls may be encountered: (a) using a formation volume factor (FVF) estimated from empirical correlations without checking the estimated bubblepoint pressure against reservoir pressure and (b) using GOR's and/or FVF's from laboratory data without adjusting for (field) separator conditions.

As noted in Sec. 3.3.2, an FVF estimated from an empirical correlation should be validated by using the companion correlation to determine if the estimated bubblepoint pressure is equal to or less than observed reservoir pressure. For example, in a case audited by the author, an FVF of 3.25 RB/STB had been used to estimate OIP. It was determined that, in this case, the in-house software available to the project engineer provided a correlation for the FVF, but not for the bubblepoint pressure. The test data used to estimate the FVF was used by the author in several correlations to estimate the bubblepoint pressure. In every case, the estimated bubblepoint pressure was significantly greater than the observed reservoir pressure. Thus, it was apparent that the test GOR used by the project engineer reflected the presence of free gas and could not be used to estimate FVF.

FVF's determined from differential liberation should always be adjusted to reflect field separator conditions. While failure to do so for low-GOR, black oil may not result in serious error, failure to do so for high-GOR volatile oils will result in substantial error.

For example, the following data were reported by a commercial laboratory for a volatile oil in Mississippi:

	Laboratory Procedure				
	Differential Liberation	0 psi Flash	Two-Stage Flash*		
			50-ST	200-ST	400-ST
R_{sb}, scf/STB	2,395	1,224	1,069	1,046	1,073
B_{ob}, RB/STB	3.129	2.041	1.898	1.885	1.900
γ_{API}, °API	42.7	50.3	52.6	52.8	52.4

* Notation "50-ST" indicates flash from bubblepoint pressure to 50 psi separator to stock tank, etc.

For reservoir engineering calculations using these data, the solution GOR and FVF should be adjusted to reflect actual or anticipated field separator conditions. For two-stage flash, the

---

[45] Op. Cit.

laboratory data indicate optimum separator pressure of about 200 psi. If, however, an unwary engineer used the differential FVF to calculate OIP, this pitfall would lead to an underestimate of about 40%.

6.9.6 Miscellaneous

Interference and Drainage

BHP interference between two wells does not necessarily mean that the interwell volume can be drained efficiently at commercial rates. The Spraberry field, discussed in Sec. 7.5, provides a classic example. Interference was demonstrated between wells on 160-acre spacing. Accordingly, it was believed such spacing would be adequate for efficient drainage of this accumulation. In subsequent years, however, it became apparent that spacing on the order of 80 acres per well (less in some areas of the field) is necessary for efficient drainage at commercial rates.

CHAPTER 7

FIELD EXAMPLES

This chapter presents brief discussions regarding geologic and/or engineering factors that have led to uncertainties in reserve estimates. References cited provide additional information.[1]

7.1 Auk (Oil) Field, North Sea

The Auk field, discovered by Shell U.K. E&P in 1971 and placed on production in 1975, is not a typical North Sea field [Buchanan and Hoogteyling 1979]. Subsequent to exploration and delineation, the field was exploited with a total of 11 wells. The main reservoir is the Zechstein dolomite, which averages 28 ft thick in the field area. The history of the field provides an example of how geologic and reservoir uncertainty can lead to large variations in reserve estimates.

Before the decision to develop this field, the operator drilled four delineation wells, including the discovery well. The Rotliegende formation, which was the main objective, had been found to be water bearing over most of the field. The only apparently commercial reservoir was the highly permeable Zechstein dolomite. This thin zone, however, could not be distinguished on the seismic sections, and the areal extent of the accumulation was in doubt [Brennand and van Veen 1975].

The operator estimated Zechstein oil reserves to be in the range of 30 to 100 million STB, with an expectation of 60 million STB. On the basis of available geologic information, which indicated the Zechstein to be limited in size, the operator did not expect a significant waterdrive. Because the oil was highly undersaturated at initial conditions, water injection was planned from the start.

A production platform was installed in 1972, and production was initiated in December 1975 from one of the development wells. From observation of this well's performance, it soon became apparent that reservoir pressure was not declining as rapidly as anticipated. Initiation of water injection was postponed pending further analysis. Buchanan and Hoogteyling [1979] observed that "it was not apparent whether the small pressure drop was caused by a larger oil accumulation, a strong natural waterdrive, or a combination of the two."

One of the wells, near a fault bounding the west side of the reservoir, began producing significant water only 6 months after being placed on production. It seemed apparent that the reservoir was in communication with a strong aquifer. The operator conducted extensive

---

[1] As a "third party" analyst, the author is not privy to the most current information developed by the operators of these fields, which might warrant revision to the material presented here. Any look-back analysis, like those presented here, must be considered a snapshot in time.

computer simulation studies to identify the source of the water. This work led to the hypothesis that the Zechstein was in communication with the underlying, waterbearing Rotliegende dune sands through a fault bounding the west side of the reservoir. The hypothesis was confirmed by water breakthrough in a second well near the west-bounding fault.

Still another surprise confronted the operator. The highest well on the structure began producing significant water shortly after it was placed on production, an occurrence not predicted by the computer simulation model. After extensive pulse testing, the operator concluded that there was a high-permeability channel linking the upstructure well to the water encroaching into the downstructure areas of the reservoir.

By Spring 1978, the operator had drilled eight development wells. Five of these wells were Zechstein producers, two were Rotliegende completions; the other was a dry hole. Buchanan and Hoogteyling [1979] noted that "The estimated ultimate recovery from these (Zechstein) wells is about 45 million barrels...thus, the initial reserve estimate of 60 million barrels in the Zechstein...can be met only if the northern or southern extensions are proved." The authors were referring to possible extensions to the Zechstein that had been suggested by a new seismic survey.

In reviewing the history of this field through 1978, the authors observed "every well yielded some surprise...plans are firm only until the next well result or latest production performance."

By 1988, the operator had made three additional completions in a small Rotliegende reservoir on the southeast side of the field, sidetracking from wellbores no longer needed for the initial completions. At that time, estimated ultimate recovery was estimated to be about 93 million STB [Trewin and Bramwell 1991], which approached the discovery high estimate of 100 million STB.

In reviewing reservoir performance through 1988, Trewin and Bramwell [1991] noted that "accurate reserve estimates are difficult to achieve...the Zechstein has produced more oil in parts of the field than the estimated STOIIP."[2]

An additional surprise seems apparent from the U.K. Dept. of Trade and Industry "Brown Book" [1996], which reported an apparently revised estimated initial ultimate recovery of 17.1 million tonnes (128 million STB). This reflects a significant increase from the operator's 1988 estimate and is more than twice the initial expectation. The increase apparently is attributed to several factors, including: (a) exploitation with horizontally completed wells, (b) infill drilling of the main field area, and (c) development of Auk North by extended-reach, multilateral drilling from the main platform [Follows 1997, Frazer 1998].

7.2 Leduc D-3 (Oil) Pool, Alberta

The Leduc D-3 pool, discovered in 1947 by Imperial Oil Co., is the largest reservoir in the Leduc field and one of the largest fields in Canada [Horsfield 1958]. The reservoir provides an example of how heterogeneities—thin low-permeability zones—were favorable to reservoir performance and how uncertainties in porosity led to continued upward revisions in estimates of OIP.

Covering an area of about 22,000 acres, the reservoir initially contained an oil column about 38 ft thick overlain by a gas cap up to 158 ft thick. The ratio of initial gas-cap volume

---

[2] This comment appears at odds with the authors' estimated ultimate recovery efficiency from the field of only 18% of OIP, shown in their Auk field data summary.

to initial oil-column volume, *m*, was approximately 1.4. Development was on 40-acre spacing. Gravity of stock-tank oil is 39°API. The oil column is completely underlain by a regional aquifer that is in pressure communication with other fields in the D-3 trend. The reservoir rock has been described as a "biohermal" carbonate. Horsfield [1958] reported that "The porosity consists mainly of vugs interconnected through the dense matrix by small crevices and fractures."

Reportedly, the vugular nature of the reservoir rock led to difficulties in recovering and analyzing cores, which led to the development of special techniques. The initial estimate of OIP, 307 million STB, was based on an average porosity of 8% and an average water saturation of 15% [Horsfield 1958]. However, this estimate was revised upward to 354 million STB "because of a higher estimated porosity based on...performance" [Wellings 1975]. This estimate was revised again in 1983 to 386 million STB, which reportedly was based on 10% porosity, 14% average water saturation, and a slightly larger initial formation volume factor than that used in the 1958 estimate [Energy Resource Conservation Board[3] (ERCB) 1984].

The reservoir has produced by a combination of gas-cap expansion and waterdrive [Horsfield 1958]. The aquifer has been supplemented by water injection into a single well since 1955. It is remarkable that this well has been able to inject up to 65,000 BWPD with "no deterioration in injection capacity" [Horsfield 1958]. Supplemental gas injection into the gas cap was initiated during 1961.

In 1975, estimated ultimate recovery was 70% of OIP [Wellings 1975]. In 1985, however, this estimate was revised to 65% of OIP [ERCB 1987]. Through 1992, cumulative oil recovery was 64.5% of OIP. This unusually high recovery efficiency from a relatively thin oil column between a gas cap and bottomwater is attributed mainly to a very high ratio of horizontal to vertical permeability [Wellings 1975]. Well-completion practices apparently contributed to this unusually high recovery efficiency. Before 1970, more than 90% of the wells were completed in a common 5-ft interval about 20 ft above the initial oil/water contact (OWC). Subsequently, recompletions were made over a common 2-ft interval about 23 ft above the initial OWC; later recompletions were made in a common 1-ft interval.

At the time of the last published study of this field [Wellings 1975], the oil column had been reduced to about 8 ft by a combination of gas cap expansion and aquifer encroachment. Individual wells, however, were still capable of producing up to 150 BOPD. From careful monitoring of the advance of the gas cap and the aquifer, a "flushing efficiency" of 80% was estimated [Wellings 1975].

Recovery efficiency from the Leduc D-3 Pool is compared with that from another D-3 Pool, in the Redwater field, which is in the same trend, about 50 miles northeast of Leduc. The Redwater D-3 Pool also produces with a strong bottomwater drive but did not have an initial gas cap, as did the Leduc D-3 Pool [Willmon 1967]. The Redwater D-3 also was developed on 40-acre spacing. Average initial oil-column thickness at Redwater, however, was 101 ft, or approximately 2.6 times that at Leduc. Estimated ultimate recovery from the Redwater D-3 is 62% of OIP [ERCB 1987]. Through 1992, cumulative oil recovery was 61.5% of OIP.

The following table compares properties of the two reservoirs and estimated ultimate recovery [Horsfield 1958, Willmon 1967, ERCB 1987]:

---

[3] In 1995 the Alberta ERCB was merged with other agencies and became part of the Alberta Energy and Utilities Board (EUB).

Pool	Porosity (fr)	Permeability (md)	Oil Viscosity (cp)	Ultimate Recovery (fr OIP)
Leduc	0.100	1000	0.78	0.65
Redwater	0.065	500	2.7	0.62

7.3 Little Creek (Oil) Field, Mississippi

The Little Creek field is in Mississippi (southeastern U.S.). The field was the location of a highly successful peripheral (linedrive) waterflood between 1962 and 1978. The Little Creek waterflood is an example of the high volumetric sweep efficiency possible in a relatively uniform sand. There were, however, difficulties in evaluating the reservoir because of diagenetic clay lining the sandstone pore walls.

Discovered by Shell Oil Co. in 1958, the Little Creek field produced from a complex of remarkably uniform Lower Tuscaloosa (Cretaceous) fluvial point bar sandstones, which formed a stratigraphic trap, draped across a low-dip, structural nose. Occurring at a depth of approximately 10,750 ft, the reservoir was developed on 40-acre spacing. Initial productive area was about 6,300 acres; average net pay was approximately 29 ft. The reservoir oil was highly undersaturated at initial conditions; bubblepoint pressure was 2,050 psi compared with initial reservoir pressure of 4,850 psi. Stock-tank oil gravity of was 39°API [Cronquist 1968].

The Lower Tuscaloosa sands in this area typically contain diagenetic clay coating the pore walls. Because of this clay coating, the sands contain high irreducible water saturation; 50 to 60% is not uncommon. In addition, the clay coating causes problems in obtaining representative porosity and residual oil saturation data; special core-handling techniques must be used.

In 1960 the reservoir was almost completely defined by drilling. Using volumetric methods, OIP was estimated to be 112 million STB. It was anticipated the reservoir would produce by expansion of the undersaturated reservoir oil and solution-gas drive; primary recovery efficiency was estimated to be 22.5% of OIP.[4]

It was recognized that fluid injection would be necessary to supplement natural reservoir energy to maintain reasonable production rates and improve ultimate recovery. There were the "traditional" concerns about not being able to develop an oil bank by waterflooding sands with high interstitial water saturation. Thus, attention first was directed to gas injection.

It was estimated that reinjection of 80% of produced gas would increase ultimate recovery efficiency to approximately 31%. This estimated recovery efficiency was lower than the 35% recovery efficiency observed in another gas-injection project in a Lower Tuscaloosa reservoir in the area. The lower recovery efficiency calculated for Little Creek was attributed to less favorable gas/oil relative permeability characteristics measured in the Little Creek cores.[5]

Determination of waterflood potential was hampered by conflicting laboratory measurements of residual oil saturation and questions about well productivity. For example, laboratory data indicated waterflood residual oil saturations ranging from 6 to 25% of pore volume. Also, it was known that production of water at a nearby Lower Tuscaloosa reservoir had caused losses in well productivity of as much as 60%.

---

[4] This estimate is consistent with the theoretical calculations shown in Table 3.1.
[5] Bruist, E.H.: *Economic Analysis of Gas Injection, Little Creek Field, Mississippi*, Shell Oil Co., New Orleans [1958].

The major causes for conflicting data on waterflood residual oil saturations were wettability changes that occurred during core handling; the problem was attributed to the clay coating on the pore walls. Additional cores were taken under carefully controlled conditions, and waterflood susceptibility tests were run to establish floodability. As a result of this work, it was determined that waterflooding was feasible; ultimate recovery efficiency was estimated at 57% of OIP, or 63.9 million STB.

Reservoir performance under waterflood was excellent. Fill-up was achieved within less than 1 year after flood operations started. Flood fronts advanced uniformly across the field at rates ranging from 0.7 to 12 ft/D [Cronquist 1968].

After flood operations had been under surveillance for several years, revised, detailed mapping was undertaken. The initial estimate of OIP was revised downward from 112 to 102 million STB. Also, the initial estimate of ultimate recovery efficiency was revised downward, from 57 to 47% of OIP, or 47 million STB. The following factors led to these revisions:[6]

- Net pay determination in the initial estimate was based on "mechanical ground rules" established to determine participation factors for individual tracts, which did not permit realistic interpretation and led to an unrealistically high estimate of total net sand volume.

- Porosity used in the initial estimate was corrected downward from 0.244 to 0.234 to account for compaction.

- Conformance efficiency was revised downward from 81 to 75% on the basis of waterflood results elsewhere in similar depositional environments.

Later in the life of the flood, however, it became apparent that overall volumetric sweep efficiency was substantially greater than anticipated, apparently exceeding 90%. To account for the observed flood-out performance, however, residual oil saturation had to be higher than the average of 15% used in the 1963 calculations. On the basis of new laboratory data, average residual oil saturation was estimated to be 21%, which was consistent with the higher volumetric sweep efficiency [Cronquist 1968]. The net effect of these revisions resulted in estimated ultimate recovery of 46 million STB.

Actual (1978) ultimate recovery from the Little Creek field was 46.8 million STB; the detailed work done by the operator in 1963 and 1966 proved to be remarkably accurate. The actual volumetric sweep efficiency was calculated to be more than 90% [Cronquist 1968].

Of interest is a 31-acre tertiary miscible pilot with high-pressure carbon dioxide, which was conducted by Shell during 1974–1977. The pilot area, one-quarter of an inverted ninespot, was confined on two sides by the sand/shale pinchout and on the other two sides by five water-injection wells. Approximately 3.4 Bscf of CO_2 was injected into a single well; 124,000 bbl of oil was recovered from three wells in the pilot area. Reportedly, "A significant volume of previously immobile oil, trapped during the waterflood...of this field (was) displaced from the pilot area." [Thurber 1978]. Appendix F provides additional information on tertiary miscible flooding in this field.

---

[6] Brock, J.A., Werren, E.G., and Mardick, M.L.: *Review. Performance Prediction and Plan of Operation, Little Creek Waterflood Unit, Lincoln and Pike Counties, Mississippi*, Shell Oil Co. New Orleans [1963].

7.4 Palm Valley (Gas) Field, Australia

Discovered in 1965 in the Amadeus basin of central Australia, the Palm Valley (gas) field provides an example of the difficulties in estimating reserves (and deliverability) for a heterogeneous, fractured (sandstone) reservoir with a low-permeability matrix.

Described as a combination structural-stratigraphic trap, the field occupies an arcuate anticline, about 20 miles long. Because of very rugged topography, seismic coverage is sparse, and the structure is not well defined along the western and eastern flanks. The gas column—trapped in fractured, low-permeability Ordovician sandstones—grosses 1,600 ft. Core porosities range from 2 to 4%; matrix permeability averages 0.1 md. [Do Rozario 1990].

Sustained production—delayed because of lack of a market and uncertainties about long term deliverability—began in 1983. Cumulative production through 1995 was 79 Bscf, or approximately 8% of the operator's "minimum" estimate of gas initially in place (GIP).

The history of published reserve estimates, summarized below, is discussed briefly in the ensuing paragraphs.

Year	Wells	GIP (Bscf)	Ult. Rec. (Bscf)	Type/Bases for Estimates	Reference
1970	2	3000	Not rptd.	Volumetric	Do Rozario [1990]
1976	3	48	38	Geologic mapping and transient pressure analysis	Strobel *et al.* [1976]
1985	3	305[a] 896[c]	224[b]	Computer simulation	Sabet and Franks [1985]
1990	7	845[d] 912[e] 158[f]	680[d] 734[e] 127[f]	Computer simulation of 88 months field pressure-production history plus two interference tests	Aguilera *et al.* [1990]
1995	10	Not rptd. Not rptd Not rptd	295[d] 59[e] 419[f]	Computer simulation of data available through September 1995	Magellan press release [15 March 96]

a. Geometric mean of 262 and 356 Bscf; classified here as proved.

b. Estimated cumulative production after 20 years. Authors did not report estimate of ultimate recovery. Do Rozario reported initial proved reserves were estimated at 325 Bscf.

c. Geometric mean of 726 and 1106 Bscf; classified here as proved plus probable.

d. Proved.

e. Incremental probable.

f. Incremental possible.

The 1970 estimate of GIP in the above table is the geometric mean of 1,000 and 9,000 Bscf, which is the range of the volumetric estimates reported by Do Rozario [1990]. Estimated productive area ranged from 78 to 121 sq miles; other parameters used in these estimates were not reported. Estimates of GIP made in 1974, after a third well had been drilled, ranged from 5,000 to 10,000 Bscf [Do Rozario 1990].

The 1976 estimate of GIP was based on transient-pressure analysis (TPA) and geologic mapping. From TPA, effective gas porosity, $\phi_g(1 - S_{wi})$, was estimated to be 0.0022 of the gross fractured interval, which was estimated to be 75 ft. From geologic mapping, the productive area was estimated to be 54 square miles (34,560 acres). Thus, the volumetric factors that would have been used in Eqn. 3.3a (Sec. 3.2.2) are

$$G_{Fi} = 43,560 \text{ (cf/acre-ft)} \times 0.0022 \times 34,560 \text{ acres} \times 75 \text{ ft/ } 0.0052 \text{ (rcf/scf)}$$
$$= 48 \text{ Bscf.}$$

Ultimate recovery was based on an assumed 80% recovery efficiency.

In reviewing the 1976 work, Sabet and Franks [1985] observed that "their [Strobel et al. 1976] model matched the pressure observed in one well and failed to match the pressure observed at the second, more distant, well."

The 1985 estimates of GIP were based on computer simulation with essentially the same permeability, thickness, and compressibility as the Strobel et al. [1976] work, except that the estimated productive area had to be increased to 228,000 acres to match the pressure observed in both wells. From their studies, Sabet and Franks [1985] reported that there was no basis for the downdip limit used in the Strobel et al. [1976] work and concluded that the "reservoir contains between 262 and 356 Bscf GIP in the fracture system and the higher-permeability matrix rock...it is quite possible total GIP ranges between 726 and 1,106 Bscf."

The 1990 estimate of GIP was based on computer simulation with a dual-porosity model to history match (a) field pressure/production history from 1983 through 1990 and (b) two interference tests, one conducted in 1974 and the other in 1986. Ultimate recovery— classified as proved initial reserves by the authors—was based on an assumed abandonment pressure of 600 psi, which is 20% of the initial reservoir pressure.

In the author's review of the 1990 estimates of incremental probable and incremental possible reserves, it is noted that the upside potential reflected by these classifications included incremental GIP but not incremental improvements in recovery efficiency.

Late in 1993, it was reported that during approximately 40 months of sustained production the performance coefficients of four wells had undergone a continuous decline [Aguilera et al. 1993]. The performance coefficients—equivalent to the U.S. Bureau of Mines (USBM) 'C" factor—had been determined by conventional isochronal testing. The authors attributed the decline to possible permeability reductions caused by compaction of the fracture system as reservoir pressure was reduced. The Aguilera et al. [1993] work cast doubt on Sabet and Franks' [1985] work, where it was estimated that field deliverability (three wells) could be sustained at about 32 MMscf/D for 20 years. In this context, it is noted that production during 1995 averaged only 19.7 MMscf/D from six wells.

Aguilera et al. [1993] did not publish details of the magnitude of the decrease. However, from graphs included in their paper for the four wells reported, the "C" factors after about 50 months of sustained production are estimated here to be from approximately 25 to 70% of initial. It is suggested here that—in addition to compaction of the fracture system—an additional factor contributing to the decline might have been expansion of each well's drainage volume deeper into the very-low-permeability matrix.

The 1995 estimate was commissioned after increasing concerns about long-term deliverability. Reportedly, the reserves were "determined entirely by simulation." Other details of the study were not released.

In reviewing the history of estimates since this field was discovered, it is noted that estimates of GIP varied by a factor of more than 50:1; proved reserves, by a factor of almost 20:1. These large disparities can be attributed, to a large degree, to a lack of subsurface control. The lack of significant production is a contributing factor. In this context, it is noted that only about 8% of the operator's estimated GIP had been produced after more than 10 years of sales. Given the complexities of this field and the relatively immature stage of depletion, it is expected that, with additional drilling and production maturity, additional reserve revisions will occur. As of September 1995, a major international consulting firm estimated proved ultimate reserves of 172 BScf—approximately 60% of the 1995 estimate released by Magellan [1995].

7.5 Spraberry (Oil) Field, West Texas

The Spraberry field provides an example of how poor-quality reservoir rock has been detrimental to the recovery efficiency of oil, despite continuing improvements in exploitation technology. Also, the field history demonstrates that pressure communication between widely spaced wells does not necessarily mean that the interwell area can be effectively drained at commercial rates.

Discovered in 1947, the Spraberry field has been described as "the world's largest uneconomic oilfield" [Handford 1981]. OIP has been estimated at 9.4 billion STB [Galloway *et al.* 1983]. The field covers over 700,000 acres; more than 8,300 productive wells have been drilled.[7] Waterflooding operations have been conducted for more than 25 years with a variety of innovative techniques [Byars 1970, Elkins *et al.* 1968, Guidroz 1967]. Through 1992, cumulative production was 709 million STB, which is about 7.5% of the Galloway *et al.* [1983] estimated OIP. Ultimate recovery has been estimated to be approximately 814 million STB [*Oil and Gas J.* 1993], or about 8.6% of the Galloway *et al.* [1983] estimated OIP.[8] As discussed in subsequent paragraphs, continued infill drilling and improved hydraulic fracturing techniques may increase this slightly.

The field produces from the Spraberry and Dean formations (Permian) which comprise a 1,000-ft-thick section of sandstones, siltstones, shales, and limestones; the top of the interval is from 6,300 to 7,200 ft deep. The structure is predominantly monoclinal, dipping westward about 50 ft/mile. The trap is stratigraphic; the productive sandstones and siltstones pinch out against shelf-edge carbonates to the north and east.

The productive zones include the upper and lower Spraberry and the Dean sandstones, which have been described as basin-filling submarine fan systems [Handford 1981]. Reservoir rock quality is very poor; porosity averages 11%; permeability, 1 md; and water saturation, 35%. The reservoir oil was approximately 460 psi undersaturated, compared with initial reservoir pressure of 2,350 psi. Stock-tank-oil gravity averages 37°API.

Although the productive interval is naturally fractured, most wells had to be hydraulically fractured to obtain satisfactory productivity [Christie and Blackwood 1952]. Individual wells initially averaged 250 BOPD, but production rates decreased rapidly. The rapid decline in productivity was attributed to the initially undersaturated reservoir oil, the rapid drainage of the fracture system, the subsequent slow influx of oil from the very-low-permeability matrix, and the relative inefficiency of solution-gas drive in the highly fractured reservoir rock.

Shortly after the field was discovered, Sohio Petroleum Co., one of the major operators in the field, undertook extensive studies to understand the reservoir better and to improve the economics of exploiting this accumulation. In a paper by Elkins [1953] in which he reported a comprehensive program of interference testing, it was concluded that "a well can deplete an area of at least 160 acres in the Spraberry as efficiently as could many wells in the same area"

As a result of this work, most of the field was developed initially on 160-acre spacing. Since initial development, however, significant additional reserves apparently have been attributed to the Spraberry by infill drilling to 80-acre spacing. For example, in 1979, before significant infill drilling, ultimate recovery was estimated to be 569 million bbl (API 1980)

---

[7] As might reasonably be expected in an accumulation this large, stratigraphy, petrophysics, and productive characteristics of individual wells vary widely. Thus, statements here should be considered of a general nature.

[8] A recovery efficiency of 8.6% of OIP is slightly less than might be estimated from Table 3.1 for this accumulation.

compared with about 814 million bbl estimated in 1992. By 1988, approximately one-third of the producing wells were on 80-acre spacing. Since that time, selective infill drilling on 80-acre spacing has continued with very successful results. One operator in the area estimates ultimate oil recovery from 80-acre infill wells at better locations to be about 80,000 bbl. This estimate is compared with an observation by Skov,[9] based on a statistical analysis of published data, that "over the last 15 or 20 years...wells...have averaged very nearly 60,000 bbl per well."

From their geologic studies, Tyler and Gholston [1987] concluded that the Spraberry is extremely heterogeneous and has several different sub-environments of deposition. It was suspected that interwell facies boundaries "compartmentalize" the reservoir, resulting in interwell stratigraphic traps that have been drained only partially by the then existing wells.

Continuing improvements in hydraulic fracturing have been another contributor to increased recovery of oil from this accumulation. Initially—early to mid-1950's—treatments involved small "hydrafractures" for cleanup. By the mid to late 1950's, however, operators were experimenting with re-fracture treatments involving about 20,000 gal of fracture fluid carrying about 30,000 lbm 20/40-mesh sand [Barba and Cutia 1992]. By the late 1980's, operators reportedly were treating with about 10 times these volumes [Blauer et al. 1992], with treatments this size apparently being warranted by increased oil recovery.

Given the experience at Spraberry [Elkins 1953], it seems apparent that interwell areas are not necessarily drained efficiently at commercial rates because wells are in pressure communication. This may be attributed to several factors, including: (a) reservoir heterogeneity/compartmentalization, (b) low transmissibility in the interwell area, and (c) a dual-porosity system, with relatively high-transmissibility fractures and low-transmissibility matrix.

7.6 Turtle Bayou "AA" (Gas) Reservoir, Louisiana

The "AA" sand is one of the major reservoirs in the Turtle Bayou field, a multi-horizon gas field in the Middle Miocene trend on the U.S. gulf coast. Discovered by Shell Oil Co. in 1949, most of the field was essentially depleted by 1990.[10] The history of the "AA" sand reservoir is an example of: (a) the importance of accurately determining the initial pressure and carefully monitoring the early production behavior of geopressured gas reservoirs and (b) the uncertainties caused by conflicting volumetric data and pressure performance.

The "AA" sand, which occurs at a subsea depth of approximately 11,380 ft at Turtle Bayou, was geopressured; the initial reservoir pressure gradient was 0.63 psi/ft.

The initial volumetric estimate of GIP—made in 1952, before the field went on sustained production—was 36 Bscf. Because the sand was initially geopressured, it was assumed that—consistent with conventional wisdom at the time—the reservoir would produce by pressure depletion. Accordingly, the recovery efficiency was estimated to be 90% of GIP.

In 1963, the operator conducted a major geologic and engineering review of this field. The nature of the drive mechanism in the "AA" Sand was uncertain but was then believed to be edgewater drive. Based on volumetric methods, GIP and ultimate reserves were estimated to be 24 and 16 Bscf, respectively. At the time of the 1963 review, it was noted that a definite gas/water contact (GWC) had not been established; the downdip limit was based on a cased-

---

[9] Personal communication, 1995.

[10] Subsequent to the author's 1984 paper, deeper pool exploratory efforts in this field by Shell resulted in the successful completion of additional producers. This activity, however, does not affect the conclusions offered here, which refer to a shallower reservoir.

hole test in a flank well that produced gas and water. It also was reported that a deeper well had been logged slightly resistive—although not conclusively commercial—suggesting the possibility of an accumulation larger than mapped [Cronquist 1984].

In 1984, by which time most of the other reservoirs in the field had been abandoned, it become apparent that: (a) the "AA" sand was producing by pressure depletion, not edgewater drive, and (b) the GIP was about 145 Bscf—almost six times the 1963 estimate [Cronquist 1984].

The size of the discrepancy between the 1963 and 1984 reserve estimates prompted a re-evaluation of the 1963 data, which led to the following observations:

- Initial reservoir pressure had not been measured; it had been estimated.

- The first BHP survey was not made until after the reservoir had been on production for about 9 months (cumulative production was 1.2 Bscf); it was a "static" survey for which the well had been shut in only 3 hours.

- Shut-in time for subsequent surveys varied from 5 to 72 hours.

- The first pressure-buildup survey was not made until about 6 years after the reservoir had been placed on production (cumulative production at that time was 8.8 Bscf).

Data from BHP surveys taken after short shut-in times were omitted, and linear regressions were run on the remaining p/Z vs. G_p points to determine which data set gave the best fit. A correlation coefficient of 0.995 was observed for one data set, which extrapolated to an estimated GIP of 157 Bscf [Cronquist 1984].

Comparing results of the "look-back" analysis of the 1963 data with those of the 1984 data—157 vs. 145 Bscf GIP—it might be concluded that the "agreement" between the two estimates of GIP was fortuitous. However, it does illustrate the need for accurate measurement and careful analysis of reservoir pressures early in the life of geopressured reservoirs.

As noted, a cased-hole formation test was the basis for the estimated initial GWC for this reservoir. Given current knowledge about gas percolation into setting cement—especially in abnormally pressured reservoirs—it seems apparent the water production on that test was the result of inadequate zonal isolation.

CHAPTER 8

RESERVES ANALYSIS, FIELD DEVELOPMENT, AND PERFORMANCE MONITORING[1]

"Surveillance is a continuous process of generating opportunities for optimizing existing assets. "
Jitendra Kikani[2]

8.1 Background

Estimating reserves is not merely a periodic, statutory exercise in routinely calculating and reporting company assets, no matter how important those functions may seem. Estimating reserves is an essential element of investment planning and resource management for every prudent operator. Estimating reserves begins with the identification of a drillable prospect, continues as the prospect is developed and placed on production, and thereafter as warranted by well and/or reservoir performance, new geologic data, improved technology, and/or changing economic conditions.

Faced with the high costs and geologic uncertainties in the North Sea in the 1970's and the highly uncertain economic environment of the late 1980's, many operators developed a team approach to optimize the appraisal, development, and operation of newly discovered major oil and/or gas accumulations. Richardson [1989], Thakur [1992], and others have provided an overview of this approach to reserves estimation, which involves teams of experienced geologists and engineers. These teams:

- plan data acquisition,
- analyze and integrate the information into a consistent interpretation of the accumulation,
- investigate various development, operating, and economic scenarios, and
- implement exploitation tactics and strategy.

As discussed by Neidell and Beard [1984], Robertson [1989], and others, modern seismic methods have played an increasingly important role in reservoir definition and management. Thomas [1986] and others have discussed the role of computer simulation in formulating initial development plans and subsequent exploitation programs. The overall effort might be considered reservoir characterization under static/dynamic conditions.

---

[1] This section is only a brief overview of a topic that has—since the late 1970's—gradually changed the oil industry's upstream modus operandi and merits an entire monograph. Early pioneers in this effort include Joe Richardson (Exxon) and Bob Sneider (Shell).
[2] SPE Distinguished Lecture, Houston, March 2000.

8.2 Well Siting and Development Plans

The degree of planning required for prudent exploitation of an oil and/or gas accumulation depends on:

- location of the accumulation with respect to infrastructure,
- operating conditions, especially with respect to climate, physical environment, and, if offshore, water depth,
- degree of geologic complexity—i.e., structural and/or stratigraphic framework,
- type of accumulation(s)—i.e., heavy oil, black oil, volatile oil, retrograde gas, wet gas, dry gas,
- reservoir drive mechanism,[3]
- production method, and
- stage of maturity.

These items are discussed briefly in the following paragraphs.

Good planning includes identification of commercially feasible exploitation scenarios and associated costs and benefits, all of which may impact well spacing, economic limits, and minimum commercial reserve volumes. For example, infill drilling from 40- to 20-acre spacing—which might be commercially feasible with the recovery of an additional 5% of OIP in a west Texas waterflood—would not be commercially feasible in a North Sea water flood. A small accumulation that could be developed separately in west Texas might be commercial in the North Sea only if a tie-in to nearby facilities is feasible.

In remote or frontier areas, good planning mandates examination of the impact of each feasible development scenario on the costs and logistics of providing supplies and support services to the area. In contrast, in areas with an established oilfield infrastructure—west Texas, for example—good planning may require little more than notifying the appropriate vendors and service companies of the next well location.

Under harsh operating conditions that may preclude year-round activity—the Beaufort Sea, for example—efficient exploitation mandates preparation of detailed timetables for each stage of the operation. Because conditions that may require cessation of operations are difficult to predict in these areas, operating timetables must include contingency plans.

The degree of geologic complexity, the type of reservoir fluids in the accumulation, and the reservoir drive mechanism have a major impact on well-drainage area, on the spacing required for efficient exploitation, and on the design and sizing of production facilities. For offshore fields, platform space and cost—and the logistics of these operations—impose major constraints on the positioning and number of wells and on the production rates that can be handled safely.

Uncertainties in geology and reservoir drive mechanism mandate a high degree of flexibility in facilities design, especially for offshore locations. Buchanan and Hoogteyling [1979] and Kingston and Niko [1975] have provided detailed discussions of these aspects of development planning for the Auk and Brent fields (North Sea), respectively. Haugen *et al.* [1988] have provided a case history of the Statfjord field (North Sea). This case history illustrates how the operator's original development plans were refined on the basis of field performance observed through an extensive cased-hole-logging program and computer

---

[3] Consistent with the scope of this work, the discussion here focuses on reservoirs produced by primary drive mechanisms.

simulation by use of progressively more performance detail and more sophisticated computer-simulation models.

In areas where drilling and producing operations are at a relatively mature stage—the U.S. Gulf Coast, for example—there tends to be less uncertainty in the planning operation for known geologic trends and moderate water depths. In these areas the geologic and reservoir conditions are reasonably well understood. The economic and operating parameters that impact economic limits and minimum reserve size can be estimated with a reasonable degree of confidence. The converse is true for areas in the early stages of exploration—at this writing, the Timor Sea (Australia), for example—where there are substantial uncertainties in all aspects—geologic, engineering, and economic.

8.3 Economic Analysis of Rate and Spacing Scenarios

Efficient exploitation of an oil and/or gas accumulation involves a balance between:

- well drainage area,
- well production capacity,
- reservoir production capacity,
- per-well reserves,
- well cost,
- operating costs,
- manpower needs,
- cost of production facilities,[4]
- capacity of production facilities, and
- market conditions.

The first four of these parameters will be controlled by reservoir geology and reservoir drive mechanism. In newly developing areas, none of these four parameters will be known with a reasonable degree of certainty until there has been a period of sustained production and operating experience.

The problem of optimizing reservoir exploitation is less difficult for most onshore areas, where wells and facilities can be added or modified without severe economic penalty as needs become apparent. On offshore fields, however, high initial costs for platform construction and installation, limited space for production facilities and wells, and high costs to add wells or modify facilities mandate detailed economic analysis of various rate and spacing scenarios. In some scenarios, it may be economically feasible to include contingency well slots in initial construction. Hillier *et al.* [1978] have provided a detailed discussion of considerations in development planning for the Forties field (North Sea).

8.4 Managing Data Flow in Newly Developing Fields

Systematic collection, interpretation, and cataloguing of geophysical, geologic, engineering, operating, and performance data are essential for the proper management of an oil and/or gas field or production complex. Specific types of data are project dependent, and development of a detailed list is not feasible. However, broad guidelines were provided in Sec. 1.1.2, and DeSorcy [1979] has provided additional guidelines. Specific examples of data collected for

---

[4] Costs of sophisticated automation for such facilities need to be balanced against costs of manpower.

two widely separated offshore areas, and the use of these data for reservoir analysis and surveillance are provided by Johnson [1988] and by Edwards and Behrenbruch [1988].

Bankhead [1962] discussed the handling of geologic and engineering data in multipay fields and the coordination necessary between geologic and engineering staffs to ensure smooth operations. Cronquist and Hodgson [1971] provided an example of the integration of well and reservoir performance data into a computerized management information system. Since those early efforts, substantial improvements in computer technology have facilitated cost-effective compilation of databases that include vast amounts of both numeric and graphical geologic, engineering, and performance data.

8.5 Performance Monitoring

After initiation of sustained production from an oil and/or gas accumulation, it is vitally important to monitor well and reservoir performance. This is especially important for complex stratigraphic-structural settings offshore. Depending on circumstances, such monitoring may include:

1. regularly scheduled well tests,
2. periodic bottomhole pressure (BHP) measurements,
3. casing surveillance,
4. time-lapse (4D) seismic,
5. infill drilling, and
6. periodic updates of geologic interpretation and reservoir simulation.

Various aspects of such monitoring are discussed in subsequent sections.

8.5.1 Regularly Scheduled Well Tests

The purpose of regularly scheduled well tests includes:

- detection of significant changes in flowing and/or shut-in tubing pressure,
- monitoring of the performance of artificial-lift equipment,
- depending on the reservoir fluids and/or drive mechanism, detection of:
  -first water production, or
  -breakthrough of free gas,
- depending on the reservoir fluids, detection of changes in "performance indicators" like:
  -producing gas/oil or condensate/gas ratio, or
  -producing water/oil or water/gas ratio.

In some newer fields, especially those with subsea completions, well performance is monitored continuously with permanent downhole-pressure gauges[5] and flowmeters in the production lines for each well. In most fields, however, wells are not so equipped and are produced to centralized production facilities. Performance data must be obtained by periodically routing each well through test separators. In many installations, such testing is automated. In any scenario, costs of sophisticated test equipment and frequency and duration of well tests must be weighed against costs of manpower and monetary benefits—i.e., costs vs. value.

---

[5] As discussed by Athichanagorn et al. [1999], installation of such gauges is a relatively recent development, and special processing and interpretation methods are needed because of: (a) instabilities of the equipment, (b) the very large amounts of data, and (c) the essentially uncontrolled measurement conditions.

The importance of a comprehensive program for monitoring well performance cannot be overstated. In many cases, for example, well performance has provided the first indications of unexpected geologic heterogeneities or unexpected reservoir performance.

8.5.2 Periodic BHP Measurements[6]

Periodic BHP measurements may include: (a) surveys to measure shut-in BHP in one or more wells and (b) special tests to measure transient BHP in one or more wells during one or more flow and shut-in periods. The purposes of such measurements include:

- determination of producing and/or shut-in BHP trends,
- detection of wellbore damage,
- determination of the degree of communication between wells and/or fault blocks, and
- detection of flow boundaries and/or estimation of drainage volume.

These items are discussed briefly in ensuing paragraphs.

As discussed in Sec. 4.1.1, accurate shut-in BHP history for producing wells is essential for reliable material-balance calculations; both shut-in and flowing BHP data are necessary for reliable computer simulation.

While it may be argued that detection of wellbore damage is not directly related to reserves estimation, detection and remediation of such damage is certainly related to prudent asset management. Damaged wells may not be effective drainage points, which may have a direct impact on (drilled, producing) reserves.

Pending initiation of sustained production, the degree of communication between wells and/or fault blocks may be a matter of conjecture. It is, however, vitally important to determine the degree of such communication to ensure efficient exploitation. Pressure-transient tests may be helpful in such determinations, but they seldom provide unique solutions. Historical BHP data and material-balance/reservoir simulation may provide better results. In this context, though, it is noted that, in the presence of a strong waterdrive, historical BHP data may not provide sufficient discrimination.

8.5.3 Through-Casing Surveillance

Through-casing surveillance may be designed for diagnosis and remediation of deficient wells and/or monitoring reservoir performance. Techniques may include:

- production logging of currently active well-zone completions and/or
- measurement of the fluid content and/or BHP of behind-pipe well zones, which may involve the use of
  - pulsed-neutron and related devices[7]—e.g., the Halliburton TMD-L™ and the Schlumberger RST,™ and/or
  - WFT devices to measure BHP and/or recover fluid samples—e.g., the Halliburton CHFT™ or the Schlumberger MDT.™

Production logging typically involves dynamic measurements of the flow (or injection) profiles for active well-zone completions. The procedure is especially useful for wells where zones of different permeability are open for production, but all zones may not be contributing

---

[6] For detailed presentations of the basic principles of pressure testing, the reader is referred to Earlougher [1977] and Lee [1982]. Continuing discussion of modern interpretation procedures—especially type-curve analysis—may be found in papers by Gringarten and others beginning in the early 1980's.

[7] Neither endorsement of those mentioned nor exclusion of others is intended by these references.

to production. It has been used successfully to monitor water-level rise in waterdrive reservoirs, but special interpretive techniques may be required to achieve consistent results [Ibrahim 1999]. At Prudhoe Bay, for example, induction-log monitoring of waterflood performance was facilitated with preinstalled sections of fiberglass casing.[8]

Measurement of the fluid content and/or BHP of behind-pipe well zones is important for scenarios where: (a) the degree of vertical communication between such zones cannot readily be determined by other means and/or (b) such zones may have been partially drained or invaded by extraneous fluids.

Special techniques, however, may have to be adopted in low-salinity environments, where pulsed-neutron-capture (PNC) logs may not adequately discriminate between oil and formation water. In discussing such environments in the Gippsland Basin (Australia), Mills [1993] reported successful application of a Σ/gamma-ray overlay technique to help identify water-flushed intervals.

8.5.4 Time-Lapse (4D) Seismic

By the mid-1990's, advances in seismic technology made it feasible to monitor fluid movements in many subsurface rock-fluid settings [King 1996]. The technology seems to be applicable in geologic settings where there are verifiable direct hydrocarbon indicators (DHI). The term "4D seismic" refers to repeated 3D seismic surveys—at successive times during the pruduction phase—that are processed to measure spatial changes in seismic amplitude that may be related to changes in fluid saturation in a producing reservoir.[9] Such changes may be indicative of:

- encroachment of an expanding gas cap,
- evolution of solution gas,
- encroachment of an aquifer, and/or
- flood-front advance in an improved-recovery project.

As noted by several authors [Hoover *et al.* 1999, Kristiansen and Christie 1999] this technology might be used to:

- improve understanding of reservoir behavior,
- help delineate reservoir subfacies and/or fluid-flow barriers,
- avoid premature breakthrough of water and/or free gas, and
- locate new wells to drain bypassed oil and/or free gas.

For example, time-lapse seismic was included in the analysis of an initially geopressured[10] light-oil reservoir produced by waterdrive, offshore Louisiana [Hoover *et al.* 1999]. As a result of this work, it was concluded that a high-permeability subfacies was preferentially swept by the encroaching aquifer.[11] Such detailed observations by the authors were facilitated by an integrated analysis of log, core, performance, and seismic data, integrated with a 3D full reservoir simulation study [Ashbaugh and Flemings 1999]. As demonstrated by this work, optimal results are obtained from time-lapse seismic if such surveys are conducted in conjunction with 3D reservoir simulation.

---

[8] Arlie Skov, personal communication, 2000.
[9] Detailed discussion of this rapidly developing technology is beyond the scope of this work, and the interested reader is referred to current papers in the geophysical literature (e.g., *The Leading Edge* and *Geophysics*). Hoover *et al.* [1999] have provided a comprehensive list of references.
[10] Geopressured reservoirs are discussed in Sec. 6.5.
[11] This observation is consistent with reports by other authors, Hartman and Paynter [1979], for example.

Fundamental to this technology is the detection of small changes in seismic amplitude from one survey to the next that can be attributed to changes in reservoir fluid saturation. Such changes must be significantly greater than the background noise that is an inherent component of all seismic data. To satisfy this condition, seismic surveys should be highly repeatable, which is difficult to achieve. One possible solution to this problem for offshore fields is "permanent" emplacement of seismic sensors in the seafloor. Kristiansen and Christie [1999] report the apparent success of such an approach for the Foinaven field, offshore Shetland Islands (Scotland).[12]

Despite best efforts to reduce background noise, 4D seismic data must be carefully processed, typically by use of statistical techniques, to distinguish between such noise and useful signals [Burkhart et al. 2000]. Different techniques are in use, and the engineer should be cautious in accepting the results from such processing without a clear understanding of the limits and corroboration from other data and studies. Depending on circumstances, it might be appropriate to seek an alternate interpretation.

Historically, oilfield seismic data has involved "P" waves—compressional waves that vibrate in the direction of wave propagation. "S" waves—shear waves that vibrate perpendicular to the direction of wave propagation—typically have been processed out of the seismic data before analysis. In contrast to "P" waves, however, "S" waves do not propagate through fluids. This distinction makes "S" waves potentially useful for reservoir characterization. It does, though, cause operational problems, as "S" waves do not propagate through the ocean and, therefore, cannot be recorded using arrays of towed streamers, as is typically done in marine seismic surveys. In the early 1990's, however, "S" waves were recorded by use of ocean-bottom sensors, and this technology—called marine 4-C—is under continuing improvement [Caldwell 1999]. New processing algorithms also are needed. To date, applications to reservoir characterization have been limited but successful. Although still embryonic, the acquisition and processing technologies are being improved rapidly and are expected to be powerful tools for reservoir characterization and monitoring.

8.5.5 Exploitation Drilling

In the context of this discussion, the term "exploitation drilling" includes wells drilled after the initial round of appraisal/development drilling, to obtain additional subsurface information and/or to increase drainage efficiency. Initial well siting is seldom optimal, especially in stratigraphically and/or structurally complex accumulations. After initiation of sustained production and development of well-performance data, it frequently becomes apparent that the accumulation is more heterogeneous than initially expected and that additional exploitation wells are required.

For example, it was recognized shortly after discovery in 1973 that the Dunbar field (offshore Shetland Islands) was quite heterogeneous and that, after initial development, additional wells probably would be required to properly characterize and exploit this accumulation [Bigno et al. 1997].

Fields discovered and developed before the advent of modern logs may be especially suitable candidates for modern exploitation drilling. The Esperson Dome field (Texas), discovered in 1939, produced 21°API oil from a series of unconsolidated, shaly sands. Well spacing was (then standard) 40 acres, and the reservoirs produced with strong waterdrive (at an unfavorable mobility ratio, because of the heavy oil). Wells typically went to high WOR's

---

[12] Other aspects of survey repeatability are discussed in the cited paper.

shortly after completion and had been on pump at low rates for many years before field performance was reviewed. The available subsurface and performance data were inadequate to characterize the reservoir. The solution was to drill several appraisal/exploitation wells at infilling locations and run a modern log-and-core program to evaluate saturation conditions after many years of production. It became apparent that, owing to the unfavorable mobility ratio, most of the oil had been under-run by the encroaching aquifer,[13] and significant volumes remained in place. Over the next several years, the operator undertook an infill-drilling program that resulted in significant increases in production rate and ultimate recovery.

8.5.6 Periodic Updates of Geologic Interpretation and Reservoir Simulation

Depending on circumstances, updates of geologic interpretation and/or reservoir simulation may occur at any stage in the appraisal, development, and/or production of an accumulation.

For example, the Ubit field (offshore Nigeria) had been on production some 25 years before the operator undertook a major review. This review resulted in: (a) drilling 37 wells with horizontal-completion legs (with an additional 20 planned), (b) increasing production by 110,000 BOPD, and (c) increasing estimated ultimate oil recovery by 500 thousand STB. Discovered in 1968, and described as "the largest producing single reservoir in (offshore) Nigeria," Ubit contained estimated OIP of 2.1 billion STB [Clayton *et al.* 1998]. The field was initially developed with only "scanty" 2D seismic data and under regulations that required 800-m spacing. It produces from an (initially) 160-ft thick oil column overlain by a gas cap that (initially) varied from 50 to 550 ft thick. Drive mechanism has been gas-cap expansion, and many of the wells produced with high GOR for the previous 25 years.

The operator's review included: (a) a 3D seismic survey, (b) petrophysical re-evaluation, and (c) full-field 3D reservoir simulation.[14] As a result of this review, the operator: (a) developed an integrated geologic model of the field, (b) concluded that the reservoir was a common hydraulic unit, and (c) undertook a successful program to redevelop the field.

In contrast to the mature stage review of the Ubit field are the early stage, ongoing reviews at the Hibernia field (offshore Newfoundland). The field was discovered in 1979 and delineated by nine additional wells during the subsequent five-year period. During the appraisal period and before initiating development drilling in 1997, the operator ran two 3D seismic surveys. The field is complexly faulted, and the main producing interval is a "compartmentalized, fluvial deltaic system" [Sydora 1999]. Estimated OIP is 3 billion STB; estimated ultimate recovery is 615 million STB. Performance is continuously monitored; all wells are equipped with permanent downhole-pressure gauges, wellhead-pressure gauges, and flowmeters. The operator's 3D model and reservoir simulator are updated with log and core data on a well-by-well basis as development continues. One of the most important problems facing the operator is accurate mapping of small faults that act as partial barriers to flow—and estimating the transmissibility of such faults for reservoir simulation. This is a problem in many oil/gas fields, and reconciliation of production performance with the extant geologic/engineering model is a vital element of reservoir surveillance.

---

[13] This phenomenon, called "Dietz tonguing," was discussed in Sec. 3.8.1.
[14] This brief summary cannot do justice to the extensive work reported by the operator, and the reader is referred to the wealth of detail provided by Clayton *et al* [1998].

CHAPTER 9

PROBABILISTIC ESTIMATION AND CLASSIFICATION OF RESERVES[1]

"There is nothing uncertain about reality; it is our vision of that reality that is uncertain."

Andre Journel [1994]

9.1 Background

Experienced reservoir engineers know that uncertainty exists in geologic and engineering data and, consequently, in the results of calculations made with these data. The degree of uncertainty in most reservoir engineering calculations, however, usually is not quantified.

As noted in Sec. 1.3.2, reserve estimates historically have been deterministic—i.e., "single valued"—with the degree of uncertainty indicated by terms like proved, probable, or possible. Additional information about the degree of uncertainty has been conveyed by describing producing category and development status—e.g., producing, behind pipe, or not developed, as discussed in Sec. 1.3.5 and illustrated by Table 1.3.

The term "deterministic," as used in this work, reflects current industry usage. Strictly speaking, however, deterministic implies a unique answer to a given set of facts and natural laws. Reserve estimation involves both uncertain "facts" and ill-defined natural laws. Reserve estimates are not unique and can hardly be considered deterministic in the classical sense.

In geologic settings and operating areas where the industry has substantial experience and in fully developed, mature fields—situations considered to have a relatively low degree of uncertainty—deterministic estimates of reserves usually are considered acceptable.

However, for new geologic settings (e.g., coalbed methane) and in new operating areas (at the time, the North Sea) the industry developed probabilistic procedures[2] to estimate and classify reserves. These procedures have been used to quantify the degree of uncertainty in, and potential range of, reserves attributed to risky ventures [Dhirr *et al.* 1991, Keith *et al.* 1986].

---

[1] This chapter focuses on reserves attributed to *known* accumulations. Estimation of "reserves" for exploratory prospects—the term "reserves" being used advisedly in this context—merits a separate monograph. In this context, the reader is referred to pioneering work by Arps and Roberts [1958], Grayson [1960], Kaufman [1963], Newendorp [1975], and Harbaugh *et al.* [1977]. The reader also is referred to the AAPG's 1992 treatise *The Business of Petroleum Exploration* and to Otis and Schneidermann [1997], who provide an extensive bibliography of more recent work.

[2] The term "procedure" is used here, rather than the term "method," to avoid confusion with the historically used terms "volumetric method" and "material balance method.' Either deterministic procedures or probabilistic procedures may be used for the analogy method, the volumetric method, or the performance method.

9.2 Statistical Considerations

9.2.1 Introduction

Discussion and comparison of deterministic and probabilistic procedures must involve basic statistical terms.[3] Statistics deals with—among other topics—sets of samples from populations and provides methods to characterize these sets as measures of the nature of the populations. For example, core and/or log data for a reservoir may be considered sample sets from which some of the characteristics of the reservoir—the population—may be estimated.

9.2.2 Theoretical Frequency Distributions

To facilitate computation and comparison, it is convenient to approximate data sets with theoretical frequency distributions. Frequency distributions are called probability density functions, or "pdf's." In Sec. 2.3.2, for example, it was noted that the frequency distribution of ultimate recoveries for a set of fields in a common geologic setting could be approximated by a log normal pdf.

From a statistical point of view, a specific data set—measurements of porosity in a well-zone, for example—may be described by several attributes. Some of these attributes may be illustrated with a histogram, **Fig 9.1,** which shows the number of values that fall in subsets—also called classes or bins—that describe the entire sample set.[4]

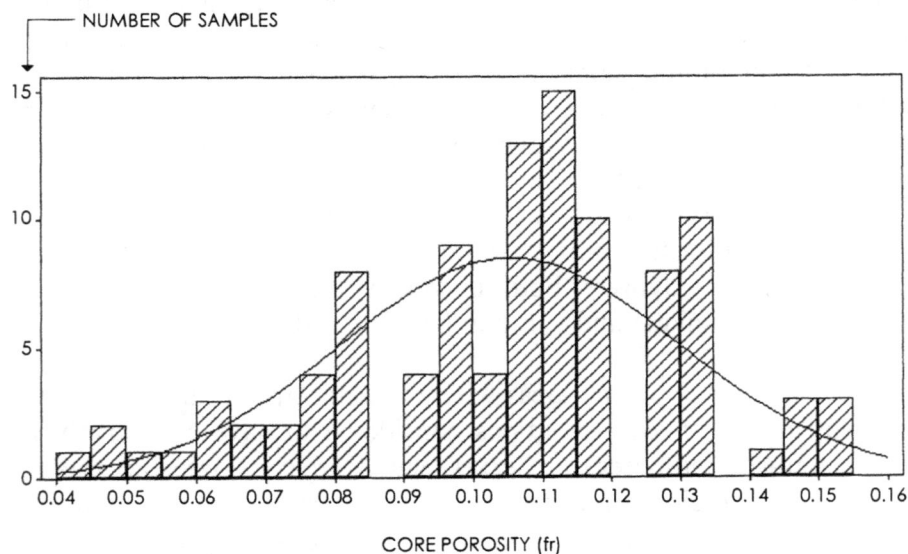

Fig. 9.1—Histogram illustrating statistical distribution of a set of data.

A histogram may be fitted with a smooth curve—a pdf—shown by the solid line on Fig. 9.1. The area under the curve defines the domain of 100% probability of occurrence of all the samples *and by inference, the sampled population.*[5]

---

[3] The brief discussion in this section cannot be considered any more than a brief introduction to the statistical concepts behind probabilistic procedures. The interested reader is referred to texts by Newendorp [1975], McCray [1975], Megill [1984], or Jensen *et al.* [1997].

[4] A histogram may be used to represent the distribution of data for either: (a) a discrete variable—e.g., point scores for a number of tosses of two dice, or (b) a continuous variable—e.g., porosity from a set of sidewall samples.

[5] The area under a segment of such a curve defined by a range of "x" values represents the probability of occurrence of the range. The area under such a curve defined by a single "x" value is infinitesimally small, thereby implying a vanishingly small probability of occurrence of a single value of "x."

Different types of theoretical pdf's, illustrated by **Fig. 9.2,** may be used to approximate specific observations of nature—for example, porosity, permeability, irreducible water

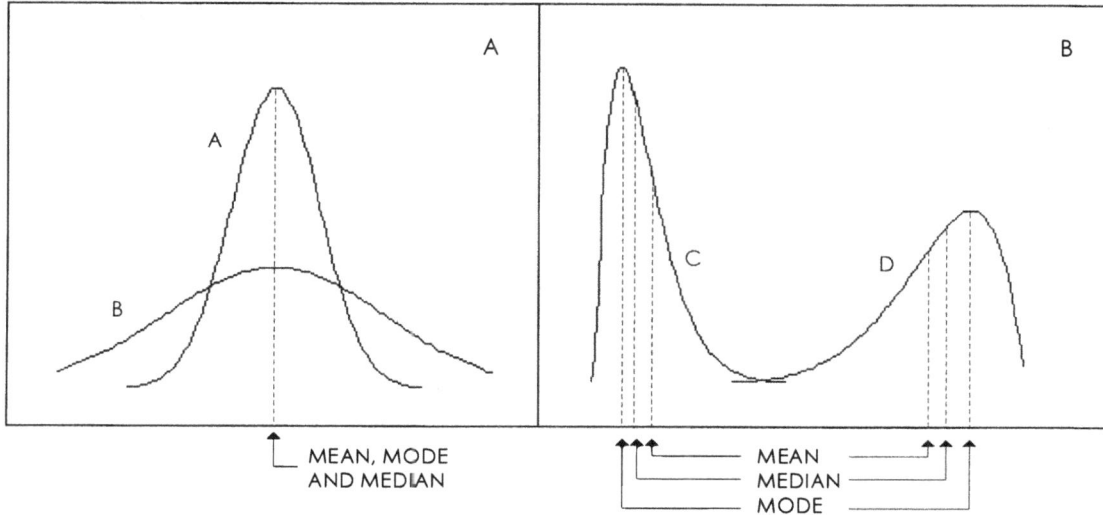

Fig. 9.2—Panel A: Symmetric distributions, A and B, illustrating different standard deviations. Panel B : Asymmetric distributions, C and D, illustrating positive and negative skew, mean, mode, and median.

saturation, and/or net pay. Such distributions may be symmetric or skewed. Pdf's have several attributes[6] that are relevant to probabilistic calculations:

- the arithmetic average, mean, or expected value,[7]
- the most likely value, or mode,
- the median, or 50th percentile,
- the degree of variation, or standard deviation, and
- the nature of the distribution of values about the mode, or type and degree of skew.

Curves "A" and "B" (Panel A) illustrate distributions where values are symmetric about the mean. Curve "B" exhibits the same type of symmetry as Curve "A," but the variation about the mean—the standard deviation—is larger than that of Curve "A." Curves "A" and "B" are each "normal" or Gaussian[8] distributions. The mean, or expected value, of a data set is the arithmetic average of all the unweighted values in the set. The mode is the value with the highest probability of occurrence. The median is the value that corresponds to the 50th percentile—50% of the samples have a smaller value; 50%, a larger value. The standard deviation is a measure of the degree of variation exhibited by the sample set. The mean, the mode, and the median of a specific symmetric distribution are the same.

Curves "C" and "D" (Panel B) illustrate skewed distributions. The mean, the mode, and the median are identified on both curves. Skewness, or skew, is a measure of the nature of the

---

[6] Although such attributes are discussed here in the context of pdf's, which imply continuous variables, they also pertain to discrete variables.

[7] The term "value," used here in a statistical cortext, should not be confused with the term "value," as used in a financia context. The terms "value" and "quantity" are used here interchangeably in a statistical context.

[8] The term "Gaussian" is used here to describe a normal distribution to avoid confusion with the term "log normal."

variation. Curves with the mode less than the mean are said to exhibit positive skew; those with the mode greater than the mean, negative skew. The skew of curve "C" is positive; that of curve "D," negative.[9] These curves are beta distributions,[10] which may be used to approximate the pdf's of data sets that are skewed.

The curves in Fig. 9.2 are unimodal; the presence of more than one "maximum" in a data set may be indicative samples from more than one population.

As discussed in Sec. 2.3.2, the distribution of initial reserves, N_{pa}, for a set of wells in a given geologic trend usually may be approximated by a log normal pdf. Such distributions exhibit positive skew, as illustrated by Curve "A" in **Fig. 9.3,** which is the pdf of N_{pa}.

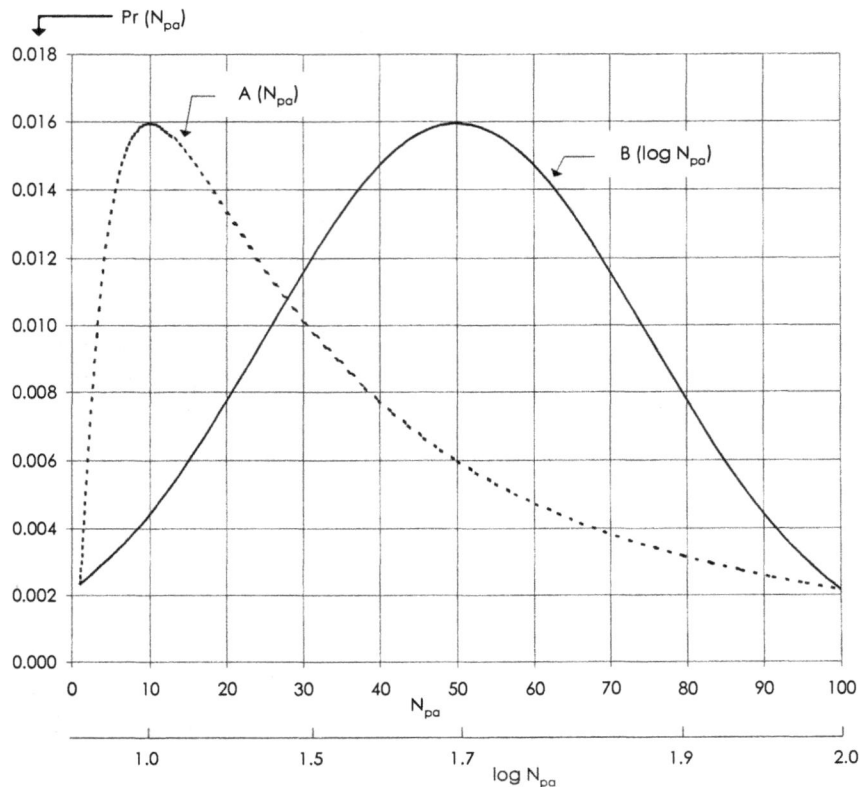

Fig. 9.3—Frequency distributions of N_{pa} and log(N_{pa}).

If the pdf of the *logarithms* of the values—log(N_{pa})—can be approximated by a Gaussian pdf, like Curve "B" (which is plotted vs. the lower x-axis), the pdf of Curve "A" is said to be log normal.

Cumulative pdf—also called cumulative density function, or cdf—is determined by plotting the cumulative sum of probabilities vs. the x-axis. **Fig. 9.4** shows a triangular pdf—often used to approximate skewed distributions—and the corresponding cdf, the short-dash curve labeled "ΣPr(X)." Cumulative pdf—usually expressed in decimal form—is numerically

---

[9] Some authors characterize the skew of Curves "C" and "D" as "left" and "right," respectively, focussing on the "hump," rather than the "tail." Others take a contrary view. Usage of the terms "negative" and "positive" to characterize skew in this work is consistent with statistical theory.

[10] Beta distributions are discussed in Sec. 9.5.5.

equal to percentile. The complement of a cdf—the solid curve labeled "Pr ($X_a \geq X$)"—is called an "expectation" curve. An expectation curve (EC) indicates the probability that the actual value, X_a, will equal or exceed the estimated value, X. With reference to Fig 9.4, for example, there is 10% probability that the actual value of X will equal or exceed 73. As discussed in Sec. 9.3.3, EC's may be used to classify reserves that have been calculated using probabilistic procedures.

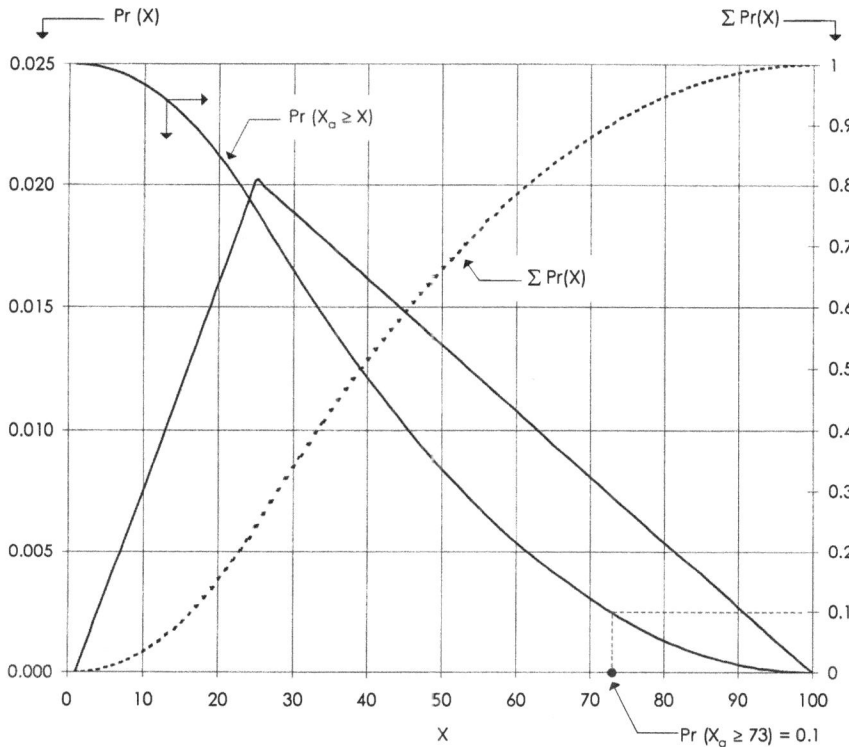

Fig. 9.4—Triangular frequency distribution illustrating cumulative frequency and expectation.

Unfortunately, the term "expectation" is used to express different, although related, quantities. One such quantity is the mean value of a distribution; the other, the probability that the actual value of a distribution will equal or exceed the estimated value. To avoid confusion, the term "expectation," used here without modifiers, means the expected quantity, or mean, of the variable X. The phrase "expectation curve" or "expectation plot" means the complement of the cumulative probability of the variable X. The phrase "expectation at P90," for example, will be used to designate 90% probability that the actual quantity will equal or exceed the estimated quantity,[11] as discussed in the previous paragraph.

9.2.3 Confidence, Expectation, Percentile, Risk, and Uncertainty

The term "confidence" frequently is used when the term "expectation" might be more appropriate. The distinction is illustrated by **Fig. 9.5,** which shows a pdf (the solid curve) and the corresponding cdf (the short-dash curve) of a normally distributed random variable, X.

---

[11] Strictly speaking, "expectation at P90" implies 90% probability that the actual quantity will be *between* the estimated quantity and the maximum estimate, which ignores the possibility that the actual quantity could *exceed* the maximum estimate.

The cross-hatched area under the solid curve between X_L and X_U is 80% of the total area under the curve. Accordingly, it might be stated "The best estimate for X is X_m, and there is 80% confidence that the true value of X is between X_L and X_U." Or, "With 80% confidence, $X = X_m \pm \Delta x$."

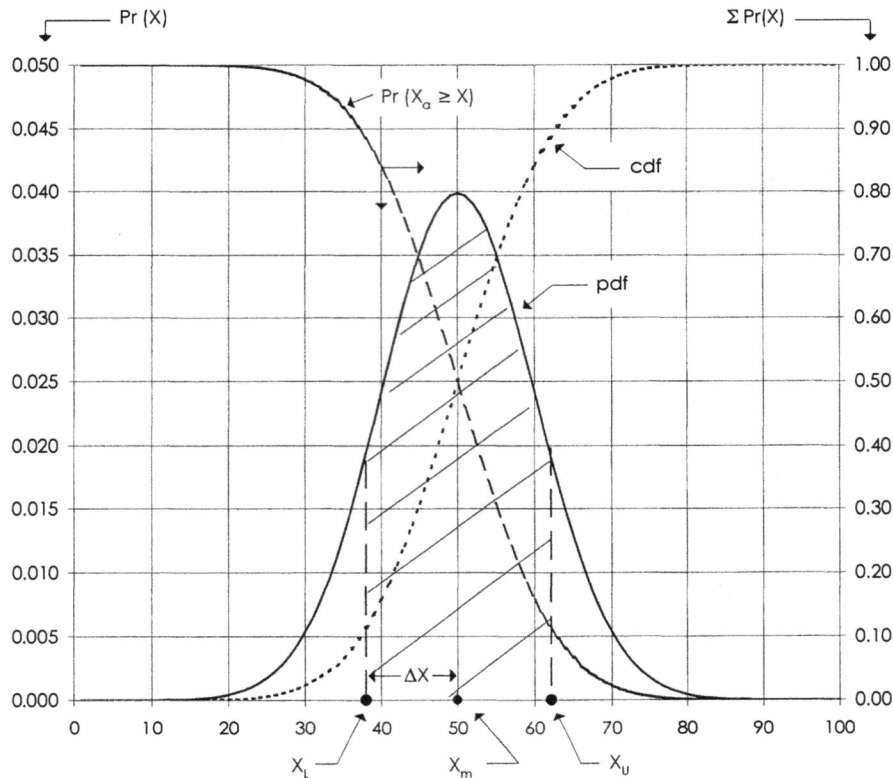

Fig. 9.5—Frequency distribution illustrating confidence interval and expectation.

The concept of confidence intervals is used in subsequent sections where the calculated upper and lower bounds at 80% confidence are used to define the 90th and 10th percentiles, respectively, of estimates based on a least squares regression. In this context, reference is made to Sec. 9.8.2, 9.9.2, and 9.11.

As discussed in previous paragraphs, expectation differs from confidence, as it refers to the probability that the actual value, X_a, will *equal or exceed* the estimated value of X. With reference to the long-dash curve of Fig. 9.5, it may be stated "There is 90% probability that X_a is equal to, or greater than, X_L." Or, "9 times out of 10, X_a will be equal to, or greater than, X_L."

The terms "percentile" and "expectation" frequently, albeit erroneously, are used interchangeably. As noted previously, percentile is numerically equal to cumulative frequency. For example, 90% of the values in a distribution are less than the 90th percentile—i.e., the 90th percentile of a cdf is defined by a cumulative frequency of 0.90. Expectation, in contrast, is the complement of cumulative frequency. For example, there is 10% chance that values in the distribution will exceed the 90th percentile.

The terms "risk" and "uncertainty" frequently are used interchangeably. Although it is a minor point, the terms convey slightly different, although related, meanings. Uncertainty implies a lack of precision. Depending on context, risk implies exposure to potential injury or loss. It also is used, however, to denote the probability of occurrence of a discrete outcome. In one context, for example, it might be stated, "Due to geologic uncertainty, drilling ventures are subject to the risk of monetary loss." In another context, it might be stated: "In this trend, the risk of a dry hole is (estimated to be) 0.75." In a related context, the term "risk" appears in the phrase "risk-weighted," applications of which are discussed in Sec. 9.5.1 and 9.7. In this work, the term "risk" is used in the context of either risk of monetary loss, or risk of geologic failure, the latter context being used in Sec. 9.5.1.[12]

9.2.4 The "Best Fit" Dilemma

Regression analysis frequently is used to "model" or analyze the relation between various types of data; for example:

- The relation between permeability, k, and porosity, ϕ, typically is modeled by a linear relation between $\log(k)$ and ϕ.
- Linear regression of p/Z vs. G_p from a volumetric gas reservoir may be used to estimate gas initially in place and/or reserves.
- Trends of production rate vs. time (or vs. cumulative production) are subject to regression analysis to determine the "best fit" and estimate reserves.
- Correlations developed from regression analysis (like those discussed in Sec. 3.6) may be used to estimate recovery efficiency.

As discussed in standard texts on statistics, linear regression analysis (LRA) is based on the assumption that error(s) in the dependent variable may be approximated by a normal distribution. The model resulting from LRA is a best fit through the means of the error distributions of the dependent variable. The assumption that the errors in the dependent variable are normally distributed leads to the conclusion that the best fit represents an estimate of the median—or P50—values of the dependent variable.

In the absence of data to the contrary, most engineers consider that the "best fit" calculated from LRA represents "reasonable certainty" and apply the results of such calculations to estimate proved reserves. For some applications, however, it may be appropriate to estimate and classify reserves based on probabilistic calculations. In a sense, LRA might be considered a probabilistic calculation that results in a P50 estimate. Accordingly, it might be concluded[13] that "best fit" estimates from LRA should be classified as "proved plus probable," rather than "proved."

9.3 Procedures Described

9.3.1 Background

Deterministic procedures for estimation and classification[14] of reserves (ECR) attributable to known accumulations were developed during the years when E&P activities of the oil and gas industry involved areas with relatively low monetary/reserve risk. Such areas included onshore and shallow waters of the continental shelves that had an established infrastructure. In such settings, newly discovered accumulations were developed and evaluated one well at a

---

[12] Rose [personal communication, 2000] makes a further distinction and uses the term "chance" to denote confidence in something happening; thus, it might be stated "Given good subsurface data there is better than an 80% chance this offset will be commercially successful."

[13] Robinson [1999] posits such fits be classified as "expected quantity."

[14] As noted in Sec. 1.1.2, reserve estimates should always be qualified to indicate the degree of uncertainty, hence the linkage in this section of the terms "estimation" and "classification."

time, without the need to install field-size platforms and/or associated facilities prior to development drilling and production.

Wells were placed on production shortly after completion and were production tested as development proceeded. In such scenarios, commerciality and extent of accumulations were determined on a well-by-well basis. During these years, a well might have been considered a "commercial unit," and ECR may be said to have been on a "well-at-a time"—rather than on a "field-at-a-time"—basis.[15]

In subsequent years, upstream activities expanded into deeper waters of the continental shelves and into new—frequently remote—geologic areas. Substantial investments generally were required for major facilities to process and transport oil and/or gas to market before development drilling and/or sustained production could be initiated. It was not feasible to determine commerciality and extent of such accumulations one well at a time. Operators had to evaluate uncertainties in risk capital and reserves at field level, rather than at well level. The emphasis shifted from risking well-size increments of capital to risking field-size increments. In such circumstances, a need was perceived to quantify the degree of uncertainty and upside potential, which led to increasing application of probabilistic procedures to ECR.

9.3.2 Deterministic Procedures

SPE-WPC [1997] definitions state, "The method of estimation is called deterministic if a single best estimate of reserves is made based on known geological, engineering and economic data." Expanding on this concise definition, ECR is considered here to be deterministic if it involves:

1) use of the best estimate of the value of each input parameter to calculate[16] oil and/or gas initially in place (O/GIP)[17] and/or reserves and

2) classification of the calculated quantities as proved, probable and/or possible, based on:

   a) *professional judgments* regarding
      -geologic setting of subject accumulation,
      -stage of development,
      -quality and quantity of geologic and engineering data,
      -degree of uncertainty in the interpretation of such data,
      -applicability of, and degree of uncertainty reflected by, analogy data,
      -operational and economic scenario, and
   b) organizational, corporate, statutory, or other, guidelines for classification of reserves, depending on the purpose of the estimate.

In the context of these definitions, the term "best estimate" is open to interpretation. Some statisticians consider the best estimate of the value of a population to be the arithmetic average, or mean value, of the samples from the population [Natrella 1963]. Keith *et al.* [1986], however, report using the median value. Elliott [1995] noted that industry practice typically involves *perceived* measure(s) of central tendency as the best estimate(s) for various input parameters.

---

[15] SPE-WPC [1997] reserve definitions include the terms "undrilled locations" and "direct offset," as discussed in Appendix B. The historic emphasis on locations and tracts is a reflection of the nature of the ownership of subsurface minerals onshore U.S. where—unless unitized—most accumulations are overlaid by numerous tracts, each with different ownership.

[16] Even though the term "calculate" is used here, the reader is reminded that the results must be considered estimates.

[17] Depending on the purpose of the estimate, calculations of O/GIP may be reported as adjuncts to estimates of reserves.

In the absence of statistical analysis, the so-called best estimate of the value of each input parameter may be the mean, mode, or median of the actual pdf of the parameter. More likely, it will be some other, undefined, value. Thus, the deterministic procedure typically does not address whether best estimates are appropriate measures of the actual statistical distributions of the input parameters and, consequently, of the calculated quantities.

For analogy estimates, like those discussed in Sec. 2.0, or for analytical calculations, like those discussed in Sec. 3.0, the best estimate of the value of each input parameter may be based on a combination of analogy, statistical analysis, and/or professional judgment.[18]

For performance based estimates, like those discussed in Sec. 4.0 and Sec. 5.0, input "parameters" will be historical pressure/production data, with best estimate(s) being based on statistical analysis and/or professional judgment.

Depending on the nature of the accumulation—stratigraphic vs. structural—and on the relative amount, type and quality of geologic control—geophysical vs. subsurface—reserves may be estimated for single wells, reservoir segments, separate fault blocks, and/or reservoirs in-toto.

Where accumulations are primarily stratigraphic and/or where volumetric mapping is not feasible, it may be appropriate to make reserve estimates for single wells. Such estimates may be based on a combination of petrophysical/test data and analogy.

Reserve estimates for offset and outstepping wells typically are based on analogy, including statistical analysis, given sufficient data. Such reserves may be classified as proved, probable, or possible, depending on a number of factors, including depositional environment of the reservoir rock, log characteristics at drilled locations, proximity to dry holes, and structural relief. Professional judgment and local experience of the reserve estimator(s) are significant factors in such classifications.

Where accumulations are primarily structural, O/GIP may be determined, initially, by volumetric mapping. Reserves in different areas of such accumulations may be assigned different classifications, depending on:

- nature of the reservoir rock and fluid,
- expected drive mechanism,
- nature of internal faulting,
- possibility of stratigraphic compartmentalization, and
- areal/structural location and spacing of existing and planned drainage points.

One method for allocating reserves of different classifications to individual wells in a reservoir is discussed in Sec. 3.8.1.

Irrespective of the geologic setting, as sufficient historical performance data become available, reserves may be based on a combination of geologic and production data. In any event, classification of the calculated quantities as proved, probable, and/or possible reserves will be consistent with guidelines promulgated by the agency for whom the estimates are prepared and/or company policy.

9.3.3 Probabilistic Procedures

SPE-WPC [1997] definitions state, "The method (of estimation) is called probabilistic when the known geological, engineering and economic data are used to generate a range of

---

[18] Such judgment might involve assigning a 40-acre drainage area to an oil well.

estimates and their associated probabilities." Expanding on this concise definition, ECR is considered here to be probabilistic[19] if it involves:

1) use of a range or pdf of values for each input parameter to calculate a corresponding range or pdf of estimates of O/GIP and/or reserves and

2) classification of various quantities within the calculated range as proved, probable and/or possible, based on the cdf of the calculated quantities, or other considerations discussed below.

For analogy estimates (see Sec. 2.0), or for analytical calculations (see Sec. 3.0), the range or pdf of each input parameter may be based on a combination of analogy, statistical analysis, and/or professional judgment.

For performance-based estimates (see Sec. 4.0 and Sec. 5.0), the input "parameters" will be historical pressure/production data, with ECR being based on an appropriate statistical procedure. In this context, reference is made to the discussion in Sec 9.2.4, regarding "best fits."

Depending on the purpose of the estimate, ECR using probabilistic procedures may include:

- drilled and undrilled areas, separately,

- drilled and undrilled fault segments, separately,

- proved areas as defined by U.S. SEC (or other applicable) regulations,

- gas cap and/or oil column, separately,

- the accumulation in-toto,

- tracts with different ownership, separately, or

- wells, separately and/or in the aggregate, for a development drilling program.

These entities are defined here as "probabilistic domains." For example, U.S. SEC guidelines specify, "In the absence of information on fluid contacts, the lowest known structural occurrence of hydrocarbons controls the lower proved limit of the reservoir." For estimates to be consistent with this guideline, the downdip limit of the probabilistic domain would be defined by lowest known hydrocarbons (discussed in more detail in Sec. 9.7.6).

In cases where significant areas of a new discovery have yet to be drilled, there may be considerable uncertainty regarding fluid content and commerciality of undrilled areas and/or fault blocks. In such situations, it might be appropriate to use probabilistic procedures for the drilled and the undrilled areas separately.[20] McKay and Taylor [1979] have discussed such an approach for fields in the Gippsland basin (Australia).

As noted, probabilistic classifications usually are based on the cdf's of the calculated quantities. Depending on the purpose of the estimate, these quantities may include O/GIP and/or reserves.[21] The cdf is expressed as an EC (discussed in Sec. 9.2.3). As illustrated by Fig. 9.6, probabilistic classifications[22] typically specify the following:

---

[19] Other procedures have been used to estimate and classify reserves that may be considered probabilistic. Thus, these statements should not be considered limiting.

[20] Under U.S. SEC, and other, guidelines, proved *undeveloped* reserves are reported separately from, but in addition to, proved (developed and undeveloped) reserves.

[21] For example, in situations where an operator is investigating various exploitation scenarios for a newly discovered accumulation, it might be appropriate to classify certain quantities of oil and/or gas as "proved initially in place." For each specific exploitation scenario, various quantities might be classified as "proved (not drilled) reserves."

[22] Caution! Probabilistic notation and concepts discussed here should not be confused with those used by many exploration geologists; see for example, Rose [2000]. In that work probabilities refer to, among other considerations, *prospect reserves* distributions or *field size* distributions. In that context $P_{90\%}$ (Rose's notation) refers to *ultimate* reserves in a *yet to be*

Proved—there is at least 90% probability that:

(a) the calculated reserves actually will be recovered, or, if appropriate, the quantities calculated initially in-place actually will be realized, *or*

(b) the actual reserves will equal or exceed the calculated quantities, or, if appropriate, the actual quantities initially in place will equal or exceed the calculated quantities.

Proved plus probable[23]—there is at least 50% probability that:

(a) the calculated reserves actually will be recovered, or, if appropriate, the quantities calculated initially in place actually will be realized, *or*

(b) the actual reserves will equal or exceed the calculated quantities, or, if appropriate, the actual quantities initially in place will equal or exceed the calculated quantities.

Proved plus probable plus possible[24]—there at least 10% probability that:

(a) the calculated reserves actually will be recovered, or, if appropriate, the quantities calculated initially in place actually will be realized, *or*

(b) the actual reserves will equal or exceed the calculated quantities, or, if appropriate, the actual quantities initially in place will equal or exceed the calculated quantities.

For the EC shown in **Fig. 9.6,** proved, proved plus probable, and proved plus probable plus possible quantities would be 17, 39, and 73, respectively.[25]

SPE-WPC [1997] reserve definitions notwithstanding, there is not global consensus regarding the expectation level to define each reserve classification, especially proved reserves. The expectation level for proved reserves varies from one entity to the next, in the range 75 to 95%.[26] There seems to be consensus at 50% for proved plus probable reserves. For proved plus probable plus possible reserves, the range varies from 5 to 15%.

The foregoing definitions reflect a "cumulative" interpretation of the EC plotted in Fig. 9.6. An "incremental" interpretation is shown by the short dotted lines, which approximate the area under the EC with three rectangles with increments on the "x" axis defined by the foregoing classifications. Thus, for the EC in Fig. 9.6, it might be said that:

• the probability of recovering proved reserves, Pr(Pv), is 0.97,

• the probability of recovering *incremental* probable reserves, Pr(Pb), is 0.70, and

• the probability of recovering *incremental* possible reserves, Pr(Ps), is 0.27.

Despite emphasis in SPE-WPC [1997] definitions on a cumulative approach to probabilistic ECR, in some situations an incremental approach might be more appropriate. Such situations involve undrilled and/or unproved reserves where incremental expenditures and/or regulatory approval are required to bring such reserves on production. Specific guidelines for these situations—identified in Sec. A.3.1 and A.3.2 and discussed in Appendix B—are extracted for the reader's convenience in the following paragraphs.

---

discovered accumulation. *After discovery,* reserves in such an accumulation might be estimated and classified using the procedures discussed here.

[23] The term "proved *plus* probable" should not be interpreted to mean that (in a probabilistic classification) an entity must be assigned *both* proved *and* probable reserves, a point discussed in subsequent paragraphs of this section and in Appendix B.

[24] As above, the term "proved *plus* probable *plus* possible" should not be interpreted to mean that (in a probabilistic classification) an entity must be assigned *both* proved plus probable *and* possible reserves.

[25] The abbreviations "1P," "2P," and "3P," respectively, are commonly used for these classifications.

[26] Petronas, for example, reportedly uses 85% probability to define proved reserves [Karra *et al.* 1995].

For probable reserves, SPE-WPC [1997] provide the following guidelines:

(1) reserves anticipated to be proved by normal step-out drilling where sub-surface control is inadequate to classify these reserves as proved,

(2) reserves in formations that appear to be productive based on well log characteristics but lack core data or definitive tests and which are not analogous to producing or proved reservoirs in the area,[27]

(3) incremental reserves attributable to infill drilling that could have been classified as proved if closer statutory spacing had been approved at the time of the estimate,

(4) reserves attributable to improved recovery methods that have been established by repeated commercially successful applications when (a) a project or pilot is planned but not in operation and (b) rock, fluid, and reservoir characteristics appear favorable for commercial application,

(5) reserves in an area of the formation that appears to be separated from the proved area by faulting and the geologic interpretation indicates the subject area is structurally higher than the proved area,

(6) reserves attributable to a future workover, treatment, re-treatment, change of equipment, or other mechanical procedure, where such procedure has not been proved successful in wells which exhibit similar behavior in analogous reservoirs, and

(7) incremental reserves in proved reservoirs where an alternative interpretation of performance or volumetric data indicates more reserves than can be classified as proved.

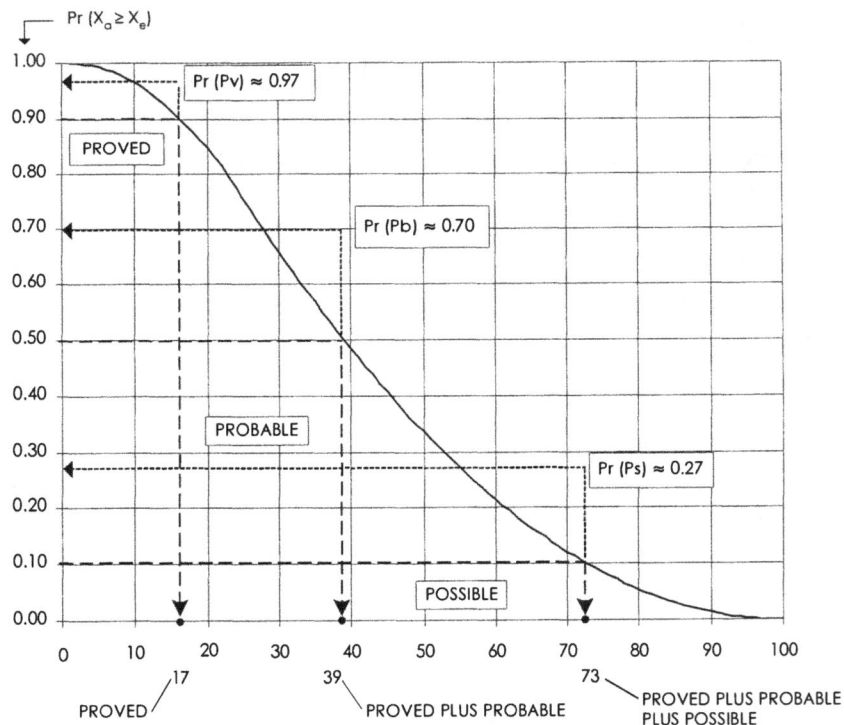

Fig. 9.6—Expectation curve for estimated reserves (arbitrary units) illustrating probabilistic classification scheme.

---

[27] Such reserves might, for example, be attributed to behind pipe zones to which proved reserves cannot be assigned.

For possible reserves, SPE-WPC [1997] provide the following guidelines:

(1) reserves which, based on geological interpretations, could possibly exist beyond areas classified as probable,

(2) reserves in formations that appear to be petroleum bearing based on log and core analysis but may not be productive at commercial rates,

(3) incremental reserves attributed to infill drilling that are subject to technical uncertainty,

(4) reserves attributed to improved recovery methods when (a) a project or pilot is planned but not in operation and (b) rock, fluid, and reservoir characteristics are such that a reasonable doubt exists that the project will be commercial,

(5) reserves in an area of the formation that appears to be separated from the proved area by faulting and geological interpretation indicates the subject area is structurally lower than the proved area.

SPE-WPC [1997] definitions state, "Proved reserves can be categorized as developed or undeveloped." Depending on the degree of geologic complexity and the stage of development, however, it may not be realistic to assign 90% probability to reserves calculated for undrilled areas or fault blocks, even if they are in close proximity to areas classified as proved. Although an EC might be calculated for such reserves, it might be appropriate to consider the results to be conditional, the condition being the drilling of a commercially successful offset or outstepping well. Conditional reserves are discussed in Secs. 9.5.1 and 9.7.

Quantities calculated and classified as illustrated by Fig. 9.6 that are subject to operational and/or other uncertainties also might be considered conditional. For example, reserves calculated for behind pipe zones may be subject to drainage and/or mechanical uncertainties.

In such circumstances, it might be appropriate to multiply the y-axis of the EC by the (subjective) probability of geologic and/or operational success, as discussed in Sec. 9.7.

9.3.4 Proved Reserves and Reasonable Certainty—The U.S. Record

Historically, estimates of proved reserves have been qualified by the phrase "reasonably certain." The inclusion of probabilistic procedures in SPE-WPC [1997] reserve definitions has led to scrutiny of this phrase in an attempt to relate "reasonable certainty" to a numerical confidence or probability. Results of a recent survey of the U.S. industry indicate substantial divergence of opinion regarding interpretation of the phrase "reasonable certainty."[28] A review of historical definitions published by various agencies in the U.S. and industry practices regarding proved reserves might shed light on this matter.

In 1946, API-AGA[29] published the first annual *Proved Reserves of Crude Oil, Natural Gas Liquids and Natural Gas.* In annual publications over the subsequent 33 years, API-AGA set standards for estimating proved reserves that became industry benchmarks. Terminology during this period, however, was modified from time to time. In 1946—as today—proved reserves included both drilled and undrilled areas. Drilled (proved) reserves were limited to those recoverable from "production systems now[30] in operation." Undrilled (proved) reserves were qualified by the phrase "every reasonable probability they will produce when drilled."

In 1960, however, API-AGA qualified proved reserves as those that could be produced "beyond reasonable doubt." Undrilled (proved) reserves were attributed to areas that were

---

[28] Most authors consider that the term has no quantitative meaning.
[29] American Petroleum Inst. and American Gas Assoc.
[30] Underlines here (not present in the original publication) emphasize critical terms and phrases.

"virtually certain of productive development." Of special interest to this discussion is the API-AGA caution that proved reserves "represent strictly technical judgments, and are not knowingly influenced by policies of conservatism or optimism."

Beginning in 1964, "reasonable certainty" was substituted for the phrase "beyond reasonable doubt." The phrase "virtually certain of productive development" was retained for reserves attributed to undrilled areas.

Terminology for (proved) reserves attributed to undrilled areas was modified beginning in 1966, with "can be reasonably judged as economically productive." The phrase "not knowingly influenced by policies of conservatism or optimism" was dropped from the definitions.

API-AGA published their last annual report in 1979. U.S. DOE/EIA (EIA) began publishing an annual report in 1977, the two-year overlap being intended to ensure continuity with the API-AGA reports. The descriptive terminology in the EIA reports has been essentially the same as that used in the API-AGA 1979 report.

What conclusions regarding industry practice for estimating proved reserves might be inferred from this history? The API-AGA caution "not knowingly influenced by policies of conservatism or optimism" seems to imply a neutral, unbiased, estimate The historical record of aggregate revisions to previous estimates (ARPE) of U.S. domestic proved reserves since 1977 might provide insight regarding industry practice. These data, compiled from EIA Annual Reports, which include only proved reserves, are plotted in **Fig. 9.7.**

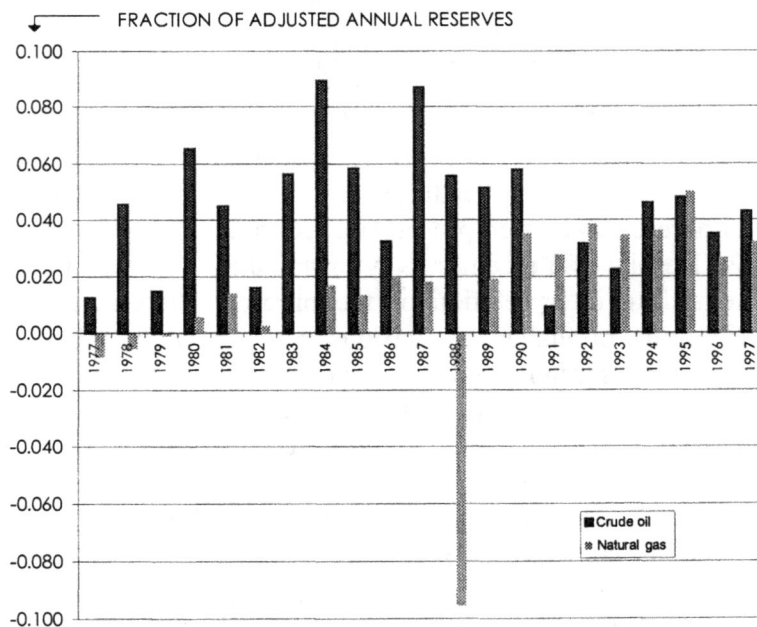

Fig. 9.7—Annual revisions to U.S. crude oil and dry gas proved reserves (Data by courtesy U.S. DOE/EIA).

On this figure, ARPE are plotted as fraction of adjusted annual reserves (FAAR). FAAR is based on year end estimated reserves, less additions during the year attributed to extensions and discoveries of new fields and/or reservoirs.[31] Two substances are plotted, crude oil and dry natural gas.

FAAR's for crude oil during this period vary from 0.009 to 0.090. There is no correlation between FAAR and average wellhead price, and no trend vs. time. The average FAAR for crude oil between 1977 and 1997 is 0.044. It is significant that FAAR's for this entire period are positive.

A comparable history is observed for dry natural gas. With the exception of five years— 1977-1979, 1983, and 1988—FAAR for the period examined have been positive. There is no correlation between FAAR and average wellhead price. The large downward revision in 1988 is attributed to reclassification of almost all the gas cap reserves at the Prudhoe Bay field (Alaska) due to lack of market. Ignoring the 1988 revision, the average FAAR for dry natural gas between 1977 and 1997 is 0.018. In contrast with the FAAR for oil, however, there is a clear, increasing trend of FAAR for natural gas vs. time. It is suspected that this trend is attributable to gradual improvements in the fracture technology of low permeability reservoirs and increasingly deep compression of major, low pressure reservoirs in the U.S. midcontinent (the Hugoton area.)

From this history, it seems apparent that—for U.S. fields—industry estimates for *proved* reserves have been *slightly* conservative. Had estimates been "not knowingly influenced by policies of conservatism or optimism," it seems apparent that FAAR would have been more or less evenly distributed between positive and negative values.

It also is significant that, in the aggregate, revisions were rather small. The operators' estimates of proved reserves or crude oil and natural gas (before annual revisions) averaged only 0.044 and 0.018 (respectively) less than observed. This history seems to imply that— during the period examined—aggregate industry estimates were only slightly less than the means of actual reserves—slightly biased on the conservative side.

Estimates with a slightly conservative bias are significantly different from estimates that are likely to be equaled or exceeded 90% of the time—i.e., significantly different from P90 estimates.

It seems apparent that—during the period analyzed—U.S. industry practice led to estimates of proved reserves that were only slightly less than actual mean values[32] and not to estimates that reflect the 10th percentile—not to P90 estimates. Accordingly, it is concluded here that the term "reasonable certainty" has been interpreted by the U.S. industry as an estimate of the mean value, which is consistent with the earlier guideline "not knowingly influenced by policies of conservatism or optimism."

9.3.5 Which Procedure—Deterministic or Probabilistic?[33]

The relative appropriateness of each procedure depends on the purpose of the estimate, the degree of geologic and/or engineering uncertainty, and the level of monetary risk. While probabilistic procedures have been advocated to identify upside potential and downside risk, an equally important purpose is quantification of the degree of uncertainty.

---

[31] "Fraction of annual adjusted reserves" is not part of EIA lexicon; it was developed for this work.

[32] This observation is consistent with that of Keith, Wilson, and Gorsuch [1986] who noted "little evidence to suggest that reserves calculated...as proved...are significantly different from those...at 50 per cent certainty using a probability method."

[33] It is estimated here that over 90% of reserve estimates currently made by the industry in North and South America are based on deterministic procedures. The state of the art n probabilistic procedures is quite embryonic. Thus, it is expected that probabilistic procedures described here will be expanded and modified over the next several years, as the industry develops a more extensive methodology.

In situations involving a significant degree of uncertainty and/or monetary risk, probabilistic procedures generally are considered more appropriate than deterministic procedures. Such situations[34] include:

- newly discovered offshore accumulations, especially those in sparsely developed areas, where there must be significant capital expenditures in advance of development and production,

- deep, onshore accumulations in geologically complex settings, especially those in high pressure, high temperature environments,

- accumulations where massive hydraulic fracturing is required to establish commercial production, but fracturing costs are substantial and the results are subject to considerable uncertainty, and

- improved oil recovery projects, especially where the method being evaluated has not met with repeated commercial success.[35]

For fully developed accumulations in a mature stage of production where there are no expectations of significant additional investment, there seems to be little benefit in making probabilistic estimates of reserves. If, on the other hand, a significant additional investment is being considered and the results from such investment are uncertain, a probabilistic approach might be warranted.

Consider, for example, a reservoir producing by water drive in which the wells, although flowing, are on a decline trend because of increasing WOR's and the consequent decreasing flow of total liquids. Depending on the operational scenario, installation and operation of artificial-lift equipment might represent significant incremental monetary expense. Long-term response to artificial lift might be uncertain, leading to uncertainty in the financial return. Such a scenario might be amenable to a probabilistic analysis.

Audits of probabilistic estimates are not as straightforward as audits of deterministic estimates.[36] For example, a probabilistic EC may be the result of several complex algorithms, each involving several pdf's for input data. The reasonableness of each such pdf, the possibility of dependencies,[37] and the validity of the algorithm(s) must be tested by the auditor. Probabilistic computations in the P90 and P10 ranges are not as robust as those in the P50 range[38] and may need to be adjusted. The auditor must have access to the computer input files and computational algorithms to make an acceptable audit and/or reasonable adjustments.

The effectiveness of third party audits also may be limited by the nature of the probabilistic domain:

- If the domain covers the entire accumulation, there may not be any direct links between estimated quantities and specific areas of the accumulation.

- It may be difficult to distinguish between, or assess treatment of, geologic uncertainties vs. engineering uncertainties.

---

[34] The "situation list," of course, tacitly includes scenarios where probabilistic procedures are mandated by company policy or the agency for which the estimate is prepared.

[35] Such projects are discussed briefly in Appendix F.

[36] The 1977 SPE report *Standards Pertaining to Estimating and Auditing of Oil and Gas Reserve Information*, which is reproduced in Appendix D, was prepared before that organization recognized probabilistic procedures. Currently, there is not a comparable document to cover such procedures; but revisions to the 1977 report are in committee.

[37] The impact of dependencies on probabilistic calculations is discussed in Sec. 9.5.4.

[38] This point is discussed in more detail in Sec. 9.5.5.

Some operators utilize a "scenario approach" to analyze the range of uncertainty in estimates of O/GIP and/or reserves. Typically, this methodology involves three separate scenarios or estimates:

1. an estimate based on the minimum (most conservative) values of pertinent parameters, which is considered "proved," or "1P,"

2. an estimate based on the most likely values of pertinent parameters, which is considered "proved plus probable, or "2P," and

3. an estimate based on the maximum (most liberal) values of pertinent parameters, which is considered "proved plus probable plus possible," or "3P."

Although seemingly a direct way to treat uncertainty, the scenario approach has several limitations:

1. Computation of the "1P" estimate tacitly assumes that all of the minimum values of the pertinent parameters occur in the same outcome, which might result in an unrealistically low estimate (with a very small probability of occurrence).

2. For the same reason, the computation of the "3P" estimate might result in an unrealistically high estimate (with a very small probability of occurrence).

3. The degree of uncertainty in the estimates is neither quantified nor related, one to the other.

9.4 Treatment of Uncertainties

Identification and quantification of uncertainties is a key aspect of probabilistic procedures. As discussed in this section, some uncertainties may be attributed to data and may be quantifiable with statistical analysis. Other uncertainties, however, may be attributed to geology and/or interpretive procedures; these may not be readily amenable to statistical analysis.

9.4.1 Types of Uncertainties

Two broad categories of uncertainties are identified here:

- Category I uncertainties, which are related to geologic and/or engineering *data* in drilled areas and/or fault blocks and the measurement accuracy[39] and interpretation of such data, and

- Category II uncertainties, which are related to the geologic *scenario* in undrilled areas and/or undrilled fault blocks.

Category I uncertainties include:

- gross rock volume in the drilled area(s),

- net-to-gross pay ratio(s) and spatial variation thereof,[40]

- in-situ properties of the rock-fluid system, including
  -petrophysical properties and
  -PVT properties,

- location of fluid contacts, if any,

- spatial distribution of permeability,

---

[39] Uncertainties attributed to measurement error are, as a practical matter, typically ignored in probabilistic analyses.

[40] The net-to-gross pay ratio in an oil column may, in general, be expected to differ from that in the gas cap, if any. Depending on the depositional environment(s) of the reservoir rock, net-to-gross pay ratios may vary substantially from well to well and/or between different zones in the same well(s).

- nature and degree of principal heterogeneity,
- degree of reservoir compartmentalization,
- drainage area(s) of individual wells, and
- recovery efficiencies of oil, gas and/or condensate.

Category II uncertainties—some of which might be considered risk factors— include:

- presence (or absence) and relative volumes of O/GIP,
- presence (or absence) of commercially productive reservoir rock, and
- areal extent of commercial accumulation (if any).

Depending on the quality and quantity of data, some Category I uncertainties may be quantifiable—i.e., amenable to statistical analysis. Such uncertainties—identified here as engineering uncertainties—include porosity, water saturation, PVT properties, recovery efficiencies, net-to-gross pay ratios, and the drainage areas of individual wells.

Some Category I uncertainties, however, may be attributable to the interpretive procedure(s) used for data analysis,[41] rather than the data. Examples include log analysis and transient pressure analysis (TPA):

- Log analysis to calculate effective porosity and water saturation in shaley sands involves selection of an appropriate petrophysical model; each model typically yields different results from the same data set; differences may be substantial.
- TPA may be used to estimate well drainage area, reservoir configuration, or other properties, which typically is done by fitting type curves; the results from type-curve fitting, however, are dependent on the model used to generate the type curves.

Interpretive uncertainties like these are not readily quantifiable and typically are resolved by professional judgment.

Other uncertainties—in both Category I and Category II—are identified here as geologic uncertainties. Such uncertainties are not readily quantifiable or amenable to statistical analysis and are subject to professional judgment. For example, depending on the geologic setting, volumetric mapping may be highly subjective. Depending on the amount of geologic data that is available and the contouring algorithm, different—but equally valid—interpretations may be made for the same area.

9.4.2 Sampling and Analysis

Quantification of many Category I engineering uncertainties is, basically, a problem of sampling and statistical analysis. Populations of concern, sampling density and frequency, and analysis methods depend on circumstances. For example, data requirements and analysis procedures for reserve estimates during the static phase—before sustained production is initiated—differ from those of the dynamic phase—after sustained production is initiated. Of concern during the static phase are the spatial distributions of initial reservoir properties that are needed to estimate O/GIP and reserves.[42] Of concern during the dynamic phase are temporal/spatial observations of well behavior that help to characterize interwell volume(s) and reservoir performance.

---

[41] Elliott [1994] calls this "modeling uncertainty."
[42] The influence of other factors on reserves is recognized; however, these are not sampling problems in the context of this discussion.

To estimate initial reservoir properties—characterize a reservoir during the static phase—initial pressure and temperature must be measured, and rock and fluid samples (or measurements) must be taken at various spatial locations in the reservoir. Adequate sampling and proper assessment of the nature of subsurface populations, however, are far more complex than implied by this brief statement.[43]

For example, determination of porosity may involve statistical analyses of different types of logs and cores, each of which may be of varying quality and degree of completeness. Procedures to resolve discrepancies between these two disparate types of data tend to be subjective and object oriented. Comparable problems are encountered in the determination of other volumetric factors, including initial water saturation and net pay.[44]

Determination of the properties of reservoir fluids and reservoir fluid contacts may involve analyses of conflicting data from logs, RFT's, well tests, and fluid samples. Collaboration between various specialists is essential. There are no standard procedures for such analyses, and the accuracy of the initial characterization will be dependent on data quality and the ingenuity of the characterization team.

During the dynamic phase, the temporal and spatial pressure/production behavior of individual wells and fluid movements in the reservoir must be monitored. Such monitoring is an essential element of prudent reservoir management, as discussed in Sec. 8.0. Given the volume of data generated during this phase of exploitation and the subjective nature of performance analysis, it is conjectural whether a generalized probabilistic approach can be developed. Some aspects of this problem, however, are discussed in Sec. 9.11.

It has been recognized that the data utilized by geologists and engineers are infinitesimal samples[45] from very large populations that typically exhibit considerable spatial variation. Geologic and engineering literature repeatedly have stressed that, even if every well were cored in a typical reservoir, on the order of 1/1,000,000 of the total reservoir volume would have been sampled. Similar observations may be made for sampling a reservoir by well logging or by fluid sampling, which might involve 1/1,000 of the total reservoir volume. Because of such limitations, core and log data alone may not provide adequate reservoir characterization, especially regarding the nature of the pdf's of critical parameters.

The problem of reservoir characterization is exacerbated by errors in laboratory analysis of cores and fluid samples,[46] by sampling disparities between log and core data, and frequently by non-representative mathematical models used for log and transient pressure analysis. Such factors—rarely quantified—are typically ignored, which contributes to the overall level of uncertainty that, as a consequence, is generally underestimated.

9.4.3 Frequency Distributions in Nature

Data validation and determination of the pdf's of input parameters used in mathematical models are among the first steps in a probabilistic analysis. Such a probabilistic analysis might involve, for example, estimating O/GIP and recovery efficiency. As discussed in Sec. 3.2, input for such an analysis might include:

---

[43] Geostatistical mapping and stochastic reservoir modeling—both subjects of considerable importance in the assessment of heterogeneous reservoirs—are beyond the scope of this work. The interested reader is referred to Davis [1986] and Yarus and Chambers [1994].

[44] Probabilistic estimation of net pay is discussed in Sec 9.9.2.

[45] In the absence of obviously spurious data, such samples historically have been accepted without systematic treatment of measurement error. Accordingly, variation between samples has been attributable to reservoir heterogeneity, ignoring the possibility that part of the variation may be attributable to such error.

[46] See Sprunt et al. [1990], discussed in Sec. 3.4.

- petrophysical data—e.g., ϕ, S_w,
- fluid data—e.g., R_{si} and B_{oi},
- volumetric data—e.g., A and h, or V_t and R_{ng}, and
- recovery efficiency of oil and solution gas—e.g., E_{Ro}, and E_{Rg}.

In the early stages of delineation and development, rarely are data sufficient to define the pdf's of these parameters. Accordingly, subjective judgments must be made regarding the minimum, the maximum, and the nature of the pdf[47] within these bounds for each of these parameters.[48]

Despite the paucity of data that typifies the early stages of development, some general observations can be made regarding the type of pdf's that reasonably can be expected for various volumetric parameters.

Parameter	Type of Frequency Distribution(s)
Porosity	Typically Gaussian, but may exhibit slight skew, either positive or negative; tendency towards positive skew and log-normal distributions in Texas carbonates and negatively skewed beta distributions in Texas sandstones; data typically exhibit covariance[49] with initial water saturation [Holtz 1993].
Interstitial water saturation	Typically exhibits negative skew, but may approach a symmetric distribution, or a distribution with positive skew.
Net pay	Typically exhibits positive skew, with log normal being a good approximation;[50] may exhibit covariance with minimum (cutoff) porosity and/or interstitial water saturation.
Permeability[51]	Typically exhibits positive skew, with log normal being a good approximation; nature of the distribution may be controlled by the type of depositional environment and/or post-depositional diagenesis; typically, covariant with porosity.
OIP (STB/ac)[52]	Generally exhibits positive skew, which ranges from 0.22 to 4.7 in Texas reservoirs [Holtz 1993].
Recovery efficiency	Typically exhibits positive skew, with log normal being a good approximation; may exhibit covariance with porosity, irreducible water saturation, permeability, net-to-gross pay ratio, and/or net pay, depending on drive mechanism.

---

[47] Fitting input data sets to theoretical pdf's is not a mandatory step for input to PC simulation programs. Several such programs facilitate customizing pdf's to fit the data. Thus, depending on circumstances, it may be preferable to utilize this feature, rather than "forcing" the data sets to fit the theoretical pdf's provided by the program manufacturer.

[48] A distinction might be made between *subjective* probability analysis–that based on professional judgments regarding the pdf's of input parameters–and *objective* probability analysis–that based on statistical analysis to determine the pdf's of input parameters. The reader is referred to Megill [1979] for further discussion of this point.

[49] Covariance is a statistical measure of the degree of correlation between data sets. Stochastic variables exhibiting a high degree of covariance may not be independent, and caution should be exercised in treating them as independent variables in probabilistic calculations. This point is discussed further in Sec. 9.5.4.

[50] Squire [1996] reported that, due to omission of thin, non-commercial zones from Santos' data base, a Pareto distribution was a better fit to the pdf of net pay.

[51] Although not directly involved in volumetric estimates of reserves, permeability may be an important correlating parameter in the selection of capillary pressure and relative permeability functions.

[52] Oil initially in place (stock tank barrels per acre); please see Notation for complete list of abbreviations.

Initial well potential	Typically exhibits positive skew, with log-normal being a good approximation; may exhibit covariance with porosity, permeability, net pay and/or type of stimulation, if any.

As noted by the foregoing summary, log-normal distributions may be reasonable approximations to many states of nature. Log-normal distributions, however, are defined by $0 < X < \infty$. In contrast, pdf's of input parameters used in ECR are finite and greater than zero. Thus, if log normal distributions are used to approximate input parameters, they should be truncated at the observed—or reasonably expected—minimum and maximum quantities of such parameters. Otherwise, simulations may be unrealistic, as discussed in Secs. 9.5.2 and 9.6.2.

The pdf's of reservoir parameters used in probabilistic calculations should represent *reservoir average* values, not the spatial distributions of these parameters. Input data for Eq. 3.1a through Eq. 3.3b—ϕ_o, S_{wo}, and h_{no}, for example—are reservoir averages. The question, however, is, average of what? Given the paucity of subsurface data typical of the early stages of reservoir appraisal, it is impossible to determine the correct average for such parameters, much less the pdf's of such averages. One approach to this problem is to use the pdf's of the subsurface samples—log and core data—to approximate the pdf's of reservoir averages. This procedure may be the only reasonable approach, even though the variances of such pdf's may be greater than the variances in the reservoir averages. As a practical matter, however, the "tails" of the sample pdf's have very low probabilities of occurrence, resulting in insignificant distortion of the simulated pdf's.

As discussed in standard texts on probability and statistics, the pdf of the product of two or more pdf's tends to approach log normal as the number of multipliers increases. This theoretical observation has been cited by many authors to support the view that the pdf's of reserve estimates should be log normal. In support of this view, reference is made to observations by Kaufman [1963] and others that pdf's of reserves attributed to sets of fields in comparable geologic settings tend to be log normal. This, however, is a separate issue. Each phenomenon is the result of two distinctly different processes. Probabilistic estimates of reserves for a specific field are generated by multiplication of pdf's, and the pdf of the product should—in theory—be log normal. The process whereby nature generates and entraps oil and/or gas is the result of many sub processes, all of which are poorly understood. It seems hardly likely, however, that nature multiplies pdf's of porosity, hydrocarbon saturation, net pay, productive area, and recovery efficiency.

9.5 Probabilistic Procedures

Three procedures[53] for probabilistic computation are discussed here: (1) probability tree, (2) parametric method, and (3) simulation. Other procedures have been discussed [Grayson 1960, Newendorp 1975, and McCray 1975].

To facilitate comparison, each of the three procedures discussed here involves the same, rather simple model, which was adopted from a classic publication by one of the industry pioneers in probabilistic analysis, Robert Megill. These procedures, however, may be used to estimate O/GIP and/or "reserves"[54] for any level—prospects, wells, tracts, and/or fault blocks.

---

[53] Other procedures have been used that may be considered "probabilistic," for example, the "scenario" method discussed briefly in Sec. 9.3.5; thus, this discussion should not be considered limiting.

[54] It is recognized that some writers might limit usage of the term "reserves" to "known accumulations." Usage here, however, is consistent with, albeit informal, current industry usage. (Such transgressions are flagged in this work with quotes.)

9.5.1 Decision, or Probability, Tree

Background

Decision trees were developed in the 1960's to facilitate analysis of complex, sequential decision processes. They also have been used for systematic analysis of the potential outcomes resulting from sequences of probabilistic occurrences and have been called stochastic decision trees. Application of such trees to the oil and gas industry has been discussed by Grayson [1960], McCray [1975], Newendorp [1975], Megill [1979, 1984], Murtha [1995], Swinkels [2001], and others. The term "probability tree" is adopted here, being considered more appropriate to the application.

A probability tree is a simple procedure for probabilistic calculation and is appropriate when there is little basic data and the nature of the pdf's of the data are unknown. In this context, probability trees might be considered subjective.

Although the procedure is simple and subject to some limitations, a probability tree is a powerful method to evaluate sequences of probabilistic occurrences and the resulting outcomes.

Example Calculation

Probability trees have been illustrated by, for example, Megill [1979] with his "School Prospect," from which the following example is taken.[55]

Consider a prospect in a trend of proved gas accumulations. Based on data compiled from discoveries on this trend, the following estimates have been made for area, net pay, and recovery factor—called "outcomes" in probability theory—and the probabilities of occurrence of each outcome, *assuming a successful well*:

	Outcomes[56]		
	Min (Prob)*	ML (Prob)*	Max (Prob)*
Average net pay (ft)	50 (0.3)	100 (0.5)	190 (0.2)
Productive Area (acres)	800 (0.3)	1000 (0.5)	1400 (0.2)
Recovery Factor (Mscf/AF)	700 (0.3)	1000 (0.6)	1500 (0.1)

* Megill's minimum, most likely, and maximum estimates do not correspond to P90, P50, or P10 estimates.

For each of the potential outcomes,[57] the sum of the probabilities of occurrence equals 1.0. The probability of a dry hole on this trend was estimated to be 0.71. Thus, the probabilities in this table are called "conditional probabilities"—the "condition" is a successful well, for which the probability is estimated to be 0.29.[58]

Table 9.1 shows the probability tree for this prospect in spreadsheet format.[59] In this model, each outcome is assumed to be independent of the others. Thus, it is assumed that there is no dependency[60] between net pay (NETPAY) and productive area (AREA) or between net pay and recovery factor (RECFAC).

---

[55] Discussion of Megill's "School Prospect" might seem dated. However, his work is an important contribution to industry literature and helps to illustrate several important points regarding probability analysis.
[56] The terms "minimum" and "maximum" used in this section reflect Megill's [1979] terminology—and that of many geologists—and differ from those defined in Sec. 9.5.2 (Eq. 9.2).
[57] There is, of course, a continuum of potential outcomes, rather than just minimum, most likely, and maximum. Methods to treat this aspect of the problem are discussed in Sec. 9.5.2 and 9.5.3.
[58] The potential application of conditional probabilities to classification of reserves is discussed in Sec. 9.7.
[59] PC programs to perform such analysis are available, for example, Palisade's *Precision Tree,*™ Merak's *Decision Tree.*™ Reference to these products does not imply endorsement. The spreadsheet format is used here to illustrate several aspects of probability tree analysis that are not readily apparent from commercial output formats.
[60] The implication of such dependencies is discussed in Sec. 9.5.4.

TABLE 9.1—SPREADSHEET FOR DECISION TREE ANALYSIS OF MEGILL'S [1979] SCHOOL PROSPECT

Input Parameters

	Min	Pr(min)	ML	Pr(ML)	Max	Pr(max)
NetPay (feet)	50	0.3	100	0.5	190	0.2
Area (acres)	800	0.3	1000	0.5	1400	0.2
RecFac (Mscf/AF)	700	0.3	1000	0.6	1500	0.1

Probability Tree   Calculations

NETPAY	Pr(NP)	AREA	Pr(A)	RECFAC	Pr(RF)	Risk Factor	ULTGAS (Bscf) Unrisked	Risked
50	0.3	800	0.3	700	0.3	0.027	28.0	0.8
				1000	0.6	0.054	40.0	2.2
				1500	0.1	0.009	60.0	0.5
		1000	0.5	700	0.3	0.045	35.0	1.6
				1000	0.6	0.090	50.0	4.5
				1500	0.1	0.015	75.0	1.1
		1400	0.2	700	0.3	0.018	49.0	0.9
				1000	0.6	0.036	70.0	2.5
				1500	0.1	0.006	105.0	0.6
100	0.5	800	0.3	700	0.3	0.045	56.0	2.5
				1000	0.6	0.090	80.0	7.2
				1500	0.1	0.015	120.0	1.8
		1000	0.5	700	0.3	0.075	70.0	5.3
				1000	0.6	0.150	100.0	15.0
				1500	0.1	0.025	150.0	3.8
		1400	0.2	700	0.3	0.030	98.0	2.9
				1000	0.6	0.060	140.0	8.4
				1500	0.1	0.010	210.0	2.1
190	0.2	800	0.3	700	0.3	0.018	106.4	1.9
				1000	0.6	0.036	152.0	5.5
				1500	0.1	0.006	228.0	1.4
		1000	0.5	700	0.3	0.030	133.0	4.0
				1000	0.6	0.060	190.0	11.4
				1500	0.1	0.010	285.0	2.9
		1400	0.2	700	0.3	0.012	186.2	2.2
				1000	0.6	0.024	266.0	6.4
				1500	0.1	0.004	399.0	1.6

Avg ULTGAS (Unrisked): 129
Expected ULTGAS: 101

From this table, it may be noted:

(a) Each outcome—ULTGAS—is "risk-weighted" by the product of the probabilities of occurrence of each combination of AREA, NETPAY, and RECFAC.

(b) The same (numerical) outcome may be realized by different combinations of input parameters—the value of the input parameters cannot, in general, be determined from the value of the outcomes.

(c) The expected ultimate, ULTGAS, *attributable to a discovery*[61]—101 Bscf—is the sum of the risk-weighted outcomes.

(d) The arithmetic average of the 27 possible (unrisked) outcomes is 129 Bscf.

None of the unrisked estimates in Table 9.1 are less than 28 Bscf, which some exploration geologists might consider unrealistic.[62] This, however, is a merely procedural example, with the calculated estimates being controlled by Megill's input parameters.

The results of this analysis are shown graphically in:

(a) a histogram of unrisked estimates, **Fig. 9.8,** which exhibits positive skew[63] and

(b) an expectation plot, **Fig.9.9a,** calculated (after Megill 1979) by ranking the unrisked estimates and corresponding risk factors and plotting the estimates vs. the complement of the cumulative risk factors.

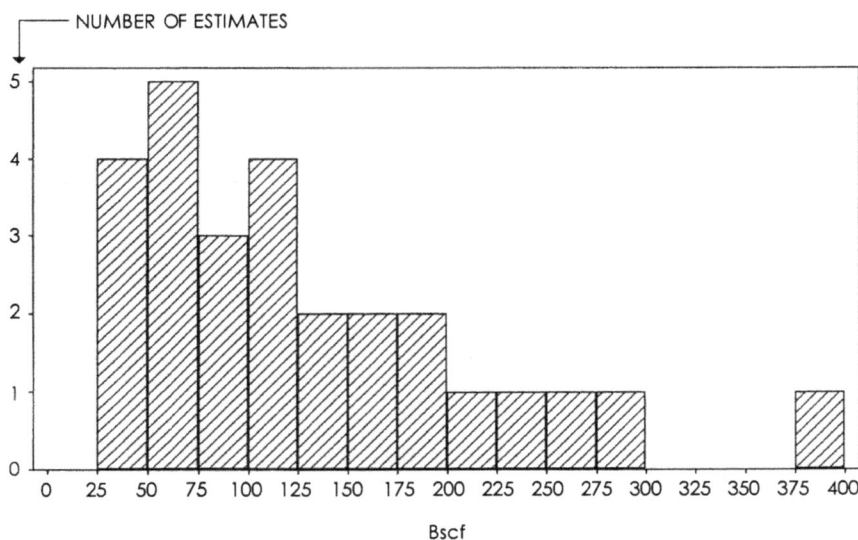

Fig. 9.8—Histogram of unrisked estimates of "reserves" calculated for Megill's School Prospect.

From Fig. 9.8, it may be noted that, *given a successful well,* the mode—the most likely outcome—is between 50 and 75 Bscf. The actual mode, computed from the data in Table 9.1, is 70 Bscf.

From Fig. 9.9a, it may be noted that the median outcome (the 50th percentile) is approximately 80 Bscf. Thus, *given a successful well,* there is 50% probability that the actual ultimate gas recovery will equal or exceed 80 Bscf.

Decision matrices are an alternate method to calculate expected quantity values of outcomes, given a set of discrete events and related probabilities. The procedure, which was discussed by Elliott [1994], is mathematically equivalent to that used in probability trees. The abbreviated matrices, however, make it difficult to visualize the chain of logic and the relations between events.

---

[61] The expected "reserves" for the decision to drill would be 29 Bscf (0.29 x 101).
[62] Rose, personal communication, 2000.
[63] The pdf of these estimates is approximately log normal.

9.5.2 Parametric Method

Background

The so called "parametric method" was developed during the late 1970's as an alternative to Monte Carlo simulation (discussed in Sec. 9.5.3) which, at the time, typically involved considerable programming effort and frequently required unacceptably long computation time. Because of its simplicity, however, the method continues to be used, [64] even in today's environment of commercially available Monte Carlo simulation software and high-speed PC's.

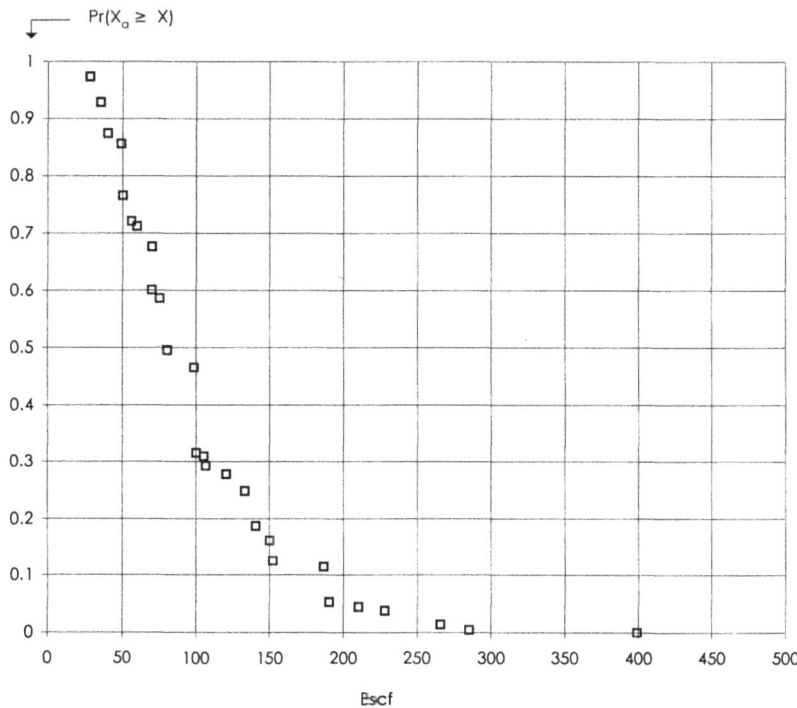

Fig. 9.9a—Expectation for unrisked estimates of "reserves" calculated for Megill's School Prospect.

The term "parametric methods," apparently introduced by Davidson and Cooper [1976], refers to the fact that the method works with descriptive parameters of pdf's, rather than pdf's. The only data required are the minimum, maximum, and most likely value of each input pdf. Knowledge of the type(s) of pdf's is not required. Advantages and shortcomings of the method vis-a-vis Monte Carlo simulation have been discussed by Smith and Buckee [1985]. Applications have been described by Capen [1992], Mireault [1994], and Otis and Schneidermann [1997].

The procedure also has been called the "three-point method," the "method of moments," and the "Warren method."[65] Although initially developed to evaluate exploration prospects, the method also may be used for many projects where a probabilistic approach is warranted. It has been used, for example, to calculate project economics [Mireault 1994].

---

[64] The procedure is used by Fekete Associates (Calgary) for their *Ca$hPot* software and by Chevron [Otis and Schneidermann 1997] for prospect analysis. Neither endorsement of these authors nor exclusion of others is intended by these references.

[65] During his tenure with the Gulf Oil Corporation, Joe Warren developed the method to estimate "reserves" attributable to exploration prospects.

Theory

For statistically independent random variables, it has been shown:

- The mean of the product of a set of pdf's equals the product of the means of the individual pdf's.
- The variance of the product of a set of pdf's equals the product of the variances of the individual pdf's.
- The pdf of the product of two or more pdf's tends to be log normal.

As discussed in Sec. 9.2.2 and 9.4.3, pdf's characteristic of oil and/or gas reservoirs typically are unimodal and approximately log normal, but may exhibit skew ranging from slightly negative to slightly positive. The means and variances of such distributions may be estimated using algorithms developed by Pearson and Tukey [1965] and by Keefer and Bodily [1983]. Accordingly, the mean of a pdf of, for example, porosity, may be estimated by[66]

$$m_1(\phi) \approx [(\phi)_{MIN} + 0.95(\phi)_{ML} + (\phi)_{MAX}]/2.95 . \dotfill (9.1)$$

The variance may be estimated by

$$m_2(\phi) \approx m_1(\phi)^2 + [(\phi)_{MAX} - (\phi)_{MIN}]^2/10.56 , \dotfill (9.2)$$

where $(\phi)_{MIN}, (\phi)_{ML}, (\phi)_{MAX}$ = minimum, most likely and maximum values of porosity,[67]
m_1, m_2 = the first and second moments—the mean and variance, respectively, and
$m_1(\phi)^2$ = square of the first moment of porosity.

With reference to Sec. 3.0, for a probabilistic estimate of (the mean value of) OIP, N_i, Eq. 3.1a could be written[68] as

$$m_1(N_i) = 7758 m_1(\phi_o) m_1(S_o) m_1(A_o) m_1(h_{no})/B_o . \dotfill (3.1c)$$

The second moment of the distribution of N_i—the variance—is the product of the second moments of the input distributions,[69] or

$$m_2(N_i) = 7758^2 m_2(\phi_o) m_2(S_o) m_2(A_o) m_2(h_{no})/B_o^2 . \dotfill (3.1d)$$

The method assumes that the pdf of N_i is log normal. The variance of the natural logarithm of (the pdf of) N_i is calculated as

$$\delta^2 = \ln[m_2(N_i)/m_1(N_i)^2] . \dotfill (9.3)$$

The value of N_i at any percentile of the distribution, $PcX(N_i)$, may be calculated as

---

[66] The discussion here follows Mireault [1994]. Other authors (cited here) follow different procedures to characterize input distributions and, consequently, use different algorithms.

[67] "Minimum" and "maximum" values for these equations should be Pc05 and Pc95, respectively.

[68] For this example, it is assumed that uncertainty in the formation volume factor is too small to warrant representation as a probability distribution.

[69] The variance of a constant is the square of the constant.

$$PcX(N_i) = Pc50(N_i)exp[\sigma z(x)] , \dotfill (9.4)$$

where $Pc50(N_i)$ = 50th percentile of N_i,
   σ = standard deviation of the distribution—the square root of the variance, $\sqrt{\delta^2}$, and
   $z(x)$ = number of standard deviations the Gaussian variable deviates from the median at a specified percentile; see, for example, Snedecor and Cochran [1980] Table A3.
and where

$$Pc50(N_i) = m_1(N_i)exp(-0.5\delta^2) . \dotfill (9.5)$$

As discussed in Sec. 9.2.3, expectation is the complement of percentile. Thus, an EC may be generated by plotting $(1.0 - PcX/100)$ vs. $PcX(N_i)$.

Example Calculation

Table 9.2 is an example of these calculations. To facilitate comparison with the work in Sec. 9.5.1, the input data were taken from Megill's School Prospect. Those data were modified, however, to approximate the input requirements of Eq. 9.1 and 9.2.

TABLE 9.2—SPREADSHEET ILLUSTRATING PARAMETRIC METHOD

	Min	ML	Max	m_1	m_2
NETPAY	20	100	250	124	20,317
AREA	500	1,000	1,700	1,068	1,276,521
RECFAC	500	1,000	1,600	1,034	1,183,502
$m_1(N_i), m_2(N_i)$				1.366E+8	3.069E+16
δ^2				0.498	
$Pc50(N_i)$				1.06E+08	
$Pc84.1(N_i)$				2.157E+8	

The EC for calculated reserves is shown on **Fig 9.9b,** which includes the results from Sec. 5.1 (Fig 9.9a). It may be noted:

- The limited number of outcomes computed with the probability tree results in poor definition of the expectation trend compared to that computed with the parametric method.

- At high expectations (greater than about 0.8) results of the two procedures are approximately the same; at lower expectations, however, results of the parametric method become increasingly greater than those from the probability tree.

The increasing disparity between the two methods at low expectations is due to the parametric method's (Eq 9.4) computing an (unbounded) log-normal distribution of estimates. As noted in Sec. 9.4.3, the log-normal distribution is bounded by $0 < X < \infty$. Thus, unless the results calculated by Eq. 9.4 are bounded by a "reasonableness test," unacceptably high values may inadvertently be included in subsequent analyses. Results from the parametric method are contrasted with those calculated by a probability tree, which are bounded by discrete, rather than continuous, input parameters.

From these results, albeit limited, it seems apparent that results from Eq. 9.4 should be reviewed critically before completing an analysis using the parametric method.

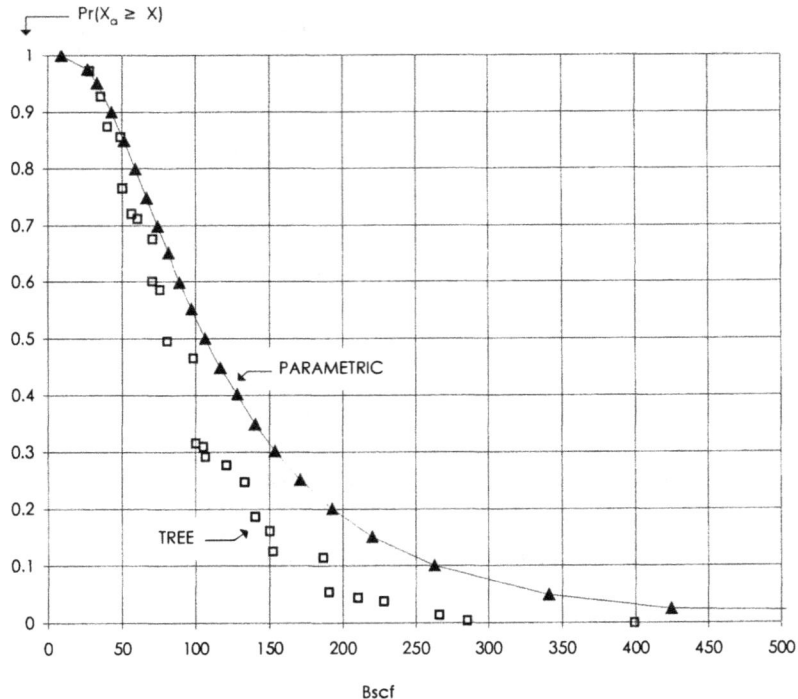

Fig. 9.9b—Expectation for unrisked estimates of reserves calculated for Megill's School Prospect using probability tree and parametric methods.

Swanson's Mean

Readers may notice the similarity between the procedure outlined above to estimate the first moment, or mean, of a pdf—Eq. 9.1—and the procedure to estimate Swanson's mean, which was introduced in published literature by Megill [1984]. Swanson's mean (SM) is an approximation to the mean value of a log normal distribution and may be calculated as

$$SM = 0.3 \times P90 + 0.4 \times P50 + 0.3 \times P10 . \quad\quad\quad\quad (9.6)$$

As noted by Megill [1984], SM is a reasonable approximation to the true mean of a log normal distribution, provided the distribution is not too highly skewed. For progressively more skewed distributions, SM approaches the median value of the distribution. The following guidelines, offered by Megill [1984], show the ratio of SM to the true mean, X_m, for several values of the ratio of P10 to P50:

P10/P50	SM/X_m
3.0	0.97
5.0	0.90
7.0	0.81
9.0	0.72
15.0	0.55

P10/P50 ratios between approximately 2.0 and 6.0 are typically observed in prospect evaluations. Ratios between approximately 6.0 and 15.0 typically are observed in basin evaluations [Megill 1984].

Swanson's mean has been used to quantify expected outcomes from drilling prospects analogous to nearby developed areas. However, quantification of expected outcomes with only the estimated mean values of reserve distributions ignores the full range of potential reserve outcomes. Depending on circumstances, such a procedure may lead to over simplified economic analysis and/or misdesign of facilities.

Graphical Methods

A variation of the parametric method involves multiplication of selected percentiles from (log normal) pdf's of input parameters to estimate the (log normal) pdf of O/GIP and/or reserves [Capen 1992 and Rose 1992]. These procedures, essentially graphical, are still used by many exploration and development geologists. In recent years, however, there has been a trend towards adaptation of these methods to PC's.

9.5.3 Simulation

Background

Simulation, also called Monte Carlo[70] simulation (MCS), was developed at the Harvard Business School [Hertz 1964] to help analyze business problems, for which there typically are no analytic solutions. In brief, the procedure involves:

1. determination of the relation(s) between a set of independent random variables and one or more dependent random variables,
2. repetitive random "sampling" of the pdf's of the independent variables, and
3. repetitive calculation to generate pdf's of the dependent variable(s).

MCS is most appropriate when the pdf's of each of the independent variables can be reasonably quantified. Depending on the complexity of the problem and the nature of the input distributions, thousands of iterations may be required to yield consistent results.[71] Depending on the complexity of the algorithms, the dependencies between input parameters, and the number and complexity of output files, MCS may require PC's with high processing speed and considerable memory.

Example Calculation

To provide a comparison with methodologies discussed in previous sections, Megill's School Prospect is analyzed in this section using MCS. For this problem, MCS involved the following steps:

1. Establish the algorithm(s) to calculate the dependent variable(s). For this discussion, initial gas reserves—ULTGAS—are considered a stochastic dependent variable (as in Sec. 9.5.1), with the same algorithm:

$$ULTGAS = NETPAY \times AREA \times RECFAC$$

---

[70] Strictly speaking, the phrase "Monte Carlo simulation" refers to a sampling procedure, rather than a computational procedure. Consistent with nearly universal usage, however, "MCS" is used here to refer to any computational process that involves iterative sampling of input pdf's and repeated calculation of an object function.

[71] Reasonable consistency can be achieved by running sufficient iterations to yield a mean standard error (MSE) on the order of 1%. MSE indicates the probability that the calculated mean (CM) differs from the true mean (TM); i.e., there is a probability of 68% that CM = TM ± MSE.

2. It is assumed for this discussion that there are no dependencies between the independent variables in the algorithm, an aspect discussed in Sec. 9.5.4.

3. As done by Megill [1979], discrete probabilities used in the tree diagram were approximated by continuous triangular distributions shown in **Fig. 9.10**.

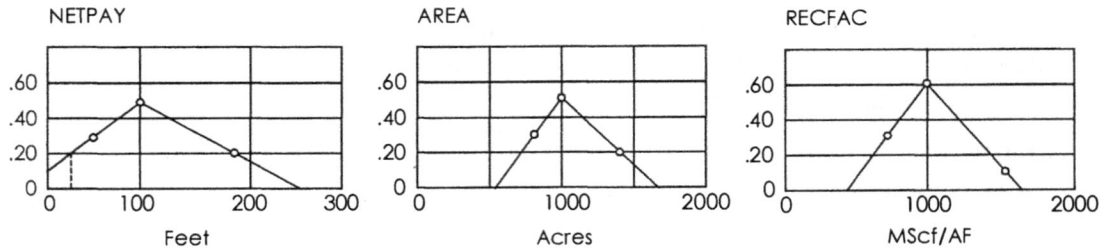

Fig. 9.10—Triangular distributions of input parameters for Megill's School Prospect.

Regarding Fig. 9.10, note that:

a)   the distribution for net pay was truncated at a minimum of 20 net ft, thereby avoiding the problem of generating negative values during the simulations and

b)   the areas under each of the triangular distributions were normalized to 1.0.

4.   Compute cdf's for each of the independent parameters, as shown on **Fig. 9.11.**

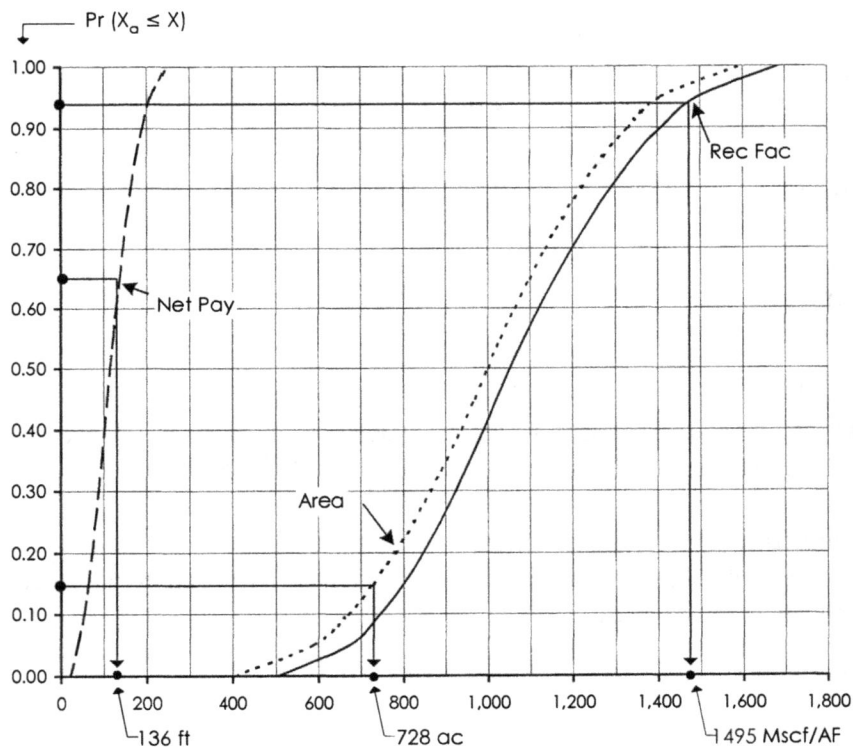

Fig. 9.11—Cumulative probability distributions of input parameters for Megill's School Prospect.

5. Generate a random number between 0.0 and 1.0 and determine the value for NETPAY that corresponds to this cumulative probability. If the random number were 0.65, as illustrated by Fig. 9.11, the corresponding value for NETPAY would be 136 ft.[72]

6. Generate a new random number between 0.0 and 1.0 and determine the value for AREA that corresponds to this cumulative probability. If the random number were 0.15, as illustrated by Fig. 9.11, the corresponding value for AREA would be 728 acres.

7. Generate a new random number between 0.0 and 1.0 and determine the value for RECFAC that corresponds to this cumulative probability. If the random number were 0.95, as illustrated by Fig. 9.11, the corresponding value for RECFAC would be 1495 Mscf/AF.

8. Using the values for AREA, NETPAY and RECFAC determined in Steps 5 thru 7, compute a value for ULTGAS which, for this iteration, would be 148 Bscf. Call this ULTGAS(1).

9. Repeat Steps 5 through 8, computing ULTGAS(2), ULTGAS(3), ULTGAS(4), ULTGAS(5) . . . ULTGAS(n), until enough values have been computed to define the pdf of ULTGAS.[73]

To obtain repeatable results, several hundred (or more) iterations generally will be required; more complex algorithms may require thousands of iterations.

Random sampling of the input data distributions was done using Latin hypercube sampling, rather than Monte Carlo sampling. Latin hypercube sampling is designed to recreate the variation in each input variable by stratifying the input probability distribution of each variable. It is a more efficient technique for statistical simulation than Monte Carlo sampling and provides greater assurance that the entire range of each input variable is sampled with fewer iterations than would be required for Monte Carlo sampling.

The results of MCS are shown on **Fig. 9.9c,** which includes the results from the previous two sections. Comparing the MCS and Parametric expectation curves, it may be noted that MCS resulted in slightly less skew than the parametric method. This might reasonably be expected, as (for this example) MCS sampled triangular distributions, which did not replicate the degree of skew of the log normal distribution generated by the parametric method.

9.5.4 Dependencies

In the example calculations in the previous sections, dependencies between input parameters were ignored. Depending on the nature of the algorithm, however, ignoring such dependencies in probabilistic calculations may lead to significant error, especially at the "tails" of the distribution.

Two broad types of dependencies may be identified:

- those between input variables used to calculate the pdf(s) of O/GIP, or reserves, for a specific accumulation and
- those between calculations of O/GIP, or reserves, for accumulations that are aggregated.

The first type of dependency is discussed in this section; the second, in Sec. 9.12.

Calculations in Sec. 9.5 assumed that the input parameters at each step in the analysis were independent. In the School Prospect, for example, it was assumed that there was no relation between net pay and productive area or between net pay and recovery factor. In some cases, however, thicker pay sections may be associated with larger productive areas and/or

---

[72] As implied by the discussion in Sec. 9.2.2, C.65 is the probability that NETPAY is *less than* 136 ft.
[73] In this example, after 10,000 iterations, the mean standard error was calculated to be less than 1%.

larger recovery factors. There may not be valid statistical correlations, and dependencies may be intuitive, rather than demonstrated. If such dependencies are ignored, however, probabilistic calculations might lead to erroneous ECR.

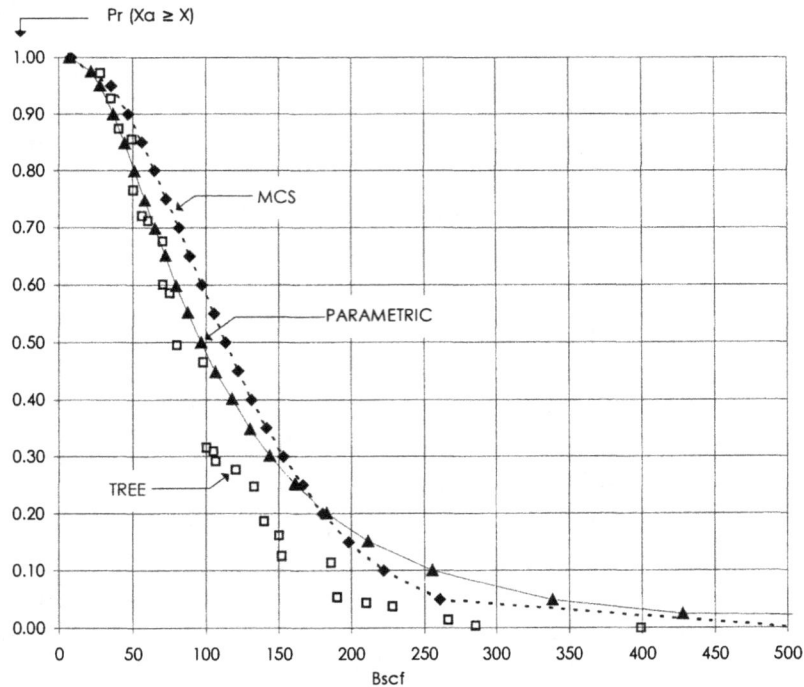

Fig. 9.9c—Expectation plots for unrisked estimates of "reserves" calculated for Megill's School Prospect using decision tree, parametric method, and MCS.

Suppose, for example, for the trend analyzed in the previous section there was a dependency suspected between average net pay and recovery factor and between average net pay and productive area. Such dependencies can be handled with a probability tree by assuming the distribution of the suspected dependent input variable to be a function of the value of the independent input variable. For example, for low values of NETPAY the distribution of RECFAC might be skewed positively; for high values, negatively. Such a procedure, however, leads to cascading levels of subjectivity and may not be suitable for more complex algorithms. Such dependencies cannot readily be handled with the parametric method without restructuring the algorithm(s), which is a major shortcoming. MCS provides a more direct method to handle dependencies. The procedure and an example are discussed in the ensuing paragraphs.

Spearman Rank Correlations

The presence of dependencies can be determined by calculating Spearman rank correlations (SRC) between pairs of input parameters. SRC is a statistical procedure to test for correlation between two data sets based on the relative *rankings* of the elements in the data sets, rather than on the *value* of the elements. The procedure may be used irrespective of the nature of the pdf's of the data sets. For computational details, the reader is referred to, for example, Snedecor and Cochran [1980]. The SRC ranges from (-)1.0, which indicates perfect negative correlation, to (+)1.0, which indicates perfect positive correlation.[74] Negative correlation

---

[74] Caution! The SRC—denoted as r_s—is not the same as the linear correlation coefficient—denoted as r—that would be

typically is observed between, for example, porosity and irreducible water saturation; positive correlation, between porosity and permeability. The probability of a (statistical) correlation between two data sets is a function of the magnitude of the SRC and the number of samples in the data set. For example, for data sets with more than about 20 samples, an SRC greater than ABS(±0.5) indicates greater than a 95% probability that there is a statistically valid correlation between data sets.

If there are sufficient data, SRC's between pairs of input variables may be computed using standard statistical software or modules in PC programs[75] for risk analysis like *Crystal Ball*™ or *@Risk*.™ If data are insufficient for such computations, sensitivity analysis might be appropriate using a reasonable range of assumed, or intuitive, SRC's for each suspect pair of variables.

Example Calculation

The influence of dependencies on Megill's School Prospect was examined by using MCS and assuming an SRC of 0.5 between AREA and NETPAY and between RECFAC and NETPAY. By so doing, the sampling procedure for the dependent input parameters was automatically adjusted (by the software) so that the dependent input parameters were sampled at a rank comparable to that at which the independent input parameters were sampled.

Results of this procedure are shown in **Fig. 9.12,** which also shows results from ignoring possible dependencies.

For each of the two scenarios, P10, P50, and P90 estimates of ULTGAS and the skew of the pdf of ULTGAS are tabulated below:

Dependencies	ULTGAS			
	P10	P50	P90	Skew
Ignored	222	114*	47	1.02
Considered	281	116	36	1.18

* The difference between the P50 estimate calculated for this example and that calculated using a probability tree is attributed to input of continuous variables vs. input of discrete variables.

It may be noted that, for this case, consideration of dependencies has a negligible effect on P50 estimates, but increases the skew of the distribution, which results in larger P10 estimates, and smaller P90 estimates.

In general, consideration of dependencies will result in greater spread between the calculated 90th and 10th percentiles than if such dependencies were not considered. A higher degree of dependency generally will result in a greater degree of spread. In this context, it is noted that the foregoing calculations assumed an SRC of 0.5. Had a larger SRC been assumed, the spread would have been larger. The relation between estimates at a specified percentile, however, is a function of the probability distributions of the input parameters and the nature of the dependencies and, thus, cannot be determined a priori. The influence of dependencies on the results of MCS is also discussed in Sec. 9.12.2.

This brief discussion is intended to serve as a "flag," rather than a comprehensive treatment of dependencies.[76] In a more comprehensive model, for example, it might be appropriate to include pdf's of porosity and net-to-gross pay ratio in the algorithm to calculate

---

calculated from a linear regression between the same two data sets.
[75] Reference to these products does not imply endorsement.
[76] For a more comprehensive (albeit somewhat outdated) discussion, please see Newendorp [1975].

the distribution of O/GIP. In this same model, it might be appropriate to calculate recovery efficiency, which may be expected to exhibit covariance with net-to-gross pay ratio and porosity, among other factors.

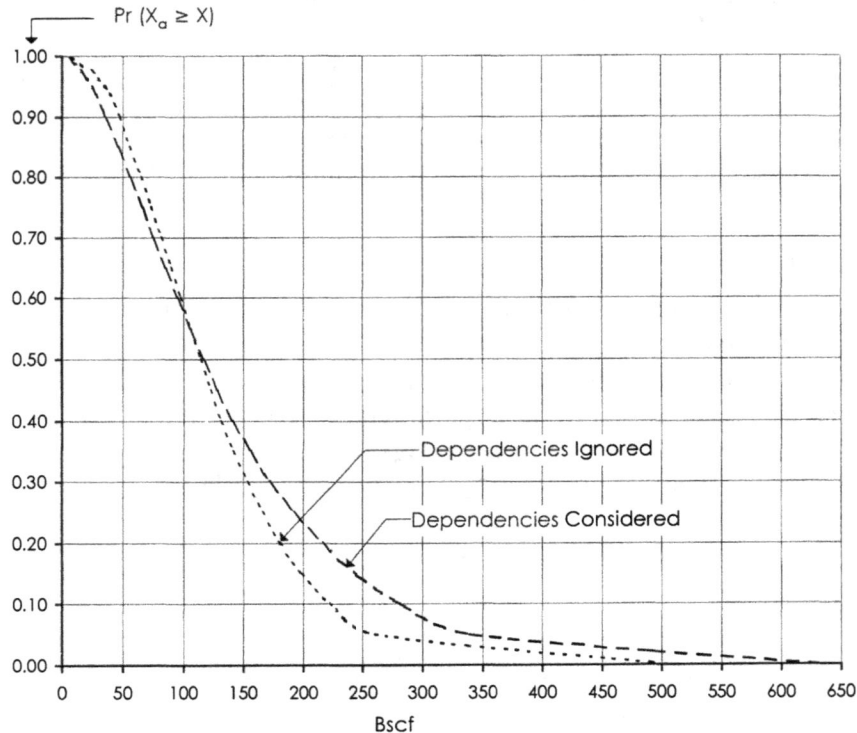

Fig. 9.12—Expectation curves for unrisked estimates of "reserves" calculated for Megill's "School Prospect" with and without dependencies.

As noted in Sec 9.4.2, significant dependencies may exist between many parameters used in estimating reserves. The available data, however, may not be sufficient to determine the degree of dependency with reasonable confidence. Depending on circumstances, it may be desirable to adopt a "what if?" approach regarding treatment of suspected dependencies. By so doing, the significance of such dependencies may be investigated, which may warrant more extensive analysis.

9.5.5 Which Procedure?

Irrespective of which procedure is used for probabilistic calculations, it should be recognized that the pdf's of calculated quantities are controlled by the pdf's of the input parameters and the treatment of dependencies. Small variations in the pdf's of the input parameters may have a significant influence on the tails of the calculated pdf's. Thus, although calculations in the P50 range may be robust, those in the P90 and P10 ranges may not.

Probability Tree

Probability trees (PT) provide a simple, easily understood procedure to estimate the expected value of a set of outcomes—the expected value of O/GIP and/or reserves. Also, the procedure may be used to *approximate* the pdf of such outcomes, as demonstrated by Megill [1979]. Special computer software is not required. Any PC spreadsheet may be programmed for PT analysis, as demonstrated in Sec. 9.5.1. One of the major advantages of a PT is the ease of visualizing the logic and computation process, an advantage not attributed to MCS.

Although a PT is simple and can provide useful insights, there are limitations, including:

- poorly defined pdf's of the object variables caused by discrete (rather than continuous) input variables,
- difficulty in handling dependencies between outcomes in complex situations, and
- cascading complexity if the analysis involves more than minimum, most likely, and maximum outcomes.

At each step of PT analysis like the School Prospect, only three (discrete) outcomes were considered: minimum, most likely, and maximum. In nature, a continuum of outcomes is possible between the minimum and maximum. In cases where the data are insufficient to define continua of values, or to quantify dependencies between successive outcomes, a PT may be adequate.

Parametric Method

The parametric method has some of the same advantages as the probability tree method:

- any PC spreadsheet may be programmed for the method, and
- the procedure may be used to approximate the statistical distribution of the object function.

Disadvantages include:

- the need to understand a complex statistical basis,
- difficulties in handling dependencies between input parameters,
- inability to input truncated distributions, and
- assumption of log normality of the output pdf's.

Simulation

In contrast with probability tree and parametric methods, simulation facilitates handling complex algorithms with numerous input parameters exhibiting a variety of pdf's and potential dependencies. The ready availability of PC programs—previously referenced—and the apparent wide acceptance of the method facilitate effective communication between users.

Drawbacks to simulation include:

- need to have sufficient input data to characterize the pdf's of input parameters,
- data analysis and treatment of input pdf's tend to be subjective,
- requirement for considerable expertise to handle the rather complex PC programs,
- need for a PC with high processing speed and considerable memory,
- the view—by some—that the procedure is done inside a "black box," and
- implied precision of the results.[77]

Despite these drawbacks, simulation appears to be the most widely used procedure for probabilistic ECR and is used in subsequent sections of this work.

---

[77] It has been claimed, for example, that probabilistic procedures lead to more *accurate* estimates; it is suggested here that more *informative* estimates might be a better point of view.

Triangular (and other) Frequency Distributions

As noted in Sec. 9.2.2 and 9.5.3, triangular distributions frequently are used to approximate the pdf's of input parameters. This procedure may have to be used when data for such parameters are inadequate to make reasonable approximations of the actual pdf's. In nature, however, the pdf's of many reservoir parameters are strongly skewed. Triangular distributions are a poor approximation for such distributions, as they tend to place more weight on the skewed tail than may be appropriate. Such weighting may cause unrealistic calculations in P90 and P10 ranges.

To avoid these problems with triangular distributions, some engineers use beta distributions to approximate the pdf's of input parameters. Beta distributions, $B(a,b)$, are (nominally) bounded by $0 < X < 1.0$, but may be scaled to fit any minimum-maximum range. Such distributions may used to approximate a wide variety of pdf's. They may be either symmetric or skewed, depending on the values of a and b. If $a = b = 1.0$, the distribution will be uniform. If $a = b > 1.0$, the distribution will be symmetric; as a (and b) increase, the standard deviation will decrease. The degree and direction of skew is a function of the ratio a/b. For $a/b > 1.0$, the skew will be negative, progressively more so as $a/b \gg 1.0$. For $a/b < 1$, the skew will be positive, progressively more so as $a/b \ll 1.0$.

Most commercially available PC programs for simulation provide modules to allow fitting input data to one or more theoretical pdf's. Such fits, however, may result in end points that are not realistic. Thus, the end points from each computed fit should be examined carefully and adjusted as warranted by the data. Exercise caution in using unbounded distributions to fit parameters with a bounded range, a point also discussed in Sec. 9.4.3.

Irrespective of which distribution is used, most geologists and engineers typically underestimate both the degree of uncertainty and the skew in stochastic variables.

9.6 Probabilistic and Deterministic Estimates Compared

9.6.1 Background

As discussed in Sec. 1.3.2, most reserve definitions and reserve literature have focused on deterministic procedures. Despite the increasing use of probabilistic procedures, there is little information published that compares the results of the two procedures for the same entity.[78]

9.6.2 Methodology

Using the analytical procedures discussed in Sec. 3.0 (volumetric methods) and in Sec. 5.0 (production/decline trend methods), calculations were made here to: (a) determine the influence of the pdf's of input variables on reserves estimated with both deterministic and probabilistic procedures and (b) compare the results of the two procedures.

Two simple models, volumetric and decline trend, were used. Each model was used to estimate oil reserves for wells in a reservoir expected to produce by solution-gas drive, with individual wells expected to exhibit hyperbolic declines.

Data used in the calculations are summarized in **Table 9.3.** For each model (volumetric and decline trend) three types of input pdf's were used: (a) symmetric triangular, (b) skewed triangular, and (c) log normal.

For deterministic calculations, the minimum, maximum, mean, and mode of each pdf were used for each input variable to calculate four corresponding reserve estimates. For the probabilistic calculations, MCS was used.[79]

---

[78] The following sections are a modified version of Cronquist [1991].

For probabilistic calculations using a volumetric model, the often observed dependency between porosity and water saturation was modeled by a negative SRC (-0.5) between these two variables. The influence of rock quality on recovery efficiency was modeled by a positive SRC (0.5) between porosity and recovery efficiency.

For probabilistic calculations using a hyperbolic decline trend model, a positive SRC (0.2) was assumed between initial decline rate and initial production rate. A negative SRC (-0.2) was assumed between the *b* exponent and the initial production rate. For the log-normal distributions, the *b* exponent was truncated at a maximum of 0.9.

9.6.3 Results

Results of the reserve calculations for each of the two models are summarized in **Table 9.4.** For the volumetric model, results of the four deterministic calculations of reserves (using the minimum, mean, mode, and maximum values) are shown with the corresponding distribution. For the log-normal distribution, only the mean value was used. Results of the probabilistic calculations are shown similarly, with the mean, P90, P50, and P10 values, and the skew of the three calculated distributions on the right side of the table.

For the decline-trend model, the same procedure outlined in the previous paragraph for the volumetric model was followed, and results are shown in Table 9.4 in the same logical order.

Of necessity, this work was limited in scope. However, from analytical calculations using the same input data sets like those made here, it seems apparent that:

- Deterministic estimates of reserves based on calculations using minimum or maximum values of input parameters are outside the range of reasonable probabilistic estimates—those at expectations of P90 and P10—and do not provide a reasonable basis to estimate either downside risk or upside potential.[80]

- For the volumetric model with symmetric and skewed triangular distributions of input parameters, deterministic calculations using minimum values of input parameters result in estimates of 71 and 95 MSTB, respectively, whereas probabilistic calculations result in P90 estimates of 281 and 387 MSTB, respectively.

- Deterministic estimates of reserves based on calculations using means of input parameters are numerically equivalent to probabilistic estimates of reserves at the mean value of the pdf; however, such deterministic estimates
    -may be only poor approximations to P50 estimates, and
    -may be significantly greater than P90 estimates.

- Deterministic estimates of reserves based on calculations using mode (most likely) values of input distributions
    -may be only poor approximations to P50 estimates, and
    -may be significantly greater than P90 estimates.

In summary, deterministic estimates of reserves based on analytical methods—volumetric or decline trend—cannot readily be compared to probabilistic estimates based on the same analytical method. Deterministic estimates based on either means or modes of the input data sets may be significantly greater than (probabilistic) P90 estimates using MCS of the same data sets.

---

[79] Probabilistic calculations were made using *Crystal Ball,*™ which is an add on to Microsoft *Excel.*™ Reference to these products does not imply endorsement.

[80] Comparable observations have been made by Mireault [1994] and others.

TABLE 9.3—INPUT DATA FOR VOLUMETRIC AND DECLINE TREND MODELS

Model	Distribution	Parameter											
		NetPay		Prod area		Porosity		WaterSat		FVF		RE	
		Mean	Skew	Mean	Skew	Mean	Skew	Mean	Skew	Mean	Skew	Mean	Skew
Volumetric	Symmetric triangular	25	0	80	0	0.20	0	0.25	0	1.25	0	0.25	0
	Skewed triangular	28	0.42	93	0.31	0.22	0.31	0.27	0.31	1.25	0	0.27	0.31
	Log normal	28	0.30	93	0.80	0.22	0.36	0.27	0.33	1.25	0	0.27	0.30

Model	Distribution	Parameter							
		Initial Prod Rate		Init Dec Rate		"b" Exponent		Economic Limit	
		Mean	Skew	Mean	Skew	Mean	Skew	Mean	Skew
Decline Trend	Symmetric triangular	175	0	0.20	0	0.50	0	5	0
	Skewed triangular	192	0.42	0.20	0.31	0.50	0.14	6	0.24
	Log normal	192	0.27	0.20	0.71	0.50	0.42	6	1.09

TABLE 9.4—RESULTS OF DETERMINISTIC AND PROBABILISTIC CALCULATIONS

Model	Distribution	Calculated Reserves (MSTB)								
		Deterministic				Probabilistic				
		Minima	Means	Modes	Maxima	Mean	P90	P50	P10	Skew
Volumetric	Symmetric triangular	71	465	465	1696	465	281	452	689	0.66
	Skewed triangular	95	695	465	3405	695	387	658	1090	0.99
	Log normal	NA	696	NA	NA	696	403	659	1043	0.90
Decline Trend	Symmetric triangular	494	1454	1454	8227	1454	1015	1477	2300	1.15
	Skewed triangular	454	1420	1454	6881	1420	966	1465	2285	1.20
	Log normal	NA	1415	NA	NA	1415	1014	1454	2150	1.05

In this context, reference is made to papers by Patricelli and McMichael [1994] and Nangea and Hunt [1997], who discuss an "integrated deterministic/probabilistic approach to reserve estimation." The approach discussed by these authors, apparently designed for new discoveries, involves probabilistic (volumetric) analysis of the entire accumulation, made consistent with deterministic (volumetric) analysis of the same accumulation to estimate the proved and probable components. Results of the two analyses are compared graphically to ensure consistency with corporate guidelines regarding expectation levels and risk factors for the probabilistic and deterministic procedures, respectively.[81]

9.7 Reserves: Expected, Risk-Weighted, Conditional, and Risk-Adjusted

"All animals are equal, but some animals are more equal than others."

George Orwell, *Animal Farm*

9.7.1 Background

SPE-WPC [1997] reserve definitions, which provide for both deterministic and probabilistic procedures for ECR, pose several potential problems, including:

- classification ambiguities if a comparison must be made between a deterministic and a probabilistic estimate for the same entity, and

- lack of guidelines regarding aggregation of reserve estimates.

For example, both deterministic and probabilistic estimates might be made for an undrilled fault block offsetting, and upthrown from, a proved area. As discussed in Sec. 9.3.2, the deterministic procedure typically involves using best estimates of values of input parameters to make a single estimate of reserves. Under SPE-WPC [1997] guidelines it would be appropriate to classify reserves for this fault block as probable, not drilled.[82] If reserves were estimated for this same fault block using probabilistic procedures, however, a full range of estimates typically would be generated, from P0 to P100.[83] What classification would be appropriate for the P90 estimate? The P50 estimate? Under SPE-WPC [1997] guidelines only probable reserves could be assigned to such a fault block. This point is discussed in Sec. 9.7.6.

Depending on the probabilistic domain (defined in Sec. 9.3.3) there may be conflicts between classifications based on probabilistic procedures and those based on incremental deterministic procedures. There also may be conflicts with U.S. SEC (or other applicable) regulatory guidelines. For example, the probabilistic domain for a specific reserve estimate might include the entire accumulation, including undrilled fault blocks. The reserves attributable to such blocks will, thereby, be included with those attributable to drilled fault blocks in the overall pdf of the reserve estimate. This will result in larger estimates at all expectations than would be the case if the undrilled fault blocks were included in a separate domain in accordance with the incremental guidelines discussed in Sec. 9.3.3. This type of conflict is illustrated in Sec. 9.7.6.

Terms like "expected reserves" and "risked reserves" are in common use in the industry, but SPE-WPC [1997] definitions provide no guidelines regarding applicability. Robinson

---

[81] As this work represents a corporate, rather than a generic, approach, further discussion is not offered here.
[82] The term "not drilled" is used here in preference to the SPE-WPC [1997] term "not developed."
[83] This statement tacitly assumes that the calculated pdf of reserves is not "risk adjusted," a point discussed in subsequent sections.

[1999], for example, has posited use of "expected" reserves rather than the historically used "proved" reserves.

Problems may also occur if reserve estimates based on disparate procedures must be aggregated. In this context, it is noted that SPE-WPC [1997] definitions state: "Because of potential differences in uncertainty, caution should be exercised when aggregating reserves of different classifications." Unfortunately, the definitions do not provide guidance regarding aggregation. In this context, it is noted that annual reports of proved reserves to U.S. DOE/EIA, or to shareholders under U.S. SEC guidelines, have (historically) involved arithmetic addition of component estimates. Such addition tacitly assumes the component estimates of the aggregate are mean estimates, which does not account for the upside potential.[84]

9.7.2 Expected Reserves

As noted in Sec. 9.2.2, the expected value, or mean, of a data set is the arithmetic average of the unweighted values in the set. As shown in Sec. 9.5.1, the mean also may be calculated as the sum of the risk-weighted values in the set. The mean of input and output pdf's is routinely calculated by commercial MCS software.

For the EC shown on Fig. 9.6, expected reserves (ER) may be approximated by

$$ER \approx 0.97 \times 17 + 0.70 \times 22 + 0.27 \times 34 = 41 \ . \qquad (9.7)$$

ER, as calculated by Eq. 9.7, is approximately equal to the sum of proved plus probable reserves, or reserves at an expectation of P50 in a probabilistic sense.

Provided that the pdf of reserve estimates is not too skewed, these are reasonable approximations to ER. If, however, there is substantial skew, these approximations become progressively less reliable.

PS-CIM [1994] recommends approximating expected reserves by

$$ER \approx P1 + Pr(P2) \times P2 + Pr(P3) \times P3 \ , \qquad (9.8)$$

where $Pr(P2)$ would normally range from 0.4 to 0.8 and $Pr(P3)$, from 0.1 to 0.4, consistent with the PS-CIM [1994] definitions of probable and possible reserves, respectively.[85] Expected reserves calculated using Eq. 9.8 may not correspond to "established reserves," historically reported to Alberta's ERCB (now EUB), which were discussed in Sec. 1.3.7.

Some companies use ER for public reporting[86] and financial planning. Based on the analysis discussed in Sec. 9.3.4, it seems apparent that U.S. operators—in the aggregate— have reported ER to the EIA.

---

[84] These comments address long standing practice in the U.S., wherein annual estimates of proved reserves to U.S. agencies have been based on deterministic procedures. It is recognized that in areas where reserves estimates are based on probabilistic procedures, Monte Carlo addition of component estimates may be used for aggregate estimates. Caution must be exercised, however, to ensure the proper treatment of dependencies between component estimates, a point discussed in Sec. 9.12.

[85] In June 1999, PS-CIM published, for industry comment, a draft of proposed revisions to the 1994 reserve definitions that might warrant different probability levels than those used in Eq. 9.8.

[86] Statoil, for example, follows this practice.

9.7.3 Risked Reserves[87]

The term "risked reserves" historically has referred to a deterministic reserve estimate that has been multiplied by the probability that the estimated reserves actually will be recovered. Expected reserves—as defined in the foregoing section—might be considered a form of risked reserves.

Applications where it may be appropriate to risk weight reserves include evaluation of mineral interests and reporting to various agencies.[88]

Historically, the industry has risk weighted deterministic estimates, using risk factors considered appropriate for the reserve classification, development status, and producing category for each such estimate. The objective of this procedure is to place such reserve estimates on a (risk) basis comparable to estimates of proved, producing reserves.

For probabilistic calculations, however, which involve a pdf rather than a single estimate, the pdf may have to be risk *adjusted*, depending on the purpose of the estimate and the probabilistic domain of the pdf. Risk *adjustment* is distinguished here from risk *weighting*, a point discussed in Sec. 9.7.7.

SPE-WPC's [1997] caution regarding aggregation might be addressed by risk weighting and/or risk adjusting the components of the aggregation, depending on the procedure used for the component estimates. Determination of consistent probability multipliers, or risk factors, however, remains an open question. Annual surveys by the Soc. of Petroleum Evaluation Engineers (SPEE) might help resolve this problem.

9.7.4 SPEE Risk Adjustment Factors

Since 1982, the SPEE [89] has conducted an annual survey to determine (among other items) "risk adjustment factors" (RAF) used by the domestic industry to evaluate oil and gas interests, primarily for the purpose of acquiring or selling properties.[90] SPEE RAF's are not used to classify reserves. To the contrary, these RAF's pertain to the probability of realization of estimates of the *present value of future net revenue* attributed to reserves of various classifications, development statuses, and producing categories. Thus, depending on circumstances, the RAF's tacitly include risks attributable to:

- actual expenditure of monies to drill or recomplete wells,

- operations associated with such activities,

- realization of estimated operating expenses, taxes, and wellhead prices,

- potential drainage of behind pipe reserves, and

- geologic and/or engineering uncertainties of reserve estimates.

---

[87] Rose [personal communication, 2000] might identify these as "chance-weighted" reserves.

[88] For example, as noted in Sec 1.3.7, the Ontario Securities Commission requires disclosure of the risk adjustment used to estimate risked probable reserves (if such are reported).

[89] The SPEE, founded in 1962, is comprised of registered Professional Engineers and AAPG Certified Petroleum Geologists involved, principally, in the evaluation of oil and gas properties. Membership has increased each year since founding and currently totals approximately 500.

[90] Over the years, these factors have been identified by the SPEE as "risk adjustments," "risk adjustment factors," "confidence factors," "acquisition adjustment factors," and, most recently, "monetary value risk factors."

Results for 1991-1998 are summarized below:

Classification, Development Status, Producing Category	Average RAF[91]	Average RAF ± 1 S.D[92]
Proved, producing	.96	1.00 - .88
Proved, shut-in	.82	.98 - .65
Proved, behind pipe	.72	.91 - .53
Proved, not drilled*	.54	.78 - .31
Probable, producing**	.28	.51 - .05
Probable, behind pipe	.26	.48 - .03
Probable, not drilled*	.20	.40 - .00
Possible, producing**	.08	.20 – .00
Possible, behind pipe	.08	.20 – .00
Possible, not drilled*	.06	.14 – .00

* These are identified as "undeveloped" by the SPEE. The term "not drilled" is preferred here, as discussed in Appendix B.
** Included for the first time in the 1997 survey.

RAF's reported by the SPEE reflect the composite professional judgment of the individual respondents, rather than a statistical analysis of actual cases. RAF's, thus, must be considered subjective, rather than objective. That notwithstanding, as shown by **Fig. 9.13,** the RAF's in the surveys for the period reported here have been remarkably consistent.[93]

Fig 9.14 is a graphical presentation of the summary data from the foregoing table. Of significance is the pronounced decrease in the average RAF for proved, not drilled, reserves, compared to that for proved, shut in, and proved, behind pipe, reserves. Of comparable significance is the decrease in average RAF's for probable and possible reserves, compared to those for proved reserves. Such decreases are a clear indication of the significantly higher uncertainty associated with estimates of future net revenue attributed to undrilled and/or non-producing reserves.[94]

From Fig. 9.14 it may be noted that, as the average RAF decreases, there is a wider diversity of opinion—a larger standard deviation—between survey respondents. Accordingly, it seems reasonable to expect a comparable diversity of estimates—more likelihood of disagreement—between evaluators of such groups of properties.

With the exception of those attributed to proved, producing reserves, average SPEE RAF's differ significantly from probability levels used to classify proved, probable, and possible reserves that were discussed in Sec. 9.3.3. This may be attributed, in significant degree, to survey respondents considering operational and economic uncertainties in addition to geologic and engineering uncertainties. In this context, the SPEE [1998] distinguishes between "reserves risk" and "monetary value risk," observing that monetary value risk factors are approximately equal to the square of reserve risk factors.

[91] These RAF's pertain to "acquisition value," rather than "loan value."
[92] S.D. = standard deviation. See Nomenclature for a complete list of abbreviations.
[93] During the period covered by these surveys, the industry used—almost exclusively—deterministic methods to estimate reserves. It is conjectural if the introduction of probabilistic methods will affect future surveys.
[94] Lower RF's for undrilled reserves are consistent with industry experience in, for example, the North Sea, where significant revisions to presanction, probabilistic (volumetric) estimates are not uncommon.

It has been observed that SPEE RAF's are more conservative than those apparently used in property acquisition. The difference may be attributed, in part, to purchasers' views that acquisition will result in more efficient operation and high confidence that undrilled probable and possible reserves will be drilled and moved to a proved, producing category.

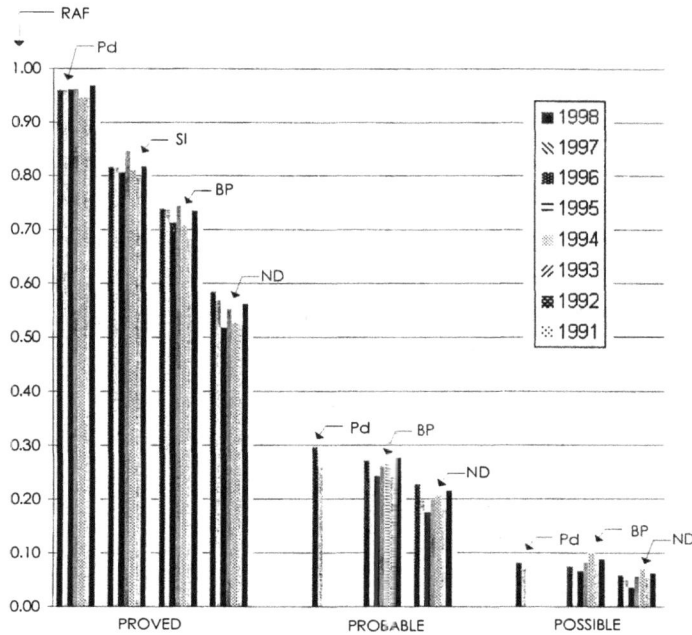

Fig. 9.13—Risk adjustment factors (RAF) reported by the SPEE between 1991 and 1998.

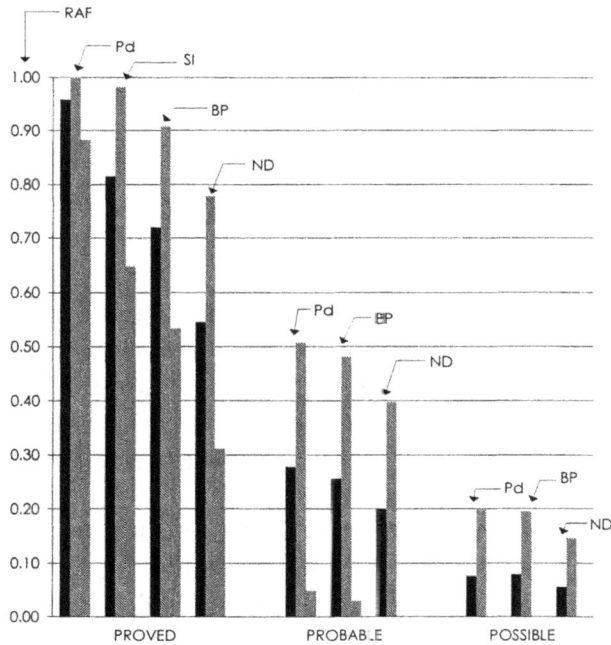

Fig. 9.14—Average RAF and average RAF ± 1 S.D.

9.7.5 Conditional Reserves

Although not formally recognized in the historic lexicon of reserve classifications, *conditional* reserves are *implied* by historic terms like "proved undeveloped reserves," "probable reserves," and "possible reserves." These terms imply "success" from either—or both—of two distinctly different scenarios, or "conditions":

a) *actions* like drilling, recompleting and/or stimulating wells or implementing improved recovery projects, and/or

b) *realizations* of better-than-expected "state(s) of nature," including
   -more efficient water drive or gravity segregation during natural depletion,
   -larger well drainage volumes, and/or
   -more efficient volumetric sweep efficiency by an improved recovery process.

For example, a partially drilled, producing accumulation with proved waterflood potential might be assigned both action- and realization-conditional reserves:

a) action-conditional reserves might include
   -proved, not drilled (primary) reserves,
   -probable, not drilled (primary) reserves,
   -proved, not developed (waterflood) reserves, and
   -probable, not developed (waterflood) reserves.

b) realization-conditional reserves might include
   -probable, producing (primary) reserves and
   -possible, producing (primary) reserves.

In deterministic ECR, the (reserve) outcomes from each action and/or realization typically are treated on an incremental basis, with each such reserve increment classified and categorized accordingly.

In probabilistic ECR, however, both action- and realization-conditional reserves might be included in the pdf's of the input parameters. This procedure makes it difficult to identify component factors and track performance.

9.7.6 An Illustration

Consider a (hypothetical) new oil discovery on a faulted anticline, **Fig. 9.15.** Well 1 encountered commercially productive sand at (-)9685 with 65 gross ft of undersaturated oil to the base at (-)9750, thereby "proving" the F2u block and establishing lowest known oil (LKO) for this fault block. Well 2 encountered 85 gross ft of wet sand, establishing highest known water (HKW) at (-)9800. The structural interpretation is based on shallow control and fair quality seismic data. Geologists estimate a 50% probability that the F2d block is in fluid communication with the F2u block, that the sand is continuous over the mapped area, and is of comparable quality to that observed in the F2u block—i.e., there is 50% probability that the F2d block is commercially productive.

Based on petrophysical data available from Wells 1 and 2, well test and fluid samples from Well 1, and analogy data from similar accumulations along trend, expectation curves of estimated ultimate oil recovery (EUOR) were calculated separately for each of the two fault blocks using MCS, Curves "A" and "B" on **Fig. 9.16.**[95]

---

[95] To simplify the discussion here, estimated ultimate recovery of solution gas is ignored.

To ensure compliance with U.S. SEC (and other) guidelines for estimating proved reserves,[96] the probabilistic domain for the proved fault block includes only the area above LKO.[97] Thus, the EC of EUOR for the F2u block, Curve "A," reflects Category I uncertainties (defined in Sec. 9.4.1) except the position of the OWC.

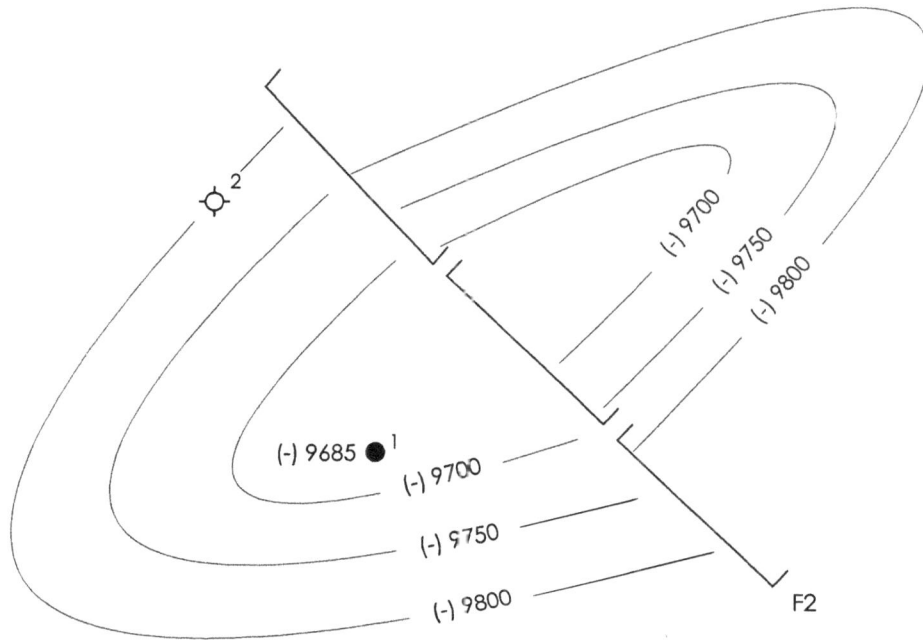

Fig. 9.15—Example of new discovery of structurally controlled accumulation.

LKO and HKW for the unproved (F2d) fault block are assumed to be the same as in the proved block. Because this fault block is unproved, it would not be included in a reserve estimate under U.S. SEC guidelines, but might be included under internal (corporate) reporting. Accordingly, the probabilistic domain for the unproved block includes the area above HKW. There is considered to be is a uniform probability of an OWC between the potential LKO and the potential HKW for that block. The EC of EUOR for the unproved block, Curve "B," reflects Category I uncertainties, including the position of the potential OWC.

The EC of EUOR for the unproved block, however, must be considered conditional, the condition being a well demonstrating commercial productivity of that block. Thus, although a "P90" estimate has been calculated, it must be considered conditional, and proved reserves cannot be assigned to this block.   A conditional EC may be risk *adjusted* by the probability of a successful outcome, which is judged for this accumulation (as discussed previously) to be 50%. Accordingly, Curve "B" is risk adjusted by multiplying the "y" axis by 0.5, resulting in Curve "C," which is the EC of *unconditional* estimates of EUOR for the unproved block. This procedure is analogous to multiplying the risk factors for each outcome in Table 9.1 by

---

[96] The U.S. SEC guideline that "the lowest known structural occurrence of hydrocarbons controls the lower proved limit" is considered here to be deterministic.

[97] For corporate planning, it might be appropriate to expand the probabilistic domain to include the area between LKO and HKW.

0.5, which reduces the expected value by 50% and is consistent with risk *weighting* a deterministic estimate.

$Pr(N_a \geq N_e)$

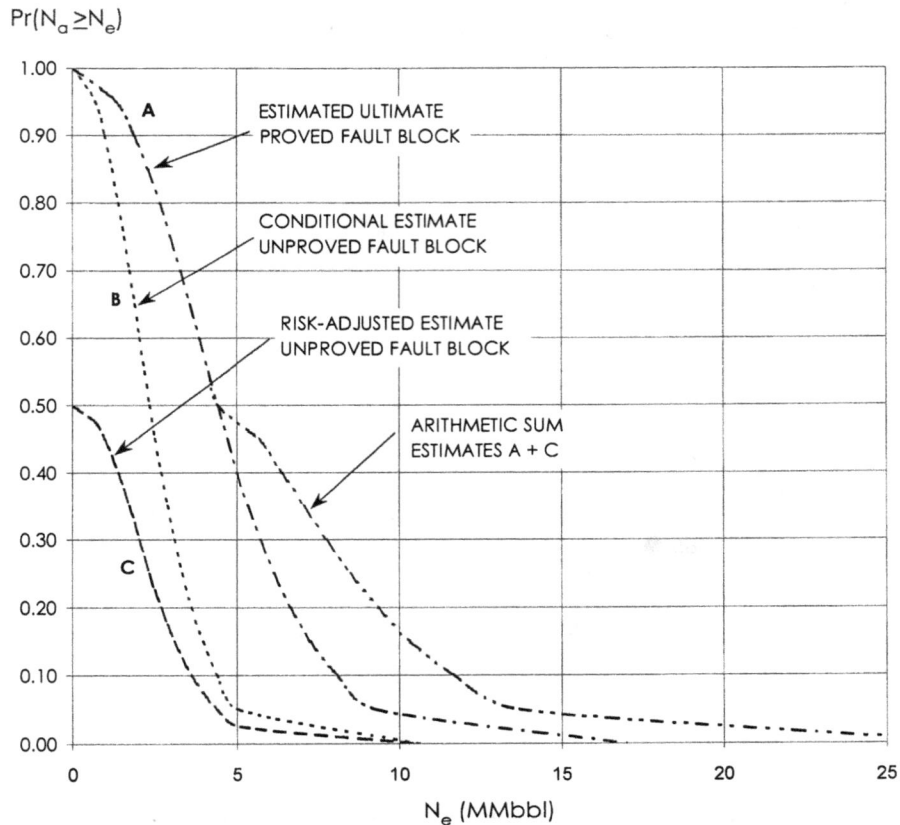

Fig. 9.16—Expectation curves of estimated ultimate recovery for proved and unproved fault blocks.

The same data sets and assumptions were used for both estimates, which are, accordingly, considered fully dependent. Thus, the EC of EUOR for the proved block is added arithmetically to the risk-adjusted EC of EUOR for the unproved block to determine the EC of EUOR for the accumulation.[98]

The following summary observations are made:

- The unconditional EC of EUOR for the proved block, Curve "A," reflects uncertainties attributed to lateral continuity of reservoir rock, petrophysical and reservoir fluid properties, recovery efficiency, etc., but not to presence of accumulation—it reflects Category I uncertainties (defined in Sec. 9.4.1) except the position of the OWC.

- The conditional EC of EUOR for the unproved block, Curve "B," reflects Category I uncertainties, including the uncertainty regarding the position of the potential OWC, but not the Category II uncertainty regarding commerciality of this block.

---

[98] Before addition of the two estimates for corporate planning, it would be appropriate to expand the probabilistic domain of the proved fault block, as discussed in the previous note. The addition here is merely to illustrate a general method to treat conditional and unconditional estimates.

- Conditional EC's should not be used to classify reserves.
- Proved (P90) and proved *plus* probable (P50) reserves may be attributed only to the proved block.
- The risk-adjusted (unconditional) EC of EUOR for the unproved block (Curve B of Fig. 9.16) reflects both Category I and Category II uncertainties.
- Unconditional EC's may be added to generate an EC for the entire accumulation.

9.7.7 Summary Comments and Proposed Descriptive Terminology

In the previous discussion a distinction was made between risk weighting (RW) a deterministic estimate and risk adjusting (RA) the EC for probabilistic estimate. As noted, each procedure leads to a comparable adjustment to expected reserves.

It also was noted that, depending on circumstances, it might be necessary to compare and/or aggregate reserve estimates based on different computational procedures or to satisfy certain reporting protocols. Descriptive terminology proposed below might help communication:

Reserve Scenario	Computational Procedure	Comparable Quantities *	Descriptive Terminology
Proved segment	Deterministic Probabilistic	Best estimate P50	Proved P50
Unproved segment	Deterministic Probabilistic	Best estimate P50	Probable** Conditional P50
Unproved segment	Deterministic Probabilistic	RW best estimate EV of RA pdf	Risk-weighted probable Risk-adjusted EV

* This table assumes the pdf of the estimates is not too skewed to preclude approximating the mean estimate with the P50 estimate.

** If the unproved segment is considered "possible," rather than "probable," comparable descriptive terminology would be appropriate.

9.8 Probabilistic Analysis of Analogy Data

Analogy data—also discussed in Sec. 2.3—may be utilized: (a) both before and after wells are drilled and (b) for both wells and/or reservoirs.

9.8.1 Predrill Analyses

In Sec. 2.3.2, it was noted that the frequency distribution of ultimate recoveries for a group of wells producing from the same geologic trend frequently may be approximated by a lognormal pdf. Provided such ultimate recoveries are statistically random and are not correlative with measurable pre-drilling parameters, the median of the pdf may be considered the (deterministic) best estimate for ultimate recovery for the *next* well to be drilled on the trend.[99] Whether such an estimate can be classified as proved depends on—among other factors—whether the next well is a direct offset to a commercial well in the objective zone and the dry-hole[100] risk on the trend. If, for example, the dry-hole risk is greater than 0.1—

---

[99] Given that the pdf of ultimate recoveries for such a group of wells typically is approximately log normal, the mean of the distribution would be biased on the high side. An estimate at the median, however, would reflect the mid-point of the pdf—there is approximately 50% probability the estimate is either too high or too low.

[100] For this (and subsequent) discussion, a dry hole is defined as a well that is not completed, or one that tests oil and/or gas at subcommercial rates with no reasonable expectation for improvement.

irrespective of location—it would not be appropriate to assign proved reserves (P90 expectation) to undrilled tracts.

Depending on circumstances, a probabilistic analysis of analogy data may be appropriate. **Fig. 9.17** is an EC generated from the data in Table 2.3.

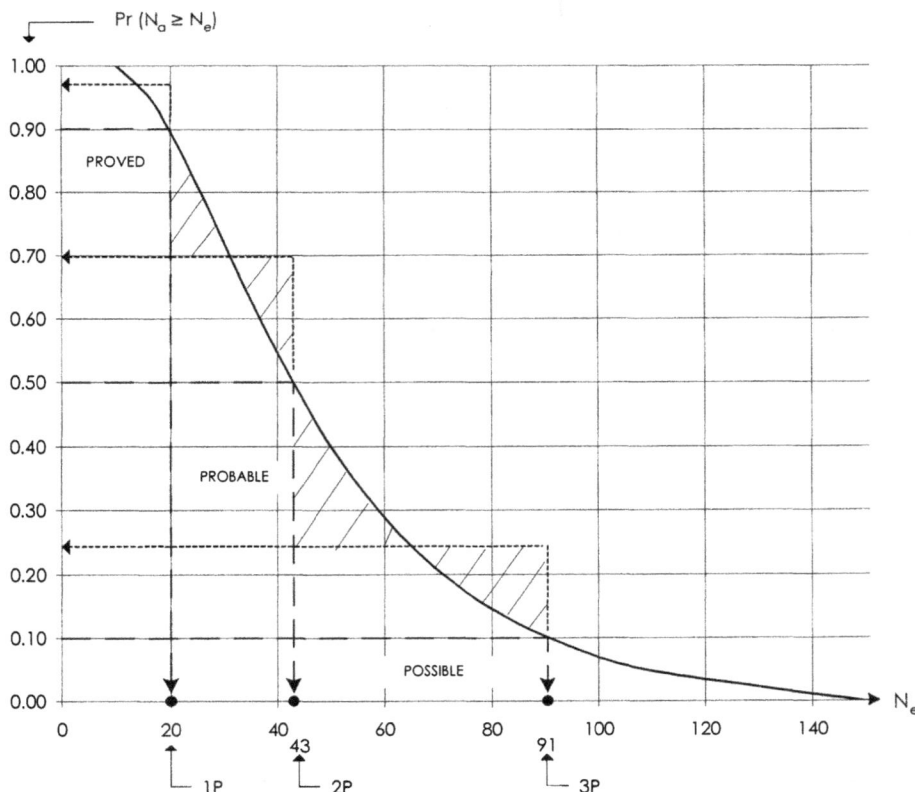

Fig. 9.17—Expectation curve for estimated ultimate oil recoveries in Table 2.3.

At expectations of P90, P50, and P10, ultimate oil recoveries for each of the next few wells in this trend are estimated to be 20, 43, and 91 Mbbl,[101] respectively. *Provided the dry-hole risk is negligibly small*, these estimates might be classified as proved (1P), proved plus probable (2P), and proved plus probable plus possible (3P) reserves, respectively.

From an incremental perspective, one might assign proved reserves of 20 Mbbl, incremental probable reserves of 23 Mbbl, and incremental possible reserves of 48 Mbbl. If appropriate for the purpose of the analysis, "risk factors" for proved reserves and for *incremental* probable and *incremental* possible reserves may be estimated, as indicated by the horizontal arrows on Fig.9.17, to be approximately 0.97, 0.70, and 0.25, respectively. ER may be estimated as 48 Mbbl (0.97 x 20 + 0.70 x 23 + 0.25 x 48). This procedure ignores the "tail" of the pdf, which would add approximately 2 Mbbl to the estimated ER, thereby equaling the integrated area under the entire EC.

Suppose, however, that operators in the trend historically have experienced a dry-hole risk of 0.2, which has remained more-or-less the same for several years, irrespective of location.

---

[101] The alert reader might note that the median estimate here (43 Mbbl) differs from that estimated for Sec. 2.3.2 (38 Mbbl). The difference is attributed to treating the data in Table 2.3 discretely compared to fitting these data with a continuous pdf.

Under these circumstances, a pre-drill location would not qualify for assignment of proved reserves if a P90 expectation is required for such assignment.

Given a drilling program to develop a series of prospects in a proven trend, it might be appropriate to apply the binomial theorem. This theorem may be used to estimate the number of successes (or failures) for a series of trials, given the probability of success for a single trial, as discussed by, for example, Newendorp [1975]. This type of analysis is generally applicable to exploration ventures and detailed discussion is beyond the scope of this work. It is conjectural if such a procedure would be acceptable to assign P90 "reserves" to undrilled prospects, even in a proved trend, a point discussed by Schuyler [1998].

9.8.2 Post-drill Analyses

In many areas, ultimate recoveries of individual wells appear to be almost completely random. Such recoveries are not correlative with measurable pre-drill parameters like seismic amplitude or with post-drill parameters like net pay or initial potential. Ultimate gas recoveries from deep gas wells in the mid continent area of the U.S., for example, exhibit such statistical scattering, with only a weak correlation with net gas. For such wells, only after several years of production history can a reliable estimate be made of GIP and ultimate gas recovery. Typically, however, reserve estimates must be made at completion, before such historical data are available.

One approach to this problem is to regress ultimate gas recovery vs. net pay for mature wells in the trend being analyzed. This is illustrated by **Fig. 9.18,** which reflects data from 33 mature, *commercial* wells competed in the Morrow sandstone of the Anardarko basin, U.S.[102] The Morrow sandstones in these wells, approximately 14,400 ft deep, apparently are meandering channel deposits in a fluvial-deltaic environment. Subject regression was forced through the origin. An unforced regression has a positive intercept of 2.9 Bscf, which, although statistically valid, is illogical.[103] Also shown for the least squares fit are the upper and lower bounds of the prediction at 80% confidence.[104] Pending sufficient production history for a performance estimate, and *unless contrary to more compelling information*, the following procedure might be followed for probabilistic estimates for new wells drilled on this trend:

Proved (P90):	Lower predicted bound, except for wells with less than 20 net ft of gas, in which case use minimum of 2 Bscf
Proved plus probable (P50):	Least-squares fit
Proved plus probable plus possible (P10):	Upper predicted bound, but only for wells with greater than 10 net ft of gas

In reviewing this procedure, it is noted that:

---

[102] These data are "best" fitted with a beta distribution (1.27, 5.66, 72.8), rather than a log normal distribution.

[103] The regression coefficient, r^2, for the unforced regression is 0.46; for the forced regression, 0.80.

[104] As implied by the discussion in Sec. 9.2.3, the upper and lower bounds of the predicted interval at 80% confidence are a reasonable approximation to the 90th and 10th percentiles, respectively. These computations were made with STATISTIX™ (www.statistix.com). No endorsement is intended by this reference. Details of these computations are beyond the scope of this work, and the reader is referred to, Snedecor and Cochran [1980].

- Many completions with less than 10 ft. of net gas were marginally commercial and were not included in the regression.

- The least-squares fit corresponds to a 640-acre drainage area with GIP = 1100 Mscf/NAF[105] and 85% recovery efficiency, which provides a "reality check" on the regression.

- Wells with less than about 20 net ft of gas with ultimate recoveries approaching the upper predicted bound on Fig. 9.18 apparently were completed in the thin edge of channel sands that were substantially thicker away from the wellbore.

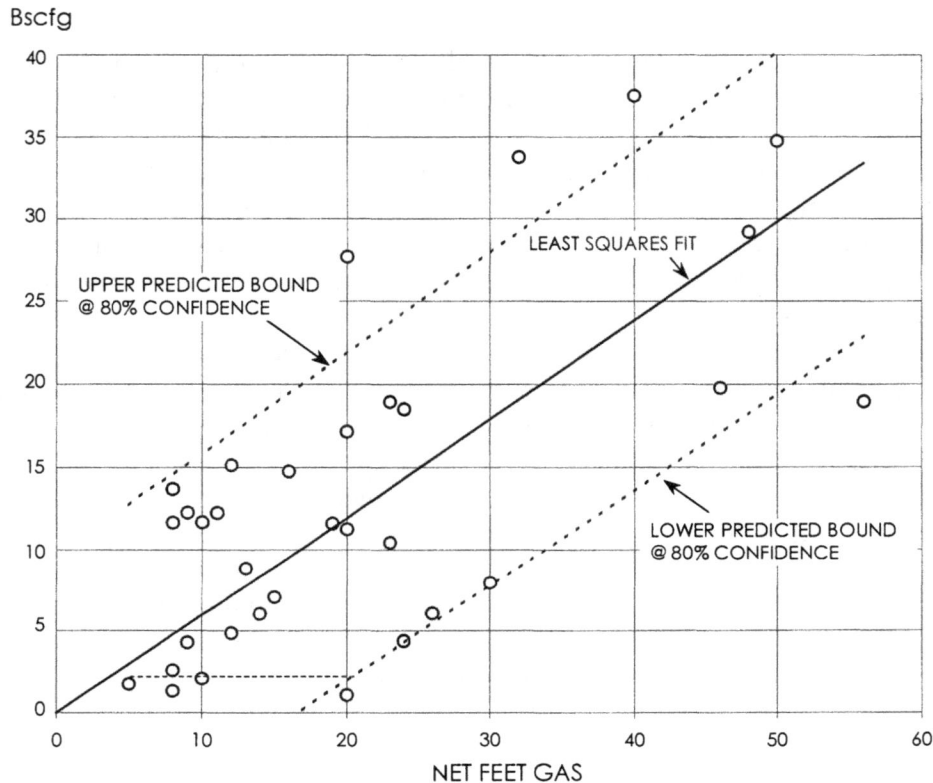

Fig. 9.18—Estimated ultimate gas recovery vs. net gas for 33 wells in U.S. Anadarko basin.

9.9 Volumetric Estimates Using Probabilistic Procedures

For a given accumulation, a volumetric estimate of O/GIP and reserves[106] includes four elements,

1. gross rock volume,
2. petrophysical properties,
3. hydrocarbon properties, and
4. recovery efficiency.

---

[105] NAF = net acre feet.
[106] It is recognized that reserves—in general—are a function of geologic, reservoir, operational and economic scenarios. To simplify this discussion, the focus here is on "technical" reserves, which ignores operational and economic factors.

The uncertainties in each of these four elements are discussed in the ensuing sections. Some uncertainties may have a greater influence on reserve estimation and project planning than others and should, accordingly, receive greater attention. For example:

- For an accumulation containing near critical fluids, the type and sizing of processing and production equipment may be strongly dependent on the PVT behavior of the reservoir fluids.

- For a low-relief oil accumulation with bottomwater, uncertainties in the relationship between oil-column thickness, net-to-gross ratio, and drainage area of individual wells may have a major influence on the drilling plan.

- For an accumulation containing highly undersaturated oil, the extent and relative strength of the aquifer may have a significant influence on the design of injection and production facilities.

9.9.1 Gross Rock Volume

Depending on the geologic setting, the quality of seismic data, and the stage of delineation, the degree of uncertainty in mapping gross rock volume (GRV) may be substantial. Robertson [2001] has provided guidelines regarding the degree of depth uncertainty that might reasonably be expected in extending seismic time-depth conversions beyond existing well control. Such uncertainty may be used to develop minimum, most likely, and maximum interpretations of top structure and GRV, as discussed in Sec. 9.10.1. There is no published information, however, regarding the statistical nature of uncertainties in GRV, and most analysts use triangular pdf's.

It seems apparent that the degree of uncertainty in GRV and, correspondingly, in O/GIP, has, in many cases, been significantly underestimated [Spears and Dromgoole 1992, Corrigan 1993].

In some cases, however, the uncertainty can be attributed to incomplete geologic definition of the accumulation, rather than to inadequate quantification of the uncertainty of the drilled areas. For example, Lia et al. [1997] presented a comprehensive analysis of the Veslefrikk field (offshore Norway] that involved geostatistical analyses to drive MCS of the reservoir aspects and computer simulations to compute potential rate-time production profiles. Yet, several years after production was initiated, the operator discovered a major horizon that had been undefined at the time of the Lia et al. [1997] study. This discovery led to an increase in expected ultimate oil recovery from 36 MMm^3 to 54 MMm^3. This increase was significantly outside the bounds of the Lia et al. [1997] study.

Given that uncertainties in GRV may be substantially larger than other uncertainties in the volumetric calculation, there might be a tendency to do only a cursory analysis of other factors. While time and/or budget constraints might seem to justify such an approach, it cannot be recommended here.

9.9.2 Petrophysical Properties

Characterization of petrophysical properties includes determination of the pdf's of porosity, water saturation, and the net-to-gross pay ratio—and the degree of dependency between these parameters. These properties are collected in a "petrophysical group," defined here as

$$\phi(1 - S_w)R_{ng},$$

where all terms are as defined in Eqs. 3.1b and 3.3b, but are considered as stochastic variables for this discussion. Potential dependencies between ϕ, S_w, and R_{ng}—discussed in subsequent sections—mandate treating these parameters as a group, rather than separately.

The degree of uncertainty in the petrophysical group is dependent on:

- lithology,
- degree of stratigraphic complexity,
- degree of post-depositional diagenesis and/or fracturing of the reservoir rock(s),
- nature of the interstitial fluids, and
- the quality and quantity of log, core, and/or test data.

Scenarios in which a high degree of uncertainty may be expected include:

- fractured, and/or vugular, carbonate reservoirs,[107]
- sandstone reservoirs comprised of finely laminated sand and shale,[108]
- low permeability sandstone reservoirs with complex mineralogy,
- sandstone reservoirs with (diagenetic) clay-coated pore walls,[109]
- shaly sandstone reservoirs with low-salinity interstitial water, and
- any reservoir with poor quality and/or incomplete log, core and/or test data.

Faced with such scenarios, a prudent operator might take appropriate steps to ensure a comprehensive evaluation program.

In many cases, uncertainties in petrophysical properties may be attributed to spatial variation in depositional environment, rather than to random variations. In such cases, it may be desirable to subdivide the reservoir into different rock regions, as supported by geological analysis, and analyze each such region separately.[110]

Porosity

Typically, porosities measured from cores are the basis for calibration of log-determined porosities. The need for such calibration, the procedures to do so, and the associated problems have been thoroughly discussed in petrophysical literature and will not be summarized here. The pdf's of core-calibrated log porosities (CCLP) may not be accurate representations of the true state of nature, however, a point not discussed in petrophysical literature. As an example, data from a case history are summarized below:

Porosity basis	Mean	S.D.	Skew
Cores	.093	.035	-0.7
NDLP	.111	.031	-0.8
CCLP (initial)	.102	.035	-0.7
CCLP (adjusted)	.097	.034	-0.9

Mean neutron-density log-porosity (NDLP) is approximately 20% larger than mean core porosity. The initial CCLP were based on standard log-analysis procedures, correlating NDLP with core porosity. The adjusted CCLP were based on correlating density porosity

---

[107] See, for example, the discussion of the LeDuc D-3 oil pool in Sec. 7.2.

[108] Examples of such reservoirs—identified as low resistivity, low contrast (LRLC)—have been compiled by the Houston and New Orleans Geological Societies [1993] and the Rocky Mountain Association of Geologists [1996].

[109] See, for example, the discussion of the Little Creek field in Sec. 7.3.

[110] Depending on circumstances, a geostatistical approach may be more appropriate; please see, for example, Yarus and Chambers [1994].

(only) with core porosity. The petrophysical procedures for both the initial and adjusted CCLP are consistent with "good practice." Mean porosity from the adjusted analysis, however, was considered more consistent with the state of nature and more suitable for MCS.

Commercial software was used to analyze each of the four porosity data sets to determine the "best" pdf for input to MCS. In each case, the analysis indicated that a Gaussian distribution was the "best" fit. A symmetric pdf (like a Gaussian distribution), however, does not account for the slight skew in the data sets, which might lead to errors in ECR.

The foregoing discussion illustrates the problems in selecting a pdf for porosity—or any other parameter—given data from a single well. The problem is compounded by the reality that pdf's over the accumulation of interest may be quite different from those observed in the first few wells.

Water Saturation[111]

Typically, initial water saturation is calculated from resistivity logs, which may be supplemented with porosity logs and other logging devices to help characterize the rock-fluid system under investigation. Occasionally, it is necessary to supplement such calculations with data from special core analysis. A detailed discussion of such procedures is beyond the scope of this work, and the reader is referred to a comprehensive body of petrophysical literature, discussed in Sec. 3.2.1.

There are several equations available to calculate S_w from log data. Selection generally depends on the lithology of the object zone. Selection of the "best" equation is subjective and introduces modeling uncertainty. One of the more general equations, which illustrates the sources of data uncertainty, is based on work by Archie [1942] and might be used in clean sandstone reservoirs,

$$S_w = [(A/\phi^m)(R_w/R_t)]^{1/n} , \qquad (9.9)$$

where A = a "constant" that may vary from about 0.7 to 1.0,

$\quad\quad m$ = cementation factor, which may vary from about 1.3 to 2.2, depending on lithology,

$\quad\quad R_w$ = resistivity of interstitial water, which should be constant in a specific reservoir,[112] but may appear to vary, depending on measurement techniques,

$\quad\quad R_t$ = true resistivity of zone, which depends on the spatial distribution of rock properties (which may vary over a wide range in the same reservoir), measurement of which depends on the borehole environment, bed thickness, and logging suite, and

$\quad\quad n$ = saturation exponent, with range comparable to m.[113]

Uncertainties in m and n may be quantified with special core-analysis data. Ideally, R_w may be measured from produced water. In the initial stages of appraisal, however, the water zone may not be encountered and tested; if it is, the water may be contaminated with mud filtrate. Theoretically, R_w may be calculated from SP logs, but response from such logs may be suppressed by shale and/or low interstitial water saturation. R_t must be determined from

---

[111] Additional comments regarding initial water saturation are provided in Sec. 3.2.1 and 3.5.6.
[112] This tacitly assumes the reservoir is not subject to hydrodynamic flow and has not been invaded by extraneous water.
[113] It should not be inferred from this statement that m and n are correlative, a point discussed by Doveton [1994].

log data, and quantification of uncertainties may be difficult. Depending on borehole environment, logging device, lithology, depth of mud-filtrate invasion, and bed thickness, substantial corrections may be needed to correct log (apparent) resistivity to true resistivity. Historically, petrophysical analyses have been deterministic, and publications regarding probabilistic approaches are sparse.

Although typically ignored in favor of a "best estimate," petrophysical uncertainties may lead to significant differences in related calculations. For example, in a major gas field in western Siberia, a plausible difference in (only) R_w caused a difference in a volumetric estimate of GIP of 20%.[114]

Net Pay—A Probabilistic View[115]

Estimation of net pay, or the net-to-gross pay ratio, is one of the major elements in volumetric mapping. As discussed in Sec. 3.5.3, however, procedures to estimate net pay tend to be highly subjective and object oriented. Generally, the procedure involves exclusion of (log) intervals of the gross rock section judged to be noncommercial, the remainder being considered net pay. Exclusion criteria typically include minimum porosity and maximum water saturation, which may involve truncation of the original pdf's of these parameters. Both porosity and water saturation are elements of the petrophysical group, defined above. Thus, depending on how net pay is determined for a specific accumulation, there may be significant dependencies between ϕ, S_w, and R_{ng}. As discussed in Sec. 9.5.4, such dependencies may need to be considered if simulation is used to calculate pdf's of O/GIP.

Minimum—or cutoff—porosity usually is based on a correlation between permeability and porosity, where cutoff porosity corresponds to the minimum permeability judged to be commercial. Depending on lithology and the quality and quantity of porosity and permeability data, such correlations may exhibit considerable scatter. Such scatter may be attributed to several factors, including: (a) the presence of more than one rock type in the data, and/or (b) spatial, rather than random, variation in the data. If there is considerable scatter in the data, investigation of these possibilities might be considered.

Fig. 9.19 shows a LSF of permeability-porosity data from a well-zone in a sandstone reservoir in Western Siberia. Also shown are the upper and lower bounds of the 80% confidence interval for the prediction,[116] which are identified as PREDICTION @ Pc90 and PREDICTION @ Pc10, respectively. These data are from a retrograde gas reservoir, and minimum commercial permeability is estimated to be 0.1 md. The lower predicted bound might be used to establish cutoff porosity for "1P" net pay, $(\phi_{co})_{1P}$; the best fit might be used for "2P" net pay, $(\phi_{co})_{2P}$; the upper predicted bound might be used for "3P" net pay, $(\phi_{co})_{3P}$.

The criteria for "commercial net pay" are considerably more complex than merely a deterministic selection of a "minimum commercial permeability." The criteria are functions of economic scenario, geologic setting, and reservoir mechanics, among other factors. Clerke [1999], for example, observed that the probability that a specific well-zone interval is net pay could be expressed as the sum of conditional probabilities of exceeding specified limits for commerciality.[117] He noted that the conditional probabilities of exceeding such limits are

---

[114] Personal communication, Rick Richardson, 1999.

[115] This section is an expansion of an informal presentation by the author at the 1999 SPE ATW on Probabilistic Assessment of Reserves, held in Dallas, 24-25 March.

[116] The considerable scatter in these data is attributed to the relative immaturity of the reservoir rock, a point noted in the next section, and the consequent poor sorting.

[117] "Complex Carbonates and Sands—Pay Examples," presentation at the SPWLA Symposium "Examining Net Pay Using Conventional and Unconventional Techniques," Houston, 18 May 1999.

determined by log-core correlation, reservoir case studies, sedimentological studies, and pressure transient analyses, among other factors. However, given corporate—and other—constraints, it is unlikely that such in-depth analysis will occur, and the determination of net pay probably will continue to remain "highly subjective and object oriented," as noted previously.

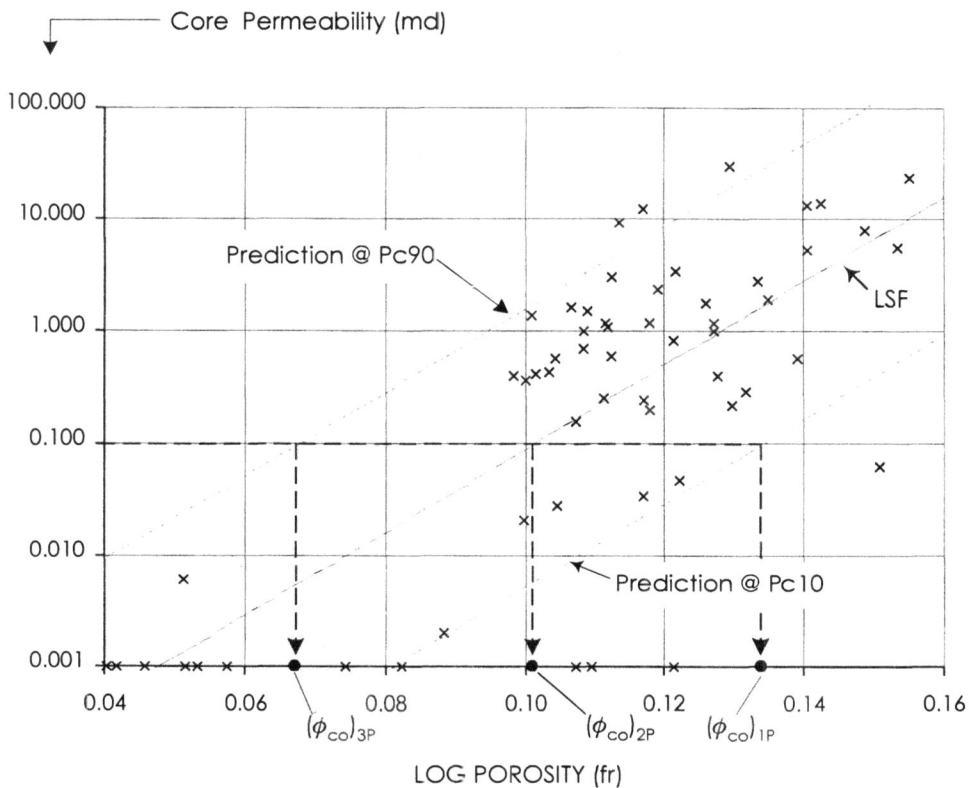

Fig. 9.19—Core permeability vs. log porosity illustrating LSF and upper and lower bounds at 80% confidence for predicted values for estimation of $(\phi_{co})_{1P}$, $(\phi_{co})_{2P}$, and $(\phi_{co})_{3P}$.

Net Pay—A Probabilistic Example

As an example of a probabilistic approach to the estimation of net pay, log and core data from a gas-bearing zone in a well in Western Siberia are analyzed here.[118] Also analyzed is the impact of the methodology on the calculation of O/GIP, which is a direct function of the petrophysical group defined previously.

Fig. 9.20 shows log-calculated porosity and water saturation through the zone of interest. With the exception of the two intervals identified as CALCAREOUS, the entire interval is gas bearing, as indicated by neutron-density log and core data. Lithology is arkosic sandstone,[119] grey, very-fine to fine-grained, slightly shaly, slightly silty, slightly calcareous.

---

[118] Depending on circumstances, procedures other than those discussed here may be more appropriate. This discussion is not intended to be limiting.

[119] Arkosic sandstones contain significant amounts of feldspar—by definition, more than 25%—which generally indicates an immature rock.

For this discussion, dimensionless cutoff porosity, ϕ_{coD}, is introduced here

$$\phi_{coD} = (\phi_{co} - \phi_{min})/(\phi_{max} - \phi_{min}) , \quad\text{...} \quad (9.10)$$

where ϕ_{co} = cutoff porosity—minimum porosity used to determine a specific classification of net pay in a well-zone,

ϕ_{min} = minimum observed porosity—lower bound of the pdf of porosity, and

ϕ_{max} = maximum observed porosity—upper bound of the pdf of porosity.

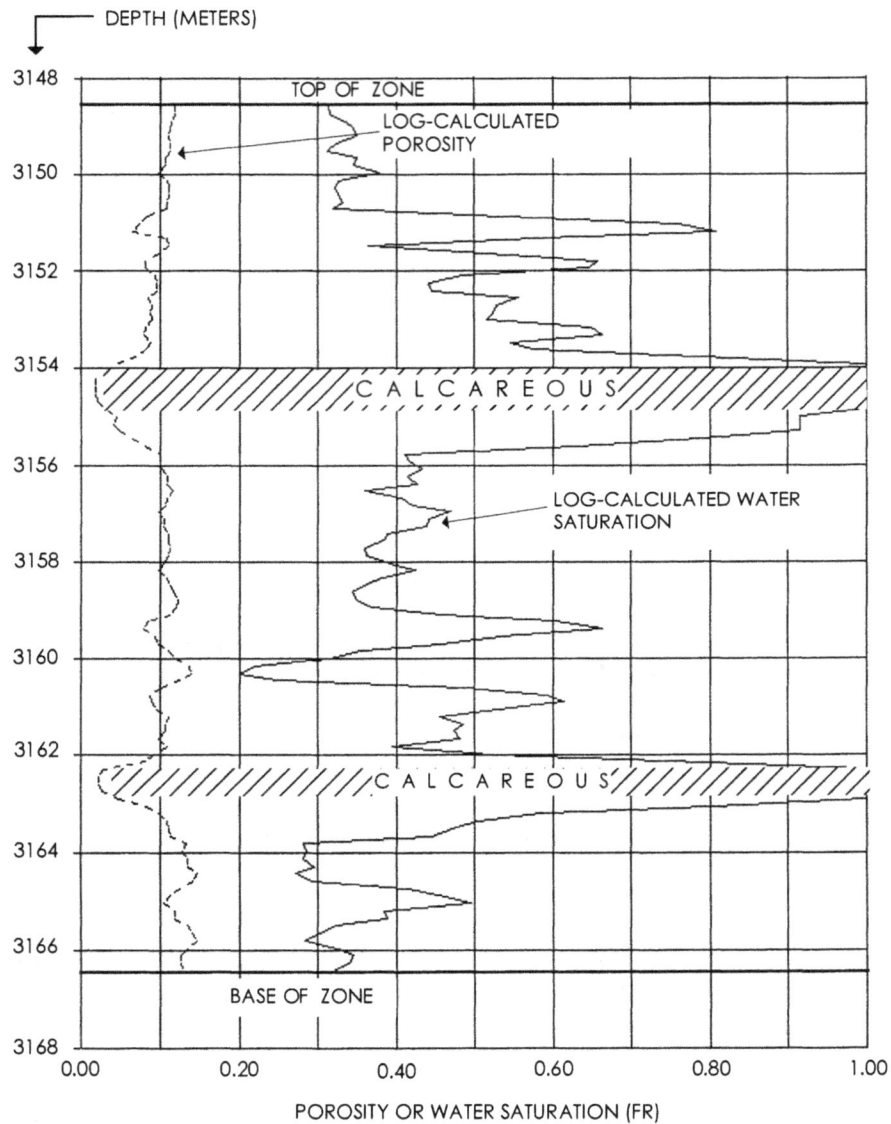

Fig. 9.20—Log-calculated porosity and water saturation vs. depth for gas zone in western Siberia.

For the gas-bearing interval shown in Fig 9.20, a continuum of net pay values was determined using ϕ_{coD} values ranging frcm "0" to "1" and maximum log-calculated water saturation of 70%. The results are shown in **Fig 9.21** as plots of R_{ng} and ϕ_W vs. ϕ_{coD}.

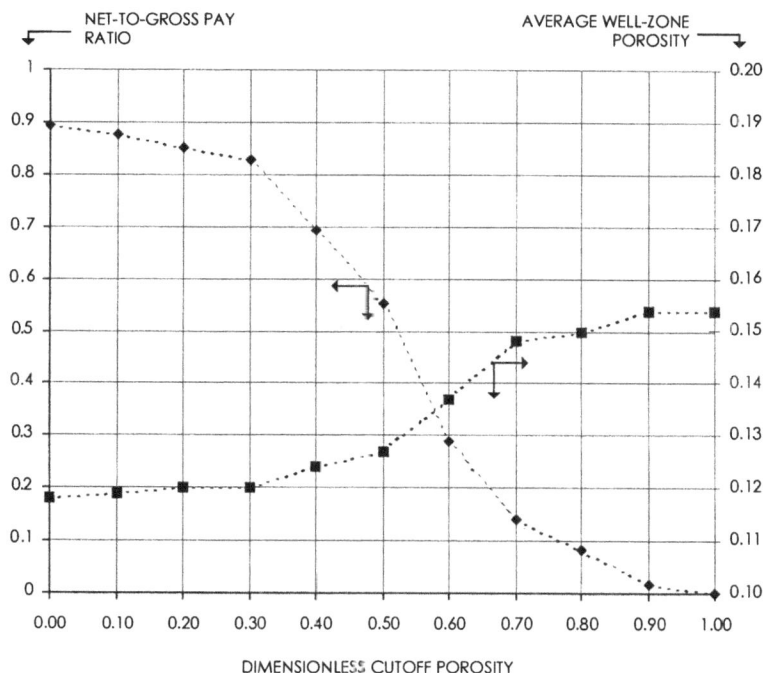

Fig. 9.21—Net-to-gross pay ratio, R_{ng}, and average well-zone porosity, ϕ_W, vs. dimensionless cutoff porosity, ϕ_{coD} .

From this figure, it may be noted:

- At ϕ_{coD} = 0.0, ϕ_W = 0.118, which is the average porosity of the original (untruncated) data set, and R_{ng} = 0.89, which reflects the zone of high water saturation at 3151 (meters) and the two calcareous intervals excluded from the gross interval.

- At ϕ_{coD} = 1.0, ϕ_W = 0.154, which is the maximum porosity of the original data set.

- The decrease in R_{ng} as ϕ_{coD} increases is not offset by commensurate increases in ϕ_W.[120]

Fig 9.21 represents one well-zone. The relation between R_{ng} and ϕ_{coD} may be expected to be different for each well-zone in a specific reservoir. Depending on circumstances, it might be appropriate to calculate a weighted average relation for a specific project.

The pdf of the petrophysical group may be expected to be affected by the dependency between R_{ng} and ϕ_{coD}. Depending on circumstances, the dependency might be significant. Accordingly, MCS was used here for two sets of calculations of the petrophysical group: (a) considering this dependency and (b) ignoring this dependency. Procedures and results are summarized in the following paragraphs.

For the dependency case, input pdf's included:

---

[120] This observation is contrary to the views of some authors, who have indicated that the decrease in R_{ng} (with increases in ϕ_{co}) is offset by increases in ϕ_W.

- water saturation defined by a Gaussian distribution fitted to log-calculated water saturations, truncated at observed minimum and assumed maximum saturations, with a calculated SRC of (-)0.88 between water saturation and porosity,
- cutoff porosity defined by a symmetric triangular distribution with minimum, most likely, and maximum values equal to $(\phi_{co})_{1P}$, $(\phi_{co})_{2P}$, and $(\phi_{co})_{3P}$, as defined by Fig. 9.19,
- R_{ng} computed dynamically from simulation of the dependency with dimensionless cutoff porosity defined by Fig. 9.21, and
- pdf of porosity defined by dynamic truncation of the initial input pdf.

For the non-dependency case, input pdf's included:

- water saturation defined the same as for the dependency case,
- R_{ng} defined by an (assumed) triangular distribution with minimum, most likely, and maximum values equal to 0.1, 0.7 and 0.9, as defined by $(\phi_{co})_{1P}$, $(\phi_{co})_{2P}$, and $(\phi_{co})_{3P}$, and Fig. 9.19; this defines the pdf "a priori," rather than computing it dynamically, and might be considered an independent "intuitive deterministic" approach, and
- pdf of porosity unchanged from the initial input pdf.

The results of these calculations are shown in **Fig. 9.22,** which includes expectation curves of the petrophysical group and the corresponding triangular distribution of R_{ng} for each of the two cases.

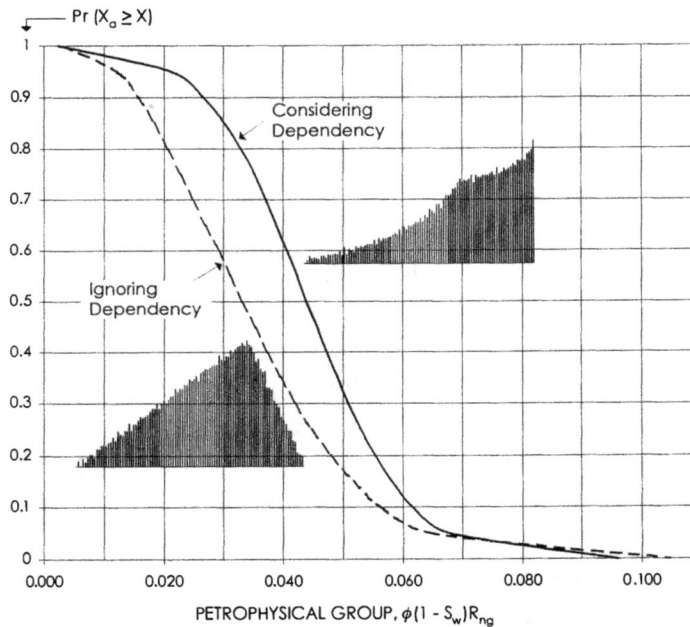

Fig. 9.22—Expectation curves for petrophysical group and pdf's for R_{ng}, considering and ignoring dependency between net-to-gross pay ratio and porosity.

Based on these calculations, albeit limited, it is noted:

- Dynamic computation of R_{ng}, rather than the "intuitive deterministic" approach, resulted in a computed R_{ng} that may be approximated by a triangular distribution defined by minimum, most likely and maximum values of 0.04, 0.90 and 0.90.

- Ignoring the dependency between R_{ng} and ϕ_{co} may lead to significant under-estimates of O/GIP; for the data examined, ignoring the dependency resulted in a P90 estimate of the petrophysical group that was 40% less than if the dependency had been considered.

It is intuitively expected that the type of dependency discussed here is common in: (a) carbonate rocks, which typically exhibit very heterogeneous porosity, and (b) in compound sandstones, which may exhibit grain-size variations ranging from silt to coarse sand. For sandstones in which shale layers are the principal element of nonproductive rock, the dependency might be insignificant.

Net pay may be a factor in unitization or other negotiated arrangements for sharing reserves in competitive accumulations. Depending on circumstances, probabilistic procedures to determine net pay, O/GIP and reserves may be appropriate. The type of dependencies identified here might have a significant influence on the calculations.

9.9.3 Hydrocarbon Properties

Hydrocarbon characterization includes: (a) determination of reservoir fluid type(s), which were discussed in Sec. 3.7.1, (b) the PVT properties of these fluids, and (c) the relative volumes of the reservoir occupied by each fluid. Uncertainties in characterization tend to be caused more by nonrepresentative fluid sampling and formation testing and by inaccuracies in reservoir pressure-temperature data, than by laboratory PVT analysis.[121] Additional uncertainty may be attributed to the inherent differences between laboratory testing processes and the flow processes that occur in the field. For sub-critical fluids, there tends to be a lower degree of uncertainty in determining fluid type and PVT behavior than for near-critical fluids. Most uncertainties must be resolved by subjective analysis of often conflicting data, as there are no standard methods.

Guidelines to estimate the relative volume of the reservoir occupied by each hydrocarbon fluid are provided in Appendix G, and an example analysis is provided in Sec. 9.10.1.

9.9.4 Recovery Efficiency

Factors[122] contributing to uncertainty in recovery efficiency (listed in the approximate order of importance) include[123]

- strength of aquifer and nature of influx,[124] if any,
- quality of the reservoir rock,
- type and degree of heterogeneity(s),[125]
- geometry of reservoir,
- location and spacing of wells, and
- hydrocarbon properties.[126,127]

---

[121] Additional comments regarding fluid sampling and well testing are provided in Sec. 3.3.2.

[122] Economic and operational factors are ignored in this discussion.

[123] The development plan may have a major influence on recovery efficiency; being under operator control, however, it is not considered an uncertainty in the same sense as the factors listed here.

[124] Please see the discussion on bottom vs. edgewater influx in Appendix I.

[125] Please see discussions regarding internal limits, Sec. 3.5.5, and heterogeneities, Sec. 3.7.

[126] The influence of reservoir fluid type on recovery efficiency was discussed in Sec. 3.6. In the context of the discussion here, the input to the equations in that section should be considered stochastic variables.

[127] In a "heavy oil" accumulation, hydrocarbon properties may become a more important factor.

The nature and complex interplay of these factors preclude statistical analysis and mandates a subjective approach to determine the most appropriate pdf's for input to MCS.

There are dependencies—both demonstrated and intuitive—between some of the factors that influence recovery efficiency and those included in the petrophysical group. Such dependencies may have a significant influence on the results of MCS:

- As noted in Sec. 3.6.5, there is a demonstrated dependency between porosity and residual hydrocarbon saturation after water encroachment.
- There is a theoretical relation between displacement efficiency, residual hydrocarbon saturation, and initial water saturation.
- There is an intuitive (positive) relation between the net-to-gross pay ratio and the volumetric sweep efficiency.

Other dependencies that may need to be considered include:

- oil reservoirs with significant gas caps in which the recovery efficiency for the (initial) solution gas may be dependent on the recovery efficiency for the (initial) free gas, depending on the degree of gravity segregation,
- bottomwater-drive reservoirs, where recovery efficiency and optimal spacing are a function of the ratio of horizontal to vertical permeability in the wells' drainage areas, which cannot be determined until the wells have been drilled, and
- oil reservoirs that are potential candidates for improved recovery, where the incremental oil attributable to improved recovery is dependent on the oil recovery during primary depletion.

9.10 Probabilistic Analysis of New Discoveries

One of the major applications of probabilistic procedures is the analysis of new discoveries, especially in areas where analogy data are sparse and development costs are significant.

9.10.1 New Discovery of Structurally Controlled Accumulations

In newly discovered, structurally controlled accumulations, depending on the quality of seismic and other data, principal uncertainties may include:

- structural dip away from areas of well control,
- location of sealing faults,
- areal variations in gross and net pay, and
- position of fluid contacts.

Two of these aspects, structural dip away from well control and the positions of fluid contacts, are discussed here in an example of problem setup.

A sand volume, or vertical distribution, curve, of the type discussed in Sec. 3.8.1, facilitates reservoir characterization for such scenarios. **Fig. 9.23** shows such a curve for a hypothetical discovery of a fault-block segment that has been appraised by three wells. Also shown is cumulative gross rock volume vs. area on top of sand. The area vs. depth curves were based on the geologist's "most likely" structural interpretation. From analysis of seismic time-depth conversions, a potential error of (±)50 ft is estimated away from well control, which would result in a potential error of (±)20 M-GAF at HKW.

Pertinent data from Fig. 9.23 are summarized below:

Well 1:   HKO   @ (-) 7,950;   LKO @ 8,025
Well 2:   HKW   @ (-) 8,050
Well 3:   LKG   @ (-) 7,825

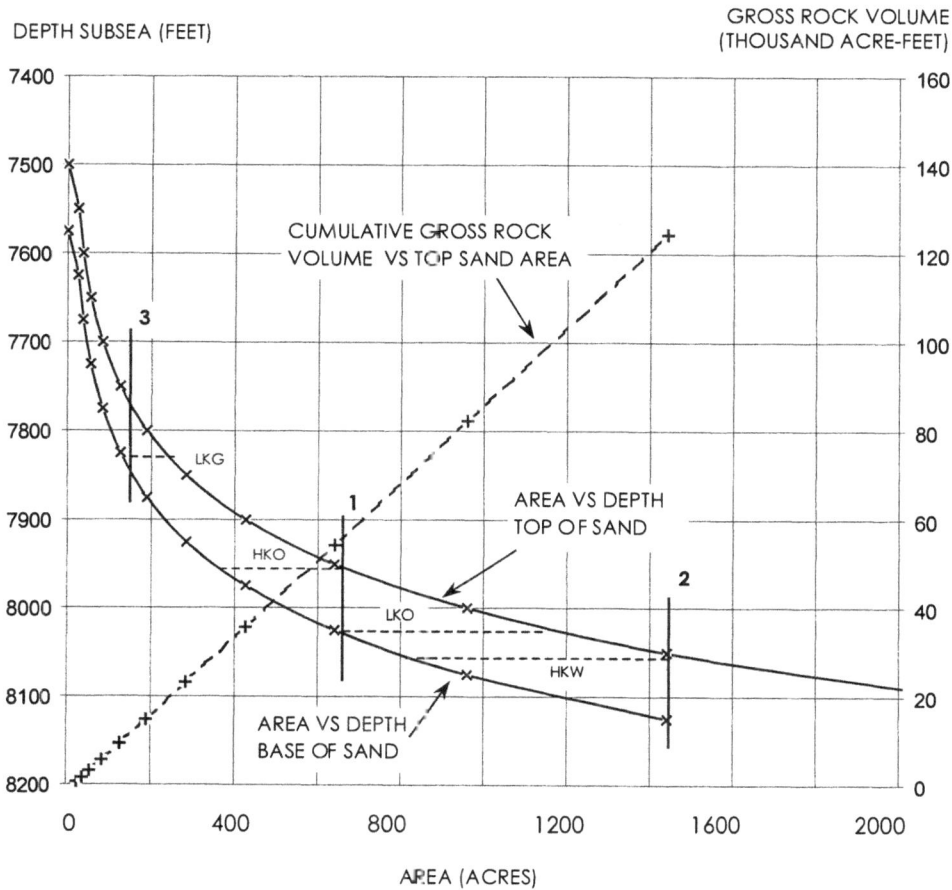

Fig. 9.23—Vertical distribution of O/GIP for hypothetical structural accumulation.

Gross rock volume (GRV) data from Fig 9.23 are as follows:

Cumulative GRV (M-GAF) Down to:		Fraction of Cum. GRV Down to HKW
LKG:	19.0	0.153
HKO:	54.3	0.437
LKO:	98.9	0.795
HKW:	124.4	1.000

The following computational algorithms may be set up to calculate the pdf's of estimated ultimate oil recovery (EUOR) and ultimate free gas recovery (EUFGR) using MCS:

$$\text{EUOR} = 7758 \times \text{GRV}_{HKW} \times \text{E}_{GRV} \times \text{F}_o \times \text{PG}_o \times E_{Ro}/B_o$$

$$\text{EUFGR} = 43560 \times \text{GRV}_{HKW} \times \text{E}_{GRV} \times \text{F}_{FG} \times \text{PG}_g \times E_{RgF}/B_g$$

where, for this scenario

GRV_{HKW} = gross rock volume down to HKW, NAF;

E_{GRV} = error function for GRV, a triangular RV defined by 0.84, 1.0, 1.16;

F_o = fraction of GRV occupied by oil;

   for estimates made under U.S. SEC guidelines,[128]

    $\text{F}_o = 0.795 - 0.437 = 0.358$;

   for estimates not required to follow U.S. SEC guidelines,

    $\text{F}_o = \text{F}_H - \text{F}_{FG}$, where[129]

      F_H = fraction of GRV occupied by hydrocarbons, a uniform RV between 0.795 and 1.000, and

      F_{FG} = fraction of GRV occupied by free gas, a uniform RV between 0.153 and 0.437;

PG_o = petrophysical group for the oil column, as defined in Sec. 9.9.2;

PG_{FG} = petrophysical group for the (free) gas column,

and other variables are as previously defined.

As they are not directly relevant to this section, potential dependencies of the type discussed in Sec. 9.5.4, and 9.9.2 are not discussed here.

9.10.2 Scenario Analysis for Major Offshore Discoveries

For offshore discoveries, estimates of initial reserves may be dependent on the exploitation plan and the economic/operating scenario, which includes:

a) number, placement, and type of wells—conventional vs. horizontal vs. multilateral,

b) method of operation—primary drive and/or fluid injection—and production rate,

c) economic scenario—development capital, future wellhead prices, and operating costs, and

d) contract terms between host government and operating group.

In such cases, estimates typically are made of O/GIP, and rate-time profiles (reserves) are calculated for various development plans to exploit O/GIP, typically using computer simulation.

In complex geologic settings, there may be considerable uncertainty regarding the type and spatial distribution of reservoir heterogeneities and the influence of such heterogeneities on ultimate recovery. Geostatistical analysis coupled with reservoir simulation has been utilized in one such setting, but with only one development and operating plan [Lia *et al.* 1997]. It seems apparent that in such geologic settings the placement of production (and injection) wells might have a significant influence on recovery efficiency. If the key heterogeneities are uncertain, however, so must also be the sub-optimal placement of wells, leading to an almost unsolvable problem.

---

[128] As noted in Appendix C, U.S. SEC guidelines set the downdip limit for proved oil reserves at LKO or OWC. It is assumed here that such guidelines (tacitly) set the updip limit for an oil column at HKO.

[129] Absent compelling data to the contrary.

9.11 Performance Estimates Using Probabilistic Procedures

As inferred by previous discussions, most applications of probabilistic procedures have been during the static phase of exploitation—during prospect appraisal, when operators typically face the greatest degree of geologic/reservoir uncertainty and monetary risk. There have been few published applications of such calculations during the performance stages of exploitation. It is recognized that the magnitude of uncertainties during the dynamic phase typically is less than that during the static phase. That not withstanding, possible approaches to probabilistic analysis during the dynamic phase are discussed in this section.

9.11.1 Volumetric Gas Reservoirs

As discussed in Sec. 4.4, GIP and reserves for a volumetric gas reservoir may be estimated with a plot of p/Z vs. G_p data. With deterministic procedures, reserves so estimated typically are classified as proved, unless there are compelling data to the contrary. Depending on circumstances, however, a probabilistic approach may be warranted. Sources of uncertainty[130] in p/Z vs. G_p data include:

(a) initial (static) reservoir pressure, p_i,

(b) gas deviation factor, Z, vs. p and T_f,

(c) cumulative gas and condensate production vs. time and calculation of reservoir "gas equivalent" of stock-tank condensate, as discussed in Sec. 4.4.1 and Appendix E.

(d) average static reservoir pressure history during the time period analyzed, and

(e) average reservoir-abandonment pressure, p_a.

Depending on circumstances, Items (d) and (e) may be subject to the largest degree of uncertainty, especially in heterogeneous, low-permeability reservoirs. Uncertainty in Item (d) also may be substantial if reservoir pressure is estimated from SITP data, rather than from bottomhole pressure surveys. Depending on reservoir gas composition and surface production facilities, Item (c) also may be subject to significant uncertainty.

In geopressured reservoirs, Item (a) may constitute a significant degree of uncertainty [Cronquist, 1984]. Item (a) may also be subject to substantial uncertainty in areas where there has been previous production.

In retrograde gas reservoirs producing at pressures below the dewpoint pressure, or in gas reservoirs with substantial nonhydrocarbon components, Item (b) may be subject to significant uncertainty, especially when there is not a PVT analysis available and empirical correlations must be used.

Irrespective of the sources of error—which may be difficult to identify—and the engineer's best efforts to omit spurious data from the calculations, the level of uncertainty may warrant a probabilistic analysis, as illustrated by **Fig 9.24**.[131] This figure shows the p/Z vs. G_p data compiled after the cumulative production of 7.3 Bscf of gas. In addition to the uncertainty in these data is the uncertainty in the static reservoir pressure at abandonment. For this reservoir, the operator's estimate for most likely p_a/Z_a is 500 psi; the maximum, 750 psi; the minimum, 250 psi.

---

[130] In the discussion following, it is assumed that the reservoir is volumetric, as discussed in Sec. 4.4, and that a linear fit of p/Z vs. G_p is the best interpretation of the available geologic and engineering data.

[131] Discussion of a comparable methodology has been provided by Elliott [1995].

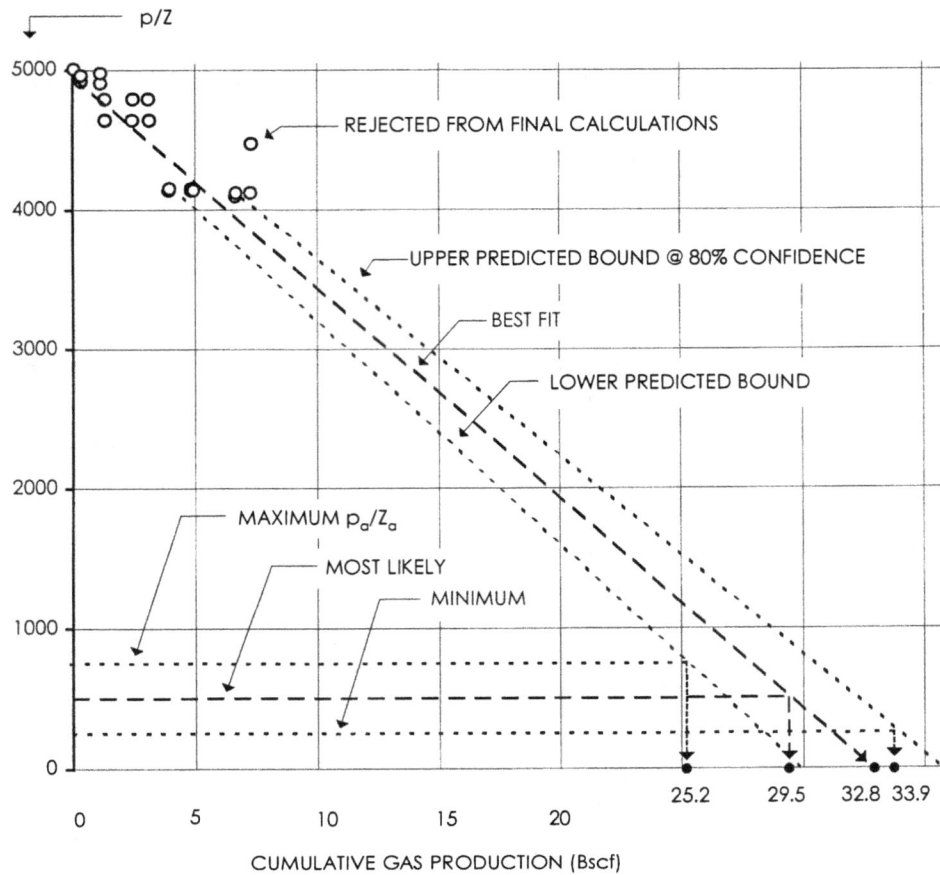

Fig. 9.24—Example *p/Z* vs. *G_p*, illustrating probabilistic method to estimate GIP and reserves.

An unweighted, unforced linear regression of all the p/Z vs. G_p data results in a correlation coefficient, r^2, of 0.747. The last data point, however, is almost three standard deviations from the mean and is considered erroneous. A regression omitting this point results in a correlation coefficient of 0.842.

The quality of the initial pressure data for this reservoir are considered to be excellent. Accordingly, the third regression was forced through the "origin," at p_i/Z_i. The correlation coefficient for this regression is 0.896. From extrapolation of the "best fit" line to $p/Z = 0$, GIP is estimated to be 32.8 Bscf.

Using standard statistical techniques, the upper and lower predicted bounds at 80% confidence may be calculated, as shown by Fig 9.24. From extrapolation of these bounds to $p/Z = 0$, "upside" GIP is estimated to be 35.7 Bscf; "downside," 29.8 Bscf. These two values are not labeled on Fig. 9.24.

It is suggested here that—in the absence of data to the contrary—the lower bound be used to established proved GIP; the best fit, proved plus probable GIP; the upper bound, proved plus probable plus possible GIP.[132] Estimates of GIP may be needed to ensure adequate

---

[132] It is recognized that—if one interprets extant definitions literally—the terms "proved, "probable," and "possible" pertain only to reserves, not to O/GIP. That not withstanding, the usage here reflects that frequently used in the application of probabilistic calculations.

drainage of the reservoir from existing wells or to determine suitability of the reservoir for gas storage.

Abandonment bottom-hole pressure in volumetric gas reservoirs, p_a, is a function of several factors, including: (a) the economic limit flow rate, (b) the pressure drawdown required to produce the well at the economic limit flow rate, and (c) the minimum wellhead flowing pressure. The economic limit flow rate for a producing entity is a function of the economic scenario anticipated at abandonment, the prediction of which is always subject to uncertainty. Pressure drawdown may be expected to vary from well to well and may exhibit considerable variation, especially in reservoirs with a high degree of permeability heterogeneity. Minimum wellhead flowing pressure —not sales line pressure—also may vary from well to well and may be influenced by the liquid gas ratio (LGR) and flowline length. While prediction of the CGR at abandonment might involve minimal uncertainty, the LGR is a function of production of extraneous formation water, which is difficult to predict.

Downside, most likely, and upside initial reserves are estimated to be 25.2, 29.5, and 33.9 Bscf, respectively,[133] as shown by Fig. 9.24. The results of the foregoing analyses, together with reserves as of the date of the estimate are summarized below.

	GIP	Initial Reserves	Cumulative Production	Reserves at Date of Estimate
Upside (Bscf)	35.7	33.9	7.3	26.6
Most likely (Bscf)	32.8	29.5	7.3	22.2
Downside (Bscf)	29.8	25.2	7.3	17.9

As discussed previously, the upper and lower predicted bounds at 80% confidence are a reasonable approximation of the 90th and 10th percentile[134] of the predicted variable, p/Z. Also, as discussed in Sec. 9.2.4, the best fit is a reasonable approximation to the 50th percentile. Accordingly, it is suggested here that—in the absence of data or reporting protocols to the contrary—upside, most likely, and downside estimates be classified as "3P," "2P," and "1P" (initial) reserves, respectively.

In view of the quality of the initial pressure data for subject reservoir, the regression was forced through the origin, p_i/Z_i. In geopressured reservoirs, however, initial pressure data might be subject to considerable error. In such cases, it might be preferable not to force the regression through the origin. Had this procedure been followed for subject reservoir, most likely GIP would have been 28.8 Bscf; upside, 32.2 Bscf; downside, 25.4 Bscf.

9.11.2 Oil Reservoirs

As discussed in Sec. 4.2, in some cases a Havlena-Odeh [1963] "tank model" material balance may be an appropriate method to estimate OIP. In such cases, consideration might be given to using linear regression to analyze the trend of data points.[135] For example, the "best fit" might be used to determine the median estimate ("2P") for OIP. The upper and lower bounds of the prediction at 80% confidence might be used to determine "3P" and "1P" OIP, respectively, provided this does not conflict with other, more compelling, data.

---

[133] This is a dry gas reservoir; thus, there are no condensate reserves.
[134] Confidence intervals and percentiles were discussed in Sec. 9.2.3.
[135] For additional discussion—and caveats—regarding this procedure, please see Carlson [1997b].

9.11.3 Performance/Decline Trend Analysis

There are several scenarios that are typified by considerable uncertainty that might be amenable to a probabilistic approach, including: (a) transient production trends, (b) poorly defined production trends from multi-well aggregates, and (c) conflicting data from performance/ decline trends.

Probabilistic approaches to decline trend analysis have been discussed by Huffman and Thompson [1994] and by Jochen and Spivey [1996]. The former authors developed methodologies to use regression analysis of production trends to estimate "1P," "2P," and "3P" reserves. The latter authors used a novel approach, called the bootstrap method, to do essentially the same thing. Both techniques, however, were developed in an academic environment, and general application of these procedures has been inhibited by lack of commercially available software. Recently, however, commercial software has become available to facilitate probabilistic analysis of decline trends,[136] and an application is discussed in the next section.

Transient Production Trends

In many producing areas, new wells in depletion drive reservoirs initially exhibit rapid declines in productivity. In many such cases, these rapid declines are caused by transient flow, which typically exhibits apparent b exponents significantly greater than 1.0.[137] Mathematical extrapolation of such trends—with b exponents significantly greater than 1.0— may lead to erroneously high estimates of reserves. Production from wells drilled earlier in these—or analogous—areas, however, may have entered the semi-steady-state phase. The more mature production trends—and indicated ultimate oil production—from these wells typically are used as models for the new wells. Usually, however, such data are too sparse for meaningful statistical analysis. Thus, the engineer typically must make judgments regarding reasonable rate-time trends and ultimate oil production for the new wells. A probabilistic approach may be appropriate.[138]

As discussed in Sec 9.2.4, the "best fit" of a trend of production data using regression analysis actually is a P50 estimate. Accordingly, reserves estimated by forward projection of such "best fits" might be classified "proved plus probable." A fit through these data that would reflect a P90—or other—estimate might be needed for certain applications. **Fig 9.25** is a semi-log, rate-time plot of oil production for a well in a low permeability, volatile oil reservoir in the U.S. with a P50 "best fit" from regression analysis.[139] The b exponent from this fit is calculated to be 1.6, which is suggestive of transient flow, consistent with the performance of other wells in this accumulation.[140] To help assess reserve uncertainty and establish appropriate spacing for future wells, P85 and P15 estimates[141] were made through these data. As discussed in Sec. 5.3.8, forward projections of fits made of the transient period should be adjusted to reflect a minimum annual decline, consistent with experience in the area.

---

[136] MICA™ Program Documentation, available at www.mcsi.com. No endorsement is intended by this reference.
[137] Hyperbolic production declines and b exponents were discussed in Sec. 5.3.2.
[138] A deterministic approach to this problem was illustrated in Fig. 1.2.
[139] This work was done using MICA™ for Windows.™ No endorsement is intended by this reference.
[140] Such wells are typically "frac-ed," which accounts for the high initial production rate and subsequent rapid decline.
[141] Estimates at P90 and P10 were considered too extreme and were rejected in favor of P85 and P15 estimates.

BOPD

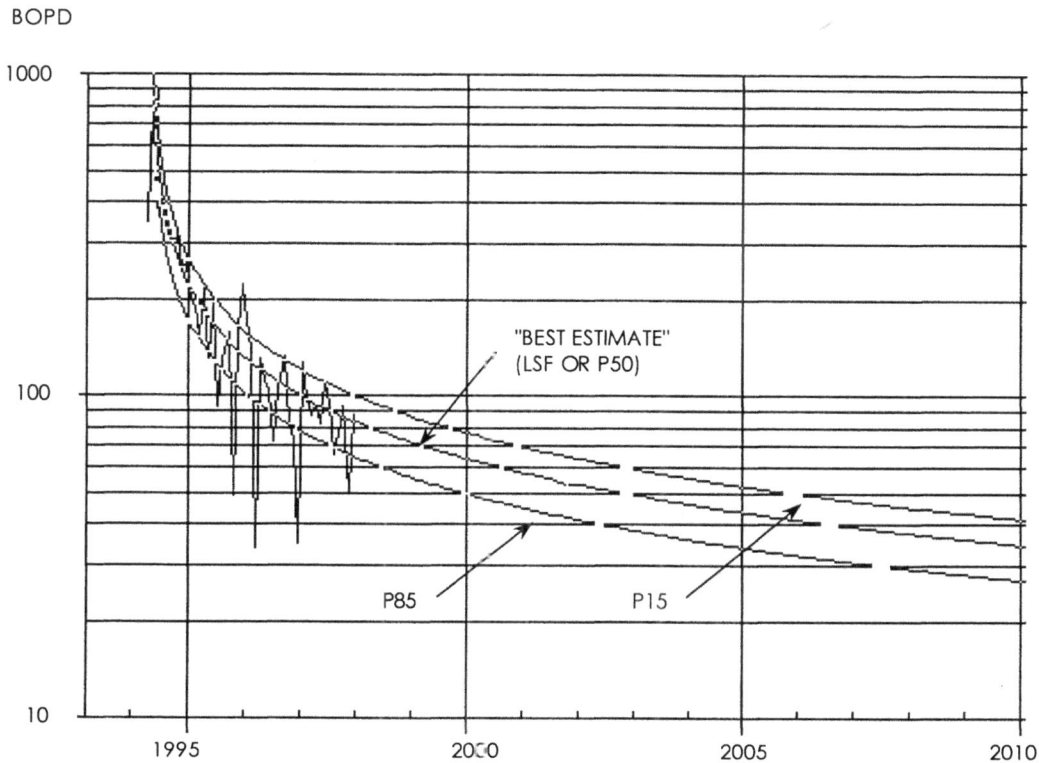

Fig. 9.25—BOPD vs. time illustrating best estimate (LSF or P50) and P15 and P85 estimates [Courtesy Molli Computer Services].

Poorly Defined Production Trends

Even in mature producing areas, decline trends may be poorly defined and subject to considerable uncertainty. Such uncertainties are likely to be observed, for example, in areas where only data on multi-well aggregates are available for analysis. This problem has been subject to considerable discussion by Purvis and others [1979, 1985, 1987, 1990], and this work was summarized in Sec. 5.1.3. A methodology like that discussed in the previous section may not be appropriate, but the findings by these authors reflect an alternate probabilistic approach to this type of problem.

Conflicting Interpretations of P/DT

Analysis of performance/decline trends[142] often requires resolution of dilemmas caused by conflicting and/or incomplete data. For example, as discussed in Sec. 5.2.1, after breakthrough,[143] fractional oil flow, f_o, for wells producing from waterdrive reservoirs typically exhibits a semilog trend vs. cumulative oil production, N_p. In some areas, however, trends of f_o vs. N_p tend to be linear, rather than semi log. In many cases, it is not possible to determine the correct trend until there has been substantial oil production after water

---

[142] Production/decline trend analysis—from a deterministic view—was discussed in Sec. 5.0. As discussed in Sec. 5.3.5, caution should be exercised in relying on mathematical analysis to the exclusion of good engineering judgment for P/DTA.

[143] A waterflooding term, breakthrough means production of first significant water and generally indicates advance of the water-oil interface into the producing well.

breakthrough. If there is a decline in oil production rate in such wells, it might be appropriate to supplement analysis of trends of f_o vs. N_p with analysis of trends of rate vs. time (q_o vs. t), and/or rate vs. cumulative production (q_o vs. N_p). Frequently, however, extrapolations of plots of f_o vs. N_p and q_o vs. N_p for the same well yield different estimates for reserves. Interpretation is further complicated by uncertainties regarding total liquid capacity of the well at high water cuts, which establishes the lower limit of the plot of f_o vs. N_p at q_{el}. In some cases, a decline in oil rate may be attributable to a combination of a decline in both f_o and total liquid production. Depending on circumstances, it might be appropriate to consider the most conservative plot indicative of proved reserves; the most optimistic, proved plus probable plus possible reserves.

9.12 Aggregation of Separate Reserve Estimates

"The total of proved reserves is not always the proved reserve of the total."

Wim Swinkels [1999]

9.12.1 Background

SPE-WPC [1997] definitions state, "Because of potential differences in uncertainty, caution should be exercised when aggregating reserves of different classifications." This caution, which first appeared in SPE [1987] definitions, referred (at that time) to aggregation of reserves estimated using deterministic procedures that had not been risk-weighted. The caution, however, is applicable to reserves estimated using either probabilistic or deterministic procedures and might be expanded to include reserves of different classifications *and* (development) status. In this context, it is noted that, under U.S. SEC guidelines, companies report "proved undeveloped" reserves separately from, and in addition to, "proved developed and undeveloped reserves."

Estimates of reserves for separate accumulations may be aggregated for various purposes including:

- reporting to shareholders and/or regulatory entities,
- purchase and/or sale of properties,
- negotiation with lending institutions,
- sales contracts, and
- corporate planning.

Depending on the purpose of the estimate, aggregations may reflect reserves for different geologic trends, states, regions, or countries. They may include wells, leases, units, reservoirs, fields, and/or concessions, each of which may be located in different geologic settings and be at different stages of exploration, development, and/or production maturity. Despite the disparate nature of the separate estimates, however, it generally it is considered that the degree of uncertainty in an aggregate estimate is less than that of the separate estimates comprising the aggregate. In the absence of a business reason to do so, however, quantification of aggregate uncertainty has not been a routine exercise. Also, the historical absence of quantification of the degree of uncertainty in separate deterministic estimates has precluded quantification of uncertainty in aggregate estimates.

Depending on requirements, aggregations may include, separately: (a) proved reserves,[144] (b) proved plus probable reserves,[145] (c) possible reserves, (d) expected reserves, and/or (e) potential additional reserves.[146]

The increasing application of probabilistic methods, in which uncertainty may be quantified, has facilitated quantification of the uncertainty in aggregate estimates.

9.12.2 Dependencies Between Separate Estimates[147]

There may be varying degrees of dependency between the separate estimates comprising an aggregate of these estimates. If there is no, or negligible, dependency between the separate estimates, the aggregate estimate may involve direct stochastic summation of the pdf's of the separate estimates. Geologic scenarios in which the separate estimates might be considered independent include:

- wells in a development drilling program where reserves attributed to each successive well are random and independent of those drilled earlier,

- widely spaced wells in a geologic trend, where there is no reasonable chance of interference, even though the wells may be in a common accumulation,

- separate accumulations in the same geologic trend or province that are not in hydraulic communication, and

- separate accumulations in different geologic trends or provinces.

Even though these conditions for independence may be satisfied, there are other conditions that may lead to dependence between estimates—e.g. use of the same seismic interpretation procedure or use of the same algorithm to estimate reserves. For example, if the API [Arps *et al.* 1967] correlation, Sec. 3.7.2, is used to estimate recovery efficiency of oil for each reservoir in an aggregate, the results might be considered partially dependent.

If there are dependencies between the separate estimates, the degree of each such dependency may have to be considered in the aggregation algorithm. Geologic scenarios in which there may be dependencies between separate reserve estimates include:[148]

- wells in a common accumulation, where the drainage areas of individual wells are treated stochastically,

- adjacent fault blocks, where the position(s) of the separating fault(s) are treated stochastically,

- stacked reservoirs, where the dip angle—and, consequently, the downdip limit of each reservoir—is treated stochastically,

- gas-cap oil reservoirs where the position of the initial gas/oil contact and/or the degree of gravity segregation during production are treated stochastically, and

- reservoirs in a common aquifer, where aquifer strength—and, consequently, recovery efficiency—is treated stochastically.

---

[144] Major operators in the U.S. report (only) proved reserves of crude oil, natural gas, and natural gas liquids annually to the EIA.

[145] Operators in Western Australia report—separately—proved (P90) and proved plus probable (P50) reserves annually to the Department of Resources Development.

[146] Operators in the U. K. report—separately—proven (P90), probable (P50) and possible (<P50) reserves annually to the Department of Trade and Industry. In addition, the U K. DTI makes estimates of potential additional reserves in "discoveries which do not meet the criteria for inclusion as possible reserves."

[147] Some of the material in this section was adopted from Carter and Morales [1998] and Swinkels [2001]. More specific credits are provided in the sections following.

[148] For a more comprehensive discussion of dependencies, please see Carter and Morales [1998].

In addition to these scenarios, there are operational situations that might lead to dependency between separate estimates. For example, if separate accumulations are produced through a common facility, the separate reserve estimates for those accumulations may be interdependent, depending on assumptions regarding facility constraints at the economic limit.

9.12.3 Aggregates of Independent Estimates

Discussed in this section is an example of a geologic/reservoir scenario in which reserves attributed to each successive well in a drilling program may be treated as independent, random events.

Fig. 9.26 is a histogram of EUOR from 30 mature wells producing from the developed part of a major oil field, offshore U.S. The break in the abscissa facilitates plotting all the data on an expanded scale. There is no correlation between the drilling sequence and the EUOR for the individual wells. Based on this observation and general knowledge about the reservoir, it seems apparent that the EUOR for each successive well in the developed area was random and not related to the outcomes from earlier wells in the same area.[149]

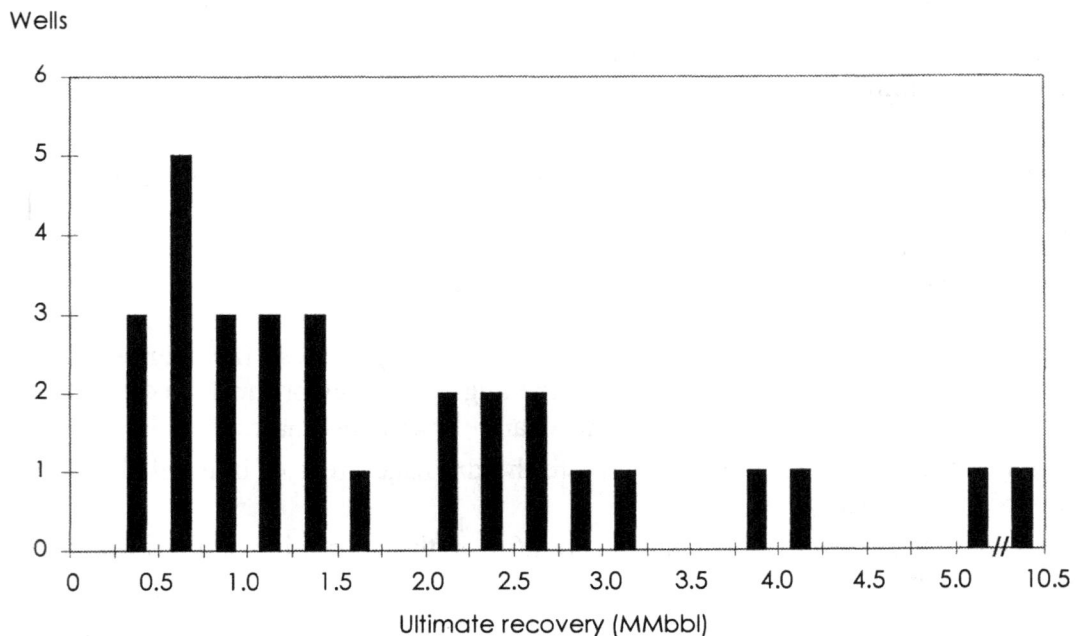

Fig. 9.26—Histogram of EUOR for 30 wells in a major reservoir, offshore U.S.

Under consideration at the time of this work was a drilling program for the then-undrilled part of the same reservoir. The reservoir characteristics of the undrilled area were considered to be analogous to those of the drilled area. Thus, the pdf of EUOR estimated for the wells in the drilled area was considered an acceptable model for the wells being considered for the undrilled area.

Statistical distribution of EUOR for wells in the developed area is shown by the expectation plot in **Fig. 9.27.** Also shown on this figure is the EC and the pdf for the beta

---

[149] Informal presentation by author during panel discussion at 1998 SPEE Annual Meeting, Monterey, California.

distribution used to fit these data,[150] which exhibit a high degree of skew.[151] From the beta distribution, the EUOR for a single well in the program may be estimated as:

Proved (P90):	90 Mbbl
Proved plus probable (P50):	1050 Mbbl
Proved plus probable plus possible (P10):	4520 Mbbl

The mean of this distribution is 1.75 MMbbl, which differs significantly from the P50 estimate, as might be expected for this highly skewed distribution.

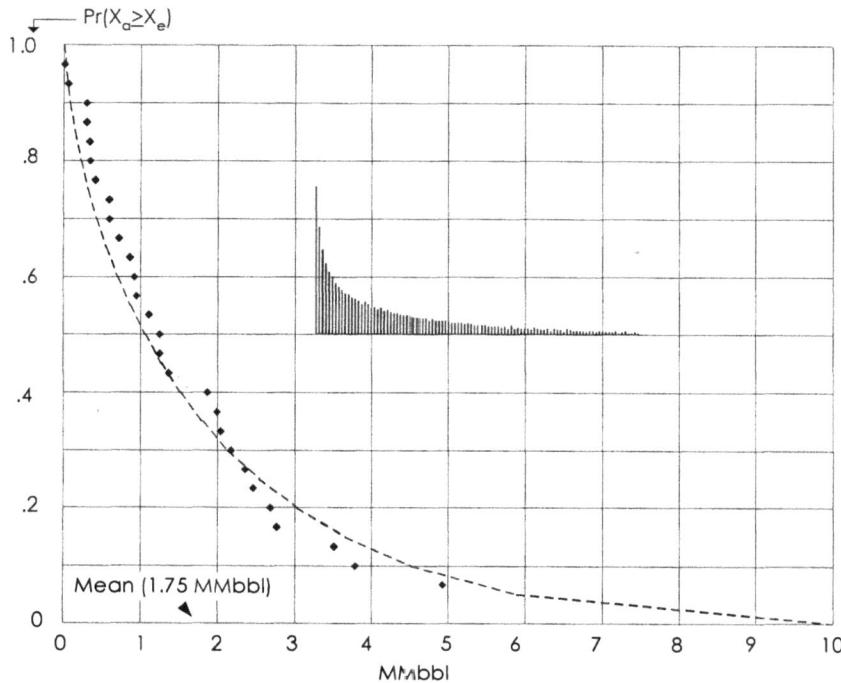

Fig. 9.27—Expectation plot, beta distribution fit, and pdf of data in Fig. 9.26.

Three-, five-, and seven-well programs were being considered for the undrilled area. EUOR and economic analysis of such programs historically has been based on simple arithmetic summation of EUOR and discounted future net revenue calculated for separate wells. Probabilistic summation, however, might provide a more informative view. Based on experience in the developed area, dry-hole risk for the undeveloped area was considered negligible. Thus, EUOR attributed to each program was estimated by successive sampling of the unrisked pdf of EUOR of the analogy wells—by probabilistic summation.[152] The results of these analyses are shown on **Fig. 9.28,** which includes expectation curves of EUOR for three-, five- and seven-well programs and the pdf's of EUOR for the three- and seven-well programs. Of interest is the shift from the strongly skewed EC of EUOR for a single well,

---

[150] A log-normal distribution did not fit these data as well as a beta distribution. The beta distribution, defined by 0.58, 6.46, 21.05, was truncated at 0.01 and 10 for MCS.

[151] The observant reader may note there are only 28 data points on Fig. 9.27, compared to the 30 wells included in Fig. 9.26. Abnormally low ultimate recovery from one well was attributed to mechanical problems and was excluded from the fit. Ultimate recovery from the other well was included in the fit, but the data point is off the scale of Fig. 9.27.

[152] This procedure, called sampling with replacement, is reasonable, provided that the total population is significantly greater than the number of samples, so that the small number of samples does not significantly affect the pdf of the initial population.

shown on Fig. 9.27, to the progressively less skewed EC's for the three- five- and seven-well programs shown on Fig. 9.28.[153]

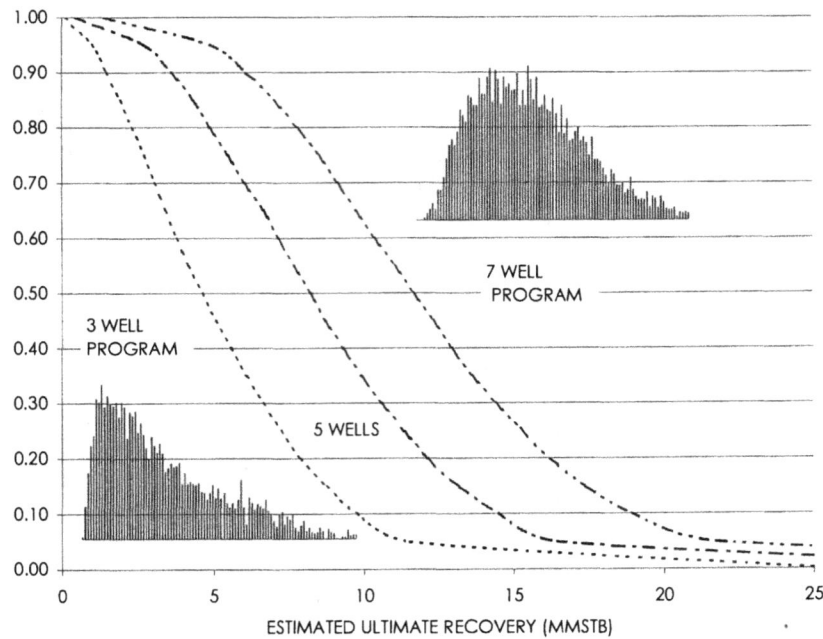

Fig. 9.28—Expectation curves (EC's) of EUOR for 3-, 5-, and 7-well programs with pdf's of EUOR for 3- and 7-well programs.

Key estimates from Fig 9.28 are tabulated below:

	Program Aggregate and (Well Average) Reserves (MMbbl) Number of Wells in Program		
	3	5	7
Proved (P90):	1.46 (0.49)	3.62 (0.72)	6.10 (0.87)
Proved plus probable (P50):	4.64 (1.54)	8.24 (1.65)	11.66 (1.81)
Proved plus probable plus possible (P10):	9.79 (3.26)	14.50 (2.90)	18.94 (2.91)
Mean:	5.24 (1.75)	8.73 (1.75)	12.22 (1.75)

Probabilistic and arithmetic summation of program reserves are compared:

	Ratio of Probabilistic to Arithmetic Summation* Number of Wells in Program		
	3	5	7
Proved (P90):	5.49	8.17	9.83
Proved plus probable (P50):	1.47	1.56	1.58
Proved plus probable plus possible (P10):	0.72	0.64	0.60
Mean:	1.00	1.00	1.00

* Small differences between ratios in this table and ratios that might be derived from data in the text are due to intermediate rounding.

---

[153] This shift is consistent with the central limit theorem of classical statistics, which posits that the pdf of the probabilistic sum of a set of pdf's approaches Gaussian as the number of pdf's in the set becomes very large.

From these calculations,[154] it is noted that:

- The probabilistic sum of reserve estimates for an aggregate of wells at P90 expectation may be significantly greater than the arithmetic sum of reserves for the individual wells at the same expectation, with the difference increasing for larger aggregates.

- The probabilistic sum of reserve estimates for an aggregate of wells at P50 expectation may be slightly greater than the arithmetic sum of reserves for the individual wells at the same expectation, with the difference increasing for larger aggregates.

- The probabilistic sum of reserves for an aggregate of wells at P10 expectation may be slightly less than the arithmetic sum of reserves for the individual wells at the same expectation, with the difference increasing for larger aggregates.

- The mean of the probabilistic sum of reserves for an aggregate of wells is the same as the arithmetic sum of the mean reserves for the individual wells—i.e., the sum of the means equals the mean of the sum.

This simple example has covered only aggregation of various multiples of a single pdf. Although not demonstrated here, it seems apparent that the degree of difference between probabilistic summation and arithmetic summation of a set of pdf's will depend on: (a) the nature and degree of skew of the individual pdf's and (b) the range of mean values represented by the set of pdf's.

Of significance is the observation that "the sum of the means equals the mean of the sum" which provides a basis for using the means of probabilistic estimates for aggregations and economic analyses.

9.12.4 Aggregates of Partially Dependent Estimates

Two procedures to aggregate partially dependent estimates have been discussed by others:

- stochastic addition of the pdf's of reserve estimates, using Spearman rank correlations to quantify the degree of dependency between estimates [Carter and Morales 1998], and

- scenario methods, using probability trees [Swinkels 2001].

Each procedure is discussed briefly in ensuing paragraphs.

Carter and Morales [1998] discussed aggregation of separate reserve estimates "within a major gas project," understood to be offshore Western Australia.[155] As noted by the authors, "correct determination of proved reserves underpins the contracted gas volumes for the project."

Potential areas[156] of partial dependency identified by the authors included:

- geologic scenario (some of which were noted here in previous sections),

- method(s) to calculate gross rock volume and GIP, and

- method(s) to estimate recovery efficiency.

---

[154] Strictly speaking, these observations must be limited to the data presented here. As a practical matter, however, generality seems to be a reasonable extension.

[155] After completion of this section, Woodside Energy Ltd. released the results of an extension to this work [van Elk *et al.* 2000].

[156] Interested readers are referred to subject paper, as the authors' treatment of this subject is too extensive to summarize here and merits careful review.

Carter and Morales [1998] work involved a comprehensive, subjective assessment of the degree of dependency between reserve estimates for each field pair. Identified dependencies for each field pair were subjectively quantified by SRC's between "0" and "1" for input to a commercial simulation program.

In a 25-field example, the authors reported that probabilistic summation with all identified dependencies resulted in 9.3% increase in P90 reserves compared to arithmetic summation of P90 reserves for the separate fields. If all the estimates had been fully independent, the increase would have been 14.8%.

Swinkels [2001] discussed use of probability trees to aggregation of separate reserve estimates, which he called a "scenario method." The method is illustrated by **Table 9.5,** which summarizes reserve estimates (arbitrary units) for three separate entities, "A," "B," and "C."

TABLE 9.5—SCENARIO METHOD FOR RESERVE AGGREGATION [after Swinkels 2001]

Entity	LoEst	P50	HiEst	Exp
A	250	458	839	512
B	187	343	629	384
C	141	257	472	288
Sums:	578	1058	1940	1184

Entity A Est	Entity A Pr[est]	Entity B Est	Entity B Pr[est]	Entity C Est	Entity C Pr[est]	Pr[sum]	Sum	RWSum
				141	0.33	0.037	578	21.3
		187	0.33	257	0.33	0.037	694	25.6
				472	0.33	0.037	909	33.7
				141	0.33	0.037	734	27.1
250	0.333	343	0.33	257	0.33	0.037	850	31.4
				472	0.33	0.037	1065	39.4
				141	0.33	0.037	1020	37.8
		629	0.33	257	0.33	0.037	1136	42.1
				472	0.33	0.037	1351	50.2
				141	0.33	0.037	786	29.0
		187	0.33	257	0.33	0.037	902	33.3
				472	0.33	0.037	1117	41.4
				141	0.33	0.037	942	34.8
458	0.333	343	0.33	257	0.33	0.037	1058	39.1
				472	0.33	0.037	1273	47.1
				141	0.33	0.037	1228	45.5
		629	0.33	257	0.33	0.037	1344	49.8
				472	0.33	0.037	1559	57.9
				141	0.33	0.037	1167	43.2
		187	0.33	257	0.33	0.037	1283	47.5
				472	0.33	0.037	1498	55.5
				141	0.33	0.037	1323	49.0
839	0.334	343	0.33	257	0.33	0.037	1439	53.3
				472	0.33	0.037	1654	61.4
				141	0.33	0.037	1609	59.8
		629	0.33	257	0.33	0.037	1725	64.1
				472	0.33	0.037	1940	72.3
Sums:	1.000					1.000		1192.6

For each entity, expectations at P90 (LoEst), P50, and P10 (HiEst) were input to the tree shown in the bottom part of this table. Probabilities for each outcome in the tree were set at 0.333, after Swinkels [2001], who considered this to reflect a "fully independent" set of estimates. The sums and corresponding probabilities for each of the 27 outcomes, together with the probability-weighted sums, are in the right three columns of this table. It may be noted that the expected value of the aggregate, 1193 units, is approximately equal to the arithmetic sum of the means (Exp), 1184 units, which is consistent with probability theory.

The same reserve estimates shown on the top of Table 9.5 were aggregated assuming a "high degree of dependency," with probabilities of each outcome taken from Swinkels [2001]. Expectation plots of the two aggregates, **Fig 9.29a,** were generated by sorting the sums and plotting vs. the complement of the corresponding accumulated probabilities.[157]

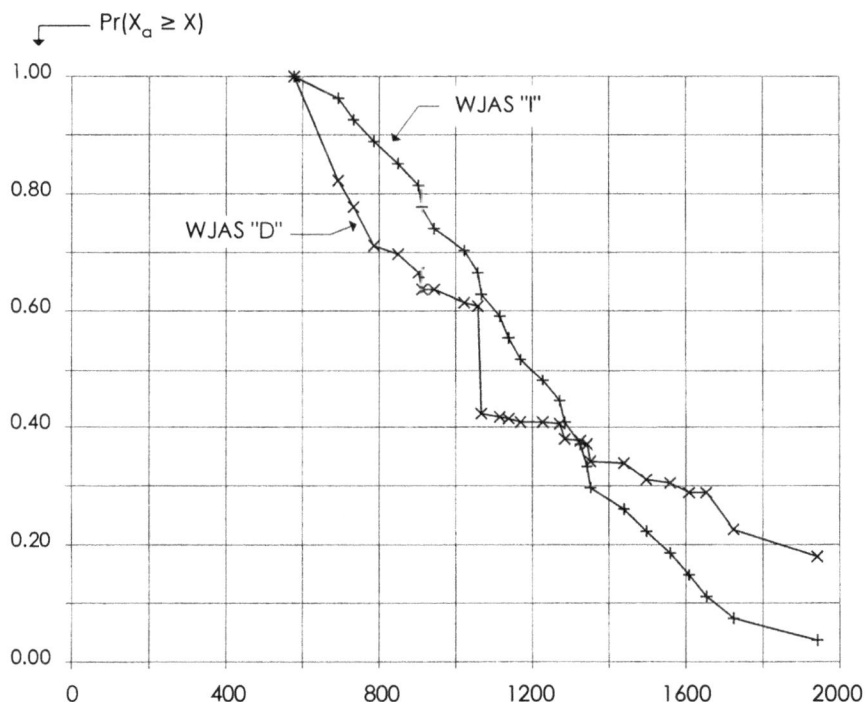

Fig. 9.29a—Expectation plots for aggregates of fully independent and highly dependent estimates of reserves [after Swinkels 2001].

For the fully independent case, identified on Fig 9.29a as WJAS "I," it may be observed that:

- At an expectation of P90, aggregate reserves are estimated to be 750 units, compared to 578 units (Table 9.5) from arithmetic addition of the separate P90 (LoEst) estimates.

- At an expectation of P50, aggregate reserves are estimated to be 1200 units, which is a reasonable approximation of the mean estimate.

---

[157] This is the same procedure that was used to generate Fig. 9.9a from Table 9.1.

- At an expectation of P10, aggregate reserves are estimated to be 1700 units, compared to 1940 units (Table 9.5) from arithmetic addition of the separate P10 (HiEst) estimates.

For the case assuming a "high degree of dependency" [Swinkels 2001], which is identified on Fig 9.29a as WJAS "D," it may be observed that, compared to the fully independent case, the expectation plot exhibits more skew—i.e., the aggregates at expectations of P90 and P10 tend to approach the arithmetic sums of the separate P90 and P10 estimates, respectively.

For a comparison between the scenario method and stochastic addition of pdf's, the same estimates used in the probability tree in Table 9.5 were used in a commercial simulation program to generate two additional aggregations. The first case assumed full independence—SRC = 0. The second case assumed complete dependence—SRC = 1. The results are shown on **Fig. 9.29b,** which includes the plots from Fig. 9.29a.

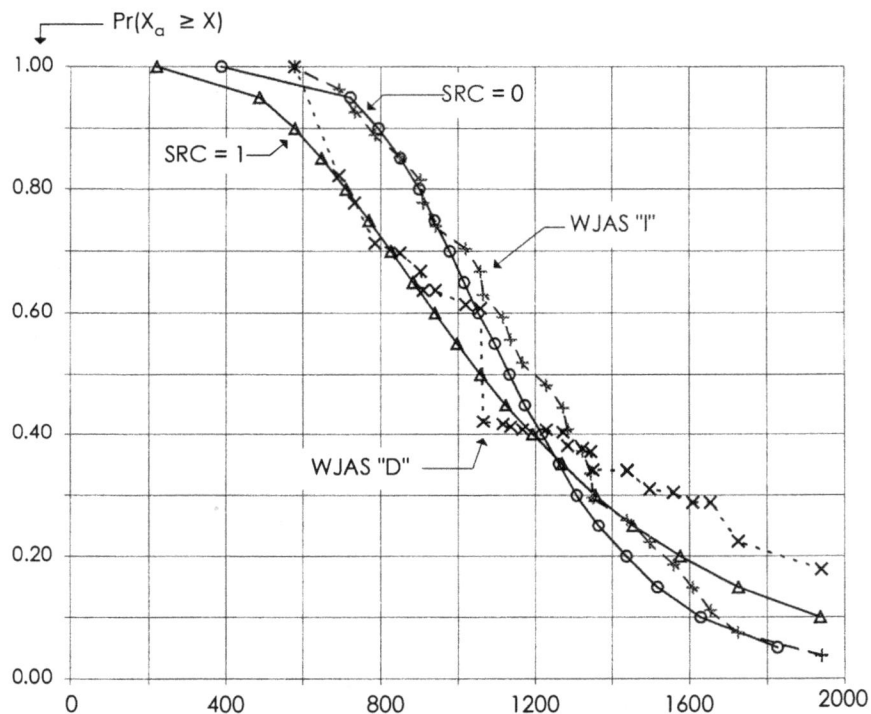

Fig. 9.29b—Expectation plots comparing scenario method and stochastic addition for aggregate estimates of reserves.

From these plots, it may be noted that:

- For the fully dependent case (SRC = 1), at expectations of P90 and P10 the aggregate estimates equal the arithmetic sum of the separate estimates at P90 (LoEst) and P10 (HiEst)—i.e., 578 and 1940 units (Table 9.5), respectively.
- Because of the discretization inherent in the scenario method, the expectation plots are not as well defined as those computed by stochastic addition.

From the work discussed here, albeit limited, and from unpublished work done by the author, it is noted that:

- Provided the data are available, stochastic addition of separate pdf's may be preferable to the scenario method to compute expectation plots of aggregate estimates.

- The scenario method may provide reasonable estimates of aggregated reserves at P90 expectations, but consideration might be given to sensitivity analysis of the probabilities estimated for the outcomes.

- In general, the pdf's of aggregates tend to be dominated by the pdf's of the upper decile of the separate estimates.

- In general, greater skew of the pdf's of the separate estimates will result in greater skew of the aggregate estimate and, consequently, more sensitivity to the degree of dependency estimated between the separate estimates.

- In any event, arithmetic addition of the means of the separate estimates will result in a statistically correct mean of the aggregate estimate, which—in the long run—may be the most meaningful statistic for financial purposes and public reporting.

9.13 Principal Points

Deterministic estimation and classification of reserves may involve one or more separate estimates for each accumulation, with classification of each estimate being based on professional judgment of the risk associated with the realization of each such estimate. Probabilistic procedures lead to a continuous probability distribution of estimates for each accumulation, with classifications being based on selected percentile rankings of estimates within each continuum.

In geologic settings and operating areas where the industry has substantial experience and in fully developed, mature fields—situations considered to have a relatively low degree of uncertainty—deterministic estimates of reserves usually are considered acceptable.

As exploitation emphasis shifted from risking well-size increments of capital to risking field-size increments, a need was perceived to quantify the degree of uncertainty in reserve estimation, which led to increasing application of probabilistic procedures.

Two principal categories of uncertainty/risk regarding volumetric reserve estimates may be identified:

- those attributed to uncertainty of the geologic and engineering data in the drilled area and

- those attributed to geologic risk regarding presence or continuity of a commercial accumulation beyond the drilled area and/or in adjacent fault blocks.

While the former category of uncertainty may be amenable to statistical analysis, the latter must be resolved with professional judgment.

Deterministic estimates of *proved* reserves based on analytical methods cannot readily be compared to probabilistic estimates of *proved* reserves, as the former typically are based on means or modes of the input data sets, which usually will result in estimates that are significantly greater than P90 estimates from the same data sets.

Probability trees, parametric methods, and MCS each have been used for probabilistic estimates. MCS, however, has become the preferred method because of wide commercial availability of generalized software and increasingly more powerful PC's, despite the need to master rather complex theoretical bases and software.

The results of probabilistic estimates are dependent of the pdf's of the input parameters, the treatment of dependencies between such parameters, and the method of aggregation of the component estimates. While such estimates may be robust in the P50 range, estimates in the P90 and P10 ranges tend to be less so. Arithmetic addition of the means of component estimates will provide a statistically correct mean aggregate estimate, which is the most useful statistic for financial planning and public reporting.

APPENDIX A

PETROLEUM RESERVES DEFINITIONS OF SOCIETY OF PETROLEUM ENGINEERS (SPE) AND WORLD PETROLEUM CONGRESSES (WPC)

A.1 Definitions

Reserves are those quantities of petroleum[1] which are anticipated to be commercially recovered[2] from known accumulations from a given date forward. All reserve estimates involve some degree of uncertainty. The uncertainty depends chiefly on the amount of reliable geological and/or engineering data available at the time of the estimate and the interpretation of these data. The relative degree of uncertainty may be conveyed by placing reserves into one of two principal classifications, either proved or unproved. Unproved reserves are less certain to be recovered than proved reserves and may be further sub-classified as probable and possible reserves to denote progressively increasing uncertainty in their recoverability.

The intent of the SPE and WPC in approving additional classifications beyond proved reserves is to facilitate consistency among professionals using such terms. In presenting these definitions, neither organization is recommending public disclosure of reserves classified as unproved. Public disclosure of the quantities classified as unproved reserves is left to the discretion of the countries or companies involved.

Estimation of reserves is done under conditions of uncertainty. The method of estimation is called deterministic if a single best estimate of reserves is made based on known geological, engineering, and economic data. The method of estimation is called probabilistic when the known geological, engineering, and economic data are used to generate a range of estimates and their associated probabilities. Identifying reserves as proved, probable, and possible has been the most frequent classification method and gives an indication of the probability of recovery. Because of potential differences in uncertainty, caution should be exercised when aggregating reserves of different classifications.

---

[1] Petroleum refers to naturally occurring liquids and gases that are predominantly comprised of hydrocarbon compounds. Petroleum may also contain nonhydrocarbon compounds in which sulfur, oxygen, and/or nitrogen atoms are combined with carbon and hydrogen. Nonhydrocarbons found in petroleum include nitrogen, carbon dioxide and hydrogen sulfide.

[2] Terms and phrases underlined here—not in the official SPE publication—are discussed in Appendix B.

Reserve estimates will generally be revised as additional geologic or engineering data becomes available or as economic conditions change. Reserves do not include quantities of petroleum being held in inventory, and may be reduced for usage or processing losses if required for financial reporting.[1]

Reserves may be attributed to either natural energy or improved recovery methods. Improved recovery methods include all methods for supplementing natural energy or <u>altering natural forces in the reservoir</u> to increase ultimate recovery. <u>Examples of such methods</u> are pressure maintenance, cycling, waterflooding, thermal methods, chemical flooding, and the use of miscible and immiscible displacement fluids. Other improved recovery methods may be developed in the future as petroleum technology continues to evolve.

A.2 Proved Reserves

Proved reserves are those quantities of petroleum which, by analysis of geological and/or engineering data, can be estimated with <u>reasonable certainty</u> to be commercially recoverable, from a given date forward, from known reservoirs and under current economic conditions, operating methods, and government regulations. Proved reserves can be categorized as <u>developed or undeveloped</u>.

If deterministic methods are used, the term reasonable certainty is intended to express a high degree of confidence that the quantities will be recovered. If probabilistic methods are used, there should be at least a 90% probability that the quantities actually recovered will equal or exceed the estimate.

Establishment of current economic conditions should include relevant historical petroleum prices and associated costs and may involve an <u>averaging period that is consistent with the purpose of the reserve estimate</u>, appropriate contract obligations, corporate procedures, and government regulations involved in reporting these reserves.

In general, reserves are considered proved if the <u>commercial producibility</u> of the reservoir is supported by actual production or formation tests. In this context, the term proved refers to the actual quantities of petroleum reserves and not just the productivity of the well or reservoir. In certain cases, proved reserves may be assigned on the basis of well logs and/or core analysis that indicate the subject reservoir is hydrocarbon bearing and is analogous to reservoirs in the same area that are producing or have demonstrated the ability to produce on formation tests.

The area of the reservoir considered as proved includes (1) the <u>area delineated by drilling</u> and defined by fluid contacts, if any, and (2) the undrilled portions of the reservoir that can reasonably be judged as commercially productive on the basis of available geological and engineering data. In the absence of data on fluid contacts, the <u>lowest known occurrence of hydrocarbons</u> controls the proved limit unless otherwise indicated by <u>definitive geological, engineering or performance data.</u>

Reserves may be classified as proved if facilities to process and transport those reserves to market are operational at the time of the estimate or there is a reasonable expectation that such facilities will be installed. Reserves in <u>undeveloped locations</u> may be classified as proved undeveloped provided: (1) the locations are <u>direct offsets</u> to wells that have indicated

---

[3] Aside from geological and/or engineering considerations, determination of correct "booking" of quantities of reserves produced under foreign concessions and/or service contracts may require competent counsel by tax accountants and/or regulatory authorities. In this context, the reader is referred to McMichael and Young [2001].

commercial production in the objective formation, (2) it is reasonably certain such locations are within the known proved productive limits of the objective formation, (3) the locations conform to existing well spacing regulations where applicable, and (4) it is reasonably certain the locations will be developed. Reserves from other locations are categorized as proved undeveloped only where interpretations of geological and engineering data from wells indicate with reasonable certainty that the objective formation is laterally continuous and contains commercially recoverable petroleum at locations beyond direct offsets.

Reserves which are to be produced through the application of established improved recovery methods are included in the proved classification when (1) successful testing by a pilot project or favorable response of an installed program in the same or an analogous reservoir with similar rock and fluid properties provides support for the analysis on which the project was based, and, (2) it is reasonably certain that the project will proceed. Reserves to be recovered by improved recovery methods that have yet to be established through commercially successful applications are included in the proved classification only (1) after a favorable production response from the subject reservoir from either (a) a representative pilot or (b) an installed program where the response provides support for the analysis on which the project is based and (2) it is reasonably certain the project will proceed.

A.3 Unproved Reserves

Unproved reserves are based on geological and/or engineering data similar to that used in estimates of proved reserves; but technical, contractual, economic, or regulatory uncertainties preclude such reserves being classified as proved. Unproved reserves may be further classified as probable reserves and possible reserves.

Unproved reserves may be estimated assuming future economic conditions different from those prevailing at the time of the estimate. The effect of possible future improvements in economic conditions and technological developments can be expressed by allocating appropriate quantities of reserves to the probable and possible classifications.

A.3.1 Probable Reserves

Probable reserves are those unproved reserves which analysis of geological and/or engineering data suggests are more likely than not to be recoverable. In this context, when probabilistic methods are used, there should be at least a 50% probability that the quantities actually recovered will equal or exceed the sum of estimated proved plus probable reserves.

In general, probable reserves may include: (1) reserves anticipated to be proved by normal step-out drilling where sub-surface control is inadequate to classify these reserves as proved, (2) reserves in formations that appear to be productive based on well log characteristics but lack core data or definitive tests and which are not analogous to producing or proved reservoirs in the area, (3) incremental reserves attributable to infill drilling that could have been classified as proved if closer statutory spacing had been approved at the time of the estimate, (4) reserves attributable to improved recovery methods that have been established by repeated commercially successful applications when (a) a project or pilot is planned but not in operation and (b) rock, fluid, and reservoir characteristics appear favorable for commercial application, (5) reserves in an area of the formation that appears to be separated from the proved area by faulting and the geologic interpretation indicates the subject area is structurally higher than the proved area, (6) reserves attributable to a future workover, treatment, re-treatment, change of equipment, or other mechanical procedures, where such procedure has not been proved successful in wells which exhibit similar behavior in analogous reservoirs, and (7) incremental reserves in <u>proved</u> reservoirs where an alternative interpretation of performance or volumetric data indicates more reserves than can be classified as proved.

A.3.2 Possible Reserves

Possible reserves are <u>those unproved reserves</u> which analysis of geological and engineering data suggests are <u>less likely to be recoverable</u> than probable reserves. In this context, when probabilistic methods are used, there should be at least a <u>10% probability that the quantities actually recovered will equal or exceed the sum of estimated proved plus probable plus possible reserves.</u>

In general, possible reserves may include: (1) reserves which, based on geological interpretations, could possibly exist beyond areas classified as probable, (2) reserves in formations that appear to be petroleum bearing based on log and core analysis but may not be productive at commercial rates, (3) incremental reserves attributed to infill drilling that are subject to technical uncertainty, (4) reserves attributed to improved recovery methods when (a) a project or pilot is planned but not in operation and (b) rock, fluid, and reservoir characteristics are such that a reasonable doubt exists that the project will be commercial, and (5) reserves in an area of the formation that appears to be separated from the proved area by faulting and geological interpretation indicates the subject area is structurally lower than the proved area.

A.4 Reserve Status Categories

Reserve status categories define the development and producing status of wells and reservoirs.

A.4.1 Developed

Developed reserves are expected to be recovered from existing wells including reserves <u>behind pipe</u>. Improved recovery reserves are considered developed only after the necessary equipment has been installed, or when the costs to do so are relatively minor. Developed reserves may be sub-categorized as producing or non-producing.

Producing

Reserves subcategorized as producing are expected to be recovered from completion intervals which are open and producing at the time of the estimate. Improved recovery reserves are considered producing only after the improved recovery project is in operation.

Non-producing

Reserves subcategorized as non-producing include shut-in and behind-pipe reserves. Shut-in reserves are expected to be recovered from: (1) completion intervals which are open at the time of the estimate but which have not started producing, (2) wells which were shut-in for market conditions or pipeline connections, or (3) wells not capable of production for mechanical reasons. Behind-pipe reserves are expected to be recovered from zones in existing wells, which will require additional completion work or future recompletion prior to the start of production.

A.4.2 Undeveloped

Undeveloped reserves are expected to be recovered: (1) from new wells on undrilled acreage, (2) from deepening existing wells to a different reservoir, or (3) where a relatively large expenditure is required to: (a) recomplete an existing well or (b) install production or transportation facilities for primary or improved recovery projects.

INTERPRETIVE COMMENTS REGARDING SPE-WPC [1997] RESERVES DEFINITIONS

B.1 Background

The terminology used in SPE-WPC [1997] definitions was taken, in large part, from definitions promulgated by SPE in 1987 [SPE 1987]. Those definitions were viewed by some as being applicable only to onshore U.S., even though they found widespread international application. Interpretive comments are provided here to facilitate continuation of international usage, which are consistent with the intent of the definitions and with international practice.

B.2 Interpretive Comments

Listed below—in the left column—are terms and phrases in SPE-WPC [1997] definitions. The terms and phrases—underlined in Appendix A—are grouped by section, as indicated by the boldface type and numbered sections below, which correspond to those in Appendix A. Interpretive comments are provided in the right column, across from each term or phrase.

These interpretive comments were developed by the author after extensive consultation, but are not to be interpreted as necessarily bearing the endorsement of SPE, WPC, or any other professional organization.

SPE-WPC [1997] Terminology	Interpretive Comment
A.1 Definitions	
commercially recovered	Interpretation of this term is quite subjective and may depend on the stage of development and production. For undrilled areas, a "prudent operator" might consider the term to imply a risk adjusted, discounted cash flow rate of return of at least 15 % per year. Depending on circumstances, less restrictive criteria might be applied. For undeveloped areas, however, such criteria typically imply a commitment for project implementation. Before reserves can be assigned to undrilled or undeveloped entities, a

judgment must be made regarding the commerciality of developing such reserves. Such judgments may be called to question, if economics are marginal at the time the reserve estimate must be made.

In international operations, for example, difficulties may arise when there is a disagreement between the host government and an operating group regarding this term. Such disagreement might preclude contract fulfillment for all or a part of an accumulation considered reserves by a host government and not by an operating group.

For producing entities, the term typically implies production of oil and/or gas for sale, irrespective of whether payout has occurred.

known accumulations

Although not formally defined, the term "known" accumulations historically has been interpreted to infer accumulations penetrated by, and appraised in, wellbore(s).

geological and/or engineering data

Historically the term "geological" has been interpreted to mean *direct* geological data like that obtained from wellbores—DST's, cores, logs, etc. *Indirect* geological data *alone*, like that inferred from seismic measurements, historically has been considered insufficient for the delineation of proved and, in some areas, probable reserves.

Although not so stated, the term "geologic" or "geological" (data) used herein includes "geophysics" or "geophysical" (data), which is consistent with dictionaries of the English language which typically define geophysics as "a branch of geology."

Given the advances in obtaining, processing and interpreting seismic data since the late 1970's, it generally is considered acceptable to utilize such data—provided it is properly calibrated to, and integrated with, wellbore data—to delineate all classifications of reserves. Robertson [2001] has provided guidelines for such application.

deterministic

This term is discussed in Sec. 9.3.2

probabilistic

This term is discussed in Sec. 9.3.3

caution should be exercised when aggregating reserves of different classifications

This phrase may seem to contradict phrases in sections describing probabilistic estimates of unproved reserves wherein—for probable reserves, for example—it is noted that "there should be at least a 50% probability that the quantities actually recovered will equal or exceed the sum of estimated proved plus probable reserves."

The primary intent of the phrase is to ensure the exercise of caution in the addition of deterministic estimates of reserves that have not been risk weighted, a subject discussed in Sec. 9.12.

Regarding summation of individual reserve estimates, there is no statement in SPE-WPC [1997] definitions regarding the level of summation to which the terms "reasonable certainty" or (for probabilistic estimates) "at least a 90% probability" should apply. The method of summation, the level of summation, and the assignment of an appropriate confidence to such summations are the responsibility of the estimator. This point is discussed in Sec. 9.12

altering natural forces in the reservoir

Such alteration might involve addition of surfactants to injection water to reduce the oil-water interfacial tension, for example, "chemical flooding."

Examples of such methods

In some countries, artificial lift is considered a method of improved recovery. This, however, is not the intent of these definitions, as it is considered that artificial lift does not affect the level of *reservoir* energy—pressure and/or gravity potential—available to drive reservoir fluids into the wellbore.

A.2 Proved Reserves

reasonable certainty

This phrase is not defined quantitatively, and interpretation is quite subjective.

Some investigators consider "*virtual* certainty" to imply a probability (or confidence) of at least 95%. As discussed in Sec. 9.3.4, industry practices over many years seem to imply "*reasonable* certainty" has been interpreted as a probability somewhat greater than 50%.

From a historical context, it is noted that in 1960 the American Petroleum Institute (API) introduced the phrase "beyond reasonable doubt" to qualify proved reserves.

Relevant to this phrasing is a 1991 ruling by the Texas Court of Criminal Appeals defining "beyond reasonable doubt" as "proof of such a convincing character that you would be willing to rely and act upon it without hesitation in the most important of your own affairs."

developed or undeveloped

The term "developed" implies "necessary equipment has been installed...to produce and transport to market." This is not a necessary condition for reserves to be classified as proved, provided there is "reasonable expectation...such facilities (will be installed) in the future."

In offshore operations, areas may have been appraised by wells that have been suspended, pending—if mechanically feasible—possible tieback to production facilities. Such areas cannot be classified as "developed," even though the degree of uncertainty is less than implied by the term "undeveloped." Thus, it is recommended that—as appropriate to reporting such reserves—this phrase be interpreted to imply that "Proved reserves may be drilled, undrilled, developed, or undeveloped."

Use of the terms "drilled" and "undrilled," as recommended above, is consistent with historical usage of these terms by the API and elsewhere in SPE-WPC [1997] definitions.

averaging period that is consistent with the purpose of the reserve estimate

For economic analysis of the development of a newly discovered accumulation, the "averaging period" might include the operator's estimate of the future time period during which development and production were anticipated.

For reporting proved reserves to the U.S. SEC, the "averaging period" would have to be the day ending the operator's fiscal year.

For estimating producing reserves, the averaging period might include the operator's estimate of the remaining years of production or the prior year of production, depending on circumstances.

For additional discussion, the reader is referred to Crossley [2001].

commercial producibility

In the context of this section, the term is interpreted to imply productivity of a drilled well (or wells) such that the operator can anticipate—at least—payout of the cost of completion.

In this case, most engineers would assign proved reserves to such wells. This does not, however, address the problem of the potential commerciality of continued development of marginal accumulations.

In cases where the commerciality of continued development is subject to doubt, engineers might assign probable, rather than proved, reserves to the undrilled areas.

area delineated by drilling

Historically, has referred to the structural and stratigraphic "limits" of an accumulation, including fluid contacts, if any, defined by interpretation of data from wells.

Seismic data historically has been used in mapping faults to delineate the structural limits of proved accumulations. Usage to delineate fluid contacts and/or lateral (stratigraphic) continuity, however, has met with limited success and has not been widely accepted.

lowest known occurrence of hydrocarbons

Historically, the term "lowest known" has referred to *direct* observations made in wellbores, for example, DST, WLFT, cores, logs.

definitive geological, engineering, or performance data

Although this phrase has not been defined, it is suggested that "definitive geological data" might include wellbore calibrated seismic data. It is also suggested that "definitive engineering data" might include RFT pressure traverses and/or capillary pressure data that might indicate a fluid-fluid contact downdip from a wellbore or RFT pressure traverses that might indicate a fluid-fluid contact between wells.

To find wide acceptance, use of such *indirect* data would have to be subject to local guidelines. (Suggested guidelines for utilization of such data are in Appendix G.)

undeveloped locations

The term "locations" reflects the historical emphasis on leases and units onshore U.S. and Canada.

The term "undeveloped" may be interpreted to mean "undrilled," depending on circumstances.

direct offsets

Arising from the "offset drilling rule" in onshore U.S., a "direct offset "refers to a well drilled on a tract (or unit) directly adjacent to a drilled tract (or unit) to protect the owner(s) of the undrilled tract (or unit) from drainage by the well on the drilled tract (or unit).

From such practice, it is inferred that a "direct offset" (well) is located a distance no greater than two well-drainage radii from the well being offset.

For international application, it might be inferred that a direct offset is a (development) well drilled no more than two well-drainage radii from the nearest (potential) producing well.

existing well spacing regulations

For onshore U.S. and Canada, "regulations" typically provide—initially—for 40- or 80-acre spacing for oil wells and 160- to 640-acre spacing for gas wells. However, such spacing may be modified subsequent to initial development, if warranted by well performance and is approved by regulatory authorities.

For international operations initial spacing regulations may be set by a development plan approved by the host government. In the absence of an approved development plan, local practice would govern interpretation of this term.

reasonably certain the locations will be developed

Depending on circumstances, some engineers may insist that—for estimates of outside-operated interests—there be an approved AFE before considering development to be "reasonably certain." In some cases, an operator's letter of intent, budgeted funds and/or a drilling schedule may be considered sufficient.

Given less certain indications of operator intent—or suspected permitting problems—reserves for such locations should be classified as unproved (probable or possible).

data from wells

The phrase was intended—in the late 1980's—to limit extrapolations of proved reserves significant distances beyond the drilled area.

This phrase might now be interpreted to infer utilization of wellbore data to calibrate seismic interpretations.

analogous reservoir

Analogous reservoirs are discussed in Sec. 2.1.

favorable production response	Depending on circumstances, a favorable *pressure* response may be considered a "production" response, for example, a "pressure maintenance" project where wells were producing at capacity at project initiation and/or were constrained by production facilities.

A.3 Unproved Reserves

future economic conditions different from those prevailing at the time of the estimate	This phrase implies that unproved reserves may be estimated assuming a plausible economic scenario different from that used for estimating proved reserves. Such differences should be clearly noted.
possible future improvements	If anticipated to have a material impact on reserve estimates, the nature and timing of such improvements should be stated.

A.3.1 Probable Reserves

those unproved reserves	Although not so stated, this phrase infers a deterministic estimate of probable reserves, which may be subjectively estimated separately from proved reserves.
more likely than not	Implies a "better than even" chance—at least a 50% chance—that the estimated quantity will be recovered
50% probability that the quantities actually recovered will equal or exceed the sum of estimated proved plus probable reserves	The use of probabilistic methods to estimate probable reserves does not mandate including proved reserves; thus, the "sum" of proved plus probable reserves may include only probable reserves.
may include	The numbered phrases following this term are intended to be *examples* of situations where reserves *may* be classified as probable; the list of examples is not intended to be limiting.
normal step-out drilling	A distinction might be made between: (a) "direct offsets," as noted in the section on proved reserves, and (b) locations drilled beyond direct offsets—identified in this section as "step outs"—to delineate a discovered accumulation.

repeated commercially successful applications

Depending on circumstances, it might be appropriate to interpret this phrase as "repeated commercially successful applications in analogous reservoirs." For additional discussion regarding this phrase, please refer to Appendix F.

proved

Incremental (probable) reserves may be attributed to proved reservoirs—either producing or nonproducing (behind pipe).

A.3.2 Possible Reserves

those unproved reserves

Although not so stated, this phrase infers a deterministic estimate of possible reserves, which may be subjectively estimated separately from proved and/or probable reserves.

less likely to be recoverable

Implies a "less than even" chance, that is, less then a 50% chance that the estimated quantity will be recovered.

10% probability that the quantities actually recovered will equal or exceed the sum of estimated proved plus probable plus possible reserves

The use of probabilistic methods to estimate possible reserves does not mandate including either proved or probable reserves; thus, the "sum" of proved plus probable plus possible reserves may include only possible reserves.

A.4 Reserve Status Categories

behind pipe

Unless a "relatively large expenditure" is required, for example, mobilizing a rig for an offshore platform, such reserves usually are considered "developed." However, some companies consider *all* behind-pipe reserves to be undeveloped. If such reserves are significant, it is recommended that appropriate notes be provided.

APPENDIX C

COMPARISON OF SPE-WPC AND U.S. SEC DEFINITIONS FOR PROVED RESERVES[1]

SPE-WPC [1997]

Proved reserves are those quantities of petroleum which, by analysis of geological and/or engineering data, can be estimated with reasonable certainty to be commercially recoverable, from a given date forward, from known reservoirs and under *current economic conditions,* operating methods, and government regulations. Proved reserves can be categorized as developed or undeveloped.

If deterministic methods are used, the term reasonable certainty is intended to express a high degree of confidence that the quantities will be recovered. If probabilistic methods are used, there should be at least a 90% probability that the quantities actually recovered will equal or exceed the estimate.

U.S. SEC [1996][2]

Proved oil and gas reserves are the estimated quantities of crude oil, natural gas, and natural gas liquids which geological and engineering data demonstrate with reasonable certainty to be recoverable in future years from known reservoirs under *existing economic and operating conditions, i.e., prices and costs as of the date the estimate is made.*

---

[1] Phrases are italicized here, not in the original publications, to emphasize significant differences between SPE-WPC and U.S. SEC.

[2] Paragraphs in this column were taken from U.S. SEC Regulation SX Rule 4-10, as of 15 July 1996. Paragraphs here are not in the original published sequence, but are grouped to correspond to similar paragraphs in SPE-WPC definitions. Subsequent to the preparation of this section the SEC released comments regarding interpretation of their definitions, which are available at www.sec.gov/offices/corpfin/acctdisc.htm (scroll down to Issues in the Extractive Industry–Definition of Proved Reserves).

Establishment of *current economic conditions should include relevant historical petroleum prices and associated costs and may involve an averaging period that is consistent with the purpose of the reserve estimate,* appropriate contract obligations, corporate procedures, and government regulations involved in reporting these reserves.

Prices include consideration of changes in existing prices provided only by contractual arrangements, but not on escalations based upon future conditions.

In general, reserves are considered proved if the commercial producibility of the reservoir is supported by actual production or formation tests. In this context, the term proved refers to the actual quantities of petroleum reserves and not just the productivity of the well or reservoir.

Reservoirs are considered proved if economic producibility is supported by either actual production or conclusive formation test.

In certain cases, proved reserves may be assigned on the basis of well logs and/or core analysis that indicate the subject reservoir is hydrocarbon bearing and is analogous to reservoirs *in the same area* that are producing or have demonstrated the ability to produce on formation test.

In certain instances, proved reserves may be assigned to reservoirs on the basis of a *combination* of electrical and other type logs *and* core analyses which indicate the reservoirs are analogous to similar reservoirs *in the same field* which are producing or have demonstrated the ability to produce on a formation test. (Interpretive Response)

The area of the reservoir considered as proved includes (1) the area delineated by drilling and defined by fluid contacts, if any, and (2) the undrilled portions of the reservoir that can reasonably be judged as commercially productive on the basis of available geological and engineering data.

The area of a reservoir considered proved includes: (A) that portion delineated by drilling and defined by gas-oil and/or oil-water contacts, if any; and (B) the immediately adjoining portions not yet drilled, but which can be reasonably judged as economically productive on the basis of available geological and engineering data.

In the absence of data on fluid contacts, the lowest known occurrence of hydrocarbons controls the proved limit *unless otherwise indicated by definitive geological, engineering or performance data.*

In the absence of information on fluid contacts, the lowest known structural occurrence of hydrocarbons controls the lower proved limit of the reservoir.

Reserves in *undeveloped locations* may be classified as proved undeveloped provided:
(1) the locations are direct offsets to wells that have indicated commercial production in the objective formation, (2) it is reasonably certain such locations are within the known proved productive limits of the objective formation, (3) the locations conform to existing well spacing regulations where applicable, and (4) it is reasonably certain the locations will be developed.

Reserves from other locations are categorized as proved undeveloped only where interpretations of geological and engineering data from wells indicate with *reasonable certainty that the objective formation is laterally continuous and contains commercially recoverable petroleum at locations beyond direct offsets.*

Reserves which are to be produced through the application of established improved recovery methods are included in the proved classification when (1) successful testing by a pilot project or favorable response of an installed program in the same or *an analogous reservoir* with similar rock and fluid properties provides support for the analysis on which the project was based, and, (2) it is reasonably certain that the project will proceed. Reserves to be recovered by improved recovery methods that have yet to be established through commercially successful applications are included in the proved classification only (1) after a favorable production response from the subject reservoir from either (a) a representative pilot or (b) an installed program where the response provides support for the analysis on which the

Reserves on *undrilled acreage* shall be limited to those drilling units offsetting productive units that are reasonably certain of production when drilled.

Proved reserves for other undrilled units can be claimed only where it can be *demonstrated with certainty that there is continuity of production from the existing productive formation.*

Reserves which can be produced economically through application of improved recovery techniques (such as fluid injection) are included in the "proved" classification when successful testing by a pilot project, or the operation of an installed program *in the reservoir*, provides support for the engineering analysis on which the project or program was based.

Additional oil and gas expected to be obtained through the application of fluid injection or other improved recovery techniques for supplementing the natural forces and mechanisms of primary recovery should be included as "proved developed reserves" only after testing by a pilot project or after the operation of an installed program has confirmed through production response that increased

project is based and (2) it is reasonably certain the project will proceed.

recovery will be achieved.

Under no circumstances should estimates for proved undeveloped reserves be attributable to any acreage for which an application of fluid injection or other improved recovery technique is contemplated, unless such techniques have been proved effective by actual tests in the area and *in the same reservoir.*

Estimates of proved reserves do not include the following: (A) oil that may become available from known reservoirs but is classified separately as "indicated additional reserves"[3], (B) crude oil, natural gas, and natural gas liquids, the recovery of which is subject to reasonable doubt because of uncertainty as to geology, reservoir characteristics, or economic factors, (C) crude oil, natural gas, and natural gas liquids, that may occur in undrilled prospects, and (D) crude oil, natural gas, and natural gas liquids, that may be recovered from oil shales, coal,[4] gilsonite and other such sources.

Developed reserves are expected to be recovered from existing wells including reserves behind pipe. Improved recovery reserves are considered developed only after the necessary equipment has been installed, or when the costs to do so are relatively minor.

Proved developed oil and gas reserves are reserves that can be expected to be recovered through existing wells with existing equipment and operating methods.

Undeveloped reserves are expected to be recovered: (1) from new wells on undrilled acreage, (2) from deepening existing wells to a different reservoir, or

Proved undeveloped oil and gas reserves are reserves that are expected to be recovered from new wells on undrilled acreage, or from existing wells where a

---

[3] U.S. DOE/EIA considers "indicated additional reserves" to be "quantities of crude oil (other than proved reserves) which may become economically recoverable from existing productive reservoirs through the application of improved recovery techniques using current technology."

[4] "...coalbed methane gas...(may)...be included in proved reserves, provided it complies in all other respects with the definition of proved reserves...where methane gas is deemed to be economically producible only as a consequence of existing Federal tax incentives...additional disclosure should be provided as to the specific quantities...that are dependent on existing U.S. tax policy" [SEC Interpretive Response].

(3) where a relatively large expenditure is required to: (a) recomplete an existing well or (b) install production or transportation facilities for primary or improved recovery projects.

relatively major expenditure is required for recompletion.

APPENDIX D

STANDARDS PERTAINING TO THE ESTIMATING AND AUDITING OF OIL AND GAS RESERVES INFORMATION

In 1977, SPE of AIME promulgated a landmark document with the same title as this Appendix. That document is reprinted here in (close to) the original format, together with the original footnotes. The reserve definitions extant at that time, which included only proved reserves, differ from those promulgated by SPE-WPC in 1997, which include proved, probable, and possible reserves. Also, those definitions did not cover probabilistic methods, as do the 1997 definitions. These differences not withstanding, the basic principles regarding estimating and auditing reserves discussed in the 1977 document are as valid today as they were then.

It is recommended here that—as warranted by circumstances—the standards regarding "proved" reserves in the 1977 document be extended by today's engineer to include not only "proved" reserves, but also "probable" reserves and "possible" reserves.

D.1 Table of Contents from SPE-AIME [1977] Report

D.2 Articles from SPE-AIME [1977] Report

ARTICLE I - THE BASIS AND PURPOSE OF DEVELOPING STANDARDS PERTAINING TO THE ESTIMATING AND AUDITING OF RESERVE INFORMATION[1]

1.1 The Nature and Purpose of Estimating and Auditing Reserve Information

Estimates of Reserve Information are made by or for Entities as a part of their normal business practices. Such Reserve Information typically may include, among other things, estimates of (i) the proved reserves, (ii) the future producing rates from such proved reserves, (iii) the future net revenue from such proved reserves and (iv) the present value of such future net revenue. The exact type and extent of Reserve Information must necessarily take into account the purpose for which such Reserve Information is being prepared and,

---

[1] These Standards Pertaining to the Estimating and Auditing of Oil and Gas Reserve Information (the "Standards") are not intended to bind the members of the Society of Petroleum Engineers (the "Society") or anyone else, and the Society imposes no sanctions for the nonuse of these Standards. Each person estimating and auditing oil and gas Reserve Information is encouraged to exercise his or her own judgment concerning the matters set forth in these Standards. The Society welcomes comments and suggested changes in regard to these Standards.

correspondingly, statutory and regulatory provisions, if any, that are applicable to such intended use of the Reserve Information.

1:2 Estimating and Auditing Reserve Information in Accordance With Generally Accepted Engineering and Evaluation Principles

The estimating and auditing of Reserve Information is predicated upon certain historically developed principles of petroleum engineering and evaluation, which are in turn based on principles of physical science, mathematics and economics. Although these generally accepted petroleum engineering and evaluation principles are predicated on established scientific concepts, the application of such principles involves extensive judgments and is subject to changes in (i) existing knowledge and technology, (ii) economic conditions, (iii) applicable statutory and regulatory provisions and (iv) the purposes for which the Reserve information is to be used.

1.3 The Inherently Imprecise Nature of Reserve Information

The reliability of Reserve Information is considerably affected by several factors. Initially, it should be noted that Reserve Information is imprecise due to the inherent uncertainties in, and the limited nature of, the data base upon which the estimating and auditing of Reserve Information is predicated. Moreover, the methods and data used in estimating Reserve Information are often necessarily indirect or analogical in character rather than direct or deductive. Furthermore, the persons estimating and auditing Reserve Information are required, in applying generally accepted petroleum engineering and evaluation principles, to make numerous judgments based upon their educational background, professional training and professional experience. The extent and significance of the judgments to be made are, in themselves, sufficient to render Reserve Information inherently imprecise.

1.4 The Need for Standards Governing the Estimating and Auditing of Reserve Information

The Society of Petroleum Engineers, a constituent society of the American Institute of Mining, Metallurgical, and Petroleum Engineers (the "Society"), has determined that the Society should adopt these Standards Pertaining to the Estimating and Auditing of Oil and Gas Reserve Information (the "Standards"). The adoption of these Standards by the Society fulfills at least three useful objectives.

First, although some users of Reserve Information are cognizant of the general principles that are applied to data bases in determining Reserve Information, the judgments required in estimating and auditing Reserve Information and the inherently imprecise nature of Reserve Information, it has become increasingly apparent in recent years that many users of Reserve Information do not fully understand such matters. The adoption, publication and distribution of these Standards should enable users of Reserve Information to understand these matters more fully and therefore avoid placing undue reliance on Reserve Information.

Secondly, the wider dissemination of Reserve Information through public financial reporting, such as that required by various governmental authorities, makes it imperative that the users of Reserve Information have a general understanding of the methods of, and limitations on, estimating and auditing Reserve Information.

Thirdly, as Reserve Information proliferates in terms of the types of information available and the broader dissemination thereof, it becomes increasingly important that Reserve Information be estimated and audited on a consistent basis. Compliance with these Standards is a method of facilitating evaluation and comparisons of Reserve Information by the users thereof.

In order to accomplish the three above-discussed objectives, the Society has included in these Standards (i) definitions of selected terms pertaining to the estimation and evaluation of Reserve Information, (ii) qualifications for persons estimating and auditing Reserve Information, (iii) standards of independence and objectivity for such persons, (iv) standards for estimating proved reserves and other Reserve Information, and (v) standards for auditing proved reserves and other Reserve Information. Although these Standards are predicated on generally accepted petroleum engineering and evaluation principles, it may in the future become necessary, for the reasons set forth in Section 1.2, to clarify or amend certain of these Standards. Consequently, the Society may, in appropriate future circumstances, determine to amend these Standards or publish clarifying statements.

ARTICLE II - DEFINITIONS OF SELECTED TERMS

2.1 Applicability of Definitions

In preparing a report or opinion, persons estimating and auditing Reserve Information shall ascribe, to proved reserves and other significant terms used therein, the definitions promulgated by the Society or such other definitions as he or she may reasonably consider appropriate in accordance with generally accepted petroleum engineering and evaluation principles; provided, however, that (i) such report or opinion should define, or make reference to a definition of, each significant term that is used therein and (ii) the definitions used in any report or opinion must be consistent with statutory and regulatory provisions, if any, that apply to such report or opinion in accordance with its intended use.

2.2 Defined Terms

The definitions set forth in this Section are applicable for all purposes of these Standards:

(a) *Entity*. An Entity is a corporation, joint venture, partnership, trust, individual or other person engaged in (i) the exploration for, or production of, oil and gas; (ii) the acquisition of properties or interests therein for the purpose of conducting such exploration or production; or (iii) the ownership of properties or interests therein with respect to which such exploration or production is being, or will be, conducted.

(b) *Reserve Estimator*. A Reserve Estimator is a person who is designated to be in responsible charge for estimating and evaluating proved reserves and other Reserve Information. A Reserve Estimator either may personally make the estimates and evaluations of Reserve Information or may supervise and approve the estimation and evaluation thereof by others.

(c) *Reserve Auditor*. A Reserve Auditor is a person who is designated to be in responsible charge for the conduct of an audit with respect to Reserve Information estimated by others. A Reserve Auditor either may personally conduct an audit of Reserve Information or may supervise and approve the conduct of an audit thereof by others.

(d) *Reserve Information*. Reserve Information consists of various estimates pertaining to the extent and value of oil and gas properties. Reserve Information may, but will not necessarily, include estimates of (i) proved reserves, (ii) the future production rates from such proved reserves, (iii) the future net revenue from such proved reserves and (iv) the present value of such future net revenue.

ARTICLE III - PROFESSIONAL QUALIFICATIONS OF RESERVE ESTIMATORS AND RESERVE AUDITORS

3.1 The Importance of Professionally Qualified Reserve Estimators and Reserve Auditors

Reserve Information is prepared and audited, respectively, by Reserve Estimators and Reserve Auditors, who are often assisted by other professionals and by paraprofessionals and clerical personnel. Reserve Estimators and Reserve Auditors may be (i) employees of an Entity itself or (ii) stockholders, proprietors, partners or employees of an independent firm of petroleum consultants with which an arrangement has been made for the estimating or auditing of Reserve Information. Irrespective of the nature of their employment, however, Reserve Estimators and Reserve Auditors must (i) examine the data base necessary to estimate or audit Reserve Information; (ii) perform such tests, and consider such matters, as may be necessary to evaluate the sufficiency of the data base; and (iii) make such calculations and estimations, and apply such tests and standards, as may be necessary to estimate or audit proved reserves and other Reserve Information. For the reasons discussed in Section 1.3, the proper determination of these matters is highly dependent upon the numerous judgments Reserve Estimators and Reserve Auditors are required to make based upon their educational background, professional training and professional experience. Consequently, in order to assure that Reserve Information will be as reliable as possible given the limitations inherent in the estimating and auditing process, it is essential that those in responsible charge for estimating and auditing Reserve Information have adequate professional qualifications such as those set forth in this Article III.

3.2 Professional Qualifications of Reserve Estimators

A Reserve Estimator shall be considered professionally qualified in such capacity if he or she has sufficient educational background, professional training and professional experience to enable him or her to exercise prudent professional judgment and to be in responsible charge in connection with the estimating of proved reserves and other Reserve Information. The determination of whether a Reserve Estimator is professionally qualified should be made on an individual-by-individual basis. A Reserve Estimator would normally be considered to be qualified if he or she (i) has a minimum of three years' practical experience in petroleum engineering or petroleum production geology, with at least one year of such experience being in the estimation and evaluation of Reserve Information; *and* (ii) *either* (A) has obtained, from a college or university of recognized stature, a bachelor's or advanced degree in petroleum engineering, geology or other discipline of engineering or physical science *or* (B) has received, and is maintaining in good standing, a registered or certified professional engineer's license or a registered or certified professional geologist's license, or the equivalent thereof, from an appropriate governmental authority or professional organization.

3.3 Professional Qualifications of Reserve Auditors

A Reserve Auditor shall be considered professionally qualified in such capacity if he or she has sufficient educational background, professional training and professional experience to enable him or her to exercise prudent professional judgment while acting in responsible charge for the conduct of an audit of Reserve Information estimated by others. The determination of whether a Reserve Auditor is professionally qualified should be made on an individual-by-individual basis. A Reserve Auditor would normally be considered to be qualified if he or she (i) has a minimum of ten years' practical experience in petroleum engineering or petroleum production geology, with at least five years of such experience being in the estimation and evaluation of Reserve Information; *and* (ii) *either* (A) has obtained, from a college or university of recognized stature, a bachelor's or advanced degree in petroleum engineering, geology or other discipline of engineering or physical science *or* (B) has received, and is maintaining in good standing, a registered or certified professional engineer's license or a registered or certified professional geologist's license, or the equivalent thereof, from an appropriate governmental authority or professional organization.

ARTICLE IV - STANDARDS OF INDEPENDENCE, OBJECTIVITY AND CONFIDENTIALITY FOR RESERVE ESTIMATORS AND RESERVE AUDITORS

4.1 The Importance of Independent or Objective Reserve Estimators and Reserve Auditors

In order that users of Reserve Information may be assured that the Reserve Information was estimated or audited in an unbiased and objective manner, it is important that Reserve Estimators and Reserve Auditors maintain, respectively, the levels of independence and objectivity set forth in this Article IV. The determination of the independence and objectivity of Reserve Estimators and Reserve Auditors should be made on a case-by-case basis. To facilitate such determination, the Society has adopted (i) standards of independence for consulting Reserve Estimators and consulting Reserve Auditors and (ii) standards of objectivity for Reserve Auditors internally employed by Entities to which the Reserve Information relates. To the extent that the applicable standards of independence and objectivity set forth in this Article IV are not met by Reserve Estimators and Reserve Auditors in estimating and auditing Reserve Information, such lack of conformity with this Article IV shall be set forth in any report or opinion relating to Reserve Information which purports to have been estimated or audited in accordance with these Standards.

4.2 Requirement of Independence for Consulting Reserve Estimators and Consulting Reserve Auditors

Consulting Reserve Estimators and consulting Reserve Auditors, or any firm of petroleum consultants of which such Individuals are stockholders, proprietors, partners or employees, should be independent from any *Entity* with respect to which such Reserve Estimators, Reserve Auditors or consulting firm estimate or audit Reserve Information which purports to have been estimated or audited in accordance with these Standards.

4.3 Standards of Independence for Consulting Reserve Estimators and Consulting Reserve Auditors[2]

Consulting Reserve Estimators and consulting Reserve Auditors, and any firm of petroleum consultants of which such individuals are stockholders, proprietors, partners or employees,

---

[2] For purposes of this Section 4.3, the term "affiliated" shall, with respect to an Entity, describe the relationship of a person to such Entity under circumstances in which such person directly, or indirectly through one or more intermediaries, controls or is controlled by, or is under common control with, such Entity; provided, however, that commercial banks and other bona fide financial institutions shall not be considered to be affiliated with the Entity to which the Reserve Information relates unless such banks or institutions actively participate in the management of the properties of such Entity.

Unless the context requires otherwise, the term "material" shall, for purposes of this Section 4.3, be interpreted with reference to the net worth of the consulting Reserve Estimators or the consulting Reserve Auditors, or any firm of petroleum consultants of which such individuals are stockholders, proprietors partners or employees.

would not normally be considered independent with respect to an Entity if, during the term of their professional engagement, such Reserve Estimators, Reserve Auditors or consulting firm:

(a) *Investments.* Either owned or acquired, or were committed to acquire, directly or indirectly, any material financial interest in (i) such Entity or any corporation or other person affiliated therewith or (ii) any property with respect to which Reserve Information is to be estimated or audited;

(b) *Joint Business Venture.* Either owned or acquired, or were committed to acquire, directly or indirectly, any material joint business investment with such Entity or any officer, director, principal stockholder or other person affiliated therewith;

(c) *Borrowings.* Were indebted to such Entity or any officer, director, principal stockholder or other person affiliated therewith; provided, however, that retainers, advances against work-in-process and trade accounts payable arising from the purchase of goods and services in the ordinary course of business shall not constitute indebtedness within the meaning of this Section 4.3(c);

(d) *Guarantees of Borrowings.* Were indebted to any individual, corporation or other person under circumstances where the payment of such indebtedness was guaranteed by such Entity or any officer, director, principal stockholder or other person affiliated therewith;

(e) *Loans to Clients.* Extended credit to (i) such Entity or any officer, director, principal stockholder or other person affiliated therewith or (ii) any person having a material interest in any property with respect to which Reserve Information was estimated or audited; provided, however, that trade accounts receivable arising in the ordinary course of business from the performance of petroleum engineering and related services shall not constitute the extension of credit within the meaning of this Section 4.3(e);

(f) *Guarantees for Clients.* Guaranteed any indebtedness (i) owed by such Entity or any officer, director, principal stockholder or other person affiliated therewith or (ii) payable to any individual, corporation, entity or other person having a material interest in the Reserve Information pertaining to such Entity;

(g) *Purchases and Sales of Assets.* Purchased any material asset from, or sold any material asset to, such Entity or any officer, director, principal stockholder or other person affiliated therewith;

(h) *Certain Relationships With Client.* Were directly or indirectly connected with such Entity as a promoter, underwriter, officer, director or principal stockholder, or in any capacity equivalent thereto, or were otherwise not separate and independent from the operating and investment decision-making process of such Entity;

(i) *Trusts and Estates.* Were trustees of any trust, or executors or administrators of any estate, if such trust or estate had any direct or indirect interest material to it in such Entity or in any property with respect to which Reserve Information was estimated or audited; or

(j) *Contingent Fee.* Were engaged by such Entity to estimate or audit Reserve Information pursuant to any agreement, arrangement or understanding whereby the remuneration or fee paid by such Entity was contingent upon, or related to, the results or conclusions reached in estimating or auditing such Reserve Information.

The independence of consulting Reserve Estimators and consulting Reserve Auditors, and the independence of any firm of petroleum consultants of which such individuals are stockholders, proprietors, partners or employees, shall not be considered impaired merely because other petroleum engineering and related services were performed (i) for such Entity or any officer, director, principal stockholder or other person affiliated therewith or (ii) in regard to any property with respect to which Reserve Information was estimated or audited; provided, however, such other services must have been of a type normally rendered by the petroleum engineering profession.

4.4 Requirement of Objectivity for Reserve Auditors Internally Employed by Entities

Reserve Auditors who are internally employed by an Entity should be objective with respect to such Entity if such Reserve Auditors audit Reserve Information relating to such Entity which purports to have been estimated or audited in accordance with these Standards.

4.5 Standards of Objectivity for Reserve Auditors Internally Employed by Entities

Reserve Auditors internally employed by an Entity would normally be considered to be in a position of objectivity with respect to such Entity if, during the time period in which Reserve Information was audited, such Reserve Auditors:

(a) *Accountability to Management.* Were assigned to a staff group which was (i) accountable to upper level management of such Entity and (ii) separate and independent from the operating and investment decision-making process of such Entity; and

(b) *Freedom to Report Irregularities.* Were granted complete and unrestricted freedom to report, to the principal executives and board of directors of such Entity, any substantive or procedural irregularities of which such Reserve Auditors became aware during their audit of Reserve Information pertaining to such Entity.

4.6 Requirement of Confidentiality

Reserve Estimators and Reserve Auditors, and any firm of petroleum consultants of which such individuals are stockholders, proprietors, partners or employees, should retain in strictest confidence Reserve Information and other data and information furnished by, or pertaining to, an Entity and such Reserve Information, data and information should not be disclosed to others without the prior consent of such Entity.

ARTICLE V - STANDARDS FOR ESTIMATING PROVED RESERVES AND OTHER RESERVE INFORMATION

5.1 General Considerations in Estimating Reserve Information

Reserve Information may be estimated through the use of generally accepted geologic and engineering methods that are consistent with both these Standards and any statutory and regulatory provisions which are applicable to such Reserve Information in accordance with its intended use. In estimating Reserve Information for a property or group of properties, Reserve Estimators will determine the geologic and engineering methods to be used in estimating Reserve Information by considering (i) the sufficiency and reliability of the data base; (ii) the stage of development; (iii) the performance history; (iv) their experience with respect to such property or group of properties, and with respect to similar properties; and (v) the significance of such property or group of properties to the aggregate oil and gas properties and interests being estimated or evaluated. The report as to Reserve Information should set forth information regarding the manner in which, and the assumptions pursuant to which, such report was prepared. Such disclosure should include, where appropriate, definitions of the significant terms used in such report, the geologic and engineering methods and measurement base used in preparing the Reserve Information and the source of the data used with regard to ownership interests, oil and gas production and other performance data, costs of operation and development, product prices, and agreements relating to current and future operations and sales of production.

5.2 Adequacy of Data Base in Estimating Reserve Information

The sufficiency and reliability of the data base is of primary importance in the estimation of proved reserves and other Reserve Information. The type and extent of the data required will necessarily vary in accordance with the methods employed to estimate proved reserves and other Reserve Information. In this regard, information must be available with respect to each property or group of properties as to operating interests, expense interests and revenue interests and future changes in any of such interests that, based on current circumstances, are expected to occur. Additionally, if future net revenue from proved reserves, or the present value of such future net revenue, is to be estimated, the data base should include, with respect to each property or group of properties, costs of operation and development, if available, product prices and a description of any agreements relating to current and future operations and sales of production.

5.3 Estimating Proved Reserves

The acceptable methods for estimating proved reserves include (i) the volumetric method; (ii) evaluation of the performance history, which evaluation may include an analysis and projection of producing ranges, reservoir pressures, oil-water ratios, gas-oil ratios and gas-liquid ratios; (iii) development of a mathematical model through consideration of material balance and computer simulation techniques; (iv) analogy to other reservoirs if geographic location, formation characteristics or similar factors render such analogy appropriate. In estimating proved reserves, Reserve Estimators should utilize the particular methods, and the number of methods, which in their professional judgment are most appropriate given (i) the geographic location, formation characteristics and nature of the property or group of properties with respect to which proved reserves are being estimated; (ii) the amount and quality of available data; and (iii) the significance of such property or group of properties in relation to the oil and gas properties with respect to which proved reserves are being estimated.

5.4 Estimating Proved Reserves by the Volumetric Method

Estimating proved reserves in accordance with the volumetric method involves estimation of oil in place based upon review and analysis of such documents and information as (i) ownership and development maps; (ii) geologic maps; (iii) electric logs and formation tests; (iv) relevant reservoir and core data; and (v) information regarding the completion of oil and gas wells and any production performance thereof. An appropriate estimated recovery factor is applied to the resulting oil in place figure in order to derive estimated proved reserves.

5.5 Estimating Proved Reserves by Analyzing Performance Data

For reservoirs with respect to which performance has disclosed reliable production trends, proved reserves may be estimated by analysis of performance histories and projections of such trends. These estimates may be primarily predicated on an analysis of the rates of decline in production and on appropriate considerations of other performance parameters such as reservoir pressures, oil-water ratios, gas-oil ratios and gas-liquid ratios.

5.6 Estimating Proved Reserves by Using Mathematical Models

Proved reserves and future production performance can be estimated through a combination of detailed geologic and reservoir engineering studies and mathematical or computer simulation models. The validity of the mathematical simulation models is enhanced by the degree to which the calculated history matches the performance history. Where performance history is unavailable, special consideration should be given to determining the sensitivity of the calculated ultimate recoveries to the data that is the most uncertain. After making such sensitivity determination, the proved ultimate recovery should be based on the selection of the most likely data encompassed within the ranges of their uncertainty.

5.7 Estimating Proved Reserves by Analogy to Comparable Reservoirs

If performance trends have not been established with respect to oil and gas production, future production rates and proved reserves may be estimated by analogy to reservoirs in the same geographic area having similar characteristics and established performance trends. Where appropriate, proved reserves may be estimated using multiples of current rates of production.

5.8 Estimated Future Rates of Production

Future rates of oil and gas production may be estimated by extrapolating production trends where such have been established. If production trends have not been established, future rates of production may be estimated by analogy to the respective rates of production of reservoirs in the same geographic area having similar geologic features, reservoir rock and fluid characteristics. If there is not available either (i) production trends from the property or group of properties with respect to which proved reserves are being estimated or (ii) rates of production from similar reservoirs, the estimation of future rates of production may be predicated on an assumed future decline rate that takes into proper consideration the cumulative oil and gas production that is estimated to occur prior to the predicted decline in such production in relation to the estimated ultimate production. Reservoir simulation is also an accepted method of estimating future rates of production. Irrespective of the method used, however, proper consideration should be given to (i) the producing capacities of the wells; (ii) the number of wells to be drilled in the future, together with the proposed times when such are to be drilled and the structural positions of such wells; (iii) the energy inherent in, or introduced to, the reservoir; (iv) the estimated ultimate recovery; (v) future remedial work to be performed; (vi) the scheduling of future well abandonments, (vii) normal downtime which may be anticipated, and (viii) artificial restriction of future production rates that is attributable to statutory and regulatory provisions, purchaser proration and other factors.

5.9 Estimating Reserve Information Other Than Proved Reserves and Future Rates of Production

A Reserve Estimator often estimates Reserve Information other than proved reserves and future rates of production in order to make his or her report more useful. Proved reserves net to the interests appraised are estimated using the Entity's net interest in the property or group of properties, or in the production therefrom, with respect to which proved reserves were estimated. The nature of the net interest of the Entity may be established or affected by any number of arrangements which the Reserve Estimator must take into account. Estimated future revenues are calculated from the estimated future rates of production by applying the appropriate sales prices furnished by the Entity or by using such other data as may be required by statutory and regulatory provisions that are applicable to such report in accordance with its intended use. Where appropriate, the Reserve Estimator deducts from such future revenues items such as (i) any existing production or severance taxes, (ii) taxes levied against property or production, (iii) estimates of future operating costs and (iv) estimates of any future development, equipment or other significant capital expenditures required for the production of the proved reserves. Such deductions normally include various overhead and management charges. For some purposes, it is desirable to subtract income taxes and other governmental levies in estimating future net revenues.

In estimating future net revenues, the Reserve Estimator should consider, where appropriate, any likely changes (i) from historical operating costs, (ii) from current estimates of future capital expenditures and (iii) in other factors which may affect estimated limits of economic production.

ARTICLE VI - STANDARDS FOR AUDITING PROVED RESERVES AND OTHER RESERVE INFORMATION

6.1 The Concept of Auditing Proved Reserves and Other Reserve Information

An audit is an examination of Reserve Information that is conducted for the purpose of expressing an opinion as to whether such Reserve Information, in the aggregate, is reasonable and has been estimated and presented in conformity with generally accepted petroleum engineering and evaluation principles.

As discussed in Section 1.3, the estimation of proved reserves and other Reserve Information is an imprecise science due to the many unknown geologic and reservoir factors that can only be estimated through sampling techniques. Since proved reserves are therefore only estimates, such cannot be audited for the purpose of verifying exactness. Instead, Reserve Information is audited for the purpose of reviewing in sufficient detail the policies, procedures and methods used by an Entity in estimating its Reserve Information so that the Reserve Auditors may express an opinion as to whether, in the aggregate, the Reserve Information furnished by such Entity is reasonable and has been estimated and presented in conformity with generally accepted petroleum engineering and evaluation principles.

The methods and procedures used by an Entity, and the Reserve Information it furnishes, must be reviewed in sufficient detail to permit the Reserve Auditor, in his or her professional judgment, to express an opinion as to the reasonableness of such Entity's Reserve Information. In some cases the auditing procedure may require independent estimates of Reserve Information for particular properties. The desirability of such re-estimation will be determined by the Reserve Auditor exercising his or her professional judgment in arriving at an opinion as to the reasonableness of the Entity's Reserve Information.

6.2 Limitations on Responsibility of Reserve Auditors

Since the primary responsibility for estimating and presenting Reserve Information pertaining to an Entity is with the management of such Entity, the responsibility of Reserve Auditors is necessarily limited to any opinion they express with respect to such Reserve Information. In discharging such responsibility, Reserve Auditors may accept, generally without independent verification, information and data furnished by the Entity with respect to ownership interests, oil and gas production, historical costs of operation and development, product prices, agreements relating to current and future operations and sales of production, and other specified matters. If during the course of the audit, however, questions arise as to the accuracy or sufficiency of any information or data furnished by the Entity, the Reserve Auditor should not rely on such information or data unless such questions are resolved or the information or data is independently verified. If Reserve Information is used for financial accounting purposes, certain basic data would ordinarily be tested by an Entity's independent public accountants in connection with their examination of the Entity's financial statements. Such basic data would include information such as the property interests owned by the Entity, historical production data and the prices, costs and discount factors used in valuations of proved reserves. Reserve Auditors should, however, review estimates of major expenditures for development and equipment and any major differences between historical operating costs and estimated future operating costs.

6.3 Understanding Among an Entity, Its Independent Public Accountants and the Reserve Auditors

An understanding should exist among an Entity, its independent public accountants and the Reserve Auditors with respect to the nature of the work to be performed by the Reserve Auditors. Irrespective of whether the Reserve Auditors are consultants or internally employed by the Entity, the understanding between the Entity and the Reserve Auditors should include at least the following:

(a) *Availability of Reserve Information.* The Entity will provide the Reserve Auditors with (i) all Reserve Information prepared by such Entity, (ii) access to all basic data and documentation pertaining to the oil and gas properties of such Entity, and (iii) access to all personnel of such Entity who might have information relevant to the audit of such Reserve Information.

(b) *Performance of Audit.* The Reserve Auditors will (i) study and evaluate the methods and procedures used by the Entity in estimating and documenting its Reserve Information; (ii) review the reserve definitions and classifications used by such Entity; (iii) test and evaluate the Reserve Information of such Entity to the extent considered necessary by the Reserve Auditors; and (iv) express an opinion as to the reasonableness, in the aggregate, of such Entity's Reserve Information.

(c) *Availability of Audit Report to Independent Public Accountants.* The Reserve Auditors will (i) permit their audit report to be provided to the independent public accountants of the Entity for use in their examination of its financial statements and (ii) be available to discuss their audit report with such independent public accountants.

(d) *Coordination Between Reserve Auditors and Independent Public Accountants.* The Reserve Auditors and the Entity's independent public accountants will coordinate their efforts and agree on the records and data of the Entity to be reviewed by each.

In the case of an audit to be conducted by consulting Reserve Auditors, it is preferable that such understanding be documented, such as through an engagement letter between the Entity and the consulting Reserve Auditors.

6.4 Procedures for Auditing Reserve Information

Irrespective of whether the Reserve Information pertaining to an Entity is being audited by consulting Reserve Auditors or Reserve Auditors internally employed by such Entity, the audit should be conducted in accordance with the following procedures:

(a) *Proper Planning and Supervision.* The audit should be adequately planned and assistants, if any, should be properly supervised.

(b) *Early Appointment of Reserve Auditors.* Where appropriate, early appointment of Reserve Auditors is advantageous to both the Entity and the Reserve Auditors. Early appointment enables the Reserve Auditors to plan their work so that it may be done expeditiously and to determine the extent to which such can be completed prior to the balance sheet date. Preliminary work by the Reserve Auditors benefits the Entity by facilitating the efficient and expeditious completion of the audit of such Entity's Reserve Information.

(c) *Disclosure of the possibility of a Qualified Audit Opinion.* Before accepting an engagement, Reserve Auditors should ascertain whether circumstances are likely to permit an unqualified opinion with respect to an Entity's Reserve Information and, if such will not, they should discuss with such Entity (i) the possible necessity of their rendering a qualified opinion and (ii) the possible remedies to the circumstances giving rise to the potential qualification of such opinion.

(d) *Interim Audit Procedures.* Many audit tests can be conducted at almost any time during the year. In the course of interim work, the Reserve Auditors make tests of the Entity's methods, procedures and controls to determine the extent to which such are reliable. It is acceptable practice for the Reserve Auditors to complete substantial parts of an audit examination at interim dates.

When a significant part of an audit is completed during the year and the Entity's methods, procedures and controls are found to be effective, the year-end audit procedure may primarily consist of an evaluation of the impact of new data. The Reserve Auditors must nevertheless be satisfied that the procedures and controls are still effective at the year end and that new discoveries, recent oil and gas production and other recent information and data have been taken into account. Reserve Auditors would not be required to retest the data base pertaining to an Entity's properties and interests unless their inquiries and observations indicate that conditions have changed significantly.

(e) *General Matters To Be Reviewed With Respect to Reserve Information.* An audit of the Reserve Information pertaining to an Entity should include a review of (i) the policies, procedures, documentation and guidelines of such Entity with respect to the estimation, review and approval of its Reserve Information; (ii) the qualifications of Reserve Estimators internally employed by such Entity; (iii) ratios of such Entity's proved reserves to annual production for, respectively, oil, gas and natural gas liquids; (iv) historical reserve and revision trends with respect to the oil and gas properties and interests of such Entity; (v) the ranking by size of properties or groups of properties with respect to estimates of proved reserves or the future net revenue from such proved reserves; (vi) the percentages of proved reserves estimated by each of the various methods set forth in Section 5.3 for estimating proved reserves; and (vii) the significant changes occurring in such Entity's proved reserves, other than from production, during the year with respect to which the audit is being prepared.

(f) *Evaluation of Internal Policies, Procedures and Documentation.* Reserve Auditors should review and evaluate the internal policies, procedures and documentation of an Entity to establish a basis for reliance thereon in determining the nature, extent and timing of the audit tests to be applied in the examination of such Entity's Reserve Information and other data and matters. The internal policies, procedures and documentation to be reviewed with respect to an Entity should include (i) reserve definitions and classifications used by such Entity; (ii) such Entity's policies pertaining to, and management involvement in, the review and approval of Reserve Information and changes therein; (iii) the frequency with which such Entity reviews existing Reserve Information; (iv) the form, content and documentation of the Reserve Information of such Entity, together with such Entity's internal distribution thereof; and (v) the flow of data to and from such Entity's reserve inventory system.

(g) *Testing for Compliance*. Reserve Auditors should conduct tests and spot checks to confirm that (i) there is adherence on the part of an Entity's internal Reserve Estimators and other employees to the policies and procedures established by such Entity; and (ii) the data flowing into the reserve inventory system of such Entity is complete and consistent with other available records.

(h) *Substantive Testing*. In conducting substantive tests, Reserve Auditors should give priority to each property or group of properties of an Entity having (i) a large reserve value in relation to the aggregate properties of such Entity; (ii) a relatively large reserve value and major changes during the audit year in the Reserve Information pertaining to such property or group of properties; and (iii) a relatively large reserve value and a high degree of uncertainty in the Reserve Information pertaining thereto. The amount of substantive testing performed with respect to particular Reserve Information of an Entity should depend on the assessment of (i) the general degree of uncertainty with respect to such Reserve Information, (ii) the evaluation of the internal policies, procedures and documentation of such Entity and (iii) the results of the compliance testing with respect to such Entity. Such substantive testing could therefore appropriately range from a limited number of tests to the complete estimation of Reserve Information with respect to a majority of an Entity's reserves.

6.5 Records and Documentation With Respect to Audit

Reserve Auditor should document, and maintain records with respect to, each audit of the Reserve Information of an Entity. Such documentation and records should include, among other things, a description of (i) the Reserve Information audited; (ii) the review and evaluation of such Entity's policies, procedures and documentation; (iii) the compliance testing performed with respect to such Entity; and (iv) the substantive tests performed in the course of such audit.

APPENDIX E

EMPIRICAL EQUATIONS AND CORRELATION CHARTS TO ESTIMATE PVT PROPERTIES OF RESERVOIR FLUIDS

E.1 Introduction

Frequently, laboratory PVT[1] data are not available when needed for reserve estimation, and empirical correlations must be used. Even when a PVT analysis is available, it is good practice to check the results against empirical correlations.

PVT correlations have appeared over many years; many, however, are not in SPE publications. Thus, they may be difficult to obtain at remote locations. Accordingly, as a convenience to the reader, charts to estimate PVT properties[2] of reservoir fluids are provided in this appendix in Sec. E.2 through E.4, corresponding to each of the three principal reservoir fluids, gas, oil, and water. Also provided are regression equations—if published—to calculate chart values. Equations previously provided in the text are referenced here for convenience.

As with any regression, the correlations discussed here represent "best fits" through specific data sets. With any such data set—especially those involving natural systems— outliers may be expected. The reader is cautioned that application of a generalized correlation to a specific case is subject to error. Accordingly, when applying such correlations to estimate PVT properties, reasonable efforts should be made to cross check available data and to seek independent corroboration.

E.2 Reservoir Gases

E.2.1 Specific Gravity

For gas wells with an initial test CGR less than about 10 STBC/MMscf, the specific gravity of the reservoir gas may be assumed to be approximately equal to the specific gravity of the separator gas. For gases with an initial test CGR greater than about 10 STBC/MMscf, if only

---

[1] Abbreviations are defined in **Notation and Abbreviations**. Terms unique to this Appendix—not usually included in standard engineering nomenclature—are defined as introduced.

[2] This appendix includes correlations to estimate PVT properties needed for volumetric and/or material balance estimates of oil and/or gas—for equations presented in this work. Not included are correlations to estimate viscosity, density, and other properties of reservoir fluids, the need for which is considered peripheral to this work, and the reader is referred to McCain [1991] for correlations for these properties. Since completion of this appendix, several additional publications regarding PVT properties of use for reserve estimation have appeared. The reader is referred to, for example, McCain *et al.* [1994].

well test data are available, **Fig. E.1** may be used to estimate the specific gravity of the reservoir gas at initial conditions [Standing 1952].[3]

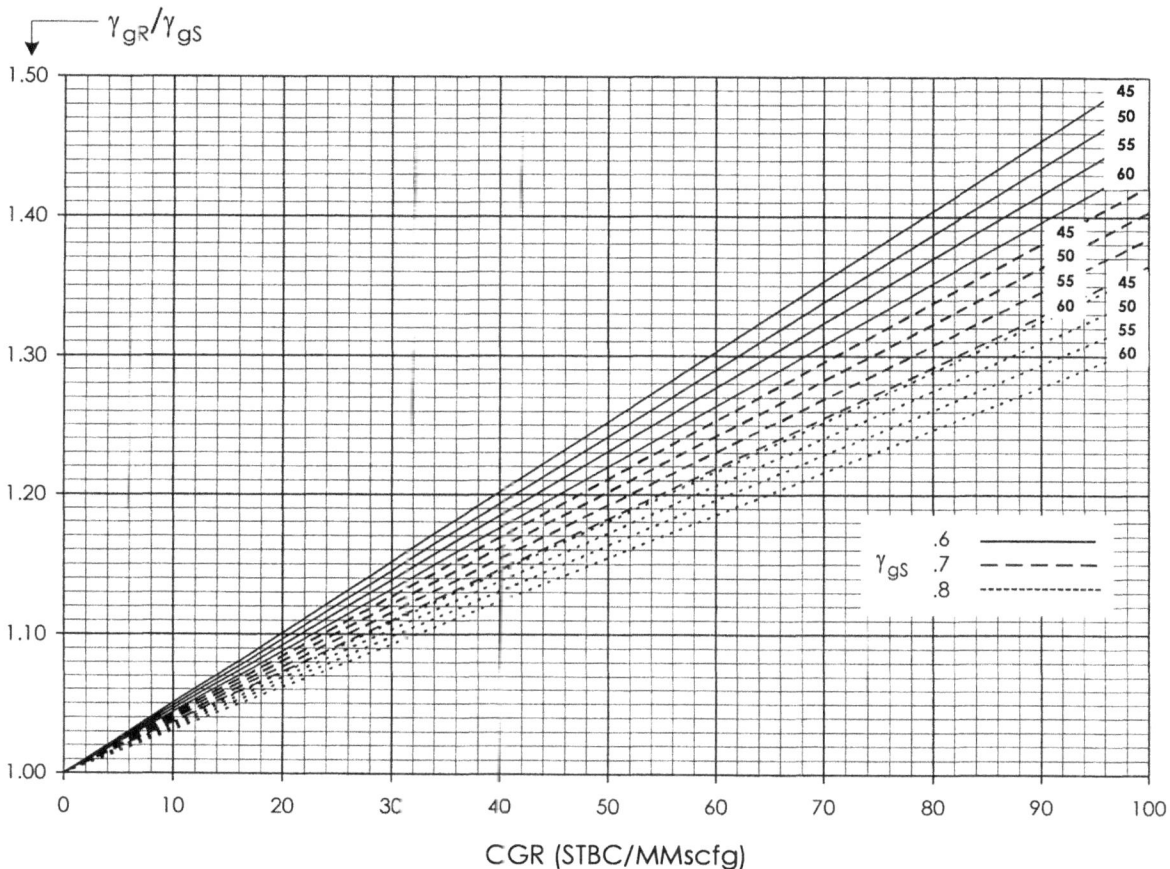

Fig. E.1—Ratio of reservoir to separator gas gravity, γ_{gR}/γ_{gS}, vs. CGR, °API, and γ_{gS} [after Standing 1952].

As an example of usage of Fig E.1, given a gas well with an initial test CGR of 65 STBC/MMscf, separator gas gravity of 0.7, and 45°API condensate, γ_{gR}/γ_{gS} would be estimated to be 1.275. Multiplying separator gas gravity by this ratio leads to estimated reservoir gas gravity of 0.89 (1.275 × 0.7 = 0.89).

Depending on accuracy requirements and on data available, it may be preferable to estimate the specific gravity of the reservoir gas using correlations developed by Gold *et al.* [1989]. These correlations account for vapors from the stock tank and gas from the second stage separator (if any) which rarely are reported during normal well testing.

The correlations may be used for either two- or three-stage separation.[4] The Gold *et al.* [1989] basic relation is

$$\gamma_{gR} = (R_1\gamma_{g1} + 4602\gamma_c + G_{fa})/(R_1 + V_{eq}), \qquad (E.1)$$

---

[3] This chart was generated from Eq. 4.21, which was discussed in Sec. 4.4.1.
[4] Two-stage separation is considered as flash through a gas/oil separator followed by flash to a stock tank. Three-stage

where γ_{gR} = specific gravity of reservoir gas (air = 1),

R_1 = GCR, ratio of gas from primary separator to stock tank condensate (scf/STBC),

G_{fa} = gas (additional) flashed from second stage separator, if present, and vented from the stock tank (scf/STBC) $\times$ (γ_g),

γ_{g1} = specific gravity of gas from primary separator (air = 1.0),

γ_c = specific gravity (water = 1.0), not API gravity, of stock tank condensate, and

V_{eq} = volume of stock-tank gas and gas from second stage separator, if present, plus "vapor equivalent" of stock-tank condensate (scf/STBC).[5]

G_{fa} and V_{eq} are dependent on the number of flashes in the separation process. For three-stage separation[6,7]

$$V_{eq} \approx 535.92 + 2.6231(p_{sp1})^{0.79318}(\gamma_{s1})^{4.6612}(^{\circ}API)^{1.2094}(T_{sp1})^{-0.84911}(T_{sp2})^{0.26987} , \quad ....(E.2)$$

$$G_{fa} \approx 2.9922(p_{sp1} - p_{sc})^{0.97050}(\gamma_{s1})^{6.8049}(^{\circ}API)^{1.0792}(T_{sp1})^{-1.1960}(T_{sp2})^{0.55367} , \quad ....(E.3)$$

where p_{sp1} = pressure at first stage separator (psia) and

$T_{sp1,sp2}$ = temperatures of first and second stage separators ($^{\circ}$F).

For two-stage separation

$$V_{eq} \approx 635.53 + 0.36182(p_{sp1})^{1.0544}(\gamma_{s1})^{5.0831}(^{\circ}API)^{1.5812}(T_{sp1})^{-0.79130} , \quad ....(E.4)$$

$$G_{fa} \approx 1.4599(p_{sp1} - p_{sc})^{1.3394}(\gamma_{s1})^{7.0943}(^{\circ}API)^{1.1436}(T_{sp1})^{-0.93466} . \quad ....(E.5)$$

For three-stage separation **Figs. E.2** and **E.3** may be used to estimate V_{eq} and G_{fa}, respectively. For two-stage separation, **Figs. E.4** and **E.5** may be used.[8]

Reportedly, these correlations were based on 234 retrograde gas samples collected worldwide. In Eqs. E.3 and E.5, p_{sc} = 14.65 psia. Appropriate adjustments should be made for other base pressures.

E.2.2 Pseudocritical Properties

Historically, **Fig. E.6** [Standing 1952] has been used to estimate the pseudocritical properties of reservoir gases. For the lines identified in Fig. E.6 as "CONDENSATE FLUIDS," the equations[9] for pseudocritical pressure and pseudocritical temperature are [Standing 1977]

$$p_{pc} \approx 706 - 51.7\gamma_g - 11.1(\gamma_g)^2 , \quad ....(E.6)$$

$$T_{pc} \approx 187 + 330\gamma_g - 71.5(\gamma_g)^2 . \quad ....(E.7)$$

---

separation involves flash through two stages of gas/oil separators prior to the stock tank.

[5] This term is comparable to the "vapor equivalent of stock-tank condensate, tank vapors and low pressure separator gas" for which a correlation was developed by Leshikar [1961], as discussed in Sec. E.2.6.

[6] Relations determined by regression analysis are identified in this work by "$\approx$" rather than "=."

[7] These correlations—identified by Gold *et al.* [1989] as "Model 2"—are for situations where stock-tank temperature is unknown—the usual case. For situations where stock-tank temperature is known, refer to Gold *et al.* [1989] "Model 1."

[8] The arrows indicate the authors' procedure to use these four figures and are not illustrative of any application discussed here.

[9] The subscript "R," included in the original equations to denote specific gravity of reservoir gas, is omitted here to be consistent with notation elsewhere in this section.

Sutton [1985] developed correlations that may provide a more accurate method to estimate pseudocritical properties, especially for high specific gravity gases

$$p_{pc} \approx 756.8 - 131.0\gamma_g - 3.6(\gamma_g)^2 , \quad\text{.. (E.8)}$$

$$T_{pc} \approx 169.2 + 349.5\gamma_g - 74.0(\gamma_g)^2 . \quad\text{... (E.9)}$$

Sutton's [1985] work included data from 634 different gases primarily from the Gulf of Mexico, Louisiana, and Texas. None of the gases contained hydrogen sulfide. Data ranges are:

Property		Minimum	Maximum
Reservoir pressure	(psia)	200	12,500
Temperature	(°F)	100	360
Z-factor	(dim)	0.748	2.147
Gas gravity	(air=1)	0.571	1.679
CO_2 (mole %)		0.01	11.86
N_2 "		0.0	2.86
C_{7+} "		0.02	14.27

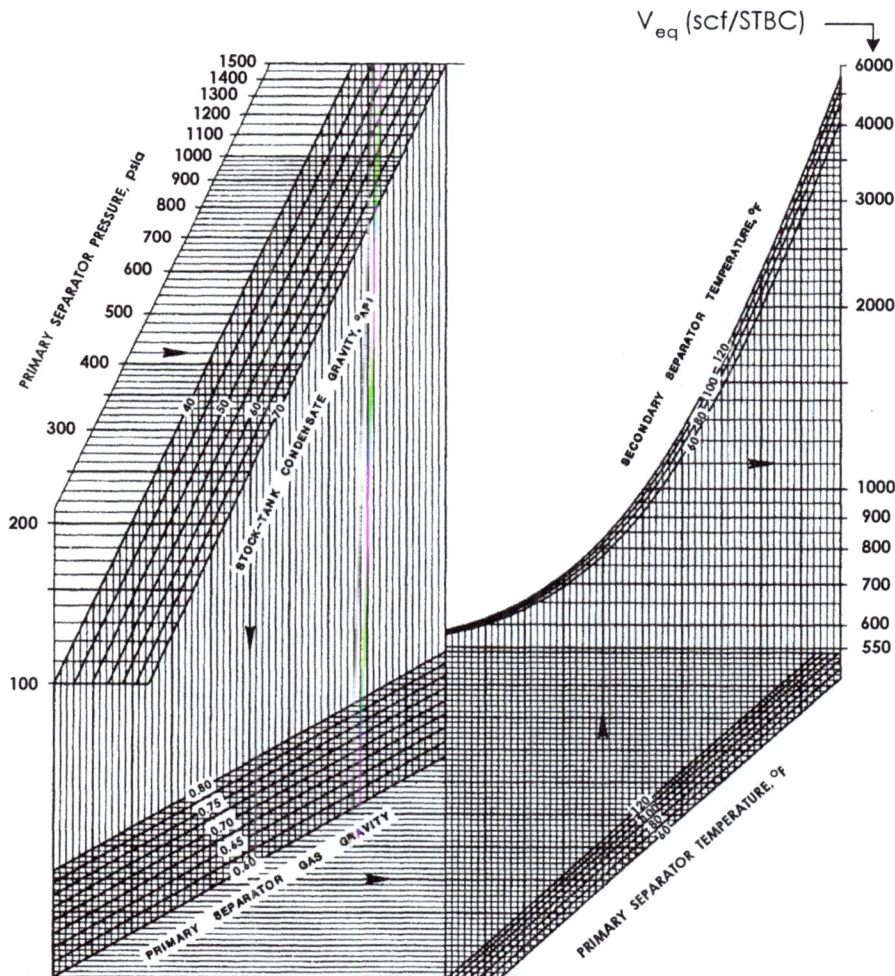

Fig. E.2—V_{eq} vs. primary separator pressure, p_{sp1}, gravity of stock-tank condensate, °API, gravity of primary separator gas, γ_{g1}, temperature of primary separator, T_{sp1} and temperature of secondary separator, T_{sp2}, for three-stage separation [after Gold et al. 1989].

Comparing Sutton's [1985] correlations with those of Standing [1977], Eqs. E.8 and E.9 vs. Eqs. E.6 and E.7, respectively, it may be noted that Sutton's correlation for pseudocritical temperature is essentially equal to Standing's. Sutton's correlation for pseudocritical pressure, however, results in estimates considerably less than Standing's, with the difference increasing as reservoir gas gravity increases. These differences may result in intolerable uncertainty in subsequent calculations. Thus, for reservoir gases estimated to have a specific gravity greater than about 0.8, a PVT analysis is recommended.

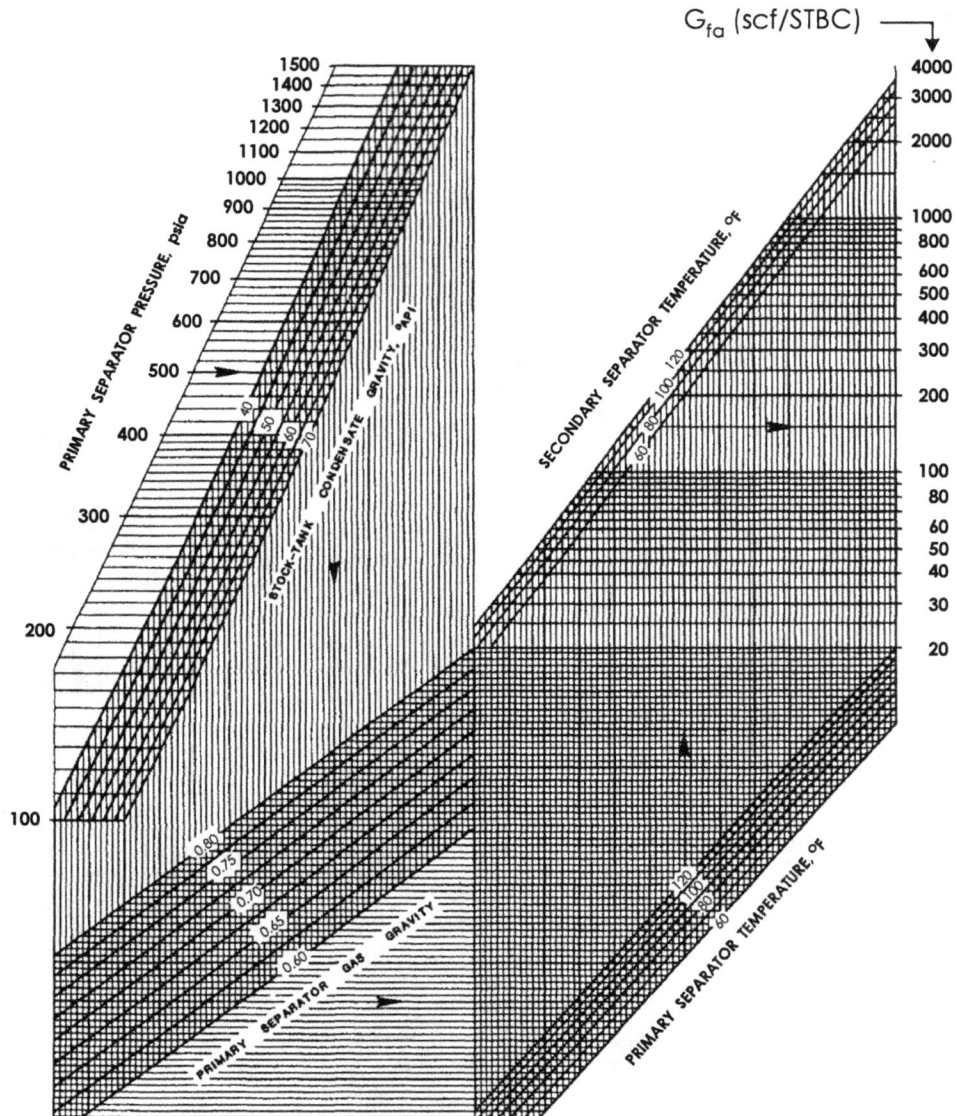

Fig. E.3—G_{fa} **vs. primary separator pressure,** p_{sp1}**, gravity of stock-tank condensate,** °**API, gravity of primary separator gas,** γ_{g1}**, temperature of primary separator,** T_{sp1} **and temperature of secondary separator,** T_{sp2}**, for three-stage separation [after Gold** *et al.* **1989].**

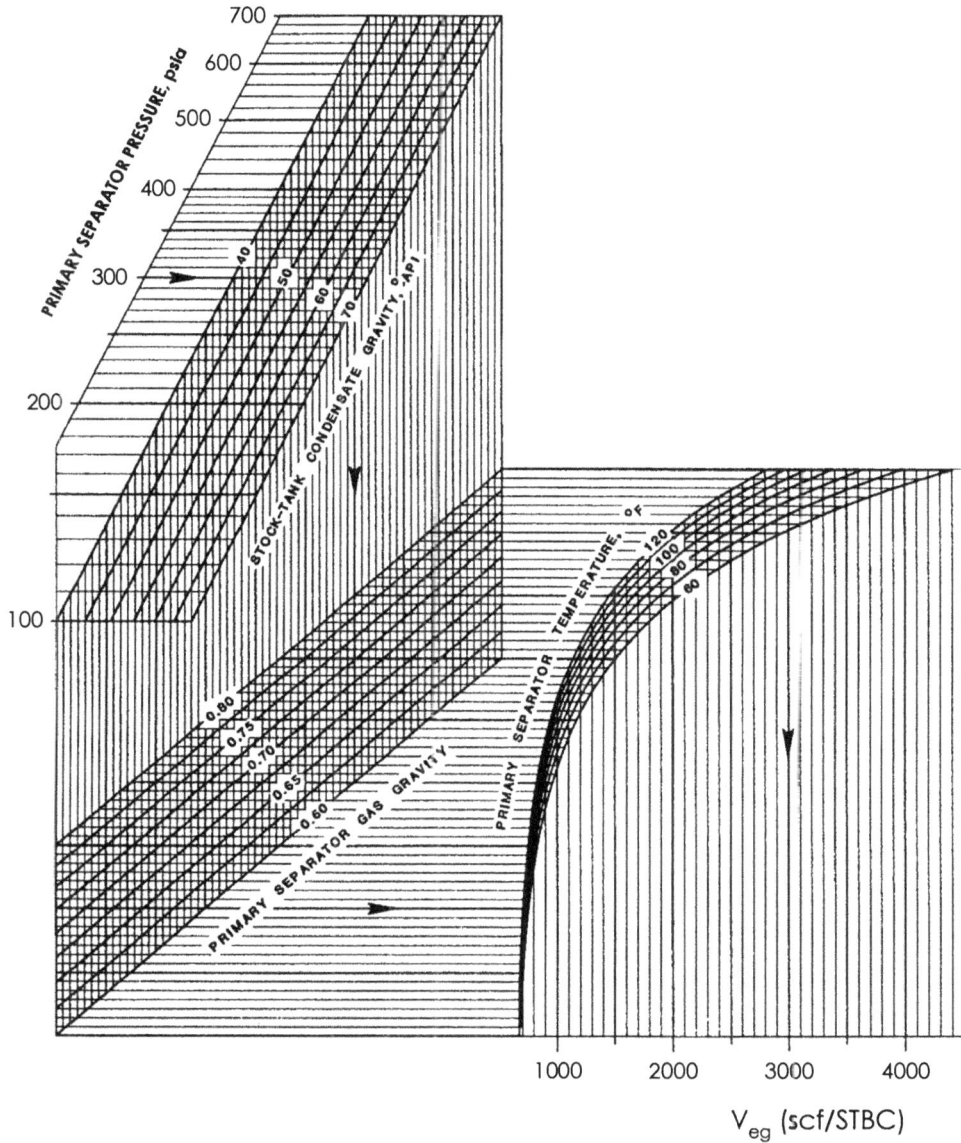

Fig. E.4—V_{eq} vs. primary separator pressure, p_{sp1}, gravity of stock-tank condensate, °API, gravity of primary separator gas, γ_{g1}, and temperature of primary separator, T_{sp1}, for two-stage separation [after Gold et al. 1989].

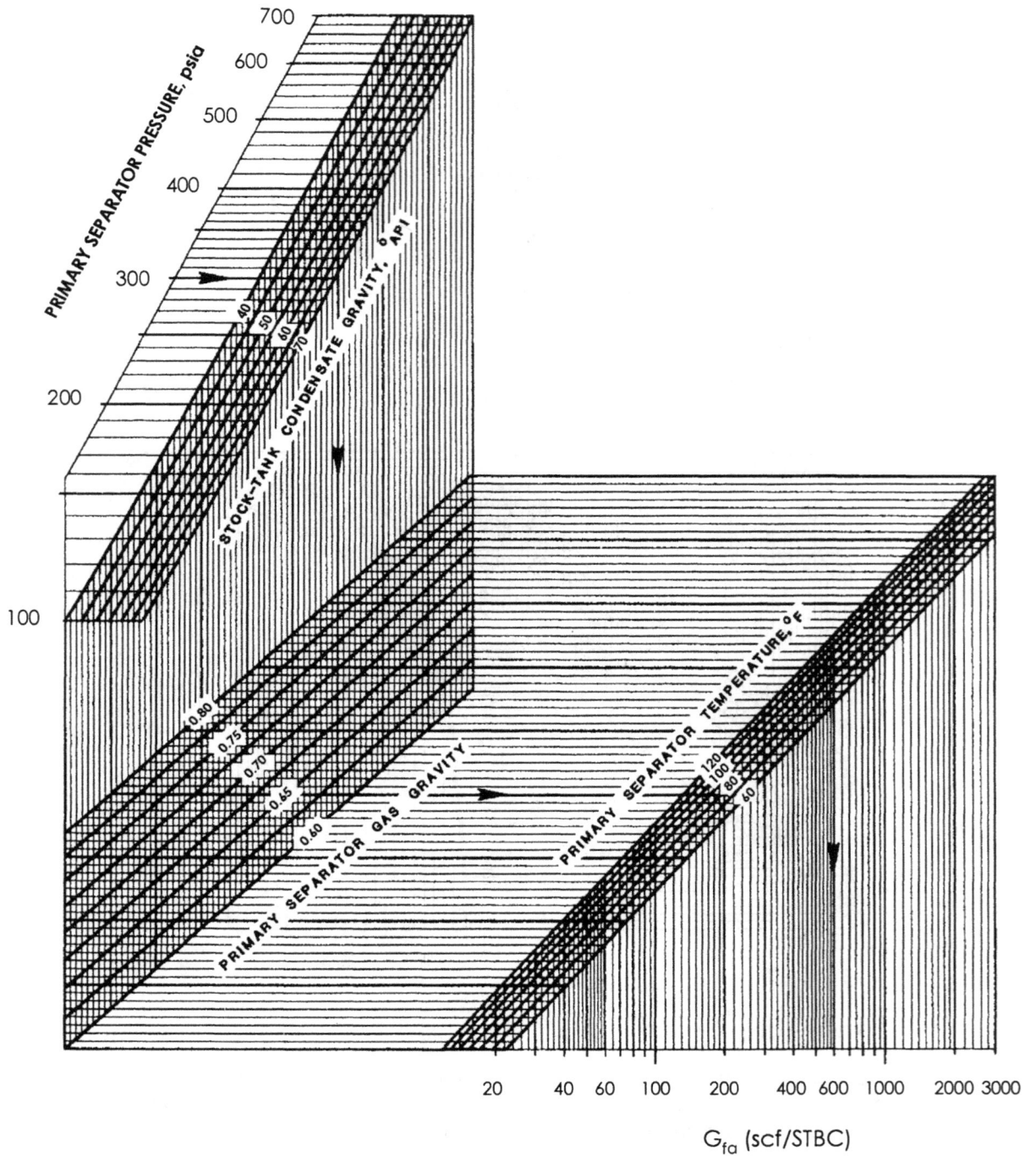

Fig. E.5—G_{fa} vs. primary separator pressure, p_{sp1}, gravity of stock-tank condensate, °API, gravity of primary separator gas, γ_{g1}, and temperature of primary separator, T_{sp1}, for two-stage separation [after Gold et al. 1989].

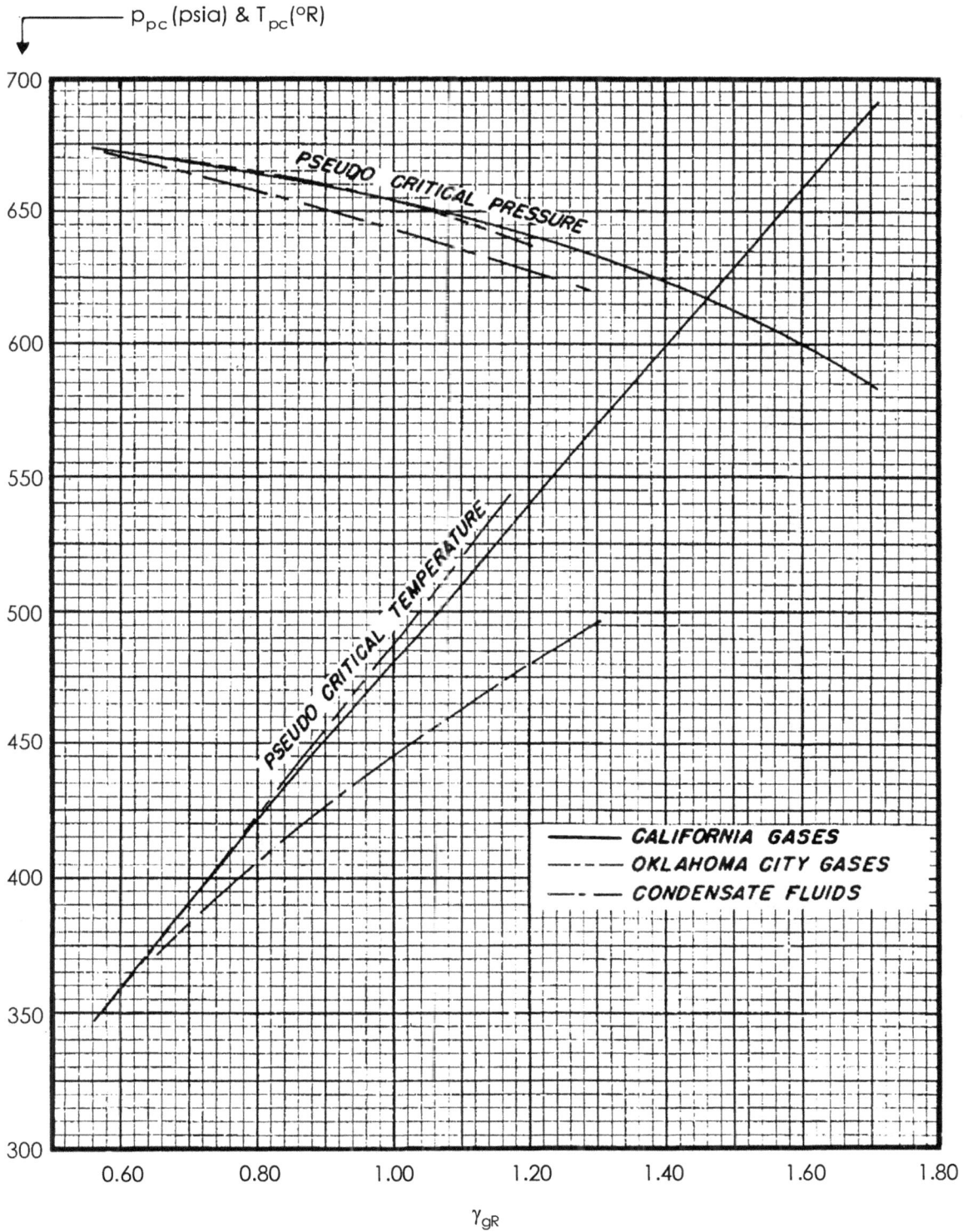

Fig. E.6—Pseudocritical pressure, p_{pc}, and pseudocritical temperature, T_{pc}, vs. reservoir gas gravity, γ_{gR} [after Standing 1952].

E.2.3 Z-Factors[10]

Gas deviation factors (Z-factors) may be estimated from **Figs. E.7** and **E.8,** which are attributed to Standing and Katz [1942] and Katz *et al.* [1959], respectively. Despite their "antiquity," these charts remain industry standard.

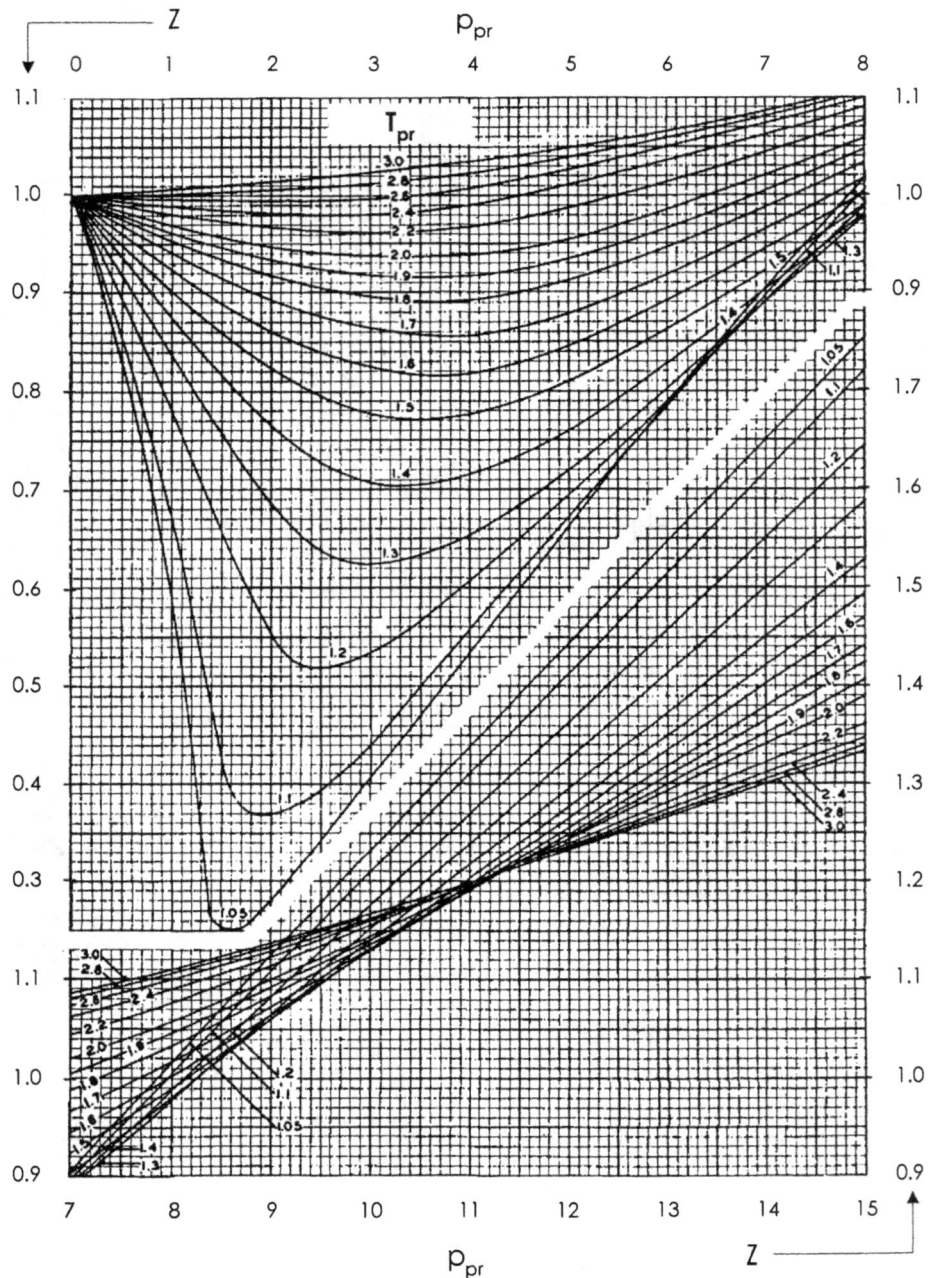

Fig. E.7—Z-factor vs. pseudoreduced pressure, *p*_{pr}, and pseudoreduced temperature, *T*_{pr}, for *p*_{pr}\<15 [after Standing and Katz 1942].

---

[10] The term "Z-factor" is used in this work in preference to the traditional term, "compressibility factor," to avoid confusion with gas compressibility, which is defined as $1/p - 1/Z(dZ/dp)$.

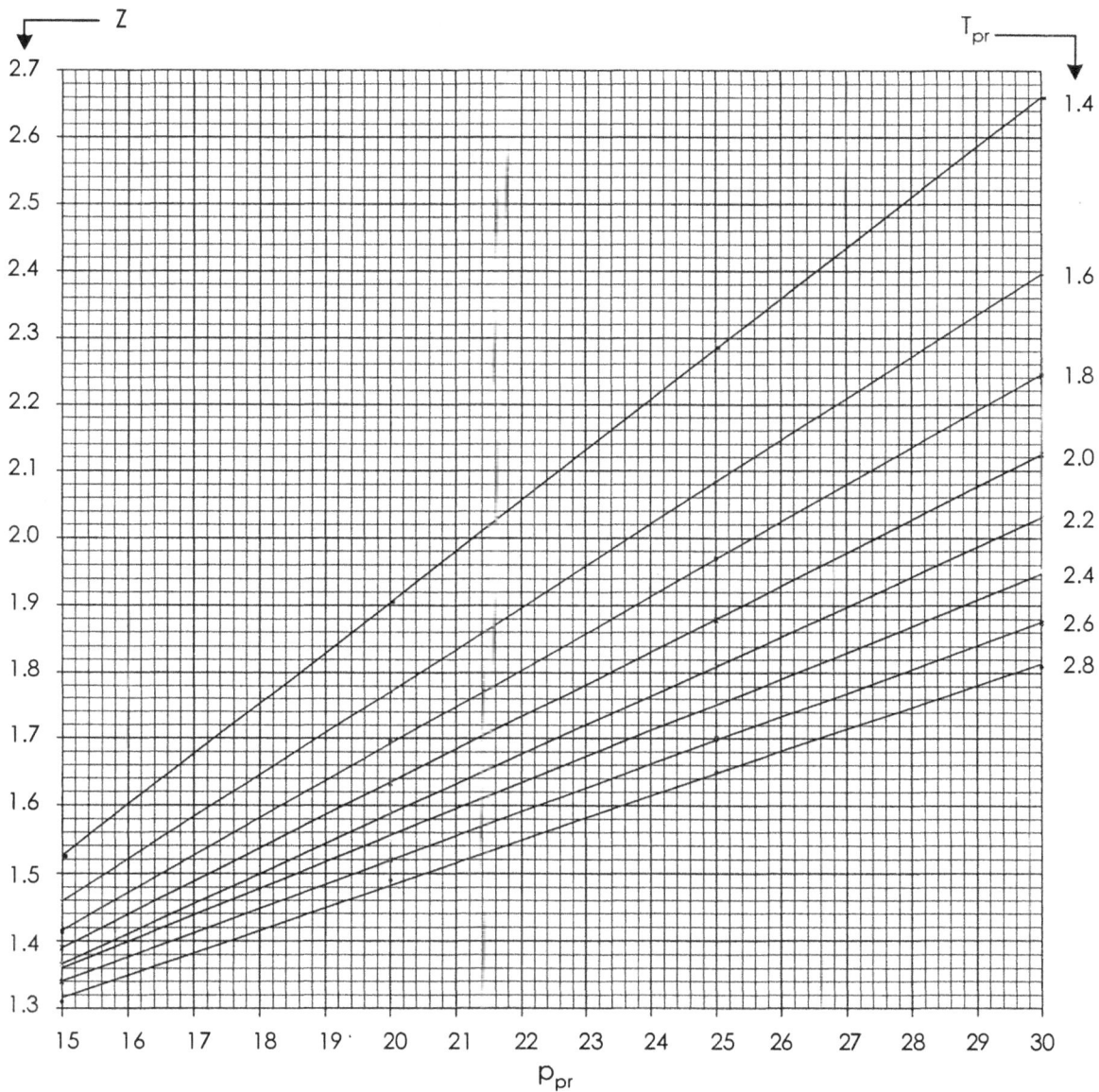

Fig. E.8—Z-factor vs. pseudoreduced pressure, p_{pr}, and pseudoreduced temperature, T_{pr} for 15< p_{pr} <30 [after Katz et al. 1959].

There have been many attempts to: (a) fit the curves on these charts (which are based on experimental data) or (b) to replicate the data (with equations of state) with varying degrees of accuracy. Reportedly [Takacs 1976], the most accurate results were obtained by Dranchuk and Abou-Kassem [1975]. These equations, however, require trial and error solution. A fit to the Z-factor charts provided by Standing [1977] is adequate for most engineering purposes

$$Z \approx A + (1 - A)/e^B + C p_{pr}{}^D, \quad\text{...} \quad (E.10)$$

$$\text{where} \quad A \approx 1.39(T_{pr} - 0.92)^{0.5} - 0.36 T_{pr} - 0.101 , \quad\text{.......................................} \quad (E.11)$$

$$B \approx (0.62 - 0.23T_{pr})p_{pr} + [0.066/(T_{pr} - 0.86) - 0.037]p_{pr}^2$$
$$+ 0.32p_{pr}^6/10^9(T_{pr} - 1) , \quad\text{...(E.12)}$$

$$C \approx 0.132 - 0.32\log T_{pr} , \text{ and } \text{...(E.13)}$$

$$D \approx \text{antilog}(0.3106 - 0.49T_{pr} + 0.1824T_{pr}^2) . \text{ ....(E.14)}$$

E.2.4 Gas Formation Volume Factors

B_{gi} in Eqs. 3.3a and 3.3b can be computed from formation temperature, initial reservoir pressure, and initial Z-factor, as shown in Eq. 4.15. However, B_{gi} and $1/B_{gi}$ may be approximated from **Figs. E.9** and **E.10** for gas gravities of 0.6 and 1.0, respectively [Arps 1962]. Interpolation may be used for other gas gravities. These charts may be used for preliminary approximations but should not be used if accurate estimates are required.

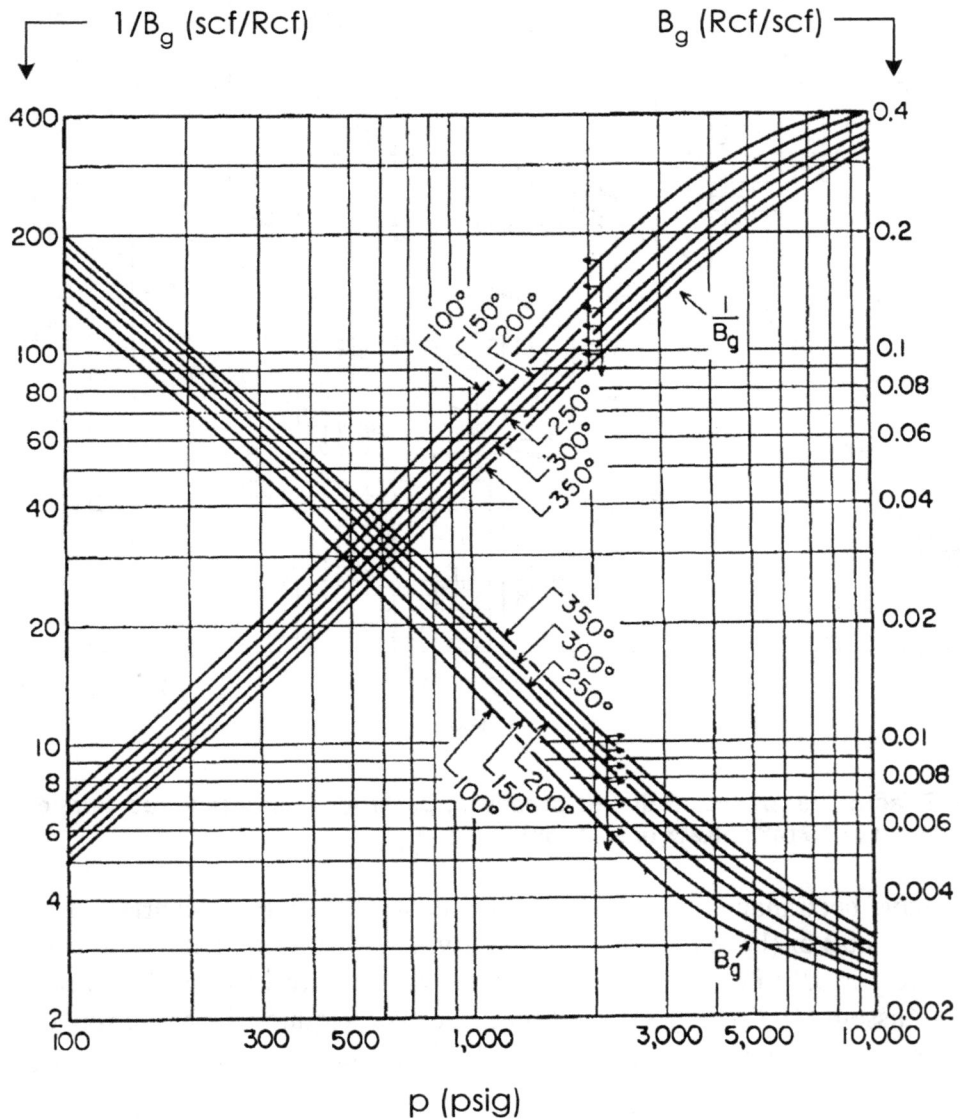

Fig. E.9—B_g and $1/B_g$ vs. reservoir pressure, p, and temperature, T_f, for 0.6 gravity gas [after Arps 1962].

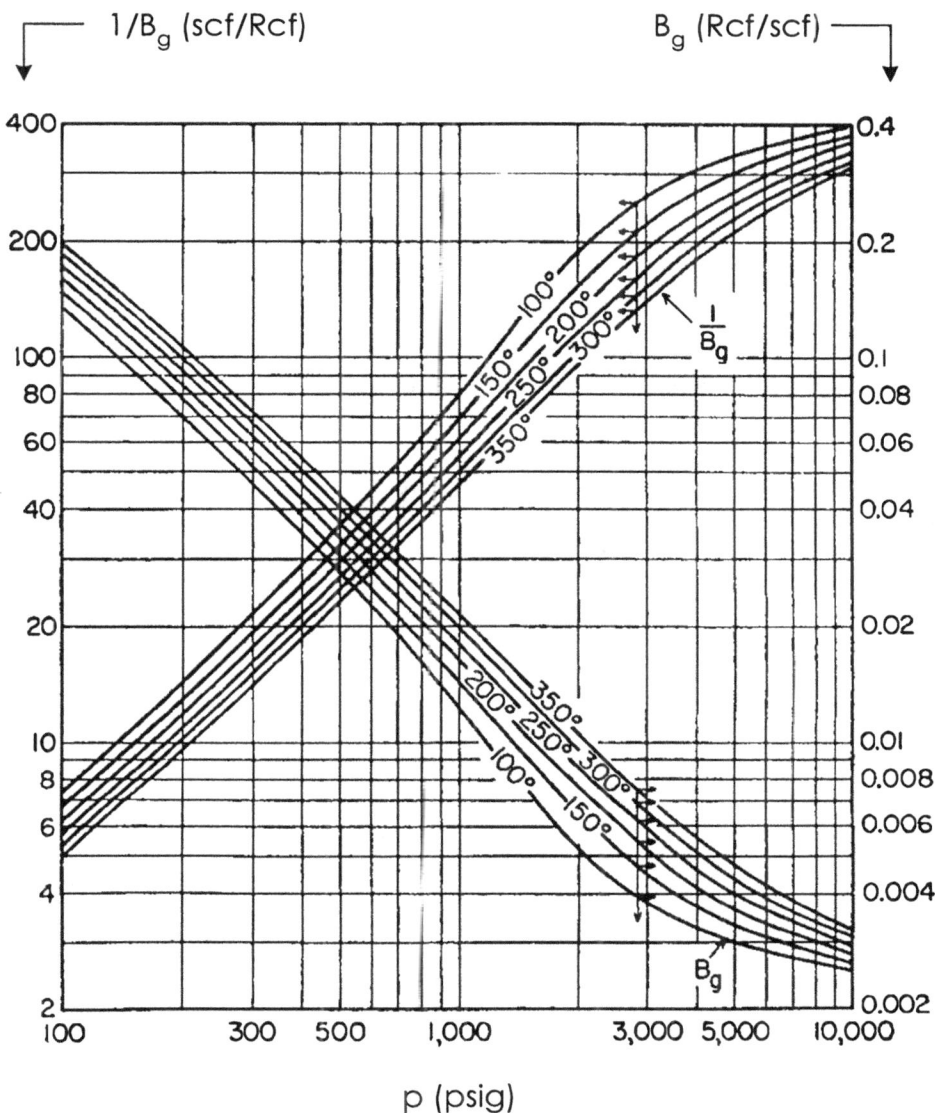

Fig. E.10—B_g and $1/B_g$ vs. reservoir pressure, p, and temperature, T_f, for 1.0 gravity gas [after Arps 1962].

E.2.5 Z-Factors for Retrograde Gases

In the absence of a PVT analysis, material balance calculations below the dewpoint pressure may be made using a correlation for the two-phase Z-factor, Z_{2P}, developed by Rayes *et al.* [1992]

$$Z_{2P} = 2.24353 - 0.0375281 p_{pr} - 3.56539/T_{pr} + 0.000829231 p_{pr}^2$$

$$+ 1.53428/T_{pr}^2 + 0.131987 p_{pr}/T_{pr} . \quad \dots\dots\dots\dots\dots\dots\dots\dots\dots (E.15)$$

Pseudoreduced pressure and pseudoreduced temperature—arguments for Eq. E.15—may be estimated using procedures outlined in Sec. 4.4.1, Eq. 4.19 and 4.20, respectively.

Eq. E.15 may be used to estimate 2-phase Z-factors if the initial heptanes-plus content of the reservoir gas exceeds about 4.0 mole %. The mole fraction of C7+ in the initial reservoir gas, y_{C7+}, may be estimated by

$$y_{C7+} \approx 0.141013\gamma_g - 0.0885119 . \quad\text{...(E.16)}$$

McCain and Piper [1994] reported that the heptanes-plus fraction of reservoir gases can be expected to be greater than about 4.0 mole % for initial test GCR's less than about 15,000 scf/STBC (CGR greater than about 67 STBC/MMscfg).

The Rayes *et al.* [1992] work included data from 131 constant composition depletion studies performed on retrograde gas samples collected worldwide. Data set properties are:

Property		Minimum	Maximum
H_2S	(mole %)	0.00	28.16
CO_2	(mole %)	0.00	63.52
N_2	(mole %)	0.01	12.74
C_1	(mole %)	19.37	94.2
C_2	(mole %)	1.95	16.66
C_3	(mole %)	0.62	12.27
iC_4	(mole %)	0.18	2.53
nC_4	(mole %)	0.25	5.02
iC_5	(mole %)	0.12	1.62
nC_5	(mole %)	0.08	2.09
C_6	(mole %)	0.14	2.04
C_{7+}	(mole %)	0.85	12.68
MW C_{7+}		108.	253.
SG C_{7+}		0.736	0.850
γ_g	(air=1)	0.644	1.557
Z_d	(dim)	0.704	1.775
p_d	(psi)	1,193.	11,844.
T_f	(°F)	94.	325.

In developing Eq. E.16 the authors used the entire data set. In developing Eq. E.15, however, the authors used data from 67 samples which were described as "relatively pure, rich gases" The initial mole % of C7+ and impurities in the 67 sample data set were GTE 4.0 and LT 5.0, respectively.

The correlation coefficient for Eq. E.15 (R^2) was 0.948; average error was 3.38%; maximum, 34.95%. Rayes *et al.* [1992] observed that the large maximum errors were associated with: (a) reservoir gases containing high concentrations of ethane and propane and (b) samples depleted to pressures less than 700 psi.

E.2.6 Vapor Equivalent of Stock-Tank Condensate Plus Tank Vapors and Low-Pressure Separator Gas

Historically, a chart prepared by Leshikar [1961] for the Texas Railroad Commission, **Fig. E.11,** has been used to estimate the gaseous equivalent of stock-tank condensate plus the associated stock-tank vapors and low-pressure separator gas (identified as V_{eq} in Sec. E.1.1).[11] Stock-tank vapors and low-pressure separator gas typically are not reported. Although this

---

[11] This term also has been called "stock-tank condensate vaporizing volume ratio," and has been noted as V_{cs}.

chart may be considered superseded by the Gold *et al.* [1989] work, it is still widely used. To the author's knowledge, a computer fit of this chart is not available in the public domain.

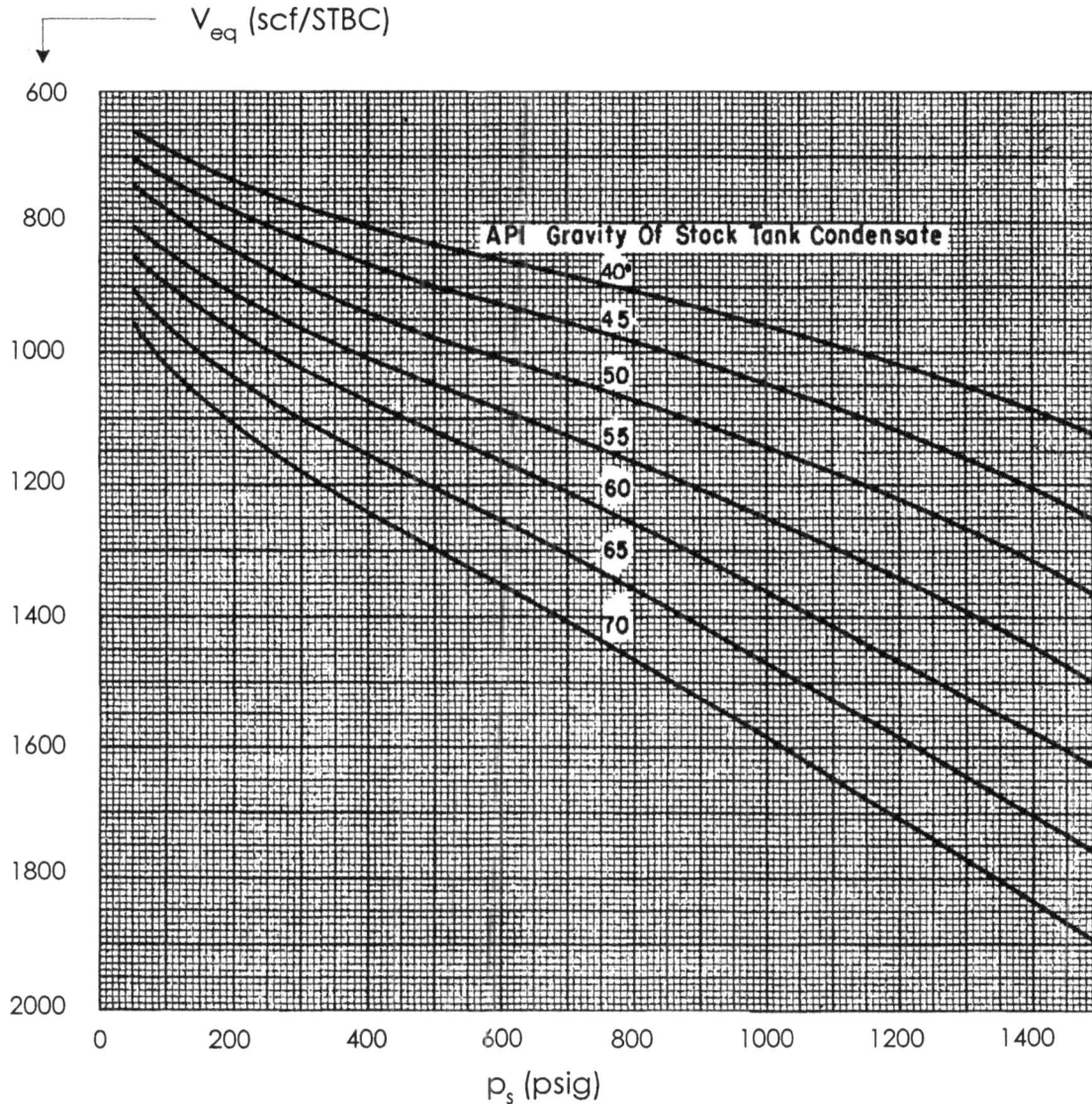

Fig. E.11—V_{eq} vs. separator pressure, p_{sp}, and gravity of stock-tank condensate, °API [after Leshikar 1961].

Depending on the data available and accuracy requirements, it may be preferable to estimate V_{eq} using correlations developed by Gold *et al.* [1989], discussed in Sec. E.2.1.

Reportedly [Gold *et al.* 1989], the correlation by Leshikar [1961] underestimates V_{eq} for high first stage separator pressures and high API gravity condensates; the absolute average error is 16%.

Irrespective of which correlation is used to estimate V_{eq}, the G_p term in Eq 4.14 is

$$G_p = G_{ps1} + V_{eq}C_p, \quad\text{..(E.17)}$$

where G_{ps1} = cumulative gas production through the first stage separator (scf).

If an acceptable estimate cannot be made of separator conditions, as required for both correlations, then the (reservoir) vapor equivalent of stock tank condensate, V_L, (only) may be estimated as

$$V_L = 133,000 \gamma_c / M_c , \dots\dots\dots\dots\dots\dots\dots\dots\dots\text{(E.18)}$$

where the constant (133,000) is based on standard conditions—60°F and 14.7 psia.

For the reader's convenience, tables published elsewhere to estimate the vapor equivalent of condensate, V_L, for a range of condensate gravities, plus a table using a correlation by Gold *et al.* [1989], are provided here:

°API	Vapor Volume (scf/STBC)		
	Craft & Hawkins	Smith	Gold *et al.*
40	Not Rep	635	575
45	684	676	648
50	752	721	717
55	814	765	783
60	870	813	845
65	930	863	904
70	Not Rep	916	960
75	Not Rep	970	1,013

Differences between values published by Craft and Hawkins [1959] and by Smith [1962] are attributed to differences in estimating the molecular weight of stock-tank condensate, M_c. Craft and Hawkins [1959] reportedly used[12]

$$M_c = 6084/(°\text{API} - 5.9) . \dots\dots\dots\dots\dots\dots\dots\dots\text{(E.19a)}$$

Smith [1962] apparently used slightly different constants, which he did not report. Gold *et al.* [1989] provided what may be the most reliable method to estimate M_c:

$$M_c = 5954/(°\text{API} - 8.8) . \dots\dots\dots\dots\dots\dots\dots\dots\text{(E.19b)}$$

which, together with Eq. E.18, was used to develop the column identified as "Gold *et al.*" in the foregoing table.

For high-pressure first stage separators and high-API-gravity condensate, the V_L values from the above table are significantly less than the V_{eq} values that would be estimated using the Gold *et al.* [1989] methodology discussed in Sec. E.2.1.

E.2.7 Dewpoint Pressure

Nemeth and Kennedy [1967] developed a "reasonably accurate" [McCain 1991] correlation to estimate dewpoint pressure. Their work was based on regression analysis of 579 dewpoint pressures from 480 different condensate systems from "most of the major oil-producing areas of the world."

---

[12] For API gravities from 40 to 60 °API, the Gold *et al.* correlation gives estimates of molecular weight from 7 to 3% greater than the Craft and Hawkins correlation.

$$\ln p_d \approx -2.0623054[y_{C2} + y_{CO2} + y_{H2S} + y_{C6} + 2(y_{C3} + y_{C4}) + y_{C5} + 0.4y_{C1} + 0.2y_{N2}]$$

$$+ 6.6259728\gamma_{C7+} - 4.4670559(E\text{-}3)[y_{C1}/(y_{C7+} + 0.002)] + 1.0448346(E\text{-}4)T_f$$

$$+ 3.2673714(E\text{-}2)(y_{C7+})(M_{C7+}) - 3.6453277(E\text{-}3)[(y_{C7+})(M_{C7+})]^2$$

$$+ 7.4299951(E\text{-}5)[(y_{C7+})(M_{C7+})]^3 - 1.1381195(E\text{-}1)[M_{C7+}/(\gamma_{C7+} + 0.0001)]$$

$$+ 6.2476497(E\text{-}4)[M_{C7+}/(\gamma_{C7+} + 0.0001)]^2$$

$$- 1.0716866(E\text{-}6)[M_{C7+}/(\gamma_{C7+} + 0.0001)]^3 + 10.746622 , \quad \text{.....................................(E.20)}$$

where y_{CO2} = mole fraction of CO_2 in the reservoir gas,
$\quad\quad$ T_f = formation temperature ($^{\circ}$R),
$\quad\quad$ M_{C7+} = molecular weight of the heptanes-plus fraction, and
$\quad\quad$ γ_{C7+} = specific gravity of the heptanes-plus fraction.

The ranges of variables in the condensate systems studied are:

Parameter		Minimum	Maximum
Dew point	(psia)	1,270	10,790
Reservoir temp. ($^{\circ}$F)		40	320
MW C7+		106	235
SG C7+		0.7330	0.8681
C_1	(mole fr.)	0.0349	0.9668
C_2	(mole %)	0.0037	0.1513
C_3	(mole %)	0.0011	0.1090
C_4	(mole %)	0.0017	0.3750
C_5	(mole %)	0.0000	0.0710
C_6	(mole %)	0.0000	0.0871
C_{7+}	(mole %)	0.0000	0.1356
N_2	(mole %)	0.0000	0.4322
CO_2	(mole %)	0.0000	0.9192
H_2S	(mole %)	0.0000	0.2986

Reportedly, the average absolute deviation of Eq. E.20 is 7.4% or 354 psi.

E.2.8 Sour Gases

For natural gases containing more then about 7 mole % carbon dioxide and hydrogen sulfide—in total—corrections must be made to the pseudocritical properties before using the "industry standard" Z-factor chart discussed in Sec. E.2.3.[13] Wichert and Aziz [1972] have shown that the pseudocriticals should be adjusted as follows:

$$T'_{pc} = T_{pc} - e , \quad \text{...(E.21)}$$

$$p'_{pc} = p_{pc}T'_{pc}/[T_{pc} + B(1 - B)e] , \quad \text{...(E.22)}$$

where T'_{pc} = adjusted pseudocritical temperature ($^{\circ}$F),
$\quad\quad$ p'_{pc} = adjusted pseudocritical pressure (psi),

---

[13]McCain [1990] reported that "the presence of nitrogen does not greatly affect the Z-factor...(which)...increases by about 1% for each 5% nitrogen in the gas."

B   = mole fraction of hydrogen sulfide, and

e = Wichert-Aziz correction factor (°R),[14] calculated as

$$e \approx 120(A^{0.9} - A^{1.6}) + 15(B^{0.5} - B^4) , \dotfill (E.23)$$

where A = sum of mole fractions of CO_2 plus H_2S.

Fig. E.12 shows the graphical relation between e and mole fraction of CO_2 and H_2S.

To use the Wichert-Aziz correlation:

1. Calculate pseudocriticals in the usual manner, using composition of the reservoir gas.
2. Determine the value of e from Eq. E.23 or from Fig. E.12.
3. Correct pseudocriticals, using Eqs. E.21 and E.22.
4. Calculate reduced pressures and reduced temperature, which become arguments for Figs. E.7 or E.8 (or the appropriate numerical approximations), to estimate corresponding Z-factors.

E.3 Reservoir Oils

E.3.1 Bubblepoint Pressure

Bubblepoint pressure at initial reservoir conditions is one of the most important parameters for oil reservoir engineering. Since publication of the Standing [1947] correlation, still an industry standard, there have been at least ten additional correlations published, including Lasater [1958], Vasquez [1976],[15] Glaso [1980], Ostermann *et al.* [1983], Obomanu and Okpobiri [1987], Al-Marhoun [1988], Asgarpour *et al.* [1989], Dokla and Osman [1992], Omar and Todd [1993], and Kartoatmodjo and Schmidt [1994].

Sample sets used by these authors, except Lasater [1958], Vasquez [1976], and Kartoatmodjo and Schmidt [1994], are from limited geographic areas.

As might be expected, the PVT properties of reservoir oils are dependent on the composition of and relative proportions of isomers in the oil. Oils generated in different geologic settings may be expected to exhibit both compositional and isomeric differences— e.g., paraffinicity and fraction of non-hydrocarbon components. Thus, correlations of PVT properties developed for oils from one geologic setting may exhibit significant differences from those for oils from a different geologic setting.

Fig. E.13 is Standing's [1947] correlation for bubblepoint pressure; regressions developed by Standing [1977] and others follow:[16]

- Standing [1977]: $p_b \approx 18.2\{\text{antilog}[0.00091T_f - 0.0125(°API)][R_{sb}/\gamma_g]^{0.83} - 1.4\}$ . ...(E.24)

---

[14] In their paper, which discussed two other methods to estimate Z-factors for sour gases, this correction term was identified as $e3$.

[15] As reported by Vasquez and Beggs [1980] and by Beggs [1987].

[16] PVT data from flash—not differential—liberation was used in these correlations. Units for T_f are °F for the correlations by Standing [1947], Vasquez [1976], Glaso [1980]; units are °R for correlations by Lasater [1958], Al-Marhoun [1988], Dokla and Osman [1992], and Kartoatmodjo and Schmidt [1994].

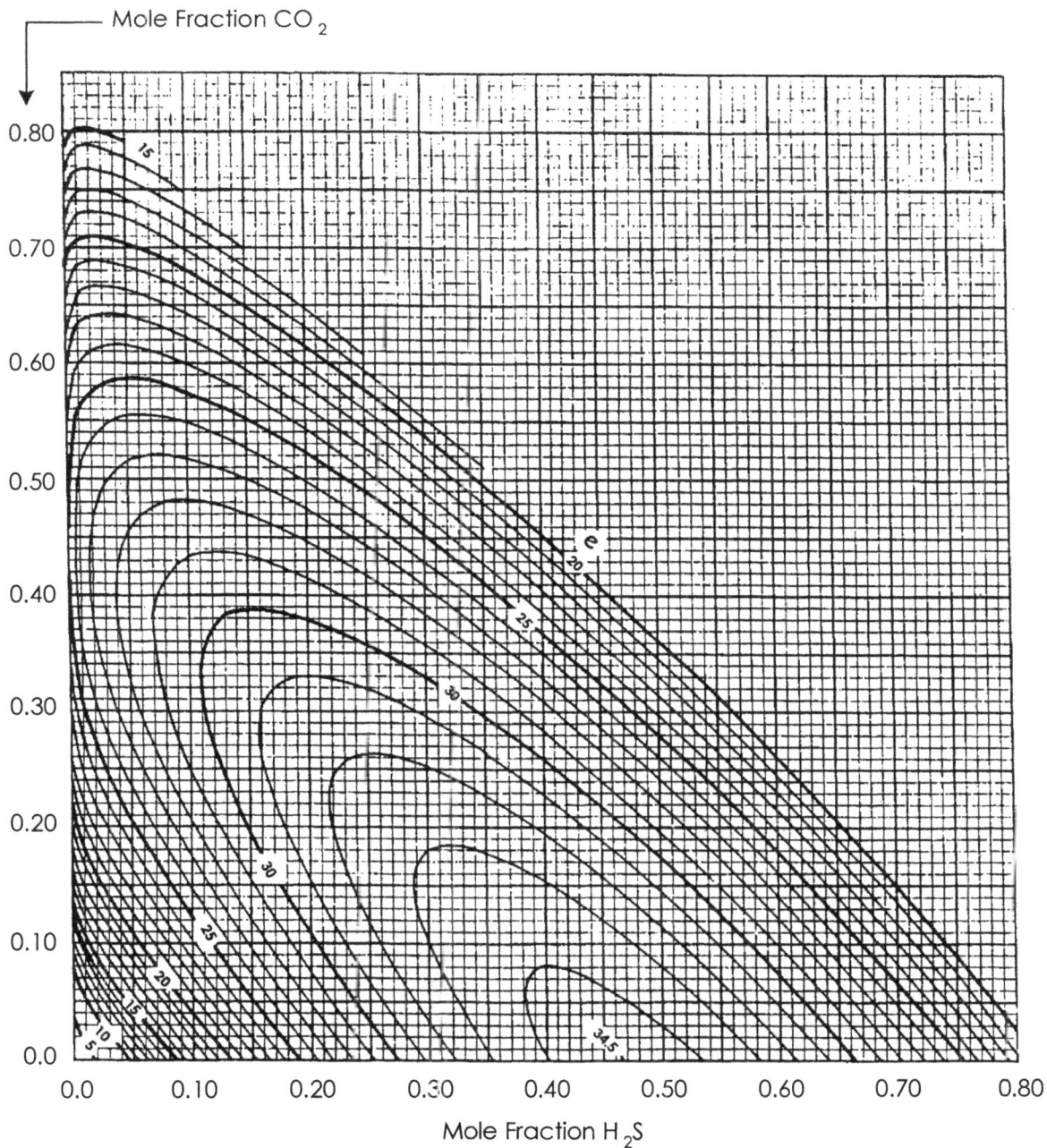

Fig. E.12—Sour-gas correction factor, e, vs. mole fractions carbon dioxide and hydrogen sulfide [after Wichert and Aziz 1972].

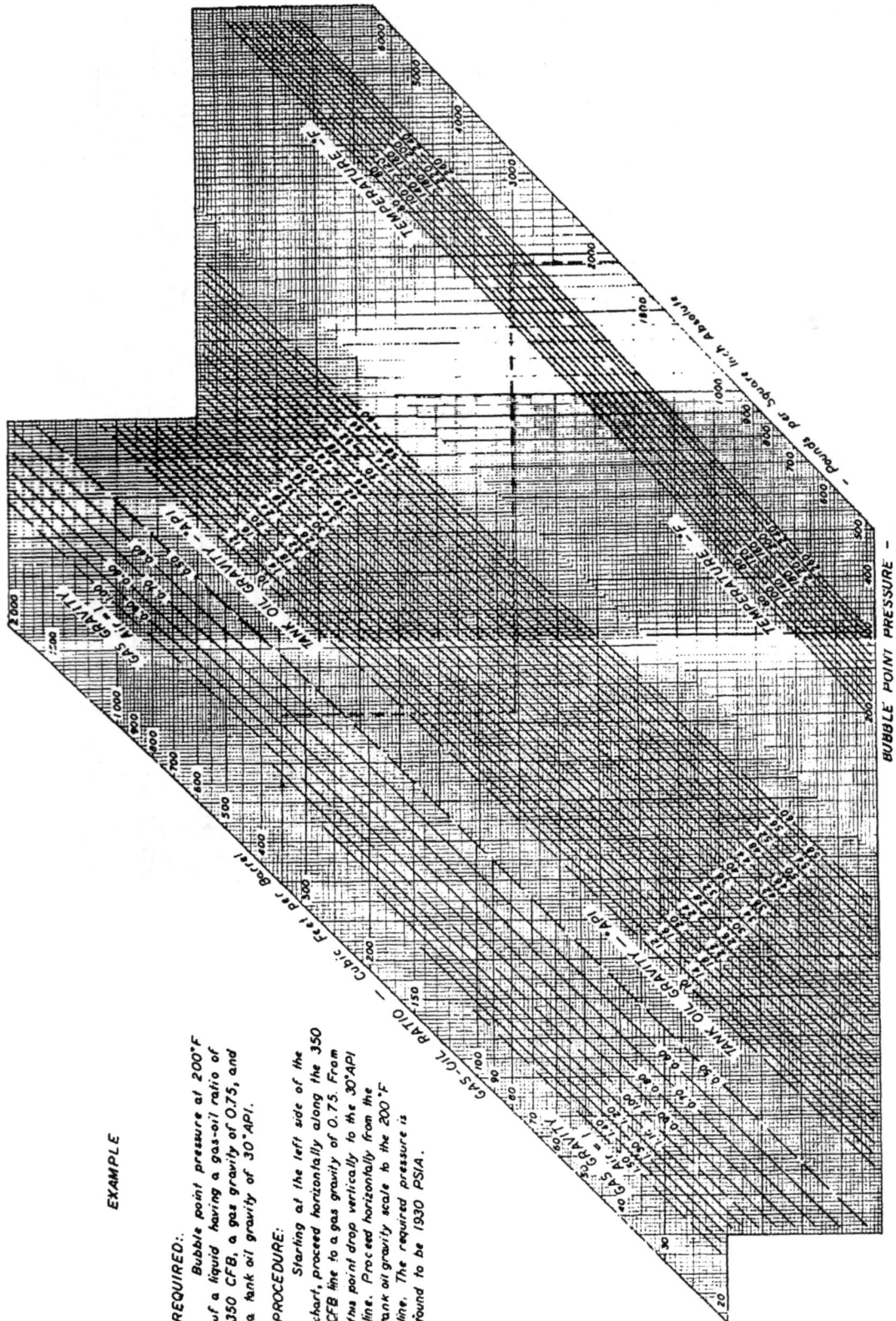

EXAMPLE

REQUIRED:

Bubble point pressure at 200°F of a liquid having a gas-oil ratio of 350 CFB, a gas gravity of 0.75, and a tank oil gravity of 30°API.

PROCEDURE:

Starting at the left side of the chart, proceed horizontally along the 350 CFB line to a gas gravity of 0.75. From that point drop vertically to the 30°API line. Proceed horizontally from the tank oil gravity scale to the 200°F line. The required pressure is found to be 1930 PSIA.

Fig. E.13—Bubblepoint pressure, p_b, vs. solution GOR, R_s, gas gravity, γ_g, gravity of stock tank oil, °API, and formation temperature T_f [after Standing 1947].

- Lasater [1958]

$$p_b \approx (p_b\gamma_g/T_f)(T_f/\gamma_g) , \quad\text{...} \text{(E.25)}$$

where $(p_b\gamma_g/T_f)$ was identified by Lasater [1958] as a "correlating factor," which is a function of the mole fraction of gas in the reservoir oil, x_g, which is calculated as

$$x_g \approx (R_{sb}/379.3)/[(R_{sb}/379.3) + 350\gamma_o/M_o] , \quad\text{..} \text{(E.26)}$$

where,[17] if $^\circ API < 40$: $\quad M_o \approx 630 - 10(^\circ API)$, or .... (E.27a)

$\quad$ if $^\circ API > 40$: $\quad M_o \approx 73,110(^\circ API)^{-1.562}$, and .... (E.27b)

where,[18] if $x_g < 0.60$: $\quad (p_b\gamma_g/T_f) \approx 0.679\exp(2.786x_g) - 0.323$, or .... (E.28a)

$\quad$ if $x_g > 0.60$: $\quad (p_b\gamma_g/T_f) \approx 8.26x_g^{3.56} + 1.95$ . ... (E.28b)

- Vasquez [1976]

$$p_b \approx \{R_{sb}/[C_1\gamma_g \exp(C_3(^\circ API)/(T_f + 460))]\}^{1/C_2} , \quad\text{...} \text{(E.29)}$$

where	Coefficient	$^\circ API < 30$	$^\circ API > 30$
	C_1	27.64	56.06
	C_2	1.0937	1.187
	C_3	11.172	10.393

Reportedly [Beggs 1987], gas gravity was observed to be a strong correlating parameter in the above correlation. Gas gravity in the correlation was that which would result from two-stage separation—100 psig separator to stock tank. Gas gravity reported from other conditions may be corrected by

$$\gamma_{gc} \approx \gamma_g[1.0 + 5.912(E{-}5)(^\circ API)T_{sp} \log(p_{sp}/114.7)] , \quad\text{...} \text{(E.30)}$$

where γ_{gc} = corrected gas gravity and

$\quad \gamma_g$ = gas gravity reported at separator temperature and pressure, T_{sp} and p_{sp} ($^\circ F$ and psia), respectively.

- Glaso [1980]

$$\log p_b \approx 1.7669 + 1.7447\log p_b^* - 0.30218(\log p_b^*)^2 , \quad\text{.......................................} \text{(E.31a)}$$

$$\text{where} \quad p_b^* \approx (R_{sb}/\gamma_g)^{0.816}T_f^{0.172}/(^\circ API)^{0.989} . \quad\text{...} \text{(E.31b)}$$

---

[17] These fits to Lasater's [1958] Fig. 2 were reported by Beggs [1987]. The reader is cautioned that these two correlations are not congruent at 40°API. Thus, depending on circumstances, it may be appropriate to average the results of these correlations if they are applied to oils with stock-tank gravity "near" 40°API. Similar comments are made for other correlations—noted in subsequent sections—which depend on API gravity.

[18] These fits to Lasater's [1958] Fig. 1—reported by Beggs [1987]—differ slightly from those reported by Standing [1952].

- Al-Marhoun [1988]

$$p_b \approx 0.00538088 R_{sb}^{\,0.715082} \gamma_g^{\,-1.87784} \gamma_o^{\,3.14370} T_f^{\,1.32657} \text{ .} \dots\dots\dots\dots\dots\text{(E.32)}$$

- Dokla and Osman [1992][19]

$$p_b \approx 8363.86 \gamma_g^{\,-1.01049} \gamma_o^{\,0.107991} T_f^{\,-0.952584} R_{sb}^{\,0.724047} \text{ .} \dots\dots\dots\dots\text{(E.33)}$$

- Omar and Todd [1993]

$$p_b \approx 18.2\{\text{antilog}[0.00091 T_f - 0.0125(^{\circ}\text{API})][R_{sb}/\gamma_g]^x - 1.4\} \text{ ,} \dots\dots\dots\text{(E.34a)}$$

where $x \approx A_1 + A_2 B_{ob} + A_3 \gamma_g + A_4 B_{ob}^{\,2} + A_5 \gamma_g^{\,2} + A_6 / \gamma_g B_{ob}$, $\dots\dots\dots\dots$(E.34b)

and $A_1 = 1.4256 \qquad A_4 = 0.04481$
$A_2 = (-)0.2608 \qquad A_5 = 0.2360$
$A_3 = (-)0.4596 \qquad A_6 = (-)0.1077$

The Omar and Todd [1993] correlation is the same as the Standing [1947] correlation, Eq. E.24, except that the exponent "0.83" in the Standing [1947] correlation has been replaced with x, defined by Eq. E.34b.

- Kartoatmodjo and Schmidt [1994]

for $^{\circ}\text{API} < 30$,

$$p_b \approx \{R_{sb}/[0.05958 \gamma_g^{\,0.7972} \text{antilog}(13.1405(^{\circ}\text{API})/T)]\}^{0.9986} \text{ ,} \dots\dots\dots\text{(E.35a)}$$

for $^{\circ}\text{API} > 30$,

$$p_b \approx \{R_{sb}/[0.03150 \gamma_g^{\,0.7587} \text{antilog}(11.2895(^{\circ}\text{API})/T)]\}^{0.9143} \text{ .} \dots\dots\dots\text{(E.35b)}$$

Kartoatmodjo and Schmidt [1994] correlations are based on gas gravity, γ_g, resulting from a 100 psi first-stage separator. They reported a dependency between separator pressure and separator gas gravity similar to that reported by Beggs [1987]

$$\gamma_{gc} \approx \gamma_{gr}[1.0 + 0.1595(^{\circ}\text{API})^{0.4078} T_{sp}^{\,-0.2466} \log(p_{sp}/114.7)] \text{ ,} \dots\dots\dots\text{(E.35c)}$$

where γ_{gc} = corrected gas gravity, adjusted for 100 psi separator, used in Eq. E.35a and Eq E.35b, where it is denoted as γ_g;
γ_g = gas gravity reported at separator temperature and pressure, T_{sp} and p_{sp} ($^{\circ}\text{F}$ and psia), respectively.

---

[19] In a published discussion of this paper, Al-Yousef and Al-Marhoun [1993] observed that "the limited data...used are not adequate to obtain an empirical correlation, and the bubble-point correlation...is not reliable to predict the behavior of UAE crudes."

Selecting the Most Appropriate Correlation

In considering which of the foregoing might be the most appropriate bubblepoint correlation for a specific application, the source, number, and characteristics of crude oil samples for these correlations might be noted:

- Standing [1947]: California; 105 samples; no nitrogen or hydrogen sulfide, less than 1 mole % carbon dioxide; bubblepoint pressures from 30 to 7000 psia; tank oil gravities from 16.5 to 63.8 °API

- Lasater [1958]: Canada;[20] western and mid-continental U.S. and South America; 137 samples; essentially free of nonhydrocarbons; UOP characterization factor approximately 11.8; bubblepoint pressures from 48 to 5780 psia; tank oil gravities from 17.9 to 57.1 °API

- Vasquez [1976]: Worldwide; 600 samples; bubblepoint pressures from 15 to 6055 psia; tank oil gravities from 15.3 to 59.5 °API

- Glaso [1980]: North Sea; 26 samples; UOP characterization factor (four samples) average 11.89; bubblepoint pressures from 150 to 7127 psig; tank oil gravities from 31.7 to 42.9 °API

- Al-Marhoun [1988]: Middle East; 69 samples; bubblepoint pressures from 130 to 3573 psia; tank oil gravities from 19.4 to 44.6 °API

- Dokla and Osman [1992]: United Arab Republic; 51 samples; bubblepoint pressures from 590 to 4640 psig; tank-oil gravities from 28.2 to 40.3 °API. See, however, Al-Yousef and Al-Marhoun [1993]

- Omar and Todd [1993]: Malaysia offshore; 93 samples from 38 fields; bubblepoint pressures from 790 to 3851 psia; tank oil gravities from 26.6 to 53.2 °API.

- Kartoatmodjo and Schmidt [1994]: Indonesia, North America, Middle East, Latin America; 740 samples; bubblepoint pressures from 15 to 6055 psi; tank oil gravities from 14.4 to 58.9 °API; the database used by these authors apparently is an augmented version of that used by Vasquez [1976]

Reportedly [Ostermann et al. 1985], Lasater's [1958] correlation is "more accurate than Standing's correlation for high gravity crudes...many engineers prefer to use Standing's [1952] correlation for crudes of API gravity below 20°."

Sutton and Farshad [1990] compared lab-determined bubblepoints with those estimated using four correlations published (as of 1984) for 31 different crude oil systems from the U.S. Gulf Coast. For systems with bubblepoint pressures greater than 1000 psia, they observed average errors ranging from 1.6 to 18.7%. The Glaso [1980] correlation gave the best results (for systems with bubblepoint pressures greater than 1000 psia), with average errors ranging from 2.2 to 11.7%.

Effects of Paraffinicity and Nonhydrocarbons

Both the paraffinicity[21] of the oil and the presence of nonhydrocarbons—nitrogen and carbon dioxide—affect the bubblepoint pressure [Lasater 1958, Jacobson 1967, Glaso 1980, Ostermann et al. 1983]. Glaso [1980] developed correlations to account for these two factors, as discussed below.

---

[20] Asgarpour et al. [1989] published PVT correlations involving oils and gases only from western Canada, which, reportedly, are more accurate—for Canadian oils—than more generalized correlations reported prior to that time.

[21] The reader is referred to Buthod [1987] for a discussion regarding the paraffin content of crude oils.

The relationship between API gravity and viscosity of a gas-free (dead) crude oil, μ_{oa}, is dependent on the paraffinicity of the oil. Glaso [1980] developed a correlation between these two parameters for oils from four North Sea fields, which have a UOP characterization factor of 11.9

$$^\circ\text{API}^* \approx \text{antilog}[3.183(\text{E}{-}11)T_f^{\,3.444}\,(\mu_{oa})^b]\,, \quad\text{...(E.36a)}$$

where $b \approx 1/(10.213\log T_f - 36.447)$ . ...(E.36b)

The API gravity of stock tank oil from regions other than the North Sea can be adjusted to a comparable paraffinicity by substituting $^\circ\text{API}^*$, calculated from Eq. E.36a, in

$$^\circ\text{API}_{cor} = (^\circ\text{API}^*)(^\circ\text{API}_F)/(^\circ\text{API}_D), \quad\text{...(E.36c)}$$

where "$^\circ\text{API}_F$" and "$^\circ\text{API}_D$" are gravities of residual oil from flash and differential liberation, respectively. "$^\circ\text{API}_{cor}$" is the proper term to use in Eq. E.31b to calculate $p_b{}^*$.

Based on laboratory analyses of a limited number of samples, Glaso [1980] developed correction factors to account for the effects of nonhydrocarbons on the bubblepoint pressure estimated for North Sea reservoir oils. These correction factors were developed to allow using the North Sea correlations for reservoirs that contain nonhydrocarbon components, which are

- for carbon dioxide

$$F_{\text{CO2}} \approx 1.0 - 693.8\,y_{\text{CO2}}T_f^{-1.553}\,, \quad\text{..(E.37a)}$$

- for hydrogen sulfide

$$F_{\text{H2S}} \approx 1.0 - [0.9035 + 0.0015(^\circ\text{API})]y_{\text{H2S}} + 0.019(45 - {^\circ\text{API}})(y_{\text{H2S}})^2\,, \quad\text{..................(E.37b)}$$

- for nitrogen

$$F_{\text{N2}} \approx 1.0 + [(-2.65(\text{E}{-}4)(^\circ\text{API}) + 5.5(\text{E}{-}3)T_f + 0.0931(^\circ\text{API})^{-0.8295}]y_{\text{N2}}$$
$$+ [1.954(\text{E}{-}11)(^\circ\text{API})^{4.699}T_f + 0.027(^\circ\text{API}) - 2.366][y_{\text{N2}}]^2\,, \quad\text{...................(E.37c)}$$

where y_{CO2}, y_{H2S}, and y_{N2} are mole fraction carbon dioxide, hydrogen sulfide, and nitrogen, respectively, in the total surface gas.

Apparently, Glaso [1980] investigated the influence of each nonhydrocarbon component separately. The influence of combinations of two or more nonhydrocarbons was not investigated. It seems unlikely that the presence of two or more nonhydrocarbons in a sample would result in a correction that was a simple multiplication of the individual correction factors.

Regarding the effects of carbon dioxide on bubblepoint pressure, Glaso [1980] observed that his experimental results differed from those of Lasater [1958]. The Glaso [1980] correlation indicates carbon dioxide causes a reduction in the bubblepoint pressure; Lasater [1958] reported an increase.

Regarding the effects of nitrogen, Glaso observed that his results differed significantly from those of Jacobson [1967].[22] Also, the Glaso [1980] correlation indicates a dependency on reservoir temperature and stock-tank gravity—factors apparently not investigated by Jacobson [1967].

In view of these discrepancies, for a significant crude oil accumulation with more than about 5% non-hydrocarbons in the separator gas, a laboratory PVT analysis is recommended.

For volatile oils, Glaso [1980] reported that a better correlation of bubblepoint pressure could be achieved by adjusting the exponent on T_f in Eq. E.31b from 0.172 to 0.130. This correction, however, apparently was based on only four samples, and PVT analysis is recommended for any volatile oil accumulation of significant size.

E.3.2 Solution Gas/Oil Ratio at the Bubblepoint

Correlations to estimate the solution gas/oil ratio at the bubblepoint have been developed by the same authors cited in Sec. E.3.1. The correlation charts and/or regression equations developed by these authors to estimate bubblepoint pressure may be solved in reverse order to estimate the solution GOR for a given bubblepoint pressure. Regressions published by many of these authors follow. Units for T_f are as noted in Sec. E.3.1, unless noted otherwise in this section

- Standing [1977]

$$R_{sb} \approx \gamma_g [p_b / 18 \text{antilog}(y_g)]^{1.204} , \qquad \text{(E.38a)}$$

where $y_g \approx 0.00091 T_f - 0.0125(^\circ API)$ and $T_f = ^\circ F$. $\qquad$ (E.38b)

- Lasater [1958][23]

$$R_{sb} \approx 132,755 \gamma_o y_g / [M_o(1 - y_g)] , \qquad \text{(E.39a)}$$

where   if $p_b \gamma_g / T_f < 3.29$, $y_g \approx 0.359 \ln(1.473 p_b \gamma_g / T_f + 0.476)$, $\qquad$ (E.39b)

if $p_b \gamma_g / T_f > 3.29$, $y_g \approx (0.121 p_b \gamma_g / T_f - 0.236)^{0.281}$. $\qquad$ (E.39c)

- Glaso [1980] [24]

$$R_{sb} \approx \gamma_g [p_b^* (^\circ API)^{0.989} / T_f^{0.172}]^{1.2255} , \qquad \text{(E.40a)}$$

where $p_b^* \approx \text{antilog}[2.8869 - (14.1811 - 3.3093 \log p_b)^{0.5}]$. $\qquad$ (E.40b)

- Vasquez and Beggs [1980]

$$R_{sb} \approx C_1 \gamma_g p_b^{C_2} \exp[C_3 (^\circ API)/(T_f + 460)] , \qquad \text{(E.41)}$$

---

[22] These comments by Glaso seem to conflict with those by Ostermann *et al.* [1983] who observed that "bubble point pressures...(in nitrogen rich crudes)...corrected for nitrogen content by both Jacobson's [1967] and Glaso's [1980] correlations...(agreed)...reasonably well."

[23] As reported by Beggs [1987].

[24] As reported by Sutton and Farshad [1984].

where the coefficients (C_1, C_2, and C_3) are the same as those used in Eq. E.29.

- Obomanu and Okpobiri [1987][25]

$$R_{sb} \approx [0.03008 p_b^{0.927} \gamma_g^{2.15} (^{\circ}\text{API})^{1.27}]/[6.4714(1.8 T_f - 460)^{0.497}] . \quad \text{............................(E.42)}$$

- Kartoatmodjo and Schmidt [1994]

for $^{\circ}\text{API} < 30$, $R_{sb} \approx 0.05985 \gamma_g^{0.7972} p_b^{1.0014} \text{antilog}[13.1405(^{\circ}\text{API})/T_f]$, ...(E.43a)

for $^{\circ}\text{API} > 30$, $R_{sb} \approx 0.03150 \gamma_g^{0.7587} p_b^{1.0937} \text{antilog}[11.2895(^{\circ}\text{API})/T_f]$ . ...(E.43b)

E.3.3 Formation Volume Factor

Correlations have been developed to estimate initial oil formation volume factor for two general cases: (1) reservoir oil initially saturated—i.e., initial reservoir pressure at bubblepoint pressure—and (2) reservoir oil initially undersaturated—i.e., initial reservoir pressure greater than bubblepoint pressure.

Reservoir Oils Initially at Bubblepoint Pressure

Correlations to estimate formation volume factors for oils initially at bubblepoint pressure have been developed by the same authors cited previously. Standing's [1947] correlation is presented in **Fig. E.14.**

Regression analyses developed by most of the authors cited in Sec. E.3.1 and E.3.2 follow. Units are as noted in Sec. E.3.1, unless otherwise noted here

- Standing [1977]

$$B_{ob} \approx 0.9759 + 0.00012[R_{sb}(\gamma_g/\gamma_o)^{0.5} + 1.25\ T_f]^{1.2} . \quad \text{...(E.44)}$$

- Vasquez and Beggs [1980]

$$B_{ob} \approx 1 + C_1 R_{sb} + C_2(T_f - 60)(^{\circ}\text{API})/\gamma_g + C_3 R_{sb}(T_f - 60)(^{\circ}\text{API})/\gamma_g , \quad \text{....................(E.45)}$$

where

Coefficient	$^{\circ}\text{API} < 30$	$^{\circ}\text{API} > 30$
C_1	4.677E-4	4.670E-4
C_2	1.751E-5	1.100E-5
C_3	(−)1.811E-8	1.337E-9

- Glaso [1980]

$$\log(B_{ob} - 1) \approx (-)6.58511 + 2.91329 \log B^*_{ob} - 0.27683(\log B^*_{ob})^2 , \quad \text{......................(E.46a)}$$

where $B^*_{ob} \approx R_{sb}(\gamma_g/\gamma_o)^{0.526} + 0.968 T_f$ . ...(E.46b)

---

[25] Database included 100 reservoirs in Nigeria. Units are kPa and $^{\circ}$K.

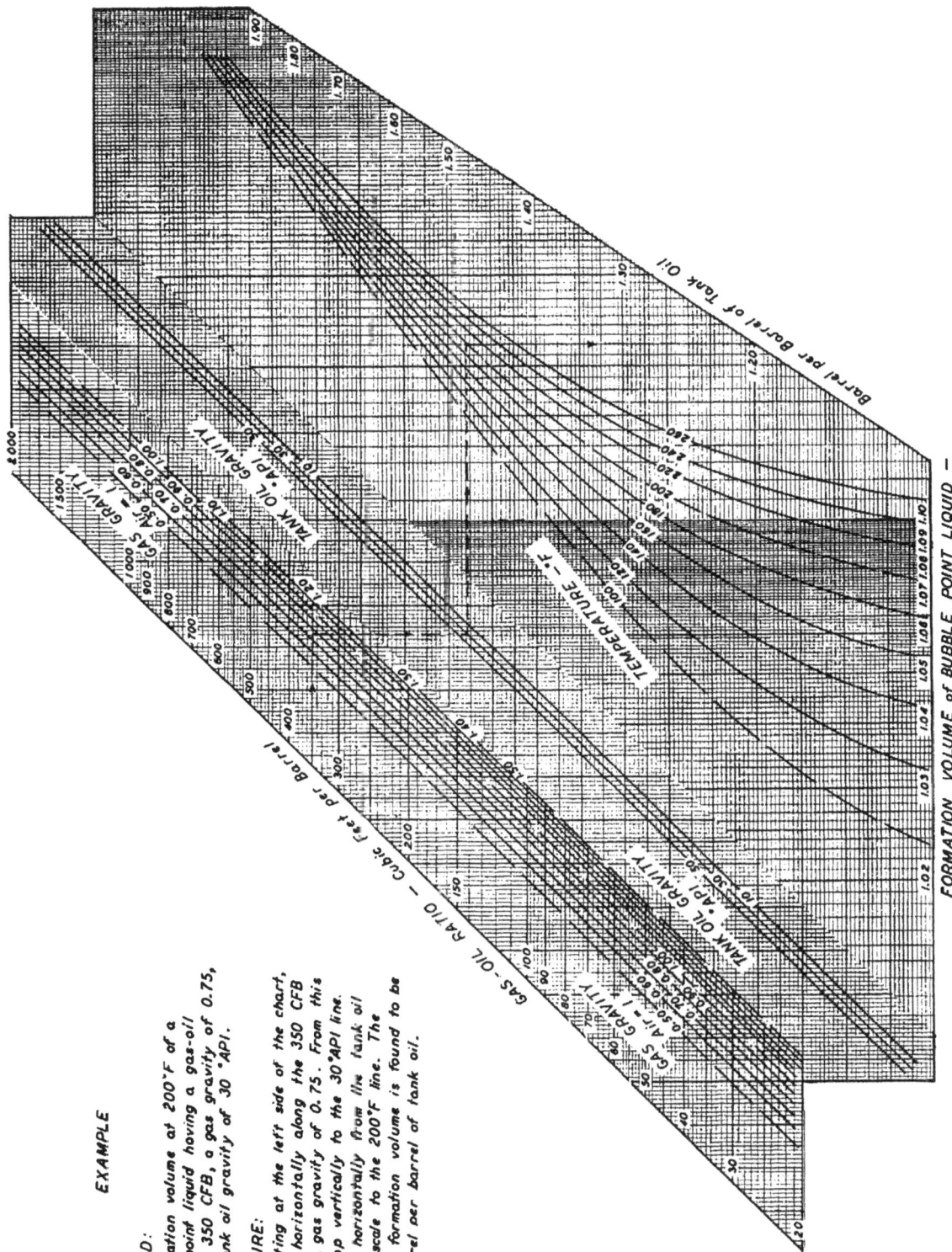

Fig. E.14—Formation volume factor, B_o, vs. solution GOR, R_s, gas gravity, γ_g, gravity of stock-tank oil, °API, and formation temperature, T_f [after Standing 1947].

- Al-Marhoun [1988]

$$B_{ob} \approx 0.497069 + 0.000862963 T_f + 0.00182594 F + 0.00000318099 F^2 , \quad \ldots\ldots\ldots (E.47a)$$

where $F \approx R_{sb}^{\ 0.742390} \gamma_g^{\ 0.323294} \gamma_o^{\ -1.202040}$ $\ldots\ldots\ldots\ldots\ldots\ldots\ldots\ldots\ldots\ldots\ldots\ldots\ldots\ldots\ldots\ldots\ldots (E.47b)$

- Obomanu and Okpobiri [1987]

for γ_o between 0.811 and 0.876,

$$B_{ob} \approx 0.3321 + 0.000788374 R_{sb} + 0.002335 R_{sb}\gamma_g/\gamma_o + 0.0020855 T_f , \quad \ldots\ldots\ldots (E.48a)$$

for γ_o between 0.876 and 0.966,

$$B_{ob} \approx 1.0232 + 0.0001065 (R_{sb}\gamma_g/\gamma_o + 1.8 T_f - 460)^{0.79} . \quad \ldots\ldots\ldots\ldots\ldots\ldots\ldots (E.48b)$$

- Dokla and Osman [1992]

$$B_{ob} \approx 0.431935(E{-}1) + 0.156667(E{-}2) T_f + 0.139775(E{-}2) M$$
$$+ \ 0.380525(E{-}5) M^2 , \quad \ldots\ldots\ldots\ldots\ldots\ldots\ldots\ldots\ldots\ldots\ldots\ldots\ldots\ldots\ldots (E.49a)$$

where $M \approx R_{sb}^{\ 0.773572} \gamma_g^{\ 0.404020} \gamma_o^{\ -0.882605} . \quad \ldots\ldots\ldots\ldots\ldots\ldots\ldots\ldots\ldots\ldots (E.49b)$

- Omar and Todd [1993]

$$B_{ob} \approx 0.972 + 0.000147 [R_{sb}(\gamma_g/\gamma_o)^{0.5} + 1.25\ T_f]^x , \quad \ldots\ldots\ldots\ldots\ldots\ldots\ldots (E.50a)$$

where $x \approx 1.1663 + 0.000762(^\circ API)/\gamma_g - 0.0399\gamma_g . \quad \ldots\ldots\ldots\ldots\ldots\ldots\ldots (E.50b)$

- Kartoatmodjo and Schmidt [1994]

$$B_{ob} \approx 0.98486 + 0.0001(R_{sb}^{\ 0.755} \gamma_g^{\ 0.25} \gamma_o^{\ -1.50} + 0.45 T_f) . \quad \ldots\ldots\ldots\ldots\ldots\ldots (E.51)$$

With reference to the discussion in Sec. 3.3.2, Fluid Data, correlations to estimate bubblepoint pressure and formation volume factor should be selected from the same data set and used together. If the bubblepoint pressure estimated from the selected correlation is equal to, or less than, estimated reservoir pressure, then the formation volume factor from the same data set may be used with reasonable confidence. If, on the other hand, the bubblepoint pressure estimated from the selected correlation is significantly greater than the estimated reservoir pressure, the presence of free gas should be suspected. Under these conditions, the formation volume factor correlation cannot provide any useful information.

Reservoir Oils Initially Above Bubblepoint Pressure

Estimation of the formation volume factor for reservoir oils that are undersaturated at initial conditions requires:

1) an estimate of bubblepoint pressure at initial conditions, p_b,

2) an estimate of the formation volume factor at bubblepoint pressure, B_{ob}, and

3) an estimate of the *average* compressibility, c_o, of the reservoir oil between bubblepoint pressure and initial reservoir pressure.

With this information, the initial formation volume factor, B_{oi}, may be calculated as

$$B_{oi} \approx B_{ob} \exp[c_o(p_b - p_i)] \dotfill \text{(E.52)}$$

Correlations to estimate p_b and B_{ob} were discussed in previous paragraphs. A method is needed to estimate c_o.

A correlation by Trube [1957] has been used for many years to estimate the compressibility of oil above the bubblepoint. To determine the parameters for this correlation, however, requires using three graphs. Regression equations for these graphs are not available in the public domain. The following correlation by Vasquez [1976][26]—based on 600 oil systems—is considered the best available [McCain 1990]

$$c_o \approx (5.0R_{sb} + 17.2T_f - 1180\gamma_g + 12.61(^{\circ}API) - 1433)/(E+5)p \dotfill \text{(E.53)}$$

Reportedly [McCain 1990], compressibilities estimated with this correlation "are generally low, by as much as 50% at high pressures. Accuracy is improved at pressures near the bubblepoint."

Subsequent to the Vasquez [1976] paper, Kartoatmodjo and Schmidt [1994] published a correlation for c_o, where

$$c_o \approx 6.8257R_{sb}^{0.5002}\gamma_g^{0.35505}T_f^{0.76606}(^{\circ}API)^{0.3613}(E-6)/p \dotfill \text{(E.54)}$$

Depending on the specific case, oil compressibilities estimated with each of these two correlations may differ substantially. In a limited number of cases investigated here, calculated compressibilities differed by a factor of almost 2:1.

McCain *et al.* [1988] published correlations to estimate the compressibility of black oils at reservoir pressures *below* the bubblepoint pressure. Some readers may be tempted to use these correlations to estimate oil compressibility *at* the bubblepoint pressure. This is inadvisable, as a sharp increase in compressibility is observed when a phase boundary is crossed—i.e., the boundary between a single phase (undersaturated) oil and the two-phase region containing saturated oil and free gas. Thus, correlations of data in the two-phase region, like that of McCain *et al.* [1988], cannot reasonably be expected to be representative of behavior in the single-phase region—i.e., at the bubblepoint.

E.4 Interstitial Water

The volume of dissolved natural gas and the formation volume factor for interstitial water—both usually ignored in material balance calculations—may become significant in calculations involving geopressured reservoirs.

---

[26] Units for formation temperature, T_f, are $^{\circ}F$.

Regarding natural gas dissolved in interstitial water, McCain [1990] has fitted the graphical correlations of Culberson and McKetta [1951], which were for methane dissolved in *fresh* water (the units for formation temperature are °F)

$$R_{sw}^* \approx A + Bp + Cp^2 , \quad\text{...(E.55a)}$$

where R_{sw}^* = solubility of methane in *fresh* water (scf/STB),

and $A \approx 8.15893 - 6.1226(E-2)T_f + 1.91663(E-4)T_f^2 - 2.1654(E-7)T_f^3 ,$ ....(E.55b)

$B \approx 1.0121(E-2) - 7.44241(E-5)T_f + 3.05553(E-7)T_f^2 - 2.94883(E-10)T_f^3 ,$ ..(E.55c)

$C \approx [-9.02505 + 0.130237T_f - 8.53425(E-4)T_f^2 + 2.34122(E-6)T_f^3$

$\qquad - 2.37049(E-9)T_f^4][E-7] \cdot$ ....(E.55d)

For *brine*, a correction to the solubility estimated from Eq. E.55a is made by

$$\log(R_{sw}/R_{sw}^*) \approx (-)0.0840655 S T_f^{-0.285854} , \quad\text{...(E.56)}$$

where S = weight % solids.

The formation volume factor of reservoir brine may be estimated by [McCain 1990]

$$B_w = (1 + dV_{wp})(1 + dV_{wT}) , \quad\text{...(E.57a)}$$

where $dV_{wp} \approx (-)1.95301(E-9)pT_f - 1.72834(E-13)p^2T_f$

$\qquad - 3.58922(E-7)p - 2.25341(E-10)p^2 ,$ ....(E.57b)

$dV_{wT} \approx (-)1.0001(E-2) + 1.3339(E-4)T_f + 5.50654(E-7)T_f^2 ,$ ....(E.57c)

where dV_{wp} = change in brine volume due to pressure,

dV_{wT} = change in brine volume due to temperature.

APPENDIX F

RESERVES ATTRIBUTABLE TO IMPROVED OIL RECOVERY METHODS

F.1 Introduction

Brief comments are provided here regarding estimation and classification of reserves attributable to improved oil recovery methods. In addition, summary data are provided for waterflood and steam injection projects—the principal methods in use today—as guidelines for such estimation. This material is intended to be a management overview; a complete treatment is beyond the scope of this work.

There are hundreds of articles on estimating reserves attributable to improved recovery methods. In addition, there are numerous books on the topic, including those by Craig [1971], Interstate Oil Compact Commission [1974], Smith [1975], Prats, [1982], Stalkup [1983], Klins [1984], National Petroleum Council [1984], Wilhite [1986], and Green and Wilhite [1998]. The National Petroleum Council's 1984 report includes an extensive bibliography that is organized by type of improved recovery process.

Waterflooding is the principal method for improved recovery of light oils. Steam injection is the principal method for heavy oils. In certain geologic, reservoir, and operational settings, carbon dioxide injection, hydrocarbon miscible flooding and nitrogen injection make significant contributions to domestic reserves; but such methods make small contribution to global reserves. Other methods, including in-situ combustion, polymer flooding, and microbial flooding, make even smaller contributions.

Waterflooding generally is effective in reservoirs with medium to light crude oils—those with stock tank gravities greater than about 20 °API. Steam injection generally is effective in reservoirs with heavy crude oils—those with stock tank gravities less than about 20 °API. Both methods are adversely affected by reservoir heterogeneities and poor operating practices.

F.2 SPE-WPC [1997] Reserves Definitions

SPE-WPC [1997] defines improved recovery as including "all methods for supplementing natural energy or altering natural forces in the reservoir[1] to increase ultimate recovery." Such methods include: "pressure maintenance, cycling, waterflooding, thermal methods, chemical flooding, and the use of miscible and immiscible displacement fluids." As used in SPE-WPC [1997] reserves definitions, the term "improved recovery" includes methods that heretofore

---

[1] In the context of these definitions, installation of artificial lift or stimulation are not considered improved recovery methods, contrary to interpretations elsewhere.

have been classified as "secondary," "tertiary," and "enhanced." The term "improved recovery" is preferred here, as there is no implication regarding sequence of application, as there is with terms like "secondary" and "tertiary."

SPE-WPC [1997] reserves definitions state (in part):

a) "Reserves which are to be produced through the application of *established*,#* improved recovery methods are included in the *proved** classification when: (1) successful testing by a pilot project or favorable response of an installed program in the same or an analogous reservoir with similar rock and fluid properties provides support for the analysis on which the project was based, and, (2) it is *reasonably certain that the project will proceed.*"*

b) "*probable** reserves may include...reserves attributable to improved recovery methods that have been established by *repeated commercially successful*,#* applications when (a) a *project or pilot is planned** but not in operation and (b) rock, fluid, and reservoir characteristics appear favorable for commercial application."

c) "*possible** reserves may include...reserves attributable to improved recovery methods when (a) a *project or pilot is planned** but not in operation and (b) rock, fluid and reservoir characteristics are such that a *reasonable doubt exists if the project will be commercial.*"*

"Established" is interpreted here to mean "repeatedly commercially successful."
* Author's italics

The foregoing definitions are somewhat ambivalent and do not cover several contingencies. Accordingly, the following additional guidelines are offered:

a) Reserves attributable to an improved recovery method may be classified as *proved* if:

-production response for a pilot or installed program in subject reservoir provides support for the analysis on which the project was based and the current economic scenario is favorable for continued operation and/or expansion, or if

-the method has been repeatedly commercially successful in reservoirs analogous to subject reservoir, a project has been budgeted[2] for subject reservoir, and regulatory approval is reasonably certain.

b) Reserves attributable to an improved recovery method may be classified as *probable* if:

-production response for a pilot or installed program in subject reservoir, although positive, does not provide definitive support for the analysis on which the project was based, but operations are going forward, pending further analysis and/or implementation of cost-saving measures, or if

-the method has been repeatedly commercially successful in reservoirs analogous to subject reservoir, a project is anticipated, and the commercial, operating, and/or regulatory scenario are conducive to such a project.

---

[2] For nonoperating interests, reserves generally should not be classified as proved until the operator has received monetary approval for the project from sufficient working-interest owners.

c) Reserves attributable to an improved recovery method may be classified as *possible* if:

  -production response for a pilot or installed program in subject reservoir is marginally commercial and generally not supportive of the analysis on which the project was based, but, operations are going forward, pending further analysis and/or implementation of cost-saving measures, or

  -even though the method has been repeatedly commercially successful in reservoirs analogous to subject reservoir, there are significant uncertainties regarding the commercial, operating, and/or regulatory scenario in subject reservoir for application of the method.

F.3 Waterflooding

Numerous analytical methods have been developed to predict waterflood performance and estimate additional oil reserves attributable to waterflooding. Most of these methods have been discussed briefly and compared by Craig [1971] and by Wilhite [1986] However, empirical methods often yield better results [Bush and Helander 1968, Guerrero and Earlougher 1962, Khan 1971, Wayland *et al.* 1971]

Numerical simulation is widely used to estimate waterflood reserves from larger reservoirs. Ideally, a history match should be made for a period of primary production before making a prediction of waterflood response.[3] However, in some areas—the North Sea, for example—prediction of incremental recovery under pressure maintenance by water injection may have to be made before primary production has commenced. In either situation, calculation of rate-time profiles and reserves are subject to uncertainty due to the industry's inability to quantify the spatial distribution of reservoir heterogeneities that influence response to water injection. Such heterogeneities include the degree of interwell communication, the degree and influence of permeability stratification, variations in water-oil relative permeability, and relative importance of gravity/viscous forces.

No matter what method is used to estimate waterflood reserves—analytical or numerical—it is good practice to check the estimate against observed response of waterfloods in analogous reservoirs in the area. In "frontier" operations, where there may not be analogous reservoirs in the area, it is advisable to check the estimate against observed response of reservoirs with similar depositional environment, fluid transmissibility (kh/μ), well spacing, and stage of depletion when flooding was initiated.

Primary and waterflood recovery efficiencies for several hundred reservoirs in the U.S. were averaged by lithology, state, and injection method by an API Subcommittee on Recovery Efficiency [API 1984], **Table F.1.**

One of the parameters shown in this table, the ratio of incremental waterflood recovery to primary recovery—the "Wf/Pr ratio"—is a rule of thumb used by some engineers to make preliminary estimates of waterflood reserves. As shown by Table F.1, however, this ratio can vary widely. Values less than about 0.5 may be attributed to heterogeneity and/or reservoir compartmentalization. Values greater than about 0.5 may be attributed to (initially) undersaturated oils and/or infill drilling, the latter being especially prevalent in West Texas' Permian Basin.

---

[3] Many of the factors that influence waterflood performance may not become apparent during primary production. Thus, even the best history match of a period of primary production is no assurance that waterflood performance can be predicted accurately.

TABLE F.1—PRIMARY AND WATERFLOOD RECOVERY EFFICIENCY FOR 230 U.S. RESERVOIRS*

Lithology	State	OIP STB/NAF	Recovery Efficiency Primary	Recovery Efficiency W'flood	Ratio Wf/Pr	Injection Method
Sandstone	California	1311	.265	.088	.332	Pattern
	Louisiana	1194	.365	.147	.403	Pattern
	Louisiana	1181	.413	.138	.334	Edge
	Oklahoma	728	.170	.106	.624	Pattern
	Texas	942	.256	.128	.500	Pattern
	Texas	897	.340	.216	.635	Edge
	Wyoming	774	.236	.211	.894	Pattern
Carbonate	Texas	388	.155	.163	1.05#	Pattern

* After API [1984]
# Please see comment in text.

The industry has extensive experience in waterflooding light oil. The method is considered here to be an improved oil recovery method that has been "repeatedly commercially successful." For practically any accumulation being considered for this method, there is an "analogous reservoir" where the method has been applied successfully. Accordingly, *provided the conditions outlined in Sec. F.2 are met*, it seems reasonable to assign proved incremental waterflood reserves based on analogy and/or volumetric analysis to any light oil reservoir, irrespective of whether there is an installed pilot or ongoing project.[4] In settings where there is economic, geologic, engineering, operating and/or regulatory uncertainty, such reserves might be classified as unproved, depending on circumstances.

F.4 Steam Injection

Steam injection includes both steamsoak (alternate injection and production cycles in single wells) and continuous steam injection in pattern operations—also called steamdrive. Although analytical models have been developed to predict heat flow in subsurface formations, most of the methods used to predict response to steam injection are semi-empirical and are based on physical model experiments and numerical simulation. A mathematical model by Myhill and Stegemeier [1978], however, apparently has been used with success.[5]

Reportedly [NPC 1984], a steamflood prediction method developed by Gomaa [1980] provides the best fit to observed performance of several large steamfloods for reservoirs with less than about 10° dip. As reservoir dip increases up to about 45°, the Gomaa methodology becomes increasingly more pessimistic, with actual results from reservoirs with about 45° dip being approximately twice what would be expected using the Gomaa method [NPC 1984].

Table F.2, compiled from data published by Farouq Ali and Meldau [1979], Prats [1982], and Matthews [1984], may be helpful in making *preliminary* estimates of steamflood reserves.

---

[4] One of the most critical factors in waterflood operations is injectivity sufficient to achieve timely fillup and displacement of oil to producing wells. This factor is not specifically mentioned in SPE-WPC [1997] guidelines and, although it might be inferred by reference to "analogous reservoirs," is frequently overlooked.

[5] W.C. Miller, personal communication, 2000.

TABLE F.2—SUCCESSFUL STEAMFLOODS

Field (Operator)[a]	Depth (ft)	Net/Grs Pay (ft)	Dip (deg)	°API	Visc. (cp)	Poros. (fr)	Perm. (md)	Oil/Stm Ratio	RecEff (fr OIP)[b]
Duri[e] (Caltex)	500	150/?		20	330	.34	1,500	.27	.60
Kern River (Getty)	900	70/?	3	14	4,000	.35	3,500	.25	.70[c]
Midway-Sunset (Chevron)	1300	260/350	10	14	1,500	.27	520	.16	.65
Mt Poso (Shell)	1800	60/75	6	16	280	.33	15,000	.21	.65[d]
South Belridge (Mobil)	1100	91/210	7	13	1,600	.35	3,000	.28	.60
Tia Juana (Maraven)	1600	125/250	4	12	5,000	.38	2,800	.34	.45
Winkleman Dome (Amoco)	1200	73/?	?	14	900	.25	600	.20	.50
Yorba Linda (Shell)	650	325/?	12	14	6,400	.30	600	.49	.55[d]

a. Due to industry consolidation, some operators may have changed since publication of cited sources.
b. More recent data suggest recovery efficiencies on the order of 0.8 for some California projects.
 Operators of some of these projects report operating costs of $4-5/bbl [Moritis 1998].
c. More recent data suggest recovery efficiency of 0.8 for this project.
d. More recent data suggest recovery efficiency of 0.7 for this project.
e. Gael, et al [1995].

There does not appear to be a statistically valid correlation between the rock-fluid parameters in this table and the indicated recovery efficiency. Prats [1982], Chu [1985], and Blevins [1990] also have published summaries on the state of the art in steamflooding.

In 1998, there were over 180 steam injection projects operating in over 7 countries, including the U..S., Venezuela, Trinidad, Indonesia, China, Germany, and Canada. Almost all were reported by the operator as being "successful" [Moritis 1998].[6]

Clearly, the industry has extensive experience in steam-flooding heavy oil. The method is considered here to be an improved oil recovery method that has been "repeatedly commercially successful." For practically any accumulation being considered for this method, there is an "analogous reservoir" where the method has been applied successfully. Accordingly, *provided the conditions outlined in Sec. F.2 are met*, it seems reasonable to assign proved incremental steam-flood reserves based on analogy and/or volumetric analysis to any heavy oil reservoir, irrespective of whether there is an installed pilot or ongoing project. In settings where there is significant economic, geologic, engineering, operating, and/or regulatory uncertainty, such reserves might be classified as probable or possible, depending on circumstances.

F.5 Other Methods

Other methods for improved oil recovery that have been, or are being, used globally include: (a) carbon dioxide flooding, both miscible and immiscible, (b) polymer flooding, (c) hydrocarbon miscible flooding, (d) nitrogen flooding, both miscible and immiscible, (e) in-situ combustion, (f) micellar-polymer, and (g) microbial flooding. [Moritis 1998, 2000]. Except for special circumstances, these methods are not considered here to be "repeatedly commercially successful." There are no reliable guidelines available for estimating reserves from these methods prior to project initiation. Each project must be evaluated on its merits.

---

[6] Moritis [2000] provided the most recent summary of worldwide IOR activity.

Accordingly, reserves attributable to such methods should be classified as "possible," pending a favorable production response from the project.

F.5.1 Carbon Dioxide Flooding

Most of the carbon dioxide flooding projects underway during 1998 were in the Permian Basin (West Texas). Almost all of these have been implemented in ongoing waterfloods in carbonate reservoirs, with the object of reducing the residual oil saturation that would be left after waterflooding. In some of these projects, operators report expecting to reduce residual oil saturation *in the CO_2 swept volume* to as low as 1 to 10% [Moritis 1998]. In some cases, data reported by the operators suggest that as much as 40% of the residual oil left by waterflooding might be recovered by carbon dioxide miscible flooding.

In addition to the carbon dioxide floods in ongoing carbonate waterfloods in West Texas, there are several post-waterflood projects being conducted in deep, low permeability sandstone reservoirs in southwestern Mississippi. Of particular interest is the Little Creek field. (Background information on this field is presented in Sec 7.3.) From data reported by the operator to the Mississippi Oil and Gas Board, it seems apparent that substantial volumes of oil left at waterflood (immobile) residual saturation have been reconnected, banked up, and displaced to producing wells by the injected carbon dioxide. Through 1997, during which commercial operations were still ongoing, an additional 8.1 million bbl oil had been produced. This is an increment of 17% of cumulative recovery by primary plus waterflood and is approximately 15% of the oil remaining in place at the end of waterflood operations.

Unlike with other injectants, the proximity of an adequate source of CO_2 and the necessary delivery infrastructure are paramount considerations.

F.5.2 Polymer Flooding

Polymer flooding generally is most effective in moderately heterogeneous reservoirs containing light oils or in reservoirs containing moderately viscous oils—less than about 100 cp. [Chang 1978, NPC 1984]. Incremental recovery attributable to polymer flooding reportedly is in the range of 1.6 to 9.7% of OIP [Chang 1978, Mack and Duvall 1984]. Lower incremental recoveries generally are associated with projects implemented in waterfloods approaching the economic limit; higher recoveries, with projects implemented early in reservoir life.[7]

Special computer programs have been developed to predict response to polymer flooding; see, for example, Mack and Warren [1984]. Many of the key parameters required for these programs, however, are difficult to determine reliably and are subject to considerable uncertainty without substantial polymer flood experience in the formation of interest. Accordingly, the predictions from computer modeling should be checked against actual results in analogous reservoirs. Comparable performance should be expected in accumulations with comparable depositional environment, Dykstra-Parsons [1950] V factor, clay content, oil viscosity, and producing WOR at project initiation.

The results of many polymer flood projects have been published [Chang 1978, Mack and Duvall 1984]. Other polymer flood projects have been identified by the NPC [1984].

---

[7] Personal communication, Ben Sloat, 1988.

F.5.3 Hydrocarbon Miscible Flooding

Most of the commercially successful hydrocarbon miscible flood projects that have been reported are in Alberta and in South Louisiana [Moritis 1998]. Operators of projects in Alberta report incremental recovery attributable to hydrocarbon miscible flooding of from 13 to 44% of OIP [ERCB 1987]. Estimated ultimate recovery in most of these projects ranges from 60 to over 80% of OIP [ERCB 1987]. The most successful of these projects in Alberta are associated with high relief carbonate "reefs," where gravity segregation results in volumetric sweep efficiencies approaching 100%. Details of such operations in the Golden Spike pool and the Rainbow Keg River F pool have been provided by Reitzel and Callow [1977] and Fong *et al.* [2001], respectively.

In one case, the operator reported conducting an enriched gas flood in a low dipping stratigraphic trap in an area waterflooded to near residual oil saturation. From computer simulation studies of the project area, it was estimated that incremental oil attributable to hydrocarbon miscible flooding would be 11.5% of OIP [Frimodig *et al.* 1988].

ESTIMATION OF DOWNDIP, UPDIP, AND LATERAL LIMITS AND CLASSIFICATION OF RESERVES IN AREAS BEYOND DIRECT WELL CONTROL IN NEWLY DISCOVERED RESERVOIRS

G.1 Introduction

Reserve definitions in the public domain provide few specifics for estimating the downdip, updip, and lateral limits for various classifications and categories of reserves if such limits are beyond direct well control. In a fully developed reservoir, this is a minor problem. In newly discovered reservoirs, however, lack of guidelines may lead to inconsistencies in estimation, classification, and categorization[1] of reserves. The objectives of this appendix are to:

a) summarize certain industry practices regarding estimating such limits,

b) discuss engineering approaches to estimating updip and downdip limits, and

c) recommend guidelines for estimating reservoir limits and classifying and categorizing reserves in areas beyond direct well control.

G.2 SPE-WPC [1997] Reserves Definitions

Regarding "proved" reserves, SPE-WPC [1997] definitions state:

"In the absence of data on fluid contacts, the lowest known[2] structural occurrence of hydrocarbons controls the proved limit unless otherwise indicated by definitive geological, engineering or performance data."

"The area of a reservoir considered proved includes...the undrilled portions of the reservoir that can reasonably be judged as commercially productive on the basis of available geological and engineering data."

---

[1] "Classification and categorization" refers to whether reserves are proved (drilled), proved (undrilled), probable (undrilled), or possible (undrilled).

[2] As observed in Appendix B, the term "known" historically has been interpreted to mean *direct* evidence of hydrocarbons, for example, flow tests, cores, or logs. *Indirect* evidence, for example, bottomhole pressure gradients, capillary pressure data, or seismic indicators, although not widely used historically, is becoming more generally accepted.

Regarding "probable" reserves, these definitions state:

"probable reserves may include...reserves anticipated to be proved by normal stepout[3] drilling where subsurface control is inadequate to classify these reserves as proved."

Regarding "possible" reserves, these definitions state:

"possible reserves may include...reserves which, based on geological interpretations, could possibly exist beyond areas classified as probable."

Regarding lateral and updip limits for assigning and classifying reserves for undrilled areas, SPE-WPC [1997] definitions make no reference to such factors as:

- geologic setting,
- fluid type and saturation pressure,
- drive mechanism, or
- well drainage area.

Yet, factors such as these may have an important bearing on the estimation and classification of reserves, especially for areas beyond direct well control. In the context of geologic setting and well drainage area, it is noted that mining industry definitions (summarized in Sec. 1.3.15) infer, for proved ore reserves, "drill hole locations...spaced closely enough to confirm continuity."

The geologic setting may be the most important factor to consider in estimating, classifying, and categorizing reserves of oil and/or gas. For accumulations characterized by a high degree of faulting, for example, piercement salt domes or overthrust structures, it may be prudent to limit proved reserves to areas in close proximity to, and/or within the drainage area of, wells demonstrated to be commercially productive. Comparable practice might be prudent for accumulations characterized by complex stratigraphy, for example, fluvial-deltaic settings. However, for many accumulations in deep marine sediments, turbidites for example, continuity has been observed over substantial areas.

G.3 Industry Practices

Industry practices regarding downdip, updip, and lateral limits for various classifications of reserves[4] reflect considerable variation, as indicated here:[5]

Entity[6]	Limit	Limits* for Various Classifications of Crude Oil Reserves		
		1P**	2P**	3P**
"A"	Updip	- - - - - - - - - - - - - - - Not stated - - - - - - - - - - - - - - - -		
	Downdip	LKO	LKO+50 ft	LKO+100 ft
	Lateral	- - - - - - - - - - - - - - - Not stated - - - - - - - - - - - - - - -		
"B"	Updip	HKO or GOC	- - - - Top of structure or GOC - - - -	
	Downdip	LKO	LKO+1GT#	HKW or spill point
	Lateral	1000 m from discovery or closest productive well	- - - - - - - - - - - Not stated - - - - - - - - - -	

---

[3] "Normal stepout" is interpreted here to mean a well drilled a distance at least two well-drainage-radii from a well (or wells) in the proven area, with reservoir delineation the primary objective.

[4] The focus here is on oil reservoirs. Similar variations are observed for gas reservoirs.

[5] This table is an illustration of the variations in practice between various entities, not a compilation of all such practices.

[6] Major international oil/gas companies.

"C"	Updip	HKO or GOC	- - - - - - - - - Not stated - - - - - - - -	
	Downdip	LKO	Halfway between LKO and HKW[##]	HKW or spill point
	Lateral	- - - - - - - - - - - - - - - Not stated - - - - - - - - - - - - - - -		

*   Provided such limits do not conflict with other data.
**   Abbreviations are defined in **Notation**.
#   Gross thickness.
Depending on circumstances, some entities might limit "2P" reserves to a distance no more than two locations downdip from the well defining LKO and "3P" reserves no more than three locations downdip from the well defining LKO.

G.4 Lateral Limits

In the context of these guidelines, "lateral" means on strike with the discovery, or nearest appraisal, well. The following guidelines—viewed here as being consistent with the intent of SPE-WPC [1997] definitions—are suggested:

1P- assign within a distance of $1.5r_{dw}$—where r_{dw} denotes the commercial radius of drainage of a well—from the discovery, or nearest, well, unless this limit conflicts with other guidelines or other geologic and/or engineering data,

2P- as above, except the recommended distance is $2.5r_{dw}$, and

3P- as above, except the recommended distance is $3.5r_{dw}$.

The commercial drainage radius of a well is difficult to determine a priori and depends on a number of factors that usually are poorly defined during the appraisal stage, including:

- petrophysical properties, especially porosity and permeability,
- initial static and abandonment flowing bottomhole pressure,
- drive mechanism,
- mobility of driven and driving fluids,
- reservoir geometry,
- reservoir stratigraphy, and
- economic limit flow rate.

In practice, wide ranges in well drainage radii may be observed, and guidelines provided here should be considered no more than preliminary estimates. For example, in thin, heavy-oil reservoirs producing by bottomwater drive, the drainage radius could be less than 100 ft. At the other extreme, the drainage radius of gas wells producing by pressure depletion from highly permeable rock might exceed 2,500 ft. Barring such extremes, and in the absence of data from analogous scenarios or compelling data to the contrary, it is suggested that well drainage radii for light oil reservoirs *initially* be assumed to be no more than 250 m; for gas reservoirs, no more than 750 m.

G.5 Initial Hydrocarbon/Water Contacts

In the absence of direct data establishing the position of an initial hydrocarbon/water contact, the position may be estimated using indirect engineering methods. Three such methods are discussed here:

1. capillary pressure methods,
2. pressure-depth methods, and
3. pressure-gradient methods.

As implied by the discussion in Sec. 3.5.2, "hydrocarbon/water contact" is a general term. It is used for simplicity in these introductory comments in lieu of the proper term, "free water level." The importance of this distinction is discussed in Sec. G.5.1.

The methods discussed here may be used—in theory—to estimate the position of either a gas/water contact or an oil/water contact. Caution should be exercised, however, in using these methods to estimate gas/water contacts, as many gas accumulations have (downdip) oil rims. Thus, unless it can be established that the gas is clearly undersaturated—i.e., that the gas is not in phase equilibrium with an oil rim—use of these methods to estimate a gas/water contact should be used advisedly. The reader is referred to Dake [1994] Exercise 2.1, where this point is discussed in detail.

The capillary pressure method is independent of the pressure-depth and pressure-gradient methods. Thus, if the necessary data are available, the two methods should be used together.

It is recognized that transient pressure analysis (TPA) has been utilized to calculate the distance from a test well to changes in transmissibility. This methodology has been used successfully to assist in mapping faults and other such discontinuities. In theory, such methodology might be used to detect a change in transmissibility due to a hydrocarbon/water contact, especially a gas/water contact. However, as observed by Earlougher [1976] "many different conditions can cause the same or similar well-test response...even very simple conditions cannot necessarily be resolved by transient testing." Thus, it is considered here that TPA—while it might *confirm* an otherwise indicated hydrocarbon-water contact—is not sufficiently definitive to be used as a primary indicator of the location of—or distance to—such contact.[7]

In the absence of definitive engineering data—or as a supplement to such data—seismic data may be used to estimate reservoir limits. However, the quality of seismic data varies widely. As a result, the interpretation of such data tends to be project specific, quite subjective, and not conducive to the development of simple guidelines. Depending on quality, properly calibrated seismic data may corroborate or refute interpretations from engineering methods, but generally should not be used without adequate subsurface control.[8]

In this context, reference is made to Robertson's [2001] summary of the state of the art regarding geophysics during the appraisal stage of new discoveries.

G.5.1 Capillary Pressure Method

Background

As noted in Sec. 3.5.6, capillary pressure, P_c, is the difference in pressure between wetting and nonwetting fluids in a porous medium. It may be expressed as a function of interfacial tension, σ, contact angle, θ, and the average pore radius, r, as [Leverett 1941][9]

$$P_c(S) = 2\sigma\cos\theta/r . \dots\dots (G.1)$$

The notation $P_c(S)$ denotes the functional relation between capillary pressure and saturation. Typically, capillary pressures are measured in the laboratory, using core[10] samples

---

[7] Recent work by Goldsberry [2000], however, may improve the industry's ability to define reservoir limits using TPA.

[8] Cases have been reported, for example, where seismic amplitude anomalies were attributed to paleo hydrocarbon/water contacts (AAPG Explorer Oct. '99).

[9] This is believed to be the primary reference for this topic. For a comprehensive discussion, the reader is referred to Bass [1987].

[10] The term "core" is not meant to be limiting, as capillary pressure also can be measured from drill cuttings.

considered representative of the reservoir being evaluated. Data reported—for each core sample—are capillary pressure vs. saturation. Laboratory fluid pairs usually are either air/brine or air/mercury. A description of laboratory procedures is beyond the scope of this work, and reference is made to Bass [1987].

Capillary pressure also may be expressed in terms of height above the free water level (FWL) and the difference in densities between the wetting and nonwetting fluids. For an oil/water system

$$P_c = Z_{FWL}(\rho_w - \rho_o)/144 \, , \quad\text{...(G.2)}$$

where Z_{FWL} = height above FWL (ft) and

                ρ_w, ρ_o = density of water, oil, at reservoir conditions (lb/cf).*

* As noted in Sec. 3.6.6, other units may be used for density, in which case the appropriate constant must be used for dimensional consistency.

Estimation of Free Water Level

As implied by Eq. G.2, the FWL is defined at zero capillary pressure, a point illustrated by Fig. 3.1, introduced in Sec. 3.5.2. In situations where the initial hydrocarbon/water contact has not been established by logging or testing, capillary pressure data may be helpful in estimating the FWL. The basic concepts were introduced by Alger, Luffel, and Truman [1989] and Smith [1991]. The method was expanded and discussed in detail by Hawkins, Luffel, and Harris [1993], and their work is summarized in the following paragraphs.[11]

Based on earlier work by Thomeer [1960], Swanson [1981], and Pittman [1992], Hawkins *et al.* [1993] developed "global" correlations between porosity, water saturation, mercury-air capillary pressure, and air permeability, as follows:

$$\log(P_{cm}) \approx (-)F_g/\ln(1 - S_w) + \log(P_{dm}) \, , \quad\text{...(G.3)}$$

where

$$F_g \approx [\ln(5.21 k_a^{0.1254}/\phi)]^2/2.303 \, , \quad\text{...(G.4)}$$

$$P_{dm} \approx 937.8/(\phi k_a^{0.3406}) \, , \quad\text{...(G.5)}$$

and P_{cm} = mercury-air capillary pressure (psia),

           P_{dm} = mercury-air displacement pressure (psia) and

           F_g = pore geometrical factor, as defined by Thomeer [1960].

Other terms are defined in **Notation.**

Eqs. G.3, G.4 and G.5 were described by Hawkins *et al.* [1993] as "a worldwide correlation for capillary pressure curves." Utilization of these equations is subject to the usual cautions regarding application of a general correlation to a specific case. If representative laboratory capillary pressure data are available, the coefficients in these equations should be adjusted to develop "local" correlations. The availability of such data would improve the reliability of this method.

---

[11] Subsequent to the Hawkins *et al.* [1993] paper, the method was (reportedly) applied successfully by Haynes [1995] for Alaska, Gulf of Mexico, and North Sea reservoirs.

In *theory,* for each (well log) level for which porosity and permeability[12] are available, Eq. G.4 is solved to calculate a value for F_g, which defines a "Thomeer [1960] curve." Eq. G.5 is solved at each such level to calculate P_{dm}. These two values, F_g and P_{dm}, and the (log-determined) value of S_w at the same level are substituted in Eq. G.3 to calculate a value for P_{cm} for that level. Eq. 3.9b, which was discussed in Sec. 3.5.6, is used to convert from mercury/air capillary pressure—the basis for the Thomeer [1960] curves—to the fluid system of interest. Finally, Eq. G.2 is used to calculate the vertical distance from the level investigated to the FWL, using the fluid densities in the reservoir being evaluated.

The basic principle is illustrated in **Fig. G.1,** which shows a good quality sand layer underlain by a poor quality sand layer in a hypothetical well that logged an LKO to the base of the poor quality sand. The presence of variable quality sand layers enhances the reliability of the technique, as more than one capillary pressure curve must be fitted in the calculations.

In *practice,* it is recognized that all the variables in these equations are subject to error. For example, the value for S_w computed from log analysis is a function of the saturation and cementation exponents. Each of these exponents is a function of lithology, which may vary over the logged interval. The "Archie equation," frequently used in logging company programs to calculate S_w, may not be the "correct" model for the log interval of interest.[13]

In a common reservoir, the four basic variables in Eqs. G.3, G.4 and G.5 (porosity, permeability, water saturation, and capillary pressure) should be related to a capillary pressure model, referenced to the FWL. Hawkins [1994] has developed a methodology whereby the depth profiles of these four basic variables can be computed and interrelated using "a single set of equations, one equation for each depth." Detailed discussion of the procedure is beyond the scope of this work, and the reader is referred to papers by Hawkins [1994] and Haynes [1995].

Comments Regarding Utilization of Capillary Pressure Data to Estimate Hydrocarbon/Water Level

The methodology discussed in the previous section is based on the following assumptions:

- The well-log interval of interest is within the part of the hydrocarbon/water transition zone (HWTZ) that exhibits a significant gradient of water saturation vs. depth.

- A correlation can be developed to relate air permeability to well log porosity.

- Laboratory capillary pressure data may be adjusted to reflect reservoir conditions.[14]

- Reservoir fluid densities do not vary significantly between the observation well(s) and the FWL.

- The FWL is a reasonable approximation to the "commercial hydrocarbon/water contact."[15]

---

[12] The permeability value, k_a, in Eqs. G.4 and G.5 is absolute—or air—permeability (md) and must be estimated independently from log-determined permeability that typically is computed by logging company interpretation programs, which assume log-calculated water saturation is irreducible water saturation.

[13] For shaly sands, for example, other equations—those attributed to Simandoux, Fertl, or Schlumberger [Asquith and Gibson 1982]—may be more appropriate.

[14] Based on limited observations, the global capillary pressure model may shift under reservoir conditions, thereby potentially obviating this assumption [Hawkins, personal communication, 1996].

[15] The phrase "commercial hydrocarbon/water contact" refers to the point within the HWTZ that defines the downdip limit of commercially recoverable hydrocarbons.

• Drainage capillary pressure curves are valid representations of the profiles of water saturation vs. height above the FWL.

These points are discussed briefly in the following paragraphs.

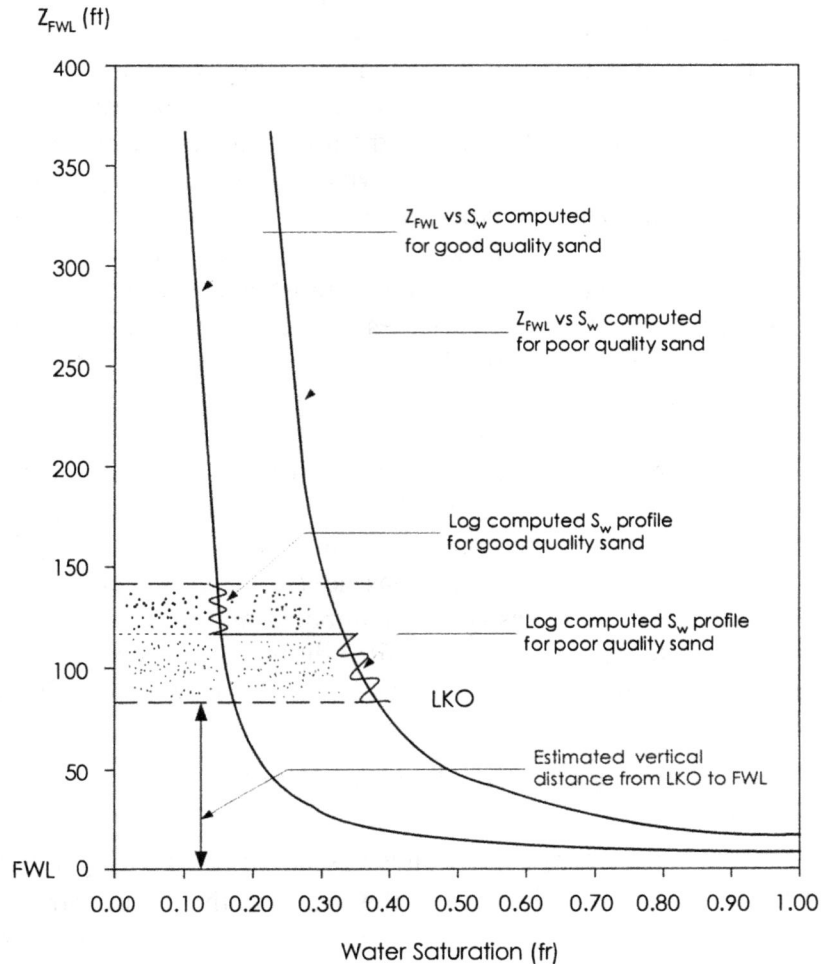

Fig. G.1—Height above FWL vs. water saturation for good and poor quality sandstone illustrating method to estimate FWL using capillary pressure data.

Hydrocarbon/Water Transition Zone

In concept, the computational procedure involves fitting profiles of water saturation vs. depth—computed from a well log—to (adjusted) global profiles of water saturation vs. depth. If the log interval of interest is in the upper part of the HWTZ, the gradient of water saturation vs. depth may be vanishingly small, and the fit between computed and global water saturation profiles—and, correspondingly, the calculated FWL—is subject to large error.

Correlation Between Air Permeability and Well Log Porosity

Depending on the degree of lithologic variation in the interval analyzed, a correlation between air permeability and log porosity may exhibit considerable scatter. This may be attributed, principally, to: (a) the fact that air permeability is a complex function of not only porosity but also irreducible water saturation and (b) the difficulties associated with correlating variables that have been measured by different—inherently disparate—procedures. A high degree of scatter also may be caused by the inadvertent inclusion of more than one rock type in the analyzed interval. In this case, provided there is a geologic basis, it might be appropriate to subdivide the analyzed interval. In any event, as discussed in Sec. 9.2.4, regression analysis of such data results in a P50 estimate, and a calculated FWL might be classified accordingly.

Conversion of Laboratory Capillary Pressure Data to Reservoir Conditions

As discussed in Sec. 3.5.6, capillary pressures determined using laboratory fluids, typically mercury/air or air/brine, must be adjusted to reflect reservoir fluids, typically gas/brine or oil/brine. The reader is referred to that section for a discussion of potential sources of error.

An additional potential source of error, also discussed in Sec. 3.5.6, is that capillary pressures measured under ambient pressure may not reflect capillary pressure behavior of the same cores under reservoir stress. As might be expected, differences are potentially more significant with less consolidated rocks and higher net overburden pressures.

Reservoir Fluid Densities

Hydrocarbon reservoir fluids—even in a common reservoir—may not exhibit the same composition at every point in the reservoir. Spatial variations in properties—especially bubblepoint—have been reported in most producing provinces in the world. The general tendency is a decrease in volatility—i.e., increase in density—with depth. The problem is especially acute in thick hydrocarbon columns in the Middle East, many of which are underlain by heavy to very heavy oil. Compositional variation with depth—not as severe as that in the Middle East—has been reported elsewhere. Examples include the Scurry Reef field (Texas), the Elk Basin field (Wyoming), and the Hilal field (Egypt).

In addition to compositional variations in the oil column, instances have been reported where there was no discernible phase boundary between a gas phase—i.e., a gas "cap"—and an oil phase in a common reservoir. Such occurrences have been reported for several fields in the North Sea, including the Oseberg field [Fjerstad et al. 1992] and the Brent field [Bath et al. 1980] and in Columbia, in the Cusiana field [Cazier et al. 1995]

In view of these occurrences, it is apparent that caution should be exercised in assuming a constant hydrocarbon gradient over a significant vertical interval in any oil and/or gas reservoir.

Commercial Hydrocarbon/Water Contact vs. Free Water Level

As discussed in Sec. G.4.2, the depth of the FWL is defined as the depth at which the capillary pressure between the wetting and nonwetting fluids is zero. The *100% water level* will be shallower than the *free water level*. The difference in these depths will depend principally on the pore size distribution of the reservoir rock. In poor quality rocks, the difference will be larger than in good quality rocks. The *producing water level*, the depth at which the reservoir will produce 100% water, will be shallower than the 100% water level. The *commercial water level*—generally considered the 50% water level—will be shallower. The difference in these depths will depend on: (a) the density contrast between the wetting and nonwetting fluids, (b) the relative permeability characteristics of the reservoir rock, and (c) the viscosity ratio. In poor quality rocks containing heavy oil, the difference will be greater than in good quality rocks containing light oil.[16]

In summary, the FWL may be a reasonable approximation to the initial hydrocarbon/water contact in reservoirs with good quality rocks containing free gas or light oil. However, in reservoirs with poor quality rocks and/or those containing heavy oil, the commercial hydrocarbon/water contact may be significantly shallower than the FWL. Thus, depending on circumstances, it might be appropriate to utilize capillary pressure, relative permeability, and fluid viscosity data to adjust from the calculated FWL to the commercial oil/water contact.

Drainage Capillary Pressure Curves

The Thomeer [1960] curves, upon which the method is based, are drainage capillary pressure curves. In some geologic settings, the initial hydrocarbon column may have shifted due to tectonic readjustment or other geologic phenomena. Depending on the degree to which capillary equilibrium has been re-established after such shifting, the Thomeer [1960] curves may not be representative of the vertical profiles of water saturation through the logged interval. Depending on the saturation history of the logged interval, imbibition capillary pressure curves may be more valid representations of such profiles.[17] Thus, in geologic settings where such shifting is a reasonable expectation, subject method should be used advisedly.

G.5.2. Pressure-Depth Method

Background

As discussed in Sec. G.4.1, at the FWL the difference between hydrocarbon and water pressures should be zero. Thus, at the FWL in an oil (or free gas) reservoir, the pressure in the oil (or free gas) phase should equal the pressure in the water phase. This observation is the basis for a procedure, called (here) the pressure-depth method, to estimate the vertical position of the FWL between two pressure observation wells.

At a gas/oil contact the gas and oil pressures should be equal. Thus, the pressure-depth method may also be used to estimate the vertical position of the gas/oil contact in a given reservoir.

The methodology to estimate the FWL for an oil reservoir is outlined in the following section.

Example Application of WFT[18] BHP Data

The following table, taken from Dake [1994], shows pressure-depth data from WFT surveys run in two different wells, identified as "A" and "B," during the appraisal of a faulted accumulation in a remote area, shown by **Fig. G.2**.[19]

---

[16] These points are illustrated in Figs. 3.1 and 3.2.

[17] A detailed discussion of drainage and imbibition phenomena is beyond the scope of this work, and the interested reader is referred to, for example, Dake [1978].

[18] The term "WFT" is used here in a generic sense and refers to devices lowered into a well on wireline that are designed to measure formation pressure and recover fluid samples. Such devices include the RFT, introduced in the mid 1970's, and improved devices, including the MDT,™ introduced by Schlumberger in 1990, and the RCI,™ introduced by Western Atlas in 1995.

[19] Fig. G.2 was adapted from Dake's Fig. 2.19.

Well "A" - wet		Well "B" - discovery	
D_{ss} (ft)	BHP (psia)	D_{ss} (ft)	BHP (psia)
6,075	2,662	5,771	2,602
6,091	2,669	5,778	2,604
6,108	2,677	5,785	2,608
6,220	2,725	5,800	2,610
6,232	2,731	5,806	2,612
		5,813	2,614

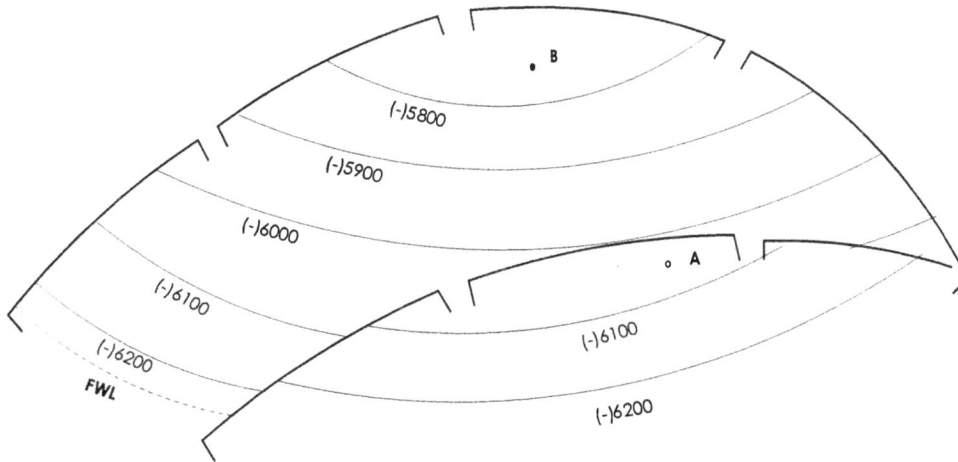

Fig. G.2—Structure on top of productive sand [after Dake 1994].

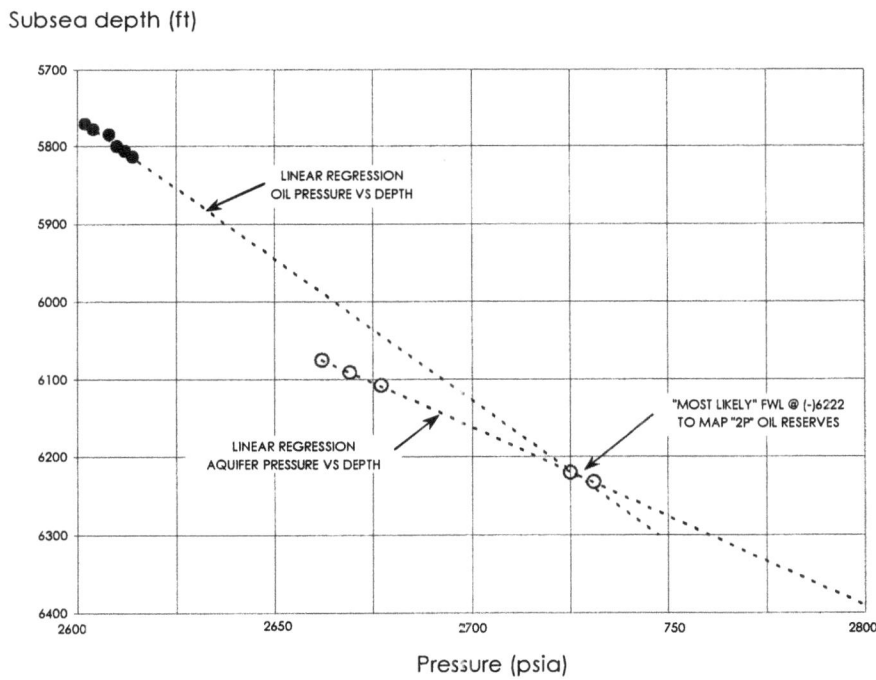

Fig. G.3—BHP vs. depth illustrating pressure-depth method to estimate the FWL [after Dake 1994].

In Well "A" the objective zone was wet, but the operator measured static BHP in porous zones at five depths between 6,075 and 6,232 ft subsea in order to determine the aquifer pressure gradient in this new area. Well "B" encountered oil to the base in a porous zone topped at 5,768 ft subsea, and the operator measured static BHP at six depths from 5,771 to 5,813 ft subsea.[20] The BHP data are plotted in **Fig. G.3,** which also shows the results from regression analyses of these data, which is discussed in the ensuing paragraphs.

The operator's structural interpretation on the top of the oil-bearing zone, Fig. G.2, shows the fault between Wells "A" and "B" dies out to the west of Well "A"—i.e., the two wells apparently are in pressure communication through a common aquifer. From examination of the pressure-depth plot, it is apparent that the east end of the fault, between Wells "A" and "B," is sealing.

For this work, a linear regression of pressure vs. depth[21] was made for each fluid with the following results:

Well "A": $p_w = 11.559 - 0.4363 D_{ss}$, ...(G.6)

Well "B": $p_o = 1,022.64 - 0.2738 D_{ss}$ . ...(G.7)

Reportedly [Dake 1994], the calculated water pressure gradient (0.436 psi/ft) was "in agreement with its salinity." Also, the calculated oil pressure gradient (0.274 psi/ft) was "confirmed by that calculated from the PVT analysis" [Dake 1994].[22]

Having verified the apparent reasonableness of the WFT-determined fluid gradients, one may use these two equations to calculate the depth of the FWL. As discussed in the previous section, at the FWL, $p_w = p_o$. Thus, equating Eqs. G.6 and G.7:

$$p_w = 11.5587 - 0.4363(D_{ss})_{FWL} = p_o = 1,022.64 - 0.2738(D_{ss})_{FWL}$$

Solving for $(D_{ss})_{FWL}$ yields:

$$(D_{ss})_{FWL} = 6,222 \text{ ft subsea}^{23}$$

The procedure outlined in the foregoing paragraphs leads to what is identified here as a "most likely"—or deterministic—estimate of the FWL. Depending on circumstances, a probabilistic approach may be appropriate, as discussed in the ensuing paragraphs.

In addition to a "best estimate," regression analysis provides useful information on uncertainties in the input data. For example, from the foregoing data the oil gradient—at 80% confidence—was calculated to be between 0.31 and 0.24 psi/ft. Depending on circumstances, this might be considered a plausible range for this prospect. Accordingly, the regression data

---

[20] Such WFT pressures generally are assumed to reflect oil pressure, which may not always be the case, as discussed by Elshahawi *et al.* [1999].

[21] Use of linear regression analysis (LRA) here is not intended to imply criticism of the graphical approach. It is, rather, intended to set the stage for a probabilistic procedure to estimate a range of potential interpretations of WFT data.

[22] Gradients reported here, calculated from regression analysis, differ slightly from those calculated by Dake [1994], who reported 0.44 and 0.28 psi/ft, respectively.

[23] Dake's—apparently graphical—interpretation of these data resulted in placing the initial oil/water contact at 6,235 ft subsea. In this context, it is noted that, if p_o at 5,800 ft is rejected as being "off trend," a solution procedure like that used here would result in a FWL at 6,248 feet subsea, which would "bracket" Dake's estimate.

may be used to calculate—at 80% confidence—the upper and lower predicted bounds for the downdip projection of oil pressure vs. depth.[24] The results of these calculations are plotted[25] in **Fig. G.4,** which includes the "best fit" interpretation from Fig G.3.

Subsea depth (ft)

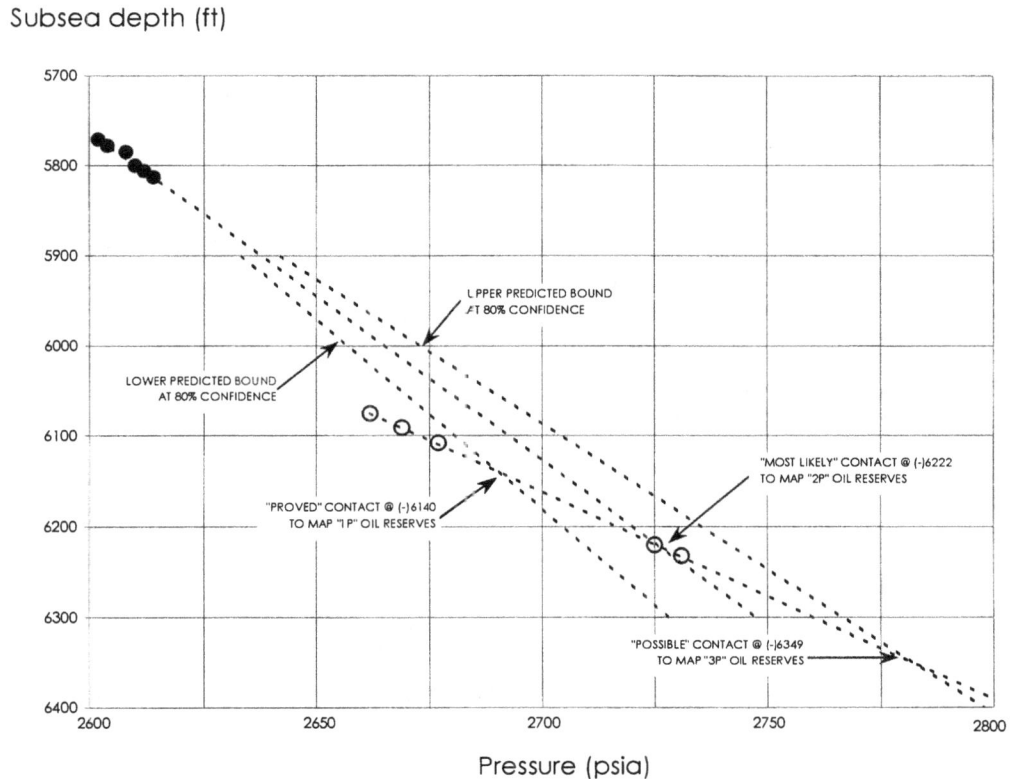

Pressure (psia)

Fig. G.4—BHP vs. depth, including upper and lower bounds of predictions at 80% confidence, illustrating pressure-depth method to estimate proved, most likely, and possible FWL's.

A procedure comparable to that used for oil zone pressures would, in general, be used for aquifer pressures. The regression coefficient for these data, however, is 0.9999, which would result in negligibly small uncertainty in the prediction—i.e., in the updip projection of aquifer pressure.

The lower bound of the prediction of oil pressure vs. depth intersects the plot of aquifer pressure vs. depth at 6,140 ft subsea; the upper bound of the prediction, at 6,349 ft subsea.

Provided there are no subsurface—or other—data to the contrary, this analysis might be used to estimate the following:

1. a *most likely* FWL at 6,222 ft subsea to map proved plus probable (2P) OIP,

2. a *proved* FWL at 6,140 ft subsea to map proved (1P) OIP, and

3. a *possible* FWL at 6,349 ft subsea to map proved plus probable plus possible (3P) OIP.

---

[24] As implied by the discussion in Sec. 9.2.3, the lower predicted bound at 80% confidence corresponds to the 10th percentile—i.e., it is a "P90" estimate; the upper predicted bound is a "P10" estimate.

[25] The upper and lower predicted bounds are slightly curved upwards and downwards, respectively, which may not be obvious from Fig. G.4.

A possible FWL at 6,349 ft subsea, however, may not be consistent with the current geologic interpretation, Fig. G.2, which reflects a spill point at about 6,300 ft subsea. Such conflicts mandate review of both the engineering and geologic data and the interpretation thereof.

Validation and Interpretation of WFT BHP Data[26]

As implied by the foregoing discussion, the hydrocarbon fluid gradients calculated from WFT pressure-depth data should be verified by: (a) comparison with those calculated from PVT data, which is preferable, or (b) empirical correlations. The aquifer fluid gradients should be verified by comparison with: (a) those calculated from salinity determined from brine samples, which is preferable, or (b) those calculated from log analysis, or (c) local aquifer gradients measured in the same stratigraphic section.

WFT pressure-depth data for each data set—typically, each fluid and well—should be analyzed by LRA to determine the least squares fits (LSF) through each data set. From these calculations, the range of fluid pressure gradients—at 80% confidence—may be compared with those determined from PVT or other data.

Consideration should be given to:

- the quality of each LSF, as calculated from LRA,
- the operational conditions under which the pressures were measured, and
- the type of WFT tools and gauges employed.

It is suggested that, for each data set—oil and water—there should be at least four valid data points and the regression coefficient, r^2, for each LSF should be at least 0.95. The range of LSF calculated pressure gradients—at 80% confidence—should be in reasonable agreement with pressure gradients determined from PVT or other data. If these criteria are met, the WFT BHP data are considered here to be "definitive" and in compliance with SPE-WPC [1997] definitions for proved reserves regarding the downdip limit.

Consideration might be given to rejection of "outliers" by use of an appropriate statistical procedure.[27] One such procedure is calculation of "residuals"—i.e., the difference between actual data and LSF calculated values. For example, the LRA of the oil pressure-depth data in this section resulted in $r^2 = 0.96$, which was considered reasonable. However, the residuals for the data points at 5,785 and 5,800 ft might have been considered too large. Had these two points been omitted from the regression, r^2 would have been 0.999, and the uncertainty in the calculated oil gradient would have been trivial.

Determination of the correct pressure-depth relation for the aquifer is essential for any such analysis to be valid. Caution should be exercised in the analysis of WFT data in accumulations exhibiting complex stratigraphy or those that are complexly faulted. In such scenarios, apparently regional aquifers may be localized by stratigraphy and/or faulting. Water-bearing zones that are apparently correlative between wells may not be common aquifers. Before regression analysis is attempted through any data set, large scale plotting and careful examination of the data are essential [Dake 1994].

---

[26] In this context, the reader is referred to Dake [1994], who has provided a comprehensive discussion of the application (and misapplication) of WFT-BHP data to estimating fluid-fluid contacts in oil and gas reservoirs.

[27] Whether or not to reject outliers is approached by many statisticians with caution; procedures (and caveats) are discussed by Snedecor and Cochran [1980].

G.5.3. Pressure-Gradient Method

Background

Depending on operating and/or geologic conditions, sets of depth-pressure measurements like those discussed in the previous section may not be available. Such conditions might include:

- adverse hole conditions,
- failure of wireline tool,
- zone(s) too thin,
- budgetary constraints, and/or
- operational oversight.

In such circumstances, a technique identified here as the pressure-gradient method might be appropriate, if necessary and sufficient data are available.

Theory and Application

The equations are based on the same principle as that applied in the previous section. If initial static BHP are available in a well in the oil zone and in the aquifer of a common reservoir, the BHP's in each of the two fluids at the FWL may be calculated by extrapolation from the observation wells as

$$(p_o)_{FWL} = p_{o1} + \rho_o(D_{FWL} - D_1) \dotfill (G.8)$$

and

$$(p_w)_{FWL} = p_{w2} - \rho_w(D_2 - D_{FWL}) . \dotfill (G.9)$$

The method is illustrated by **Fig. G.5,** which shows two pressure observation wells, Well 1 in the oil column and Well 2 in the aquifer.

Equating $(p_o)_{FWL}$ to $(p_w)_{FWL}$, Eqs. G.8 and G.9, and solving for D_{FWL} yields

$$D_{FWL} = (p_{o1} - p_{w2} - \rho_o D_1 + \rho_w D_2)/(\rho_w - \rho_o) \dotfill (G.10)$$

where

$(p_o)_{FWL}$	= BHP in the oil column at the FWL (psia),
p_{o1}	= BHP measured in Well 1 (psia),
ρ_o	= oil density at reservoir conditions (psi/ft),
D_{FWL}	= subsea depth of the FWL (ft),
D_1	= subsea depth of the pressure measurement in Well 1 (ft),
$(p_w)_{FWL}$	= BHP in the aquifer at the FWL (psia),
p_{w2}	= BHP measured in the aquifer at Well 2 (psia),
ρ_w	= water density at reservoir conditions (psi/ft) and
D_2	= subsea depth of the pressure measurement in Well 2 (ft).

Recognizing that both the pressure and the gradient data are subject to error, Eq. G-10 might be expanded to:

$$D_{FWL} = [(p_{o1} \pm \varepsilon p_{o1}) - (p_{w2} \pm \varepsilon p_{w2}) - (\rho_o \pm \varepsilon \rho_o)D_1 + (\rho_w \pm \varepsilon \rho_w)D_2]/$$

$$[(\rho_w \pm \varepsilon \rho_w) - (\rho_o \pm \varepsilon \rho_o)] , \dotfill (G.10a)$$

where ε denotes the *estimated* error in each of the data elements. Such engineering estimates may be made in lieu of statistical analysis discussed in Sec. G.4.2 and are used to establish reasonable error bars for projections of oil and aquifer BHP's.

In the absence of pressure-depth data like those used in the previous section, these equations may be used to estimate minimum, maximum, and most likely FWL's. Example data and typical engineering estimates of error are provided below:

Well	Feet Subsea	Static BHP	Gradient (psi/ft)	BHP Error	Gradient Error
Oil	5,633	2,613	.31	±3	±0.02
Water	6,328	2,865	.45	±3	±0.02

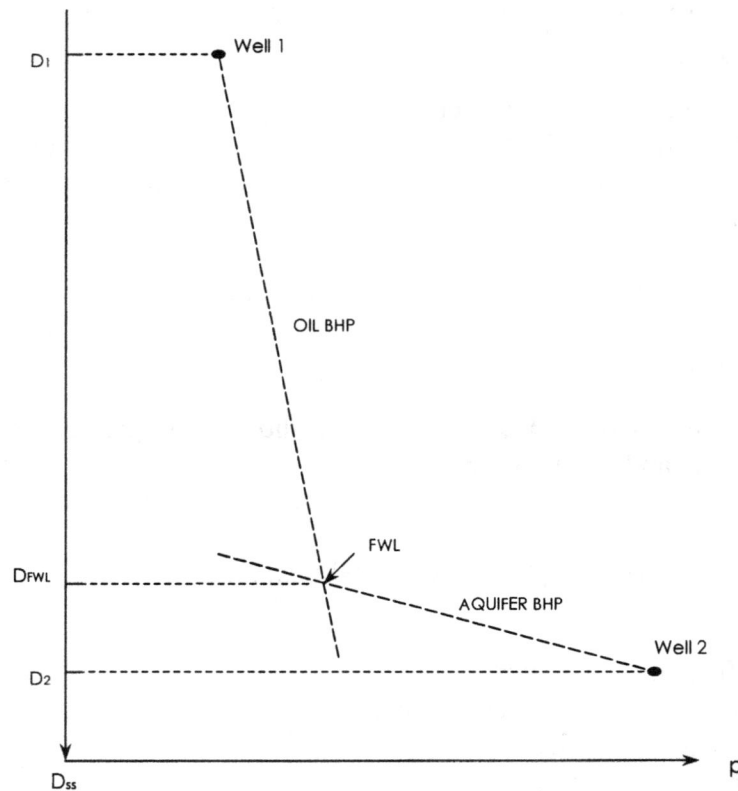

Fig. G.5—Plots of oil and aquifer BHP's illustrating pressure-gradient method to estimate FWL.

Using these data in Eqs. G.10 and G.10a provides the following estimates:

Most likely FWL: 6,067 ft subsea
Minimum FWL: 5,925 ft subsea
Maximum FWL: 6,209 ft subsea

The procedure is illustrated by **Fig. G.6.**

The minimum FWL is the intersection of the projection of estimated minimum oil pressure and the projection of estimated maximum aquifer pressure. Estimated minimum oil pressure vs. depth subsea is calculated as:

$(2,613 - 3) + (D_{ss} - 5,633)(0.31 - 0.02).$

Estimated maximum aquifer pressure vs. depth subsea is calculated as

$(2,865 + 3) + (D_{ss} - 6,328)(0.45 - 0.02).$

These projections intersect at $D_{ss} = 5,925$ ft, the minimum FWL, where estimated $p_o = p_w = 2,695$ psia.

Subsea depth (ft)

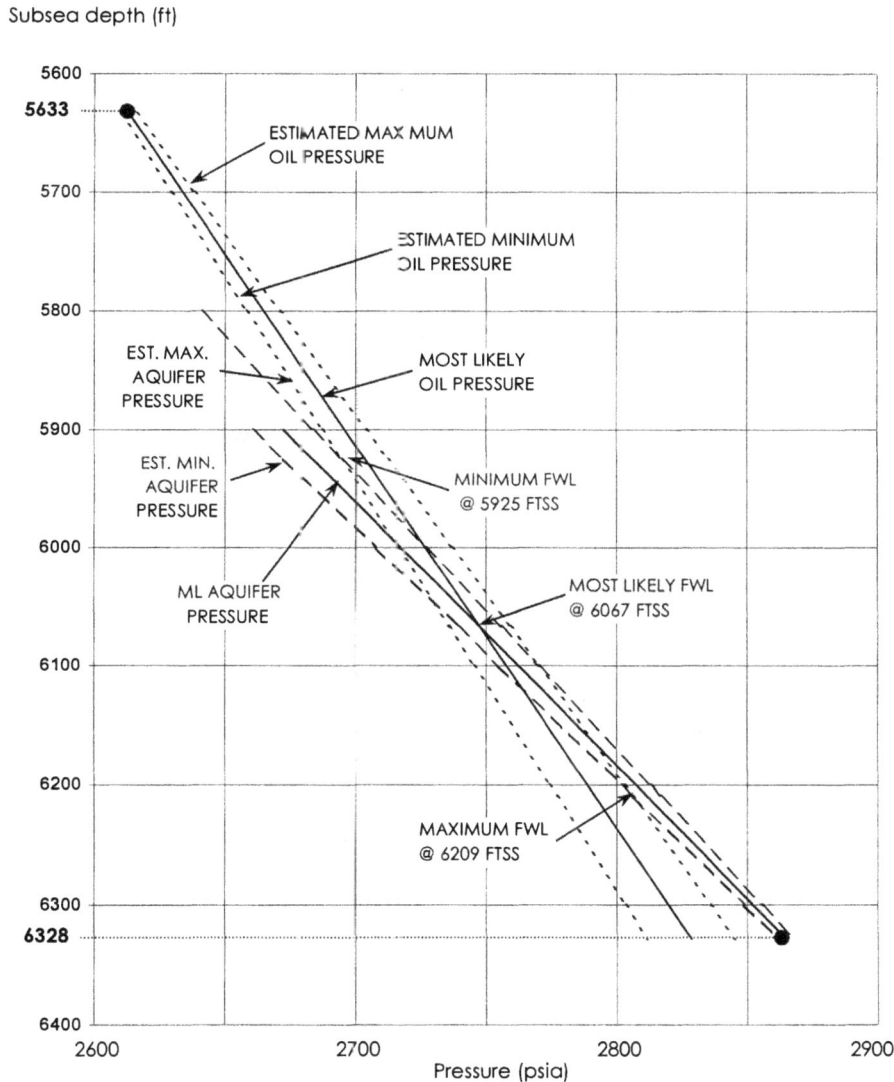

Fig. G.6—Plots of minimum, most likely, and maximum oil and aquifer BHP's illustrating pressure-gradient method to estimate minimum, most likely, and maximum FWL's.

Given the sparser data set, the pressure-gradient method is considerably less robust than the pressure-depth method and is sensitive to small errors in fluid gradients. This is especially the case for oil reservoirs where—in the absence of PVT data—it is conjectural if reliable estimates can be made. Given necessary and sufficient well test data, however, the oil gradient might be estimated using:

$$\rho_{or} = (350\gamma_{os} + 0.076R_{sb}\gamma_g)/B_{ob} , \dots\dots\dots\dots\dots\dots\dots\dots\dots\dots\dots\dots\dots\dots\dots\dots\dots\dots(G.11)$$

where ρ_{or} = density of reservoir oil at the *bubblepoint* (lb/bbl),

γ_{os} = specific gravity of stock-tank oil (dim) and

γ_g = specific gravity of solution gas (air = 1.0).

and other terms are as previously introduced.[28] Eq. G.11 is adapted from McCain's [1990] Examples 11-7 and 11-9, and the reader is referred to that work for further discussion. If the reservoir oil is undersaturated at initial conditions, corrections must be made to the bubble-point density calculated from Eq. G.11. In any event, for use in Eq. G.10, the units (lb/bbl) must be converted to (psi/ft) by multiplying by the factor:

$$(\text{lb/bbl})(\text{bbl/5.614 cf})(\text{cf/62.4 lb})(0.433 \text{ psi/ft}) = 0.00124$$

Because of the sparse data and consequent less robust calculations, this method is not here considered "definitive" in the same sense as the pressure-depth method. Accordingly, the terms minimum, most likely, and maximum are used here to describe the results of the calculations rather than proved, most likely, and possible as done with the pressure-depth method. However, estimates of minimum, most likely, and maximum FWL's may be considered reasonable *approximations* to P1, P2, and P3 limits, respectively, unless these interpretations are counterindicated by other data.

Before using any of these methods for a gas accumulation, the possibility of a downdip oil rim should be investigated. If the static BHP is at, or slightly above, the dewpoint pressure, the possibility of a downdip oil rim cannot be ruled out. Given this possibility, the pressure-gradient method might be used advisedly to estimate the position of a FWL.[29] The dewpoint pressure may be measured from PVT samples or estimated from an empirical correlation. The available correlation, that of Nemeth and Kennedy [1967], has an absolute average error of 354 psi, which may not be acceptable, depending on circumstances.[30]

G.5.4 General Application of Uncertainty Analysis to Estimate 1P, 2P, and 3P GOC and/or FWL

The methodologies discussed in Secs. G.5.2 and G.5.3 may be generalized to estimate the positions of 1P, 2P and 3P GOC's and/or FWL's. The procedure is illustrated by **Fig. G.7** and is summarized in the following table:

Case	Updip fluid	Downdip fluid	GOC or FWL		
			1P	2P	3P
GOC	Gas	Oil	C	B	A
OWC (FWL)	Oil	Water	A	B	C
GWC (FWL)	Gas	Water	A	B	C

---

[28] Comments—and caveats—are provided in Appendix E regarding the reliability of PVT correlations.

[29] If—unknown at the time of the estimate—there actually is an oil rim, then use of this method will result in an estimated FWL deeper than actual, that is, the estimated gas accumulation will be larger than actual. This error may be offset by the greater value of the oil rim.

[30] Details regarding this correlation are provided in Appendix E.

The notation in Fig. G.7 reflects that of Sec. G.4.2—that of the pressure-depth method.[31] If the pressure-gradient method is used, the best estimates of BHP and fluid gradients would be used in lieu of the least squares fits to estimate Point "B." Thus, to estimate a P2 GOC, Eq. G.10 would be modified to reflect gas and oil column pressures and gradients. To estimate a P2 FWL in a nonassociated gas reservoir, Eq. G.10 would be modified to reflect gas column and aquifer pressures and gradients. A similar procedure would be used to estimate Points "A" and "C," using Eq G.10a.

As gas is a less valuable substance than oil, the estimated 1P GOC for gas cap oil reservoirs is set at the deep end of the uncertainty range.

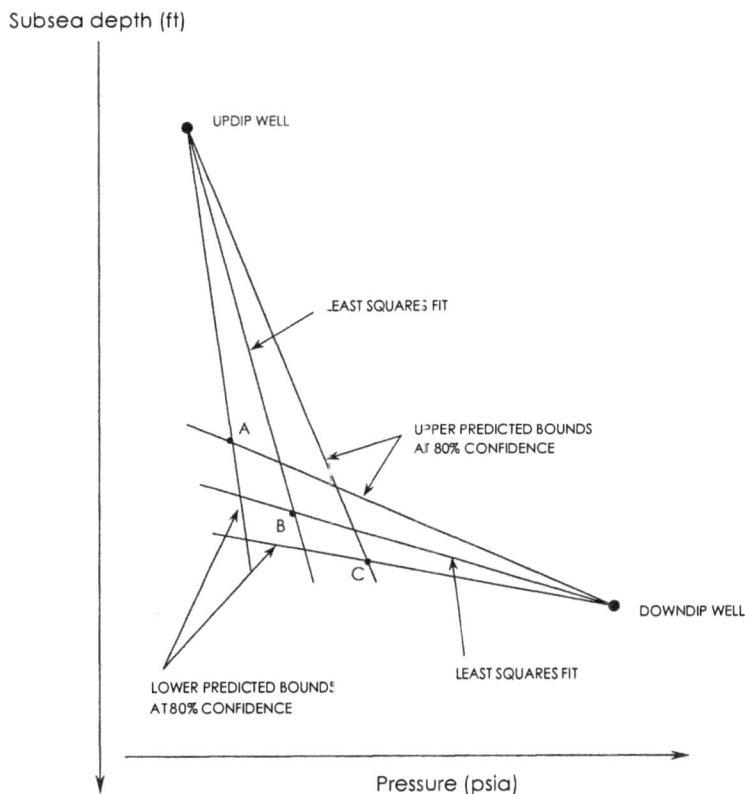

Fig. G.7—Plots of minimum, most likely, and maximum oil and aquifer BHP's illustrating pressure-gradient method to estimate minimum, most likely, and maximum FWL's

For probabilistic (volumetric) calculations of O/GIP, Points "A," "B," and "C" might be used to define the frequency distribution of gross rock volume for input to MCS. In this context, reference is made to Sec. 9.10.1.

The data scenarios discussed in the foregoing sections are simplified presentations of what may actually be encountered, and the engineer must exercise ingenuity in applying the concepts presented here. For example, there may be pressure depth data in only a single oil well and only regional data for the aquifer. In such circumstances, Eq. G.8 might be used in

---

[31] As observed for Fig. G.4, the upper and lower predicted bounds are slightly curved upwards and downwards, respectively, which may not be obvious from Fig. G.7.

combination with an equation relating regional aquifer pressure with depth, like those discussed in Sec. 3.4.3, here written as Eq. 3.7c

$$(p_o)_{FWL} = p_{ol} + \rho_o(D_{FWL} - D_1) \quad\dotfill\quad (G.8)$$

$$(p_w)_{FWL} \approx g_P D_{FWL} + p_r \quad\dotfill\quad (3.7c)$$

Equating $(p_o)_{FWL}$ to $(p_w)_{FWL}$ and solving for D_{FWL}

$$D_{FWL} \approx (p_{ol} - \rho_o D_1 - p_r)/(g_P - \rho_o) \quad\dotfill\quad (G.12)$$

Considering that equations like Eq 3.7c typically are generated from regression analysis, a FWL determined from Eq. G.11 should be considered to have an expectation no greater than P50. Further to this point, some of the initial pressures used to generate correlations like Eq. 3.7a may have been measured in the *hydrocarbon* column and may not, therefore, be reflective of initial *aquifer* pressure.

Concluding this section, it is emphasized that the methods discussed in Secs. G.5.1 through G.5.3 are to estimate the free water level (FWL), which may be significantly deeper than the commercial water level (CWL), depending on circumstances. In general, the FWL is a reasonable approximation for the CWL in good quality rocks with light-oil or free-gas accumulations. The approximation becomes progressively poorer—i.e., the CWL becomes progressively shallower than the FWL—for poor quality rocks and/or heavy oil. In such scenarios, fractional flow calculations would be appropriate to correct from the estimated FWL to the CWL, as illustrated by Fig. 3.1.

G.6 Suggested Guidelines for Estimating Position of Fluid/Fluid Contacts and Classifying Reserves in Areas Beyond Direct Well Control

G.6.1 Possible Scenarios

The following general scenarios are identified for this discussion:
1. Reservoir fluids significantly undersaturated, and either:
    a. gas discovery (Scenario 1.a) or
    b. oil discovery (Scenario 1.b).
2. Reservoir fluids at saturation pressure or saturation conditions indeterminate, and gas discovery (Scenario 2.a), or
    a. b. oil discovery (Scenario 2.b), or
    c. oil and gas in different wells apparently in the same reservoir (Scenario 2.c).

For each of these scenarios the downdip limit may (or may not) be defined by HKW or spill point (SP), which is used as a scenario modifier in next section. For example, a significantly undersaturated gas discovery with HKW defined in a downdip well is denoted here as "1.a.HKW"; if there is neither HKW or a spill point, the scenario is denoted as "1.a."

G.6.2 Recommended Guidelines

This section provides guidelines to estimate updip and downdip limits and fluid/fluid contacts for some of the scenarios identified Sec. G.6.1. These guidelines may be considered as models for other possible scenarios. Reservoir limits and/or fluid/fluid contacts estimated with such guidelines should be considered as preliminary and subject to revision, depending on subsequent subsurface geologic and/or engineering data and/or well performance data.

SCENARIO	CONTACT OR LIMIT	RESERVE CLASSIFICATION	RECOMMENDED GUIDELINE[32,33]
1.a	Updip	1P	Updip distance consistent with Sec. G.3.
1.a	Updip	2P	As above.
1.a	Updip	3P	As above or structural crest.
1.a	Downdip	1P	If data available, method discussed in Sec. G.4 3; otherwise, at a downdip distance consistent with Sec. G.3.
1.a	Downdip	2P	If data available, method discussed in Sec. G.4.3 corroborated by method discussed in Sec. G.4.1; otherwise, at a downdip distance consistent with Sec. G.3.
1.a	Downdip	3P	If data available, method discussed in Sec. G.4.3; otherwise, at a downdip distance consistent with Sec. G.3.
1.a.HKW	Downdip	1P	Depending on data available, method discussed in Sec. G.4.3 or G.4.4; otherwise, at a downdip distance consistent with Sec. G.3.
1.a.HKW	Downdip	2P	Depending on data available, method discussed in Sec. G.4.3 or G.4.4, corroborated by method discussed in Sec. G.4.1; otherwise, at half the distance between LKG and HKW.
1.a.HKW	Downdip	3P	Depending on data available, method discussed in Sec. G.4.3 or G.4.4; otherwise, at HKW.
1.b	Updip	1P, 2P, & 3P	Consistent with Scenario 1.a, except consider updip volume to be oil filled.
2.a	Updip	1P, 2P, & 3P	Consistent with Scenario 1.a.
2.a	Downdip	1P, 2P, & 3P	Consistent with Scenario 1.a.
2.b	GOC (assumed)	1P, 2P, & 3P	Updip distances to assumed GOC consistent with Sec. G.3.
2.c.HKW	Updip (gas)	1P, 2P, & 3P	Consistent with Scenario 1.a.
2.c.HKW	GOC	1P, 2P, & 3P	Depending on data available, method discussed in Sec. G.4.4.
2.c.HKW	OWC	1P	Depending on data available, method discussed in Sec. G.4.2 or G.4.3.
2.c.HKW	OWC	2P	Depending on data available, method discussed in Sec. G.4.2. or G.4.3 and corroborated with method discussed in Sec. G.4.1; otherwise, halfway between LKO and HKW.
2.c.HKW	OWC	3P	Depending on data available, method discussed in Sec. G.4.2 or G.4.3; otherwise, at LKW.
2.c.SP	Downdip	1P	Consistent with Scenario 1.a.
2.c.SP	Downdip	2P	Depending on data available, method discussed in Sec. G.4.3, corroborated with method discussed in Sec. G.4.1; otherwise, halfway between LKO and SP, unless inconsistent with Sec. G.3.
2.c.SP	Downdip	3P	Depending on data available, method discussed in Sec. G.4.3; otherwise, at SP.

---

[32] Unless these guidelines are contrary to corporate or other guidelines or are not consistent with other data.

[33] These guidelines might be considered deterministic. They might also be used to define appropriate frequency distributions of gross rock volume for MCS, discussed in Sec. 9.10.1.

REGIONAL CORRELATIONS FOR FORMATION TEMPERATURE AND INITIAL PRESSURE

H.1 Introduction

In the early stages of appraisal of a newly discovered accumulation, measured initial bottom-hole pressure and temperature data may be too sparse to provide a reliable estimate of initial conditions. In this case, regional correlations may be helpful. Provided in this appendix are correlations for initial reservoir pressure and formation temperature for several regions in the world. For extensively exploited areas where there are comprehensive public databases, like North America, correlations have been developed for specific geologic settings [Scientific Software 1975, Hitchon 1984].[1] In areas where exploitation has been less extensive, or where there are less publicly available data, the correlations are more general.[2]

Reservoir temperature and initial pressure estimated using these correlations should be considered preliminary and not a substitute for actual measurements. As observed in Sec. 3.4.1, however, measured pressures in new accumulations that are significantly less than those expected from local/regional correlations might be an indication of interference from other accumulations in the area.

H.2 Initial Reservoir Pressure

The following correlations have been developed for initial pressure in oil and/or gas reservoirs in the hydropressured interval[3] in North America:[4]

Mississippi: $p_i = 0.479\,D_{ss} + 180$ .... (H.1)

Louisiana Gulf Coast: $p_i = 0.465 D_{ss}$ .... (H.2)

Texas Gulf Coast: $p_i = 0.45 D_{ss}$ .... (H.3)

West Texas: $p_i = 0.439\,D_{ss} + 1250$ .... (H.4)

---

[1] Depending on the area, these simple, linear relations may not be valid for the entire productive interval. Shifts to higher pressure gradients with depth have been observed in many areas of the world, including the U.S. Gulf Coast and Green River basins, Romania, the North Sea, Burma, and China's Sichuan basin [Luo et al, 1994].

[2] The author has an ongoing project to improve these correlations and solicits such data from operators. Please respond to chapcron@aol.com.

[3] The hydropressure interval is that portion of the geologic column in which initial reservoir fluid pressure is hydrostatic, i.e., the initial pressure can be attributed to the weight of the overlying static column of formation water in the interval of saturation [Gary, et al. 1972].

[4] Readers may observe that—depending on the scope of their database and the statistical treatment thereof—constants in these correlations may vary slightly.

Illinois, Indiana, Kentucky, Michigan: $p_i = 0.427D_{ss} + 210$ ....(H.5)

Kansas: $p_i = 0.475D_{ss} + 375$ ....(H.6)

Oklahoma: $p_i = 0.439D_{ss} + 310$ ....(H.7)

Montana, North Dakota, Wyoming, Utah, and Colorado: $p_i = 0.471D_{ss} + 1120$ ....(H.8)

Alberta Lower Cretaceous: $p_i = 0.36D - 133$ ....(H.9)

Alberta Devonian D-2: $p_i = 0.39D - 200$ ....(H.10)

Alberta Devonian D-3: $p_i = 0.358D - 73$ ....(H.11)

Alberta Bashaw carbonate complex: $p_i = 0.527D_{ss} + 702$ ....(H.12)

Alberta Rimbey-Meadowbrook carbonates: $p_i = 0.466D_{ss} + 536$ ....(H.13)

Alberta Windfall-Swan Hills carbonates: $p_i = 0.491D_{ss} + 947$ ....(H.14)

Saskatchewan (gas pools):[5] $p_i = 0.477D_{ss} + 971$ ....(H.15)

Saskatchewan (oil pools):[6] $p_i = 0.460D_{ss} + 875$ ....(H.16)

The following correlations have been developed for initial pressure in oil and/or gas reservoirs in the hydropressured interval in other areas of the world. The reader is cautioned, the data for some of these correlations are not as complete as for those in North America. Correlations from sparse data are specifically identified.

North Sea (psia):[7] $p_i = 0.446D_{ss} + 86$ ....(H.17)

Egypt (sparse data): $p_i = 0.51D - 80$ ....(H.18)

Offshore Nigeria (sparse data): $p_i = 0.42D + 230$ ....(H.19)

Western Siberia,[8] down to about 11,000 ft: $p_i = 0.436D + 38$ ....(H.20)

Gulf of Thailand (sparse data): $p_i = 0.42D_{ss} + 62$ ....(H.21)

Venezuela:[9] $p_i = 0.445D + 107$ ....(H.22)

Offshore Vietnam (sparse data): $p_i = 0.47D_{ss} - 140$ ....(H.23)

where: p_i = initial pressure (psi)
D = depth (ft)
D_{ss} = depth subsea (ft)

---

[5] Data from Saskatchewan Division of Energy and Mines; correlation excludes pools with apparently subnormal initial pressure.
[6] Difference between Eqs. H.15 and H.16 not statistically significant.
[7] Accumulations tend to be compartmentalized. Depending on geologic setting, a generalized correlation may not be applicable to some accumulations. See, for example, Archer and Wall [1986].
[8] Data from which the author developed this correlation appeared in Littke, et al. Fig 4 [AAPG Bull., Oct. 1999]. Raw data were provided to the author by Bernhard Cramer, courtesy Urengoyneftegazgeologiya. Data cover a very large area, and local variations should be expected.
[9] Raw data provided by Anibal Martinez.

H.3 Formation Temperature

The following correlations have been developed for oil and/or gas reservoirs in North America:

Mississippi:	$T_f = 75 + 0.013D_{ss}$ .... (H.24)
Louisiana Gulf Coast:	$T_f = 74 + 0.0125D_{ss}$ .... (H.25)
North Texas:	$T_f = 60 + 0.01675D$ .... (H.26)
Oklahoma Anadarko basin:	$T_f = 66 + 0.011D$ .... (H.27)
Oklahoma deep Anadarko basin (below 21,000 ft):	$T_f = 66 + 0.014D$ .... (H.28)
Alberta:	$T_f = 35 + 0.020D$ .... (H.29)
Alberta Bashaw carbonates:	$T_f = 32 + 0.019D$ .... (H.30)
Alberta Rimbey-Meadowbrook carbonates:	$T_f = 49 + 0.017D$ .... (H.31)
Alberta Windfall-Swan Hills carbonates:	$T_f = 32 + 0.019D$ .... (H.32)
Saskatchewan (gas pools):	$T_f = 81 + 0.022D_{ss}$ .... (H.33)
Saskatchewan (oil pools):	$T_f = 89 + 0.019D_{ss}$ .... (H.34)

Nichols [1947] has published an iso-temperature gradient map of the Texas-Louisiana Gulf Coast region.

In general, formation temperature in the hydropressure interval in the U.S. may be estimated from thermal gradient maps published by the USGS [1976], using the equation:

$$T_f = T_{sa} + g_G D \, , \quad\text{.. (H.35)}$$

where T_f = formation temperature (°F),
T_{sa} = average surface temperature (°F),
g_G = geothermal gradient (°F/ft), and
D = depth (ft).

The following correlations have been developed for other areas of the world. Data for these correlations are not as complete as for those in North America.

Egypt (sparse data):	$T_f = 86 + 0.015D$ .... (H.36)
Offshore Nigeria (sparse data):	$T_f = 75 + 0.015D$ .... (H.37)
North Sea:*	$T_f = 71 + 0.015D_{ss}$ .... (H.38)
Western Siberia:*	$T_f = 13 + 0.0168D$ .... (H.39)
Gulf of Thailand (sparse data):	$T_f = 95 + 0.027D_{ss}$ .... (H.40)
Venezuela:*	$T_f = 102 + 0.014D$ .... (H.41)
Offshore Vietnam (sparse data):	$T_f = 62 + 0.017D_{ss}$ .... (H.42)

* Regional average; local variations to be expected.

APPENDIX I

BOTTOMWATER-DRIVE OIL RESERVOIRS

As observed in Sec 3.7.2 Waterdrive, there is no discussion in the API [1967] work regarding the relative efficiency of *bottom* vs. *edge* waterdrive. The nature of water encroachment into a reservoir is an important consideration in estimating reserves, and several studies of bottomwater-drive reservoirs are discussed in the following paragraphs. Note that the API [1967] results are for recovery efficiency from the *entire* reservoir. In bottomwater-drive reservoirs, recovery efficiency from *individual wells* may vary substantially, depending on well spacing and the ratio of horizontal to vertical permeability within each well's drainage area, among other factors.

Muskat [1947] established the theoretical foundation to estimate the performance of bottomwater-drive oil reservoirs. In that work, he made a clear distinction between encroachment of bottomwater due to an active aquifer and "coning" of underlying static water in a pressure depletion reservoir. In the latter case, there is a minimum production rate at which water will not be produced, the so-called "critical rate." In the case of a bottomwater-drive reservoir, however, the "critical rate" has no meaning, as water eventually will be produced by individual wells as the encroaching aquifer rises into the well completion interval. Unfortunately, in much of the published literature, the term "coning" is used to describe bottomwater encroachment, which causes confusion. The term "doming" is suggested here as being more appropriate.

Muskat's [1947] analytical treatment of bottomwater-drive reservoirs was, of necessity at the time, limited to the assumptions of: (a) unit mobility ratio and (b) zero density contrast between oil and water. For conditions other than unit mobility ratio and zero density contrast, analytical studies of bottomwater drive published by Hutchinson and Kemp [1956] and model studies published by Henley et al. [1961] and by Ciucci et al. [1966] provide a basis for a *preliminary* estimate of volumetric sweep efficiency from this type of reservoir.[1] Fig. 4 from the Hutchinson and Kemp [1956] paper and Figs. 2, 4, and 5 from the Henley et al. [1961] paper are reproduced here as Figs. I.1 through I.4, respectively.

Reserve estimates based on these figures should be considered *preliminary*, pending development of performance trends in individual wells and/or history matching by computer simulation.

---

[1] Reference to these papers might seem dated. However these studies can provide valuable insight regarding well spacing and recovery efficiency, pending obtaining reservoir data for more detailed computer simulation.

The Hutchinson and Kemp [1956] work involved analytical studies using dimensionless well spacing, A_{wD}, of 2.25, 6, and 12, where dimensionless well spacing was defined as in the Muskat [1947] work:

$$A_{wD} = (a_w/h_{to})(k_V/k_H)^{0.5} , \dots\dots\dots\dots\dots\dots\dots\dots\dots\dots\dots\dots\dots\dots\dots\dots\dots\dots\dots\dots\dots\dots\dots\dots\dots(I.1)$$

where a_w = well spacing (ft or m),
         h_{to} = initial gross oil zone thickness (ft or m),
         k_V = vertical permeability (md, average over well drainage area)[2] and
         k_H = horizontal permeability (md, average over well drainage area).

Values of A_{wD} for typical reservoir scenarios are summarized below:

Well Spacing* (a_w)	Gross Oil Zone Thickness* (h_{to})	Permeability Ratio (k_V/k_H)	Dim. Well Spacing (A_{wD})
1000	10	0.1	32
		0.04	20
		0.02	14
500	50	0.1	3.2
		0.04	2.0
		0.02	1.4

*Consistent units

In unfractured sandstone reservoirs, the ratio of horizontal to vertical permeability is in the range 10:1 to 50:1. Dimensionless well spacing less than about 5 is considered "close" and, for favorable water/oil mobility ratios, generally will result in high volumetric sweep efficiency at commercial WOR's. From the foregoing table, it may be concluded that, to achieve such spacing, interwell distance should be no more than about 10 times the gross oil column thickness.

Other parameters investigated in this study included dimensionless well penetration, Lc_D, and mobility ratio, M_{wo}, where dimensionless well penetration was defined in the same manner as in the Muskat [1947] work:[3]

$$L_{cD} = L_c/h_{to} , \dots\dots\dots\dots\dots\dots\dots\dots\dots\dots\dots\dots\dots\dots\dots\dots\dots\dots\dots\dots\dots\dots\dots\dots\dots\dots\dots\dots\dots(I.2)$$

where L_c = vertical distance, from the top down, the completion interval extends
                into the initial oil zone (ft or m) and
         h_{to} = total thickness of the oil column (ft or m).

Mobility ratio was defined consistent with the current SPE [1987] definition—the mobility of water at residual oil saturation divided by the mobility of oil at irreducible water saturation.

---

[2] The average ratio of vertical to horizontal permeability *over the drainage area of a well* cannot be determined by measurements of k_V and k_H on small samples of rock in the laboratory. In unfractured sandstone reservoirs, for example, this ratio is controlled by shale and low permeability "layers" randomly interbedded in the gross sandstone interval. For further discussion of the influence of shale layers on reservoir performance, see Richardson *et al.* [1978] or Haldorsen and Lake [1984].

[3] The notation here for dimensionless well penetration differs from that of Henley *et al.* [1961], who used b_D.

From **Fig. I.1,** which was calculated for $A_{wD} = 6$, it may be observed that, at this spacing, the parameter having the greatest influence on volumetric sweep efficiency is mobility ratio. These results indicate that, at a water/oil ratio (WOR)[4] of 20 to 1, volumetric sweep efficiency may be expected to vary from about 40% (for a mobility ratio of 2) to about 80% (for a mobility ratio of 1/3). Calculations made for $A_{wD} = 12$ indicate volumetric sweep.

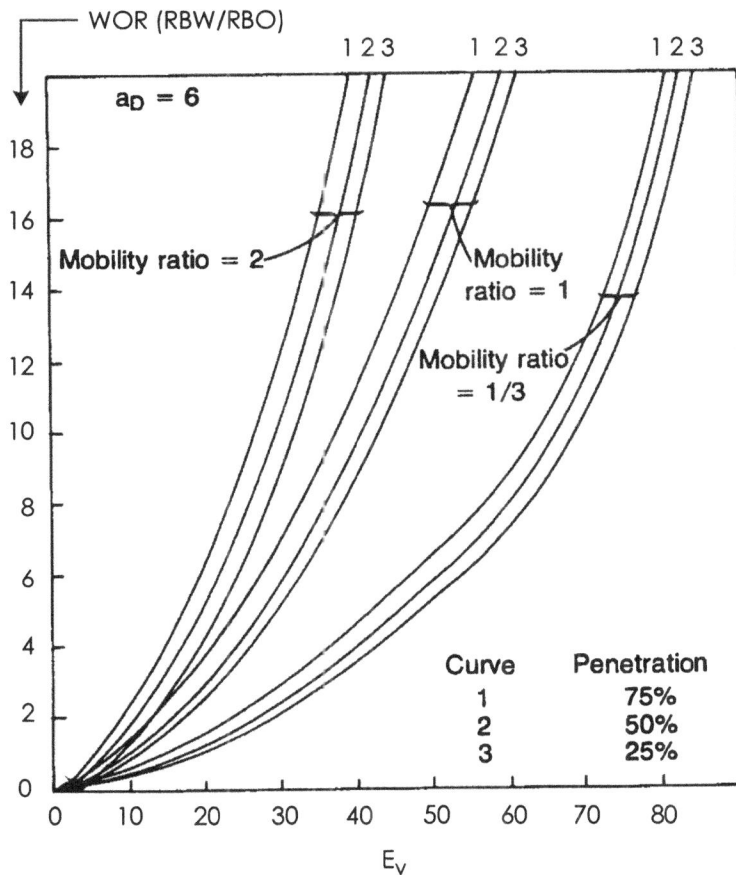

Fig. I.1—Water/oil ratio, WOR, vs. volumetric sweep, E_V, for dimensionless well spacing, a_D, equal to 6.0, illustrating how water/oil mobility ratio and penetration are estimated to influence bottom-waterdrive performance [after Hutchinson and Kemp 1956].

efficiency (at a WOR of 20 to 1) may be expected to vary from about 10 to 25% for the same variation in mobility ratio as shown in Fig. I.1. Calculations for $A_{wD} = 2.25$ indicate volumetric sweep efficiency (at a WOR of 20 to 1) of about 90% for mobility ratios of 1 and 2.

The above results were calculated assuming negligible gravity effects. Regarding this assumption, the authors observed "gravity forces can affect the performance, but in most cases any significant improvement...will be realizable only at very low rates of oil production."

The Henley *et al.* [1961] scaled model study included water/oil mobility ratios in the range 10 to 0.1, which covers oils in the medium to high API gravity range (approximately 20

---

[4] WOR in these figures is RBW/RBO and must be multiplied by the factor B_o/B_w to yield STBW/STBO.

to 40°API, respectively). Their Figs. 2, 4, and 5 are shown here as Figs. I.2, I.3, and I.4, respectively. The parameter R_3 was defined as the gravity/viscous force ratio:

$$R_3 = kg(\rho_w - \rho_o)A_w/q\mu_o , \quad\quad\quad\quad\quad\quad\quad\quad\quad\quad\quad\quad\quad\quad\text{(I.3)}$$

where k = horizontal permeability (md),
           g = gravity constant,
           ρ = density (gm/cc),
           A_w = well spacing (acres),
           q = production rate (RB/D) and
           μ_o = oil viscosity (cp).

The Henley *et al.* [1961] results are in reasonable agreement with those reported by Hutchinson and Kemp [1956] for the case where $R_3 = 0$.

The Henley *et al.* [1961] results for close ($A_{wD} = 2.0$) and wide ($A_{wD} = 12$) spacing for $R_3 > 0$ are shown in **Figs. I.2** and **I.3,** respectively. From these figures, it may be concluded—

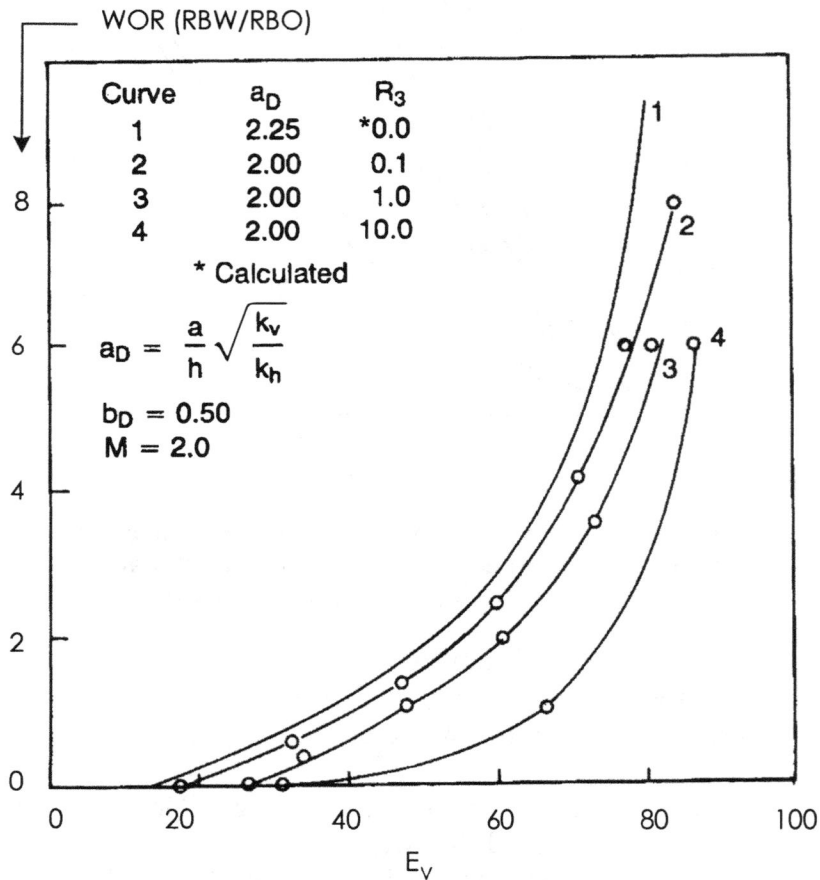

Fig. I.2—Water/oil ratio, WOR, vs. volumetric sweep, E_v, for a water/oil mobility ratio of 2.0, illustrating estimated influence of gravity-viscous force ratio, R3, on bottom-water-drive performance of closely spaced wells [after Henley *et al.* 1961].

other factors being the same—that flow rate has a more significant effect on sweep efficiency on wide spacing than on close spacing.

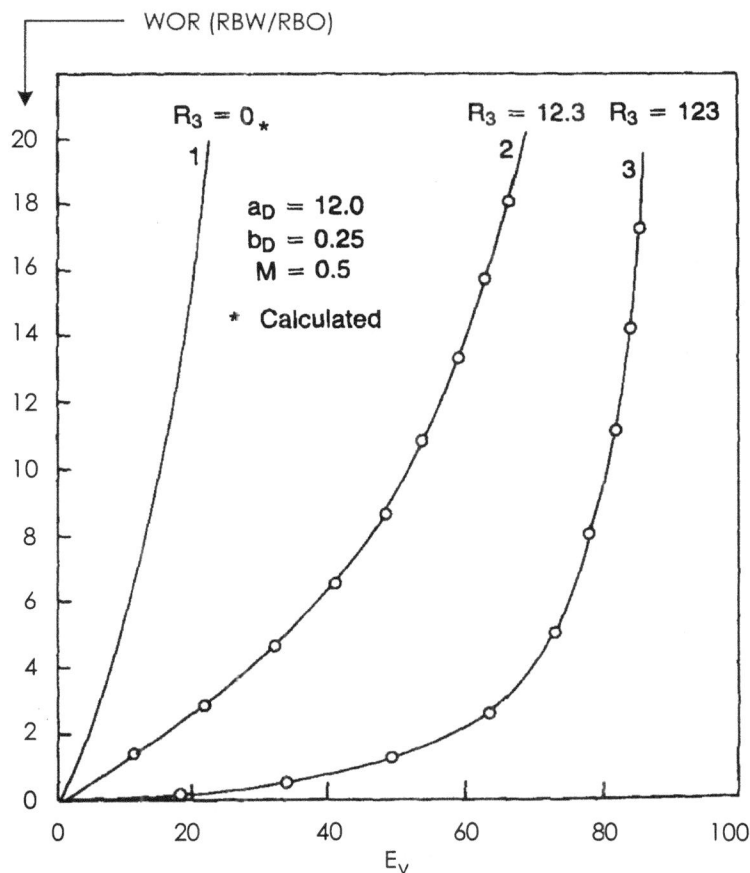

Fig. I.3—Water/oil ratio, WOR, vs. volumetric sweep, E_V, for a water/oil mobility ratio of 0.5, illustrating estimated influence of gravity-viscous force ratio, R3, on bottom water drive performance of widely spaced wells [after Henley et al. 1961].

The influence of mobility ratio, well spacing and gravity/viscous force ratio are shown in **Fig I.4**. From this figure, it may be concluded that on wide spacing (A_{wD} =12) flow rate and mobility ratio have a greater influence on sweep efficiency than on close spacing (A_{wD} = 2.0).

Ciucci et al. [1966] reported on a model study of bottomwater drive using a dimensionless well spacing (A_{wD}) of 1.6 and a mobility ratio of 200. This mobility ratio would be applicable to about 15°API crude. The model parameters in their study were defined in the same manner as in the Henley et al. [1961] study. However, direct comparison of these results with those of earlier studies is not possible, as the range of parameters used by Ciucci et al. [1966] does not overlap that used by Henley et al. [1961]. The results of the latter authors, however, are *qualitatively* consistent with those of the earlier authors. Of particular significance is the observation by Ciucci et al. [1966] that "the sweep efficiency increases remarkably as the flow rate is lowered," which corroborates the work of Henley et al. [1961].

All of the published studies cover models with the *ratio* of horizontal to vertical permeability constant throughout the model, with the tacit assumption that the permeability *level* also is constant. In nature, this type of permeability distribution is rare. Fluvial point bars typically have *lower* average permeabilities toward the top. In contrast, offshore barrier bars and beach sands typically have *higher* average permeabilities toward the top [Sneider et al. 1978]. All other factors being the same, it can be expected that wells with higher average permeabilities toward the top of the oil column will exhibit higher ultimate oil recovery efficiencies due to bottomwater drive than wells with lower average permeabilities toward the top.

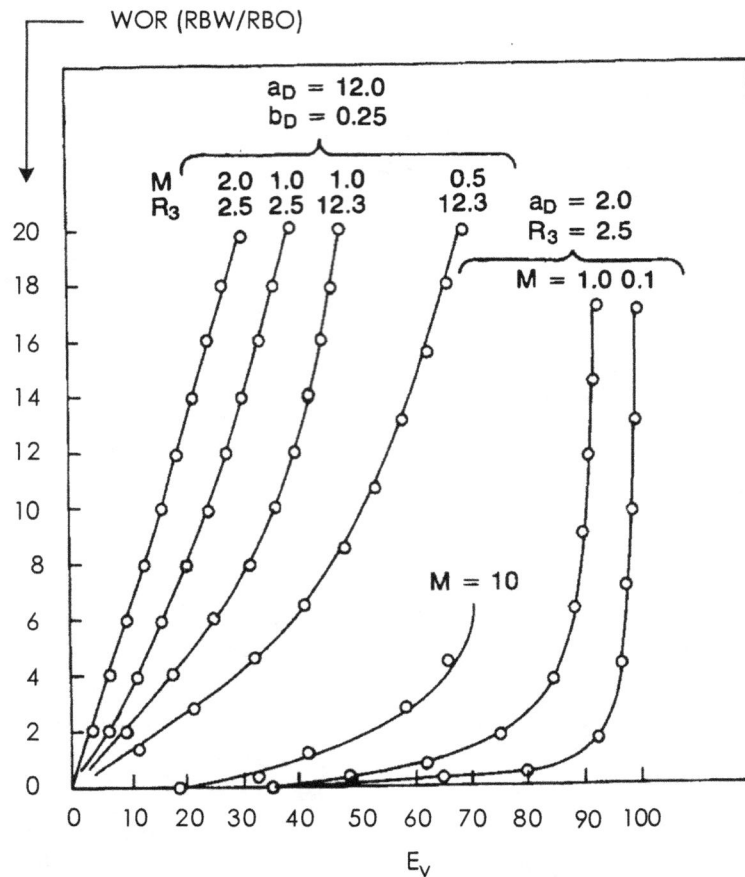

Fig. I.4—Water/oil ratio, WOR, vs. volumetric sweep, E_V, illustrating estimated influence of water/oil mobility ratio and gravity-viscous force ratio on performance of bottom-waterdrive reservoirs [after Henley *et al.* 1961].

APPENDIX J

METHOD TO CALCULATE THE COMPOSITION AND SOME PROPERTIES OF RESERVOIR GAS FROM THE COMPOSITION OF SEPARATOR FLUIDS

J.1 Background

In the absence of a PVT analysis, it may be necessary to calculate the composition and properties of a reservoir gas from the compositions and/or properties of the surface fluids— i.e., separator gas and condensate, stock-tank condensate, and stock-tank vapors. Although service companies typically perform some of these calculations, it behooves the prudent engineer to understand the general procedure. Occasionally, service company calculations may be invalid due to erroneous test data, or other reason, and the engineer may need to correct the work.

For all cases where the composition and/or properties of reservoir fluids are estimated from the composition and/or properties of the surface fluids, there must be valid field test data, including:

- condensate/gas ratio (CGR), which usually is reported as STBC/MMscf of separator gas and may be reported—if metered at the separator—as SpBC/MMscf,
- specific gravity of separator gas, stock-tank condensate, and stock-tank vapors,
- ratio of separator condensate to stock-tank condensate, and
- separator operating temperature and pressure.

While some of these data are routinely measured, for example, STBC/MMscf and gravity of stock-tank condensate, other data may be measured only when fluid samples are obtained, if at all.

Regarding availability of compositional data, there are several scenarios:

a) composition of separator gas, stock-tank condensate, and stock-tank vapors known, or

b) composition of separator gas and separator condensate known, or

c) composition of surface fluids unknown, other than specific gravity of separator gas and stock-tank condensate, a scenario discussed in Appendix E, or

d) any combination of the foregoing.

There are numerous combinations and permutations regarding data availability, and the reader is referred to McCain [1990] Chap. 7 for a detailed discussion and example calculations for several different scenarios. The procedure[1] discussed here assumes Scenario b) for compositional data.

J.2 Procedure

J.2.1 Calculation of the Composition of Reservoir Gas

Field and laboratory data are used to calculate the number of lbm-mol of separator condensate per lbm-mol of separator gas, which becomes the basis for a mathematical combination of the two fluids. In the following steps, reference is made to **Table J.1,** where Cols. "D" through "G"—Properties of Components—are taken from Eilerts [1957]. Boldface numbers (and fonts) here correspond to boldface numbers (and fonts) in Table J.1. Computational steps are:

1) calculate the number of barrels per lbm-mol of separator condensate for each component:
   -multiply the values in Col. "G" (gal/lbm-mol) by the mole-fraction of the corresponding component divided by 42—e.g., for C1, bbl/mol = 0.0869 × 5.530/42 = 0.0114 , Col. "H"
   -for C7+, gal/lbm-mol = (113/0.755) × (42/350) = 17.959, which is used to calculate bbl/mol for the C7+ component using the above formula
   -sum the individual components—thus, there are **0.3477** separator barrels of condensate per lbm-mol

2) lbm-mol per barrel of separator condensate is calculated as 1/0.3477 or 2.876

3) calculate the number of lbm-mol of separator condensate per lbm-mol of separator gas:
   -from the ideal gas law, estimate scf/lbm-mol for the separator gas, V_M, for, in this case, "standard[2] conditions" of 15.025 psia and 60°F:

   $$V_M = RT/p = 10.732 × 520/15.025 = 371.42 \text{ scf/lbm-mol,}$$

   where 10.732 is the universal gas constant.

   -from the field test data, calculate the number of separator barrels of condensate per lbm-mol of separator gas:

   $$(14.38 \text{ STBC/MMscf}) × (1.216 \text{ SpBC/STBC}) × 371.42 \text{ scf/lbm-mol}) =$$

   $$0.00649 \text{ SpBC/lbm-mol separator gas}$$

   -calculate the number of lbm-mol of separator condensate per lbm-mol of separator gas:

   $$(2.876 \text{ lbm-mol/SpBC}) × (0.00649 \text{ SpBC/lbm-mol separator gas}) =$$

---

[1] Although the basic data were adapted from McCain [1990], the procedure discussed here follows that of Craft, Hawkins, and Terry [1991]. Despite the procedural difference, the results reported here are practically identical to those reported by McCain [1990].

[2] This is the standard pressure base for Alabama, Colorado, Louisiana, Mississippi, Montana, Nebraska, New Mexico, Utah, and Wyoming. Other states use either 14.65 or 14.73 psia. U.S. DOE/EIA uses 14.73 psia.

0.0187 lbm-mol separator condensate per lbm-mol of separator gas

4) calculate the composition of the reservoir gas:
   -multiply the mole-fraction of separator condensate for each component in Col. "C" by 0.0187, Col. "I"
   -sum Cols. "B" and "I" in Col. "J" and normalize by dividing by **1.0187**, Col. "K."

J.2.2 Calculation of some Properties of the Reservoir Gas

In the absence of a PVT analysis several properties of a reservoir gas may be used to estimate other properties, as discussed in Appendix E. These properties include specific gravity, pseudocritical pressure, and pseudocritical temperature.

Specific Gravity

The specific gravity of a gas is defined with respect to the molecular weight of air, which is approximately 28.97. From Table J.1, the molecular weight of the reservoir gas, **20.86**, is calculated by summing the products of the mole fraction of each component, Col. "K," by the molecular weight of the corresponding component, Col. "D." The specific gravity is calculated as **20.86/28.97 = 0.720.**

Pseudocritical Properties

The pseudocritical properties of a gas may be computed by summing the products of the mole fraction of each component by the corresponding pseudocritical property. For example, in Table J.1 the pseudocritical pressure, **662** psia, Col. "M," is calculated by summing the products of Col. "K" and "E." The pseudocritical temperature, Col. "N," is calculated with a similar procedure.

For such calculations, the pseudocritical properties of the C7+ component may be estimated from [Kesler and Lee 1976]:

$$p_{pc} \approx \exp[8.3634 - 0.0566/\gamma_{C7+} - (0.24244 + 2.2898/\gamma_{C7+} + 0.11857/\gamma_{C7+}^2)10^{-3}T_B$$

$$+ (1.4685 + 3.648/\gamma_{C7+} + 0.47227/\gamma_{C7+}^2)10^{-7}T_B^2$$

$$- (0.42019 + 1.6977/\gamma_{C7+}^2)10^{-10}T_B^5] , \quad\text{..(J.1)}$$

$$T_{pc} \approx 341.7 + 811\gamma_{C7+} + (0.4244 + 0.1174\gamma_{C7+})\, T_B + (0.4669 - 3.2623\gamma_{C7+})10^5/T_B , \quad\text{.........(J.2)}$$

where T_B is the boiling point, which can be estimated by [Whitson 1982]:

$$T_B \approx (4.5579M_{C7+}^{0.15178}\, \gamma_{C7+}^{0.15427})^3 . \quad\text{... (J.3)}$$

TABLE J.1—CALCULATION OF THE COMPOSITION AND PROPERTIES OF RESERVOIR GAS FROM COMPOSITION OF SEPARATOR GAS AND SEPARATOR CONDENSATE

Field Data:

CGR = 14.38 STBC/MMScf
R_{sf} = 1.216 SpBC/STBC
°API = 55.9   (SG= 0.7551)

Laboratory Data and Calculations

Properties of C7+ fraction of separator condensate

S.G.= 0.794
M.W.= 113 lbm/lbm-mol

-A-	-B-	-C-	-D-	-E-	-F-	-G-	-H-	-I-	-J-	-K-	-L-	-M-	-N-
	Separator Fluids		Properties of Components				Calculations*			Reservoir Gas*			
Comp	Gas	Cond	MW	CritPress	CritTemp	gal/lbm-mol	bbl/mol	0.0187(C)	(B)+(I)	Comp	MW	CritPress	CritTemp
C1	0.8372	0.0869	16.04	666.4	342.9	5.530	0.0114	0.00163	0.8388	0.8234	13.21	548.7	282.4
C2	0.0960	0.0569	30.07	706.5	549.5	7.440	0.0101	0.00106	0.0971	0.0953	2.87	67.3	52.4
C3	0.0455	0.0896	44.10	616.0	665.6	10.417	0.0222	0.00168	0.0472	0.0463	2.04	28.5	30.8
iC4	0.0060	0.0276	58.12	527.0	734.0	12.380	0.0081	0.00052	0.0065	0.0064	0.37	3.4	4.7
nC4	0.0087	0.0539	58.12	550.6	765.2	11.929	0.0153	0.00101	0.0097	0.0095	0.55	5.2	7.3
iC5	0.0028	0.0402	72.15	490.4	828.7	13.851	0.0133	0.00075	0.0036	0.0035	0.25	1.7	2.9
nC5	0.0022	0.0400	72.15	488.6	845.4	13.710	0.0131	0.00075	0.0029	0.0029	0.21	1.4	2.4
C6	0.0014	0.0782	86.18	436.9	913.2	15.565	0.0290	0.00146	0.0029	0.0028	0.24	1.2	2.6
C7+	0.0002	0.5267	113.00	455.0	1068.0	17.959	0.2252	0.00985	0.0100	0.0099	1.11	4.5	10.5
Total	1.0000	1.0000					**0.3477**	0.01870	**1.0187**	1.0000	**20.86**	**662.0**	**396.0**

S.G.= **0.720**

*Boldface identifies values discussed in text.

NOTATION AND ABBREVIATIONS

Notation

Most notation is SPE [1993] standard, but is listed here for completeness and convenience. Other notation was developed to make equations in cited publications consistent with SPE notation. Units in this section are consistent with those used in equations in the text.[1]

$\approx$	= approximately equal
a	= nominal decline rate (1/month[i])
a_i	= initial nominal decline rate (1/month)
a_w	= distance between wells (ft)
A	= area (ac or ha)
A_g	= area of nonassociated gas reservoir or gas cap (ac or ha)
A_o	= area of oil reservoir or oil column in oil reservoir with gas cap (ac or ha)
A_w	= well spacing (ac or ha)
A_{wD}	= dimensionless well spacing
b	= hyperbolic decline exponent
b_g	= reciprocal gas formation volume factor (scf/RB or SV/RV)
B_g	= gas formation volume factor (RB/scf or Rcf/scf)
B_{gi}	= gas formation volume factor at initial conditions (RB/scf or Rcf/scf)
B_{ob}	= oil formation volume factor at bubblepoint pressure (RB/STB or RV/SV)
B_{oi}	= oil formation volume factor at initial conditions (RB/STB or RV/SV)
B_{ti}	= initial two-phase formation volume factor; oil plus dissolved gas (RB/STB or RV/SV)
B_w	= water formation volume factor (RB/STB or RV/SV)
c_e	= effective compressibility of oil, water, and pore volume (1/psi)
c_o	= oil compressibility (1/psi)
c_p	= pore volume compressibility (1/psi)
c_{pf}	= fracture "pore" volume compressibility (1/psi)
$(c_p)_H$	= pore volume compressibility measured under hydrostatic stress (1/psi)
$(c_p)_U$	= pore volume compressibility measured under uniaxial stress (1/psi)
c_{pma}	= matrix pore volume compressibility (1/psi)
c_{pn}	= pore volume compressibility at maximum net overburden pressure (1/psi)
c_{px}	= pore volume compressibility at minimum net overburden pressure (1/psi)
c_t	= total (matrix plus fractures) compressibility (1/psi)
c_w	= water compressibility (1/psi)
C	= water influx constant (bbl/psi)
C	= decline curve constant; defined as $a_i/q_i{}^b$
C_i	= condensate initially in place (STBC; usually reported as MSTBC[ii])
C_p	= cumulative condensate production (STBC)
$(C_{pa})_{pd}$	= ultimate recovery of condensate attributable to pressure depletion (STBC)
C_R	= condensate reserves; specify date (STBC)
C_{Ri}	= initial condensate reserves (STBC)
D	= effective decline rate (dim)
D	= depth (ft)
D_{ss}	= depth subsea (ft)
E	= gas expansion; term used by gas engineers (scf/Rcf)
E_c	= "expansion" of initial reservoir pore volume due to compression effects on pore volume and interstitial water (V/V/psi)
E_D	= microscopic displacement efficiency; defined as $(1 - S_{wi} - S_h)/(1 - S_{wi})$

---

[1] Despite efforts to standardize to the "SI" system of units, most oilfield usage still reflects the traditional British system. Accordingly, with the exception of key equations in Sec. 3, the latter system is used herein.

E_g = expansion of a unit volume of free gas (RB/scf)

$(E_{Rc})_{pd}$ = recovery efficiency of condensate attributable to pressure depletion (fr)

$(E_{RgF})_{pd}$ = recovery efficiency of free gas attributable to pressure depletion (fr)

$(E_{Ro})_{sg}$ = recovery efficiency of oil attributable to solution gas drive (fr)

$(E_{Ro})_{wd}$ = recovery efficiency of oil attributable to waterdrive (fr)

$(E_{Ro})_{xp}$ = recovery efficiency of oil attributable to expansion above the bubblepoint (fr)

E_t = expansion of a unit volume of oil and initially dissolved gas (RB/STB or RV/SV)

E_V = volumetric sweep efficiency (fr)

f_o = fractional flow of oil; defined as $q_o/(q_w + q_o)$

f_{wa} = fractional flow of water at abandonment, usually the economic limit; defined as $q_w/(q_w + q_{oa})$

F_g = pore geometrical factor defined by Thomeer [1960]

F_{pR} = reservoir volume of produced fluids: oil, free gas and water (RB)

g = gravitational constant

g_P = pressure gradient (psi/ft)

g_T = thermal gradient (deg/ft)

G_{Fi} = free gas initially in place (scf; usually reported as MMscf[iii])

$(G_i)_m$ = cumulative gas injected, multiple of m (dim)

G_p = cumulative gas production (scf)

G_{pa} = cumulative gas production at the economic limit; ultimate recovery (scf)

G_r = gas remaining in the reservoir, both free and dissolved (scf)

G_R = gas reserves; specify date (scf)

G_{Ri} = initial gas reserves (scf)

G_{RFi} = initial free gas reserves (scf)

G_{RSi} = initial solution gas reserves (scf)

G_{Si} = solution gas initially in place (scf)

h_n = net thickness (ft)

h_{nG} = average net thickness of free gas in a nonassociated gas reservoir or gas cap (ft)

h_{nO} = average net thickness of oil reservoir or oil column (ft)

h_t = gross thickness (ft)

h_{tG} = average gross thickness of gas column (ft)

h_{tO} = average gross thickness of oil reservoir or oil column (ft)

h_{tow} = gross thickness of oil column in a well (ft)

k = permeability (md or D)

k_g = effective permeability to gas (md)

k_H = horizontal permeability (md)

k_o = effective permeability to oil (md)

k_V = vertical permeability (md)

L_c = vertical distance well completion extends into the oil zone (ft)

L_{cD} = dimensionless well penetration; defined as L_c/h_{tow}

m = ratio of reservoir volume of initial gas cap to reservoir volume of initial oil column (dim); in petrophysics, m is the cementation exponent

$m_1(x)$ = the first moment, or mean, of the distribution of x

$m_2(x)$ = the second moment, or variance, of the distribution of x

M_c = molecular weight of condensate

M_o = molecular weight of oil

M_{wo} = water-oil mobility ratio (dim)

n = saturation exponent (petrophysics)

N_e = estimated oil reserves; generally, the abscissa of an expectation curve

N_G = gravity number

N_i = oil initially in place (STB; usually reported as MMSTB[iv])

N_p = cumulative oil production; specify date (STB)

N_{pa} = cumulative oil production at abandonment; ultimate recovery (STB)

N_r	= oil remaining in place; specify date (STB)
N_R	= oil reserves; specify date (STB)
N_{Ri}	= initial oil reserves (STB)
p_a	= reservoir pressure at abandonment (psi)
p_b	= bubblepoint pressure (psi)
p_d	= dewpoint pressure (psi)
p_H	= hydrostatic pressure (psi)
p_{Hn}	= net hydrostatic pressure (psi)
p_i	= initial reservoir pressure (psi)
p_{ob}	= overburden pressure (psi)
p_{obn}	= net overburden pressure (psi)
$(p_{on})_n$	= minimum net overburden pressure (psi)
$(p_{on})_x$	= maximum net overburden pressure (psi)
p_{pc}	= pseudocritical pressure (psi)
p_{pr}	= pseudoreduced pressure (dim)
p_r	= reference pressure; at $D=0$ or $D_{ss}=0$ (psi)
p_s	= saturation pressure; p_b or p_d (psi)
p_{sp}	= separator pressure (psi)
p_{sc}	= pressure at standard conditions (psi)
p_U	= uniaxial pressure (psi)
P_{cL}	= capillary pressure at laboratory conditions (psi)
P_{cR}	= capillary pressure at reservoir conditions (psi)
P_{cm}	= mercury-air capillary pressure (psi)
P_{dm}	= mercury-air displacement pressure (psi)
q	= production rate (STB or scf) per day, month, or year
q_{el}	= economic limit production rate (STB or scf) usually, per day
q_g	= gas production rate (scf); per day, month, or year
q_o	= oil production rate (STB); per day, month, or year
Q	= cumulative production (scf or STB)
Q_a	= cumulative production at abandonment (scf or STB)
$Q_D(\Delta t_D, r_D)$	= van Everdingen & Hurst solution to the diffusivity equation for unsteady state water influx
r_{dw}	= drainage radius of well (ft)
r_D	= dimensionless boundary; r_e/r_i (dim)
r_e	= exterior boundary (ft)
r_i	= interior boundary (ft)
r_w	= radius of wellbore (ft)
r_s	= Spearman rank correlation coefficient
R	= producing gas/oil ratio (scf/STB or m^3/m^3)
R_c	= producing condensate/gas ratio (STBC/MMscf)
R_{cd}	= producing condensate/gas ratio at dewpoint pressure (STBC/MMscf)
R_{ci}	= initial producing condensate/gas ratio (STBC/MMscf)
R_{cp}	= cumulative (produced) condensate/gas ratio (STBC/MMscf)
R_{kmf}	= ratio of matrix to fracture permeability (dim)
R_{ng}	= net-to-gross pay ratio (dim)
R_{ngG}	= net-to-gross ratio, average in gas zone (dim)
R_{ngO}	= net-to-gross ratio, average in oil zone (dim)
R_p	= cumulative produced gas-oil ratio; G_p/N_p (scf/STB or m^3/m^3)
R_{pz}	= ratio of p/z at abandonment to p/z at initial conditions; defined as $p_a z_i / p_i z_a$
R_s	= solution gas-oil ratio (scf/STB)
R_{sb}	= solution gas-oil ratio at the bubblepoint (scf/STB or m^3/m^3)
R_{si}	= initial solution gas-oil ratio (scf/STB or m^3/m^3)
R_{sw}	= gas solubility in reservoir brine
R_t	= true resistivity (petrophysics)

R_T = total GOR from entire separator train including stock-tank vapors (scf/STB)
R_w = resistivity of interstitial water (petrophysics)
$R_{\mu ow}$ = oil/water viscosity ratio (dim)
S_{fD} = dimensionless storage in fractures
S_{gc} = critical gas saturation (fr)
S_{gr} = residual gas saturation (fr)
S_h = hydrocarbon saturation, S_o or S_g (fr)
S_{om} = mobile oil saturation—defined as $(1 - S_{wi} - S_{or})$—(fr)
S_{or} = residual oil saturation (fr)
S_w = water saturation (fr)
S_{wc} = connate, or interstitial, water saturation (fr)
S_{wf} = movable (floodable) water saturation—defined as $(S_w - S_{wi})$—(fr)
S_{wG} = average water saturation in the free gas zone (fr)
S_{wi} = irreducible, not necessarily initial, water saturation (fr)
S_{wma} = matrix water saturation (fr)
S_{wO} = average water saturation in the oil zone (fr)
Δt_D = incremental dimensionless time
Δt_{el} = incremental time to reach economic limit; usually written t_{el} (months or years)
T_B = boiling point temperature (°R)
T_c = critical temperature (°R or °F)
T_{ct} = cricondentherm (°R or °F)
T_f = formation (reservoir) temperature (°R or °F)
T_{pc} = pseudocritical temperature (°R)
T_{pr} = pseudoreduced temperature (dim)
T_r = reference temperature; at D=0 or D_{ss}=0 (°R or °F)
T_s = average surface temperature (°R or °F)
T_{sc} = temperature at standard conditions (°R or °F)
V_{eq} = gas equivalent of stock-tank condensate, plus gas from stock-tank and second stage separator, if any (scf/STBC)
V_h = fraction of reservoir filled with hydrocarbons (dim)
V_{hg} = fraction of hydrocarbon volume occupied by free gas (dim)
V_M = standard cubic feet per pound mole
V_{pR} = reservoir pore volume (RB)
V_{ps} = pore volume (core) at standard conditions (cc)
V_t = gross rock volume (GAF or m^3)
V_{tG} = gross rock volume in gas zone (GAF or m^3)
V_{tO} = gross rock volume in oil zone (GAF or m^3)
W_e = cumulative water influx (RB)
W_p = cumulative water production (STB)
Y = mole fraction in gas phase
z = mole fraction in mixture
$z(x)$ = standard deviations the (Gaussian) variable deviates from the median
Z = height above a datum (ft)
Z = gas deviation factor; identified as "Z-factor" (dim)
Z_a = Z-factor at reservoir abandonment pressure (dim)
Z_{FWL} = height above free water level (ft)
Z_i = Z-factor at initial reservoir pressure (dim)

α = dip angle (deg)
Δ = change or increment
ϕ = porosity (fr)
ϕ_{co} = cutoff porosity (fr)
$(\phi_{co})_{2P}$ = cutoff porosity for proved plus probable net pay
ϕ_{coD} = dimensionless cutoff porosity

ϕ_f = fracture porosity (fr)
ϕ_{gR} = average porosity for gas reservoir or gas cap (fr)
ϕ_{ma} = matrix porosity (fr)
ϕ_{min} = minimum porosity observed in a data set (fr)
ϕ_n = porosity for the "nth" interval, usually in a well (fr)
ϕ_{oR} = average porosity for oil reservoir or oil column (fr)
ϕ_W = average porosity for a well zone (fr)
μ_g = gas viscosity (cp)
μ_o = oil viscosity (cp)
μ_{oa} = oil viscosity at atmospheric conditions; "dead oil" (cp)
μ_{ob} = oil viscosity at bubblepoint (cp)
μ_{oi} = oil viscosity at initial conditions (cp)
μ_{wi} = water viscosity at initial conditions (cp)
ρ_o = density of oil (gm/cc or lb/cf)
γ = specific gravity, not density (dim)
γ_{API} = specific gravity of oil, °API; written "°API" in some equations
γ_{gR} = specific gravity of reservoir gas (air=1)
γ_{gS} = specific gravity of separator gas (air=1)
σ = interfacial tension (dynes/cm)
σ = standard deviation (of a sample set)
σ_{ow} = oil-water interfacial tension (dynes/cm)
σ_{owL} = oil-water interfacial tension at laboratory conditions (dynes/cm)
v = Poisson's ratio (dim)
θ = contact angle (degrees)
θ_R = contact angle at reservoir conditions (degrees)

Abbreviations[2]

2D = acquisition/processing methodology for seismic data that results in (only) a two dimensional portrayal of earth structure and/or stratigraphy

3D = acquisition/processing methodology for seismic data that results in a three dimensional portrayal of earth structure and/or stratigraphy

4D = seismic data acquired during sequential 3D surveys that may be processed to portray changes in reservoir fluid saturation

1P = proved reserves; also used to designate an estimate based on the lower predicted bound of a least squares fit at 80% confidence (e.g., 1P FWL)

2P = proved (P1) plus probable (P2) reserves; also used to designate an estimate based on a least squares fit (e.g., 2P FWL)

3P = proved (P1) plus probable (P2) plus possible (P3) reserves; also used to designate an estimate based on the upper predicted bound of a least squares fit at 80% confidence (e.g., 3P FWL)

AFE = authority for expenditure
AGA = American Gas Assoc.
antilog(x) = 10 raised to the "x" power
API = American Petroleum Inst.
°API = API gravity
°API$_D$ = API gravity of residual oil from differential liberation
APPEA = Australian Petroleum Production and Exploration Assn.
atm = atmospheres
B = billion; same magnitude as "G," or giga, which is used in some areas (10^9); denoted as "bn" by some publications

---

[2] This list is provided (primarily) for readers whose primary language is not English, as a guide to often confusing engineering abbreviations!

bbl	= barrel(s)
BHP	= bottom-hole pressure; in this work, refers to static reservoir pressure
BTLPD	= barrels total liquid per day
CAOFP	= calculated absolute open flow potential (MMscfd)
cdf	= cumulative density function; the cumulative frequency of a stochastic variable; also called cumulative pdf
cm	= cubic meters (also written m^3)
cp	= centipoise(s)
C_1, C_2	= methane, ethane, etc.
C_{3+}	= propane, plus heavier fractions
CCLP	= core calibrated log porosity
CGR	= condensate-gas ratio (STBC/MMscf); general notation, with specific conditions identified with subscripts
CGR_t	= CGR measured during a test (STBC/MMscf)
CGR_p	= cumulative CGR (STBC/MMscf)
CIP	= condensate initially in place (STB or m^3)
D	= Darcies
dim	= dimensionless; corresponding subscript is "$_D$"
DDL	= downdip limit
e	= base of natural logarithms (2.718282 . . .)
E+7	= 10 raised to the 7th power (10,000,000)
E&P	= exploration and production
EC	= expectation curve; complement of the cdf
ECR	= estimation and classification of reserves
EIA	= Energy Information Administration (of U.S. Dept. of Energy)
EL	= economic limit; generally refers to production rate
ER	= expected (quantity of) reserves
EUFGR	= estimated ultimate free gas recovery
EUOR	= estimated ultimate oil recovery
EV	= expected value
$\exp(x)$	= "e" raised to the "x" power
FAAR	= fraction of adjusted annual reserves
FGIP	= free gas initially in place; may also be written as GIP (MMscf or MMm^3)
$fn(x,y)$	= function of arguments in parentheses
fr	= fraction
ft	= feet
FTP	= flowing tubing pressure (psig)
GAF	= gross acre feet
GCR	= gas-condensate ratio (scf/STBC)
GDT	= gas down to, same as LKG, subsea depth
GEC	= gas equivalent of condensate (only)
GEC+	= gas equivalent of condensate plus stock-tank vapors plus second stage separator, if any
GIP	= (free) gas initially in place (scf or m^3)
GOR	= gas-oil ratio (scf/STB or m^3/m^3))
GRV	= gross rock volume (acre-ft or m^3)
GT	= greater than (also denoted as >)
GTE	= greater than or equal to (also denoted as ≥)
ha	= hectares (10^4 m^2)
HCPV	= hydrocarbon pore volume; equal to $Ah\phi(1 - S_w)$
HKO	= highest known oil, same as "OUT," subsea depth
HKW	= highest known water, same as "WUT," subsea depth
HWTZ	= hydrocarbon-water transition zone

IOR	= improved oil recovery; used here in preference to the terms "secondary oil recovery," "tertiary oil recovery," or "enhanced oil recovery"
ISITP	= initial shut-in tubing pressure (usually psig)
kPa	= kilopascal
LGR	= liquid/gas ratio (bbl/MMscf)
LHS	= Latin hypercube sampling
LKG	= lowest known gas, same as "GDT," subsea depth
LKO	= lowest known oil, same as "ODT," subsea depth
$\ln(x)$	= natural logarithm (base e) of "x"
$\log(x)$	= logarithm (base 10) of "x"
LRA	= linear regression analysis
LRLC	= low resistivity, low contrast; generally refers to sandstone reservoirs
LSF	= least-squares fit
LT	= less than (also denoted as <)
LTE	= less than, or equal to (also denoted as ≤)
m	= meter(s)
M	= molecular weight; always subscripted, for example, M_o
M	= thousand (10^3)
MCS	= Monte Carlo simulation, including simulation using Latin hypercube sampling
md	= millidarcies
microsip	= $1/10^6$ psi
MM	= million (10^6); denoted as "mn" by some publications
MMbbl	= million (stock tank) barrels; also written as "MMSTB"
MMm^3	= million (standard) cubic meters
MMscf	= million standard cubic feet; also written as "MMcf"
MMscfd	= million standard cubic feet per day; also written as "MMscf/D," or "MMcfd"
mtn	= metric ton, or tonne, (1000 kg or 2200 lb)
NAF	= net acre feet
ODT	= oil down to, same as "LKO," subsea depth
OIP	= oil initially in place (usually STB, but may be tonnes); elsewhere, abbreviated as 'STOOIP"
O/GIP	= oil and/or (free) gas initially in place
OUT	= oil up to, same as "HKO," (subsea depth)
P1	= proved reserves; an estimated quantity of oil, gas, or condensate
P2	= probable reserves; an estimated quantity of oil, gas, or condensate; estimated separately or as an increment to proved reserves
P3	= possible reserves; an estimated quantity of oil, gas, or condensate; estimated separately or as an increment to probable reserves
P90	= 10th percentile of a quantity estimated using probabilistic procedures; denotes probability that the actual quantity will equal or exceed the estimate, i.e., $Pr(X_a \geq X_e) = 0.9$; typically considered a proved estimate; some geologists use the term to denote the 90th percentile
P50	= 50th percentile of a quantity estimated using probabilistic procedures; denotes probability that the actual quantity will equal or exceed the estimate, i.e., $Pr(X_a \geq X_e) = 0.5$; typically considered the sum of proved plus probable estimates
P10	= 90th percentile of a quantity estimated using probabilistic procedures; denotes probability that the actual quantity will equal or exceed the estimate, i.e., $Pr(X_a \geq X_e) = 0.1$; typically considered the sum of proved plus probable plus possible estimates; some geologists use the term to denote the 10th percentile
Pa	= pascals

PbBP	= probable, behind pipe
PbPd	= probable, producing (always incremental to PvPd)
PbND	= probable, not drilled (preferred here, more generally denotes proved, not developed)
Pc10	= 10th percentile (numerically equivalent to P90 as used here)
PC	= personal computer or workstation
pdf	= probability density function
P/DTA	= production/decline trend analysis
PNC	= pulsed neutron capture
$Pr(X \geq X_1)$	= probability X is greater than, or equal to, X_1
$Pr(X\|E)$	= conditional probability of X, given the occurrence of event "E"
PsBP	= possible, behind pipe
PsND	= possible, not drilled
psi	= pounds per square inch (use only if measurement conditions unspecified)
psia	= pounds per square inch, absolute
psig	= pounds per square inch, gauge
PvBP	= proved, behind pipe
PvPd	= proved, producing
PvND	= proved, not drilled (preferred here, more generally denotes proved, not developed)
PvSI	= proved, shut-in (use only if previously produced or tested)
PvWF	= proved, waiting on facilities (generally, implies previously tested)
PVT	= pressure-volume-temperature
RA	= risk adjusted; procedure used for probabilistic estimates
RBHCPS	= reservoir barrels (of) hydrocarbon pore space
RBW/RBO	= reservoir barrel(s) (of) water per reservoir barrel (of) oil
RFT	= repeat formation tester; used to identify a family of such devices, as discussed in Appendix G
RV	= random variable
RV/SV	= reservoir volume/surface volume (consistent units)
RW	= risk weighted; procedure used for deterministic estimates
scf	= standard cubic feet (of natural gas)
scfd	= standard cubic feet per day; may also be written as scf/D
SD	= standard deviation
SG	= specific gravity (^{o}API or with reference to air or water)
SGIP	= solution gas initially in place (scf)
SITP	= shut-in tubing pressure; usually reported as psig
SM	= Swanson's mean
SP	= spill point
SpBC	= separator barrel(s) of condensate
SRC	= Spearman rank correlation
STB	= stock-tank barrel(s)
STBC	= stock-tank barrel(s) of condensate
STBD	= stock-tank barrel(s) per day
STBO	= stock-tank barrel(s) of oil
TLPD	= total liquids per day; usually oil plus water, but may refer to condensate plus water
TPA	= transient pressure analysis
Tscf	= trillion (10^{12}) standard cubic feet (of gas); also written as Tcf
TVDSS	= true vertical depth subsea (feet or meters)
UDL	= updip limit
U.S. DOE/EIA	= United States Dept. of Energy-Energy Information Admin.
U.S. SEC	= United States Securities and Exchange Commission

WFT   = wireline formation tester; "WFT" is used here in a "generic" sense and
                            refers to devices lowered into a well on electrical cable that are designed to
                            measure formation pressure and recover fluid samples; devices include the
                            MDT,^{TM} introduced in 1990 by Schlumberger, and the RCI,^{TM} introduced
                            by Western Atlas in 1995
WGR   = water-gas ratio (bbl wtr/MMscf gas)
WOR   = water-oil ratio (bbl wtr/bbl oil)

---

[i] Depending on circumstances, units also may be 1/day or 1/year
[ii] Reporting of condensate also may be in MMBC, or in multiples of tonnes, depending on circumstances.
[iii] Reporting of natural gas also may be in Mcf, Bcf, or Tcf, or multiples of cubic meters, depending on circumstances.
[iv] Reporting of crude oil also may be in Mbbl or Bbbl, or in multiples of tonnes, depending on circumstances.

REFERENCES

References Cited

Aalund, L.R.: "EOR Projects Decline, But CO_2 Pushes Up Production," *Oil & Gas J.* (18 April 1988) 33.

Abdul-Majeed, G.H. and Salman, N.H.: "Statistical Evaluation of PVT Correlations—Solution Gas Oil Ratio," *J. Cdn. Pet. Tech.* (July–August 1988) 95.

Abdul-Majeed, G.H. and Salman, N.H.: "An Empirical Correlation for Oil FVF Prediction," *J. Cdn. Pet. Tech.* (November–December 1988) 118.

Abrahamson, K.A., Lindbo, K., and Vollset, J.: "Resource Classifications and Their Usefulness in the Resource Management of an Oil Company," *Proc.*, Norwegian Petroleum Society Conference, Stavanger, 6–8 December 1993; *Quantification and Prediction of Petroleum Resources,* A.G. Dore, and Larsen R. Sinding (eds.), NPF Special Publication 6, Elsevier Science Publishing Co., Amsterdam (1996) 63.

Acuna, H.G. and Harrell, D.R.: "Adapting Probabilistic Methods To Conform to Regulatory Guidelines," paper SPE 63202 presented at the 2000 SPE Annual Technical Conference and Exhibition, Dallas, 1–4 October.

Adams, D.M.: "Experiences With Waterflooding Lloydminster Heavy-Oil Reservoirs," *JPT* (August 1982) 1643; *Trans.,* AIME, **273.**

Agarwal, R.G., Al-Hussainy, R., and Ramey, H.J.: "The Importance of Water Influx in Gas Reservoirs," *JPT* (November 1965) 1336; *Trans.,* AIME, **234.**

Aguilera, R., Franks, L.N., and Au, A.D.: "Well Test Analysis of a Naturally Fractured Gas Reservoir—A Case History," *J. Cdn. Pet. Tech.* (April 1992) 46.

Aguilera, R., Au, A.D., and Franks, L.N.: "Isochronal Testing of a Naturally Fractured Gas Reservoir: A Case History," *J. Cdn. Pet. Tech.* (October 1993) 25.

Aguilera, R.: "Recovery Factors and Reserves in Naturally Fractured Reservoirs," *J. Cdn. Pet. Tech.* (July 1999) 15.

Ajufo, A.O.: "Capillary Pressure Characteristics at Overburden Pressure Using the Centrifuge Method," paper SPE 26148 presented at the 1993 SPE Gas Technology Symposium, Calgary, 28–30 June.

Al-Fattah, S.M. and Al-Marhoun, M.A.: "Evaluation of Empirical Correlations for Bubblepoint Oil Formation Volume Factor," *Saudi Aramco J. of Tech.* (Winter 1995/96) 45.

Alger, R.P., Luffel, D.L., and Truman, R.B.: "New Unified Method of Integrating Core Capillary—Pressure Data With Well Logs," *SPEFE* (June 1989) 145.

Al-Marhoun, M.A.: "PVT Correlations for Middle East Crude Oils," *JPT* (May 1988) 650; *Trans.,* AIME, **285.**

Al-Marhoun, M.A.: "New Correlations for Formation Volume Factors of Oil and Gas Mixtures," *J. Cdn. Pet. Tech.* (March 1992) 22.

Al-Yousef, H.Y. and Al-Marhoun, M.A.: "Discussion of Correlation of PVT Properties for UAE Crudes," *SPEFE* (March 1993) 80.

Amaefule, J.O. *et al.:* "Reservoir Description: A Practical Synergistic Engineering and Geological Approach Based on Analysis of Core Data," paper SPE 18167 presented at the 1988 SPE Annual Technical Conference and Exhibition, Houston, 2–5 October.

Amaefule, J.O. *et al.:* "Enhanced Reservoir Description: Using Core and Log Data To Identify Hydraulic (Flow) Units and Predict Permeability in Uncored Intervals/Wells," paper SPE 26436 presented at the 1993 SPE Annual Technical Conference and Exhibition, Houston, 3–6 October.

American Assn. of Petroleum Geologists (AAPG): *Geology of Tight Gas Reservoirs,* C.W. Spencer, and R.F. Mast, (eds.), AAPG, Tulsa, Oklahoma (1986).

American Geological Inst. (AGI): *Glossary of Geology,* M. Gary, R. McAfee, Jr. and C.L. Wolf, (eds.), AGI, Washington, DC (1972).

American Petroleum Inst. (API): *Reserves of Crude Oil, Natural Gas Liquids and Natural Gas in the United States and Canada as of December 31, 1979,* API, Washington (1980) **34.**

American Petroleum Inst. (API): *A Statistical Study of Recovery Efficiency*, API *Bull*. D14, second edition, API, Dallas (1984).

Amyx, J.W., Bass, D.M. Jr., and Whiting, R.L.: *Petroleum Reservoir Engineering*, McGraw Hill Book Co., New York City (1960) Chap.2.

Anders, E.L. Jr.: "Mile Six Pool—An Evaluation of Recovery Efficiency," *Trans.*, AIME (1953) **198**, 279.

Andersen, M.A.: "Predicting Reservoir-Condition PV Compressibility From Hydrostatic-Stress Laboratory Data," *SPERE* (August 1988) 1078; *Trans.*, AIME, **285**.

Archer, J.S. and Wall, C.G.: *Pet. Eng. Principles and Practice*, Graham and Trotman, Ltd., London (1986).

Archie, G.E.: "The Electrical Resistivity Log as an Aid in Determining Some Reservoir Characteristics," *Trans.*, AIME (1942) **146**, 54.

Arps, J.J.: "Analysis of Decline Curves," *Trans.*, AIME (1945) **160**, 228.

Arps, J.J. and Roberts, T.G.: "The Effect of the Relative Permeability Ratio, the Oil-Gravity and the Solution Gas-Oil Ratio on the Primary Recovery From a Depletion Type Reservoir," *Trans.*, AIME (1955) **204**, 120.

Arps, J.J.: "Estimation of Primary Oil Reserves," *Trans.*, AIME (1956) **207**, 182.

Arps, J.J. and Roberts, T.G.: "Economics of Drilling for Cretaceous Oil on East Flank of Denver-Julesberg Basin," *AAPG Bull.* (1958) **42**, 2549.

Arps, J.J.: "Estimation of Primary Oil and Gas Reserves," *Petroleum Production Handbook*, T. Frick (ed.), SPE, Dallas, Texas (1962) Chap. 37.

Arps, J.J.: "Engineering Concepts Useful in Oil Finding," *AAPG Bull.* (1964) **48**, 157.

Arps, J.J. *et al.*: *A Statistical Study of Recovery Efficiency*, API *Bull*. D-14, American Petroleum Inst., Dallas (1967).

Asgarpour, S. *et al.*: "Pressure-Volume-Temperature Correlations for Western Canadian Gases and Oils," *J. Cdn. Pet. Tech.*, (July–August 1989) 103.

Ashbaugh, J.P. and Flemings, P.B.: "Dynamic Reservoir Characterization of the ST295 Field (Offshore Louisiana): Reservoir Simulation, Acoustic Modeling, and Time-Lapse Seismic Refines Geologic Model and Illuminates Dynamic Behavior," Soc. of Exploration Geophysicists Workshop, Houston, 5 November 1999.

Asquith, G. and Gibson, C.: *Basic Well Log Analysis for Geologists*, American Assn. of Petroleum Geologists, Tulsa, Oklahoma (1982).

Athichanagorn, S., Horne, R.N., and Kikani, J.: "Processing and Interpretation of Long-Term Data From Permanent Downhole Pressure Gauges," paper SPE 56419 presented at the 1999 SPE Annual Technical Conference and Exhibition, Houston, 3–6 October.

Australian Petroleum Production and Exploration Assn. (APPEA): "Guidelines for Reporting Oil and Gas Reserves," Canberra, Australia (May 1995),.

Australian Petroleum Production and Exploration Assn. (APPEA): "Guidelines for Reporting Oil and Gas Reserves," Canberra, Australia (1 May 1995).

Bahorich, M. and Farmer, S.: "3-D Seismic Discontinuity for Faults and Stratigraphic Features: The Coherence Cube," *Leading Edge* (October 1995) **14**, 10, 1053.

Bailey, W.: "Optimized Hyperbolic Decline Curve Analysis of Gas Wells," *Oil & Gas J.* (15 February 1982) 116.

Bain, J.S., Nordberg, M.O., and Hamilton, T.M.: "Three-Dimensional Seismic Applications in Interpretation of Dunlin Field, U.K. North Sea," *JPT* (March 1981) 407; *Trans.*, AIME, **271**.

Bankhead, C.C.: "Processing of Geological and Engineering Data in Multipay Fields for Evaluation," paper SPE 256 presented at the 1962 SPE Petroleum Economics and Evaluation Symposium, Dallas, 15–16 March.

Barba, R.E.: "Optimizing Hydraulic-Fracture Length in the Spraberry Trend," *SPEFE* (September 1989) 475.

Barbe, J.A. and Schnoebelen, D.J.: "Quantitative Analysis of Infill Performance: Robertson Clearfork Unit," *JPT* (December 1987) 1593; *Trans.*, AIME, **283**.

Barber, A.H. Jr. *et al.*: "Infill Drilling To Increase Reserves—Actual Experience in Nine Fields in Texas, Oklahoma, and Illinois," *JPT* (August 1983) 1530; *Trans.*, AIME, **275**.

Barrufet, M.: "A Brief Introduction to Equations of State for Petroleum Engineering Applications," *Pet. Eng.* (March 1998) 81.

Barrufet, M.: "Evaluation of Standard PVT Properties From Equations of State," *Pet. Eng.* (May 1998) 49.

Barrufet, M.: "Equations of State Properties for Simulated Depletion of Gas Condensates and Volatile Oils," *Pet. Eng.* (August 1998) 49.

Barrufet, M.: "Importance of the C_7^+ Fraction in Phase Behavior Calculations," *Pet. Eng.* (October 1998) 47.

Baskin, D.K., Hwang, R.J., and Purdy, R.K.: "Predicting Gas, Oil and Water Intervals in Niger Delta Reservoirs Using Gas Chromatography," *AAPG Bull.* (March 1995) 337.

Bass, D.M.: "Analysis of Abnormally Pressured Gas Reservoirs with Partial Water Influx," paper SPE 3850 presented at the 1972 SPE Abnormal Subsurface Pressure Symposium, Baton Rouge, Louisiana, 15–16 May.

Bass, D.M.: "Properties of Reservoir Rocks," *Pet. Eng. Handbook*, H.B. Bradley (ed.), SPE, Richardson, Texas (1987) Chap. 26.

Bath, P.G., Fowler, W.N., and Russell, M.P.: "Brent Field, a Reservoir Engineering Review," paper EUR 164 presented at the 1980 SPE European Offshore Petroleum Conference, London, 21–24 October.

Begg, S.H, Carter, R.R., and Dranfield, P.: "Assigning Effective Values to Simulator Gridblock Parameters for Heterogeneous Reservoirs," *SPERE* (November 1989) 455.

Beggs, H.D.: "Oil System Correlations," *Pet. Eng. Handbook*, H.B. Bradley (ed.), SPE, Richardson, Texas (1987) Chap. 22.

Beggs, H.D. and Robinson, J.R.: "Estimating the Viscosity of Crude Oil Systems," *JPT* (September 1975) 1140.

Behrenbruch, P. and Quang Du, P.: "The Significance of Large Variations in Oil Properties of the Dai Hung Field, Vietnam," paper SPE 29302 presented at the 1995 SPE Asia Pacific Oil and Gas Conference, Kuala Lumpur, 20–22 March.

Belardo, S. and Ryan, D. (eds.): *Western Australian Oil & Gas Review,* Dept. of Resource Development, Perth, Australia (1995).

Beliveau, D., Payne, D.A., and Mundry, M.: "Waterflood and CO_2 Flood of the Fractured Midale Field, " *JPT* (September 1993) 881.

Bell, J.S. and Shepard, J.M.: "Pressure Behavior in the Woodbine Sand," *Trans.* AIME (1951) **192,** 19.

Bennion, D.W. and Griffiths, J.C.: "A Stochastic Model for Predicting Variations in Reservoir Rock Properties," *SPEJ* (March 1966) 9; *Trans.,* AIME, **237.**

Berg, O.R.: "Seismic Detection and Evaluation of Delta and Turbidite Sequences: Their Application to the Exploration for the Subtle Trap," *The Deliberate Search for the Subtle Trap,* AAPG Memoir 32, M.T. Halbouty (ed.), American Assn. of Petroleum Geologists, Tulsa, Oklahoma (1982).

Bernard, W.J.: "Reserves Estimation and Performance Prediction for Geopressured Gas Reservoirs," *J. Pet. Sci. & Eng.* (1987) No. 1, 15.

Berry, F.A.F.: "High Fluid Potentials in California Coast Ranges and Their Tectonic Significance," *AAPG Bull.* (1973) **57,** 1219.

Bigno, Y., Baillie, J., and Coombes, T.: "The Interpretation of Reservoir Pressure Data in the Dunbar Field (UKCS)," paper SPE 37758 presented at the 1997 SPE Middle East Oil Show, Bahrain, 15–18 March.

Blasingame, T.A. and Lee, W.J.: "Properties of Homogeneous Reservoirs, Naturally Fractured Reservoirs, and Hydraulically Fractured Reservoirs From Decline Curve Analysis," paper SPE 15018 presented at the 1986 SPE Permian Basin Oil and Gas Recovery Conference, Midland, 13–14 March.

Blauer, R.E., Onat, I.N., and Lemieux, M.J.: "Production Data Indicate Reservoir and Fracture Performance in the Spraberry," paper SPE 23995 presented at the 1992 SPE Permian Basin Oil and Gas Recovery Conference, Midland, 18–20 March.

Blevins, T.R.: "Steamflooding in the U.S.—A Progress Report," *JPT* (May 1990) 548.

Bourgoyne, A.T.: "Shale Water as a Pressure Support Mechanism in Gas Reservoirs Having Abnormal Formation Pressure," *J. Pet. Sci. & Eng.* (1990) No. 3, 305.

Bournazel, C. and Jeanson, B.: "Fast Water-Coning Evaluation Method," paper SPE 3628 presented at the 1971 SPE Annual Technical Conference and Exhibition, New Orleans, 3-6 October.

Brennand, T.P. and van Veen, F.R.: "The Auk Oilfield," *Petroleum and the Continental Shelf of Northwest Europe,* John Wiley & Sons, New York City (1975).

Briggs, P.J. *et al.:* "Development of Heavy Oil Reservoirs," *JPT* (February 1988) 206.

Brigham, W.E., Satman, A., and Soliman, M.Y.: "Recovery Correlations for In-Situ Combustion Field Projects and Application to Combustion Pilots," *JPT* (December 1980) 2132; *Trans.,* AIME, **269.**

Brons, F.: "On the Use and Misuse of Production Decline Curves," *API 801-39E,* API., Dallas (1963).

Brown, C.A., Erbe, C.B., and Crafton, J.W.: "A Comprehensive Reservoir Model of the Low-Permeability Lewis Sands in the Hay Reservoir Area, Sweetwater County, Wyoming," paper SPE 10193 presented at the 1981 SPE ATC&E, San Antonio, 5–7 October.

Browne, M.W.: "Scientist Suggests Some Oil Reservoirs are Refilling," *Houston Chronicle* (1 Oct 1995)

Bruns, J.R., Fetkovich, M.J., and Meitzen, V.C.: "The Effect of Water on p/Z-Cumulative Gas Production Curves," *JPT* (March 1965) 287.

Buchanan, R. and Hoogteyling, L.: "Auk Field Development: A Case History Illustrating the Need for a Flexible Plan," *JPT* (October 1979) 1305.

Bulnes, A.C.: "An Application of Statistical Methods to Core Analysis Data of Dolomitic Limestone," *Trans.,* AIME (1946) **165,** 223.

Burkhart, T., Hoover, A.R., and Flemings, P.B.: "Time-Lapse (4-D) Seismic Monitoring of Primary Production of Turbidite Reservoirs at South Timbalier 295, Offshore Louisiana, Gulf of Mexico," *Geophysics* (March–April 2000) 933.

Bush, J.L. and Helander, D.P.: "Empirical Prediction of Recovery Rate in Waterflooding Depleted Sands," *JPT* (September 1968) 933.

Buthod, P.: "Crude Oil Properties and Condensate Properties and Correlations," *Pet. Eng. Handbook,* H.B. Bradley (ed.), SPE, Richardson, Texas (1987) Chap. 21.

Byars, C.: "Parts of Spraberry Trend Fading Out," *Oil & Gas J.* (5 October 1970) 68.

Caldwell, J.: "Marine Multicomponent Seismic-Acquisition Technologies," paper OTC 10981 presented at the 1999 Offshore Technology Conference, Houston, 3–6 May.

Capen, E.C.: "A Consistent Probabilistic Approach to Reserves Estimates," paper SPE 25830 presented at the 1993 SPE Hydrocarbon Economics and Evaluation Symposium, Dallas, 29–30 March.

Capen, E.C.: "Dealing With Exploration Uncertainties," *The Business of Petroleum Exploration,* American Assn. of Petroleum Geologists, Tulsa, Oklahoma (1992) 29.

Carlson, M.R.: "What You Should Know About Evaluating Simulation Results—Parts 1 & 2. *J. Cdn. Pet. Tech.* (May and August 1997) **36,** Nos. 5 & 7.

Carlson, M.R.: "Tips, Tricks, and Traps of Material Balance Calculations," *J. Cdn. Pet. Tech.* (December 1997) 34.

Carter, P.J. and Morales, E.: "Probabilistic Addition of Gas Reserves Within a Major Gas Project," paper SPE 50113, presented at SPE Asia Pacific Oil and Gas Conference and Exhibition, Perth, Australia, 12–14 October 1998.

Casey, J. and Cronquist, C.: "Estimate GOR and FVF Using Dimensionless PVT Analysis," *World Oil* (November 1992) 83.

Cazier, E.C. *et al.:* "Petroleum Geology of the Cusiana Field, Llanos Basin, Columbia," *AAPG Bull.* (October 1995) 1444.

Chaney, P.E. *et al.:* "How To Perforate Your Well To Prevent Water and Gas Coning," *Oil & Gas J.* (7 May 1956) 108.

Chang, H.L.: "Polymer Flooding Technology—Yesterday, Today, and Tomorrow," *JPT* (August 1978) 1113.

Chen, H.Y., Posten, S.W., and Wu, C.H.: "Characterization of the Austin Chalk Producing Trend," paper SPE 15533 presented at the 1986 SPE Annual Technical Conference and Exhibition, New Orleans, 5–8 October.

Chierici, G.L., Ciucci, G.M., and Pizzi. G.: "A Systematic Study of Gas and Water Coning By Potentiometric Models," *JPT* (August 1964) 923; *Trans.*, AIME, **231.**

Chierici, G.L., Pizzi, G., and Ciucci, G.M.: "Water Drive Gas Reservoirs: Uncertainty in Reserves Evaluation From Past History," *JPT* (February 1967) 237; *Trans.*, AIME, **240.**

Chierici, G.L., Ciucci, G.M., Sclocchi, G., and Terzi, L.: "Water Drive From Interbedded Shale Compaction jn Superpressured Gas Reservoirs—A Model Study," *JPT* (June 1978) 93.

Christie, R.S. and Blackwood, J.C.: "Characteristics and Production Performance of the Spraberry," *Drill. & Prod. Prac.*, API, New York City (1952).

Chu, C.: "State-of-the-Art Review of Steamflood Field Projects," *JPT* (October 1985) 1887, *Trans.*, AIME, **279.**

Cipolla, C.L. and Kyte, D.G.: "Infill Drilling in the Moxa Arch: A Case History of the Frontier Formation," paper SPE 24909 presented at the 1992 SPE Annual Technical Conference and Exhibition, Washington, DC, 4–7 October.

Ciucci, G.M., Chierici, G.L., and Maranto, R.: "Viscous Oil Production by Water Drive," *Pet. Eng.* (August 1966) 49.

Clayton, C.A. *et al.:* "Ubit Field Rejuvenation: A Case History of Reservoir Management of a Giant Oil Field, Offshore, Nigeria," paper SPE 49165 prepared for presentation at the 1998 SPE Annual Technical Conference and Exhibition, New Orleans, 27–30 September.

Clemen, R.T.: *Making Hard Decisions,* Duxbury Press, Belmont, California (1996).

Coats, K.H.: "Reservoir Simulation," *Pet. Eng. Handbook*, H.B. Bradley (ed.), SPE, Richardson, Texas (1987) Chap. 48.

Cobb, W.M. and Marek, F.J.: "Net Pay Determination for Primary and Waterflood Depletion Mechanisms," paper SPE 48952 prepared for presentation at the 1998 SPE Annual Technical Conference and Exhibition, New Orleans, 27–30 September.

Connelly, W. and Krug, J.A.: "Evaluating Oil, Gas Opportunities in Western Siberia—Log and Core Data," *Oil & Gas J.* (23 November 1992) 97.

Connelly, W. and Krug, J.A.: "Evaluating Oil, Gas Opportunities in Western Siberia—Reservoir Description," *Oil & Gas J* (7 December 1992) 83.

Cook, A.B., Spencer, G.B., and Bobrowski, F.P.: "Special Considerations in Predicting Reservoir Performance of Highly Volatile Type Oil Reservoirs," *Trans.*, AIME (1951) **192,** 37.

Cordell, J.C. and Ebert, C.K.: "A Case History—Comparison of Predicted and Actual Performance of a Reservoir Producing Volatile Crude Oil," *JPT* (November 1965) 1291.

Corey, A.T.: "The Interrelation Between Gas and Oil Relative Permeabilities," *Prod. Monthly* (1954) **19,** 38.

Corey, A.T.: *Mechanics of Immiscible Fluids in Porous Media,* Water Resources Publishing, Highlands Ranch, Colorado (1994).

Corrigan, A.F.: "Estimation of Recoverable Reserves: The Geologist's Job," *Petroleum Geology of Northwest Europe*, J.R. Parker (ed.), Geological Soc., London (1993) 1473.

Craft, B.C. and Hawkins, M.F.: *Applied Petroleum Reservoir Engineering*, Prentice Hall, Englewood Cliffs, New Jersey (1959) 45.

Craft, B.C., Hawkins, M.F., and Terry, R.E.: *Applied Petroleum Reservoir Engineering*, second edition, Prentice Hall, Englewood Cliffs, New Jersey (1991) 11.

Craig, F.F. Jr. *et al.:* "Optimized Recovery Through Continuing Interdisciplinary Cooperation," *JPT* (July 1977) 755.

Criss, C.R. and McCormick, R.L.: "History and Performance of the Coldwater Oil Field, Michigan," *Trans.*, AIME (1954) **201,** 22.

Cronquist, C. and Hodgson, R.: "Application of a Computer Management Information System to Surveillance and Analysis of Production Behavior in Complex Fields," paper SPE 3471 presented at the 1971 SPE Annual Technical Conference and Exhibition, New Orleans, 3–6 October.

Cronquist, C.: "Waterflooding by Linear Displacement in Little Creek Field, Mississippi," *Trans.*, AIME (1968) **243,** 525.

Cronquist, C.: "Dimensionless PVT Behavior of Gulf Coast Reservoir Oils," *JPT* (May 1973) 538.

Cronquist, C.: "Volatile Oils: Fluid Characteristics, Reservoir Behavior and Production Facilities, *World Oil* (April 1979) 159.

Cronquist, C.: "Determining Recovery From Partial Water Drive Reservoirs," *World Oil* (September 1980) 107.

Cronquist, C.: "Turtle Bayou 1936-83: Case History of a Major Gas Field in South Louisiana," *JPT* (November 1984) 1941; *Trans.*, AIME, **277**.

Cronquist, C.: "Reserves and Probabilities—Synergism or Anachronism?" *JPT* (October 1991) 1258.

Cronquist, C.: "Discussion of a Comparison of Probabilistic and Deterministic Reserve Estimates: A Case Study," *SPERE* (May 1996) 135.

Crossley, D.: "Current Economic Conditions," *SPE-WPC Supplemental Vol.*, SPE, Richardson, Texas (2001).

Culberson, O.L. and McKetta, J.J.: "Phase Equilibria in Hydrocarbon-Water Systems III. The Solubility of Methane in Water at Pressures to 10,000 psia," *Trans.*, AIME (1951) **192**, 223.

Cummings, C.A. and Gentry, R.W.: "A Method for Predicting Solution Gas-Drive Production Decline," paper SPE 15021 presented at the 1986 SPE Permian Basin Oil and Gas Recovery Conference, Midland, 13–14 March.

Cutler, W.W. Jr.: "Estimation of Underground Oil Reserves by Oil Well Production Curves," Bull. 2281924, U.S. Bureau of Mines, Bartlesville, Oklahoma, (1924).

Dake, L.P.: *Fundamentals of Reservoir Engineering,* Elsevier Science Publisher, New York City (1978).

Dake, L.P.: *The Practice of Reservoir Engineering,* Elsevier Science Publisher, Amsterdam (1994).

Da Prat, G. H., Cinco-Ley, H., and Ramey, H.J. Jr.: "Decline Curve Analysis Using Type Curves for Two-Porosity Systems," *SPEJ* (June 1981) 354; *Trans.*, AIME, **271**.

Davidson, L.B. and Cooper, D.O.: "A Simple Way of Developing a Probability Distribution of Present Value," *Trans.*, AIME (1976) **261**, 1069.

Davis, J.: *Statistics and Data Analysis in Geology,* John Wiley & Sons, New York City (1986).

Davis, T.B.: "Subsurface Pressure Profiles in Gas Saturated Basins," *Elmworth—Case Study of a Deep Basin Gas Field*, Memoir 38, American Assn. of Petroleum Geologists, Tulsa, Oklahoma (1984).

De Ghetto, G., Paone, P., and Villa, M.: "Pressure-Volume-Temperature Correlations for Heavy and Extra Heavy Oils," paper SPE 30316 presented at the 1995 SPE International Heavy Oil Symposium, Calgary, 19–21 June.

De Leebeeck, A.: "The Frigg Field Reservoir: Characteristics and Performance," *North Sea Oil and Gas Reservoirs*, Graham & Trotman, London (1987).

Denny, M.J. and Heusser-Maskell, J.: "Reservoir Performance Monitoring Techniques Used in the Forties Field," *JPT* (March 1984) 457; *Trans.*, AIME, **277**.

DeSorcy, G.J.: "Estimation Methods for Proved Recoverable Reserves of Oil and Gas," *Proc.*, 10th World Pet. Cong., Bucharest, Romania (1979).

DeSorcy, G.J. *et al.*: "Definitions and Guidelines for Classification of Oil and Gas Reserves," *J. Cdn. Pet. Tech.* (May 1993) 10.

Deussen, A.: "Acre-foot Yields of Texas Gulf Coast Oil Fields," *Trans.*, AIME (1936) **118**, 53.

Dhir, R., Dern, R.R. and Mavor, M.J.: "Economic and Reserves Evaluation of Coalbed Methane Reservoirs," paper SPE 22024 presented at the 1991 SPE Hydrocarbon and Evaluation Symposium, Dallas, Texas 11–12 April.

Dickinson, G.: "Geological Aspects of Abnormal Reservoir Pressures in Gulf Coast Louisiana," *AAPG Bull.* (1953) **37**, 410.

Dietz, D.N.: "A Theoretical Approach to the Problem of Encroaching and Bypassing Edge Water," *Proc.*, Akad. van Wetenschnapper., Amsterdam (1953).

Dietz, D.N.: "Determination of Average Reservoir Pressure From Build-Up Surveys," *JPT* (August 1965) 955; *Trans.*, AIME, **234**.

Dobrynin, V.M.: "Effect of Overburden Pressure on Some Properties of Sandstone," *SPEJ* (December 1962) 360; *Trans.*, AIME, **225**.

Dodson, C.R., Goodwill, D., and Mayer, E.H.: "Application of Laboratory PVT Data to Reservoir Engineering Problems," *Trans.*, AIME (1953) **198**, 287.

Dodson, C.R. and Standing, M.B.: "Pressure-Volume-Temperature and Solubility Relations for Natural Gas-Water Mixtures," *Drill. & Prod. Prac.*, API (1945).

Do Rozario, R.F.: "Palm Valley Gas Field—Australia," *Treatise of Petroleum Geology, Atlas of Oil and Gas Fields—Structural Traps IV,* N.H. Foster and E.K. Beaumont (eds.) American Assn. of Petroleum Geologists, Tulsa, Oklahoma (1990).

Dokla, M.E. and Osman, M.E.: "Correlation of PVT Properties for UAE Crudes," *SPEFE* (March 1992) 41; *Trans.,* AIME, **293.**

Dokla, M.E. and Osman, M.E.: "Author's Reply to Discussion of Correlation of PVT Properties for UAE Crudes," *SPEFE* (March 1993) 81.

Dranchuk, P.M. and Abou-Kassem, J.H.: "Calculations of Z-Factors for Natural Gases Using Equations of State, " *J. Cdn. Pet. Tech.* (July-September 1975) **14,** 34.

Driscoll, V.J.: "Recovery Optimization Through Infill Drilling—Concepts, Analysis, and Field Results," paper SPE 4977 presented at the 1974 SPE Annual Technical Conference and Exhibition, Houston, 6–9 October.

Duggan, J.O.: "The Anderson 'L'—An Abnormally Pressured Gas Reservoir in South Texas," *JPT* (February 1972) 132.

Dutton, S.P. *et al.: Major Low-Permeability Sandstone Gas Reservoirs in the Continental United States,* U. of Texas, Austin, Texas (1993).

Dykstra, H. and Parsons, R.L.: "The Prediction of Oil Recovery by Water Flood," *Secondary Recovery of Oil in the United States,* American Petroleum Inst., New York City (1950).

Earlougher, R.C. Jr.: *Advances in Well Test Analysis,* Monograph Series, SPE, Richardson, Texas (1977) **5.**

Eaton, B.A. and Jacoby, R.H.: "A New Depletion-Performance Correlation for Gas-Condensate Reservoir Fluids," *Trans.,* AIME (1965) **234,** 852.

Edwards, K.A. and Behrenbruch, P.: "Use of Well Test Results in Oilfield Development Planning in the Timor Sea," *JPT* (October 1988) 1372; *Trans.,* AIME, **285.**

Edwardson, M.J.: "Calculation of Formation Temperature Disturbances Caused by Mud Circulation," *Trans.,* AIME (1962) **225,** 416.

Eilerts, C.K. *et al.: Phase Relations of Gas Condensate Fluids,* Monograph 10, U.S. Bureau of Mines, American Gas Assn., New York City (1957).

Elkins, L.F.: "Reservoir Performance and Well Spacing, Spraberry Trend Area Field of West Texas," *Trans.,* AIME (1953) **198,** 177.

Elkins, L.F., Skov, A.M., and Gould, R.C.: "Progress Report on Spraberry Waterflood Reservoir Performance, Well Simulation, and Water Treating and Handling," *Trans.,* AIME (1968) **243,** 1039.

Elkins, L.F.: "Uncertainty of Oil in Place in Unconsolidated Sand Reservoirs—A Case History," *JPT* (November 1972) 1315.

Elliott, D.C.: "Reserve Estimates: Uncertainty and Its Implications," *J. Cdn. Pet. Tech.* (April 1995) **34,** 4.

Elliott, D.C.: "Uncertainty and Risk in Reserves Evaluation," *Determination of Oil and Gas Reserves,* Petroleum Soc.-CIM Monograph No. 1, Calgary (1994) 266.

Ellis, D.V.: "Nuclear Logging Techniques," *Pet. Eng. Handbook,* H.B. Bradley (ed.), Richardson, Texas (1987) Chap. 50.

Elshahawi, H., Fathy, K., and Hiekali, S.: "Capillary Pressure and Rock Wettability Effects on Wireline Formation Tester Measurements," paper SPE 56712 presented at the 1999 SPE Annual Technical Conference and Exhibition, Houston, 3–6 October.

Elsharkawy, A.M. and El-Mater, D.: "Geographic Location Considered in PVT Calculation Program," *Oil & Gas J.* (22 January 1996) 36.

Energy Information Admin.: *U.S. Crude Oil, Natural Gas and Natural Gas Liquids Reserves,* DOE/EIA-0216 (88), U.S. Dept. of Energy, Washington, DC (1989).

Energy Resources Conservation Board (ERCB): *Alberta's Reserves of Crude Oil, Oil Sands, Gas, Natural Gas Liquids, and Sulphur,* Energy Resources Conservation Board, Calgary (1984, 1987, 1990, 1991).

Ershaghi, I. and Abdassah, D.: "A Prediction Technique for Immiscible Processes Using Field Performance Data," *JPT* (April 1984) 664; *Trans.,* AIME, **277.**

Ershaghi, I. and Omoregie, O.: "A Method for Extrapolation of Cut vs. Recovery Curves," *JPT* (February 1978) 203.

Ershaghi, I., Handy, L.L., and Hamdi, M.: "Application of the X-Plot Technique to the Study of Water Influx in the Sidi El-Itayem Reservoir, Tunisia," *JPT* (September 1987) 1127.

Ershaghi, I.: "Author's Reply to Discussion of a Prediction Technique for Immiscible Processes Using Field Performance Data," *JPT*, (May 1989) 559.

Fanchi, J.R.: "Analytical Representation of the van Everdingen-Hurst Aquifer Influence Functions for Reservoir Simulation," *SPEJ* (June 1985) 405; *Trans.*, AIME, **279**.

Farouq Ali, S.M. and Meldau, R.F.: "Current Steamflood Technology," *JPT* (October 1979) 1332.

Fast, C.R., Holman, G.B., and Covlin, R.J.: "The Application of Massive Hydraulic Fracturing to the Tight Muddy 'J' Formation, Wattenberg Field, Colorado," *JPT* (January 1977) 10.

Fatt, I.: "Pore Volume Compressibilities of Sandstone Reservoir Rocks," *Trans.*, AIME (1958) **213**, 362.

Felsenthal, M.: "Correlation of k_g/k_o Data with Sandstone Core Characteristics," *Trans.*, AIME (1959) **216**, 258.

Fertl, W.H. and Timko, D.J.: "How Downhole Temperatures, Pressures Affect Drilling," *World Oil* (September 1972) 45.

Fetkovich, E.J.: "Advanced Decline Curve Analysis Identifies Fracture Stimulation Potential," paper SPE 38903 presented at the 1997 SPE Annual Technical Conference and Exhibition, San Antonio, 5–8 October.

Fetkovich, M.J.: "A Simplified Approach to Water Influx Calculations—Finite Aquifer Systems," *JPT* (July 1971) 814.

Fetkovich, M.J.: "Decline Curve Analysis Using Type Curves," *JPT* (June 1980) 1065; *Trans.*, AIME, **269**.

Fetkovich, M.J. *et al.*: "Case Study of a Low-Permeability Volatile Oil Field Using Individual-Well Advanced Decline Curve Analysis," paper SPE 14237 presented at the 1985 SPE Annual Technical Conference and Exhibition, Las Vegas, Nevada, 22–25 September.

Fetkovich, M.J. *et al.*: "Decline-Curve Analysis Using Type Curves—Case Histories," *SPEFE* (December 1987) 637; *Trans.*, AIME, **283**.

Fetkovich, M.J. *et al.*: "Depletion Performance of Layered Reservoirs Without Crossflow," paper SPE 18266 presented at the 1988 SPE Annual Technical Conference and Exhibition, Houston, 2–5 October.

Fetkovich, M.J., Reese, D.E., and Whitson, C.H.: "Application of a General Material Balance for High-Pressure Gas Reservoirs," paper SPE 22921 presented at the 1991 SPE Annual Technical Conference and Exhibition, Dallas, 6–9 October.

Fetkovich, M.J., Fetkovich, E.J., and Fetkovich, M.D.: "Useful Concepts for Decline Curve Forecasting, Reserve Estimation and Analysis," paper SPE 28628 presented at the 1994 SPE Annual Technical Conference and Exhibition, New Orleans, 25–28 September.

Finley, R.J.: *Geology and Engineering Characteristics of Selected Low Permeability Gas Sandstones: A National Survey-RI 138*, Bureau of Economic Geology, Austin, Texas (1984).

Fiorillo, G.: "Exploration and Evaluation of the Orinoco Oil Belt," *Exploration for Heavy Crude Oil and Natural Bitumen,* American Assn. of Petroleum Geologists, Tulsa, Oklahoma (1987).

Fjerstad, P.A. *et al.*: "Long–Term Production Testing Improves Reservoir Characterization in the Oseberg Field," *JPT* (April 1992) 476.

Floris, F.J.T. and Peersmann, M.R.H.E.: "Uncertainty Estimation in Volumetrics for Supporting Hydrocarbon Exploration and Production Decision Making," *Pet. Geoscience* (1998) **4**.

Follows, E.: "Integration of Inclined Pilot Hole Core With Horizontal Image Logs To Appraise an Aeolian Reservoir, Auk Field, Central North Sea," *Pet. Geoscience* (March 1997) 43.

Fong, D.K., *et al.*: "Design and Implementation of a Successful Vertical Tertiary Hydrocarbon Miscible Flood," *J. Cdn. Pet. Tech.* (February 2001) 35.

Frailing, W.G.: Supplement, 1962 "Lakeview Pool, Midway-Sunset Field," *Field Case Histories, Oil Reservoirs,* Reprint Series, SPE, Richardson, Texas (1962) **4**, 41.

Frazer, F.: "Shell Applying New Research to Marginal Prospects," *Offshore* (August 1998) 46.

Frimodig, J.P., Sankur, V., and Chun, C.K.: "Design of a Tertiary Hydrocarbon Flood for the Mitsue Reservoir," *JPT* (February 1988) 215.

Gael, B.T., Gross, S.J., and McNaboe, G.J.: "Development Planning and Reservoir Management in the Duri Steamflood," paper SPE 29668 presented at the 1995 SPE Western Regional Meeting, Bakersfield, California, 8–10 March.

Galloway, W.E., *et al.: Atlas of Major Texas Oil Reservoirs,* U. of Texas, Austin, Texas (1983).

Garb, F.A.: "Property Evaluation With Hand-Held Computers: Part 9—Liquid Yield for Depletion Drive Gas Reservoirs and Gas Recovery Factors for Depletion Drive and Water Displacement Reservoirs," *Pet. Eng.* (October 1978) 72.

Garthwaite, D.L., and Krebill, F.K.: Supplement, 1962 "Pressure Maintenance by Inert Gas Injection in the High Relief Elk Basin Field," *Field Case Histories, Oil Reservoirs*, Reprint Series, SPE of AIME, Dallas, Texas (1962) **4**, 82.

Garthwaite, D.L.: Supplement, 1975 "Pressure Maintenance by Inert Gas Injection in the High Relief Elk Basin Field," *Field Case Histories, Oil and Gas Reservoirs*, Reprint Series, SPE of AIME, Dallas, Texas (1975) **4a**, 266.

Geertsma, J.: "The Effect of Fluid Pressure Decline on Volumetric Changes of Porous Rocks," *Trans.,* AIME (1957) **210**, 331.

Gentry, R.W. and McCray, A.W.: "The Effect of Reservoir and Fluid Properties on Production Decline Curves," *JPT* (September 1978) 1327.

George, C.J. and Stiles, L.H.: "Improved Techniques for Evaluating Carbonate Waterfloods in West Texas," *JPT* (November 1978) 1547.

Ghauri, W.K., Osborne, A.F., and Magnuson, W.L.: "Changing Concepts in Carbonate Waterfloods— West Texas Denver Unit Project—An Illustrative Example," *JPT* (June 1974) 595.

Gibbons, K.: "Application of Geostatistics in the Petroleum Industry," *SPE/WPC Supplemental Vol.,* SPE, Richardson, Texas (2001).

Glaso, O.: "Generalized Pressure-Volume-Temperature Correlations," *JPT* (May 1980) 785; *Trans.,* AIME, **269**.

Godec, M.L. and Tyler, N.: "The Economic Potential of Strategic Infill Drilling in the San Andres/Grayburg of South Central Basin Platform Play," paper SPE 18929 presented at the 1989 SPE Hydrocarbon Economics and Evaluation Symposium, Dallas, 9–10 March.

Gold, D., McCain, W.D. Jr., and Jennings, J.W.: "An Improved Method for the Determination of the Reservoir-Gas Specific Gravity for Retrograde Gases," *JPT* (July 1989) 747; *Trans.,* AIME, **287**.

Goldsberry, F.: "Shockwave Model Reduces Risk," *Hart's E&P* (September 2000) 43

Gomaa, E.E.: "Correlations for Predicting Oil Recovery by Steamflood," *JPT* (February 1980) 325; *Trans.,* AIME, **269**.

Grace, J.D., Caldwell, R.H., and Heather, D.I.: "Comparative Reserves Definitions: U.S.A., Europe, and the Former Soviet Union," *JPT* (September 1993) 866.

Grayson, C.J. Jr.: *Decisions Under Uncertainty: Drilling Decisions by Oil and Gas Operators,* Harvard Business School Boston, Massachusetts (1960).

Green, D.W. and Wilhite, G.P.: *Enhanced Oil Recovery,* Textbook Series, SPE, Richardson, Texas (1998) **6**.

Gruy, H.J. and Crichton, J.A.: "Plotting Pressure Drop Against Cumulative Production of Gas Fields on Log-Log Paper," *Pet. Eng.* (September 1950) B-76.

Guerrero, E.T. and Earlougher, R.C.: "Analysis and Comparison of Five Methods Used To Predict Waterflood Reserves and Performance," *Drill. & Prod. Prac.,* API (1962) -1961 28:78. New York: API.

Guidroz, G.M.: "E.T. O'Daniel Project—A Successful Spraberry Flood," *JPT* (September 1967) 1137.

Gurley, J.: "A Productivity and Economic Projection Method—Ohio Clinton Sand Gas Wells," *JPT* (November 1963) 1183.

Haldorsen, H.H. and Lake, L.W.: "A New Approach to Shale Management in Field-Scale Models," *SPEJ* (August 1984) 447; *Trans.,* AIME, **277**.

Hale, B.W.: "A Type Curve Approach to Reserves for the Big Piney Gas Field," paper SPE 9840 presented at the 1981 SPE/DOE Low Permeability Symposium, Denver, Colorado, 27–29 May.

Hale, B.W.: "Analysis of Tight Gas Well Production Histories in the Rocky Mountains,"*SPERE* (July 1986) 310; *Trans.,* AIME, **281.**

Hale, B.W. and Mahmood, A.M.: "The Impact of Measurement Error on Deliverability Projections for Tight Gas Wells," paper SPE 9837 presented at the 1981 SPE/DOE Low Permeability Symposium, Denver, Colorado, 27–29 May.

Hall, H.N.: "Compressibility of Reservoir Rocks," *Trans.,* AIME (1953) **204,** 309.

Hammerlindl, D.J.: "Predicting Gas Reserves in Abnormally Pressured Reservoirs," paper SPE 3479 presented at the 1971 SPE Annual Technical Conference and Exhibition, New Orleans, 3–6 October.

Handford, C.R.: "Sedimentology and Genetic Stratigraphy of Dean and Spraberry Formations (Permian), Midland Basin, Texas," *AAPG Bull.* (1981) **65,** 1602.

Harari, Z., Wang, S.T., and Saner, S.: "Pore-Compressibility Study of Arabian Carbonate Reservoir Rocks," *SPEFE* (December 1995) 207.

Harbaugh, J., Doveton, J., and Davis, J.: *Probability Methods in Oil Exploration,* John Wiley & Sons Inc., New York City (1977).

Harlan, L.E.: "A Method for Determining Recovery Factors in Low Permeability Gas Reservoirs," paper SPE 1555 presented at the 1966 SPE Annual Technical Conference and Exhibition, Dallas, 25 October.

Harpole, K.J.: "Improved Reservoir Characterization—A Key to Future Reservoir Management for the West Seminole San Andres Unit," *JPT* (November 1980) 2009; *Trans.,* AIME, **269.**

Hartman, J.A. and Paynter, D.D.: "Drainage Anomalies in Gulf Coast Tertiary Sandstones," *JPT* (October 1979) 1313.

Harville, D.W. and Hawkins, M.F.: "Rock Compressibility and Failure as Reservoir Mechanisms in Geopressured Gas Reservoirs," *JPT* (December 1969) 1528.

Haugen, S.A., Lund, O., and Hoyland, L.A.: "Statfjord Field: Development Strategy and Reservoir Management," *JPT* (July 1988) 863; *Trans.,* AIME, **285.**

Havlena, D. and Odeh, A.S.: "The Material Balance as an Equation of a Straight Line," *JPT* (August 1963) 896; *Trans.,* AIME, **228.**

Havlena, D. and Odeh, A.S.: "The Material Balance as an Equation of a Straight Line—Part II, Field Cases," *JPT* (July 1964) 815; *Trans.,* AIME, **231.**

Havlena, D.: "Interpretation, Averaging, and Use of the Basic Geological-Engineering Data," *J. Cdn. Pet. Tech.* (October–December 1966) 236.

Havlena, D.: "Dynamic Reservoir Data," *J. Cdn. Pet. Tech.* (July–September 1968) 128

Hawkins, J.M., Luffel, D.L., and Harris, T.G.: "Capillary Pressure Model Predicts Distance to Gas/Water, Oil/Water Contact," *Oil & Gas J.* (18 January 1993) 39.

Hawkins, J.M.: "Integrated Formation Evaluation With Regression Analysis," paper SPE 28244 presented at the 1994 SPE Petroleum Computer Conference, Dallas, 31 July–3 August.

Hawkins, M.F.: "Material Balance in Expansion Type Reservoirs above Bubble Point," *Trans.,* AIME (1955) **204,** 267.

Haynes, B. Jr.: "An Evaluation of a Method To Predict Unknown Water Levels in Reservoirs and Quantifying the Uncertainty," paper SPE 29466 presented at the 1995 SPE Production Operations Symposium, Oklahoma City, , 2–4 April.

Hazeu, G.J.A. *et al.:* "The Application of New Approaches for Shale Management in a Three-Dimensional Simulation Study of the Frigg Field," *SPEFE* (September 1988) 493.

Hefner, J.M.: "A Comparison of Deterministic and Probabilistic Reserve Estimates," MS thesis, Colorado School of Mines, Golden, Colorado (1992).

Hefner, J.M. and Thompson, R.S.: "A Comparison of Probabilistic and Deterministic Reserve Estimates: A Case Study," *SPERE* (February 1996) 43.

Henley, D.H., Owens, W.W., and Craig, F.F. Jr.: A Scale-Model Study of Bottom-Water Drives," *JPT* (January 1961) 90; *Trans.,* AIME (1961) **222.**

Hertz, D.B.: "Risk Analysis in Capital Investment," *Harvard Business Rev.* (January–February 1964) **42,** 1, 95.

Higgins, R.V. and Lechtenberg, H.J.: "Merits of Decline Equations Based on Production History of 90 Reservoirs," paper SPE 2450 presented at the 1969 SPE Rocky Mountain Regional Meeting, Denver, Colorado, 25–27 May.

Hillestad, J.G. and Goode, D.L.: "Reserves Determination—Implications in the Business World," *APEA J.* (1989) **1**, 52.

Hillier, G.R.K., Cobb, R.M., and Dimmock, P.A.: "Reservoir Development Planning for the Forties Field," paper EUR 98, *Proc.,* European Offshore Petroleum Conference and Exhibition, Graham and Trotman, London (1978).

Hitchon, B.: "Geothermal Gradients, Hydrodynamics, and Hydrocarbon Occurrences, Alberta, Canada," *AAPG Bull.* (1984) **68,** 713.

Holtz, M.H.: "Estimating Oil Reserve Variability by Combining Geologic and Engineering Parameters," paper SPE 25827 presented at SPE Hydrocarbon Economics and Evaluation Symposium, Dallas, 29–30 March 1993.

Honarpour, M., Koederitz, L., and Harvey, A.H.: *Relative Permeability of Petroleum Reservoirs,* CRC Press (1984), Reprinted by SPE, Richardson, Texas (1994).

Hoover, A.R., Burkhart, T., and Flemings, P.B.: "Reservoir and Production Analysis of the K40 Sand, South Timbalier 295, Offshore Louisiana, With Comparison to Time-Lapse (4-D) Seismic Results," *AAPG Bull.* (October 1999) 1624.

Horne, J.S., Berkoz, A., and Wood, L.R.: "Development of a Marginal Property: Petronella Field," *JPT* (June 1988) 723.

Horsfield, R.: "Performance of the Leduc D-3 Reservoir," *JPT* (February 1958) 21.

Horsfield, R.: Supplement, 1962 "Performance of the Leduc D-3 Reservoir," *Field Case Histories, Oil Reservoirs,* Reprint Series, SPE, Richardson, Texas (1962) **4,** 65.

Houston Geological and New Orleans Geological Societies: *Productive Low Resistivity Well Logs of the Offshore Gulf of Mexico,* D.C. Moore (ed.) 1993

Hoyland, L.A., Papatzacos, P., and Skjaeveland, S.M.: "Critical Rate for Water Coning: Correlation and Analytical Solution," *SPERE* (November 1989) 495.

Hubbert, M.K.: "Entrapment of Petroleum Under Hydrodynamic Conditions," *AAPG Bull.* (August 1953) 1954.

Huffman, C.H. and Thompson, R.S.: "Probability Ranges for Reserve Estimates From Decline Curve Analysis," paper SPE 28333 presented at the 1994 SPE Annual Technical Conference and Exhibition, New Orleans, 25–28 September.

Hurst, W.: "Further Discussion of Water Drive Gas Reservoirs: Uncertainty in Reserves Evaluation From Past History," *JPT* (July 1967) 965.

Hutchinson, T.S. and Kemp, C.E.: "An Extended Analysis of Bottom Water Drive Reservoir Performance," *Trans.,* AIME (1956) **207,** 256.

Ibrahim, A.A. and Mostafa, A.: "Production Logging Problem Description in October Field, Gulf of Suez," paper SPE 56650 presented at the 1999 SPE Annual Technical Conference and Exhibition, Houston, 3–6 October.

Irrgang, H.H.: "Evaluation and Management of Thin Oil Column Reservoirs in Australia," *APEA J.* (1994) **1**, 64.

Jacks, H.H.: "Forecasting Future Performance," *Reservoir Simulation,* C.C. Mattox and R.L. Dalton (eds.), Monograph Series, SPE, Richardson, Texas (1990) **13,** 99-110.

Jacobson, H.A.: "The Effect of Nitrogen on Reservoir Fluid Saturation Pressure," *J. Cdn. Pet. Tech.* (July–September 1967) 101.

Jacoby, R.H., Koeller, R.C., and Berry, V.J. Jr.: "Effect of Composition and Temperature on Phase Behavior and Depletion Performance of Rich Gas-Condensate Systems," *JPT* (July 1959) 406; *Trans.,* AIME **216.**

Jacoby, R.H. and Berry, V.J. Jr.: "A Method for Predicting Depletion Performance of a Reservoir Producing Volatile Crude Oil," *Trans.,* AIME (1957) **210,** 27.

Jansa, L.F. and Urrea, V.H.N.: "Geologic and Diagenetic History of Overpressured Sandstone Reservoirs, Venture Gas Field, Offshore Nova Scotia, Canada," *AAPG Bull.* (October 1990) 1640.

Jardine, D. *et al.*: "Distribution and Continuity of Carbonate Reservoirs," *JPT* (July 1977) 873.

Jenkins, R.E.: "Typical Core Analysis of Different Formations," *Pet. Eng. Handbook,* H.B. Bradley (ed.), SPE, Richardson, Texas (1987) Chap. 27.

Jensen, J.L.: "Discussion of Characterizing Average Permeability in Oil and Gas Formations," *SPEFE* (September 1992) 271.

Jensen, J.L. *et al.: Statistics for Petroleum Engineers and Geoscientists,* Prentice Hall Inc., Upper Saddle River, New Jersey (1997).

Jochen, V.A. and Spivey, J.P.: "Probabilistic Reserves Estimation Using Decline Curve Analysis with the Bootstrap Method," paper SPE 36633 presented at the 1996 SPE Annual Technical Conference and Exhibition, Denver, Colorado, 6–9 October.

Johnson, J.D.: "Cost-Effective Data Acquisition for the Odin Field," *JPT* (October 1988) 1316.

Johnson, J.P. and Rhett, D.W.: "Compaction Behavior of Ekofisk Chalk as a Function of Stress," paper SPE 15872 presented at the 1986 SPE European Petroleum Conference, London, 20–22 October.

Joiner, D.S. and Long, G.B.: "Retrograde GOR Behavior of Condensate Reservoirs From Well Test Data," paper SPE 7492 presented at the 1978 SPE Annual Technical Conference and Exhibition, Houston, 1–3 October.

Jones, S.C.: "Two-Point Determinations of Permeability and PV vs. Net Confining Stress," *SPEFE* (March 1988) 235.

Journel, A.G.: "Geostatistics and Reservoir Geology," *Stochastic Modeling and Geostatistics,* American Assn. of Petroleum Geologists, Tulsa, Oklahoma (1994) 19.

Jung, H.: "Reserve Definitions—Clarifying the Uncertainties," paper CIM 96-43 presented at the 1996 Petroleum Soc.-CIM Annual Technical Meeting, Calgary, 10–12 June.

Karra, S., Egbogah, E., and Yang, F.W.: "Stochastic and Deterministic Reserves Estimation in Uncertain Environments," paper SPE 29286 presented at the 1995 SPE Asia Pacific Oil and Gas Conference, Kuala Lumpur, 20–22 March.

Kartoatmodjo, T. and Schmidt, Z.: "Large Data Bank Improves Crude Physical Property Correlations," *Oil & Gas J.* (4 July 1994) 51.

Katz, D.L. *et al.: Handbook of Natural Gas Engineering,* McGraw Hill Book Co., New York City (1959) 107.

Katz, D.L. *et al.:* "How Water Displaces Gas From Porous Media," *Oil & Gas J.* (10 January 1966) 55.

Kaufman, G.M.: *Statistical Decision and Related Techniques in Oil Exploration,* Prentice-Hall Inc., Englewood Cliffs, New Jersey (1963).

Kaveler, H.H.: "Engineering Features of the Shuler Field and Unit Operation," *Trans.,* AIME (1944) **155,** 58.

Kaveler, H.H. and Tarner, J.: Supplement, 1962 "Shuler Field and Unit Operation," *Field Case Histories, Oil Reservoirs,* Reprint Series, SPE of AIME, Dallas (1962) **4,** 26.

Kazemi, H.: "Low-Permeability Gas Sands," *JPT* (October 1982) 2229; *Trans.,* AIME, **273.**

Keefer, D.L. and Bodily, S.E.: "Three Point Approximations for Continuous Random Variables," *Management Science* (1983) **29,** 5, 694.

Keelan, D.K.: "Core Analysis," *Subsurface Geology,* fourth edition, L.W. LeRoy and D.O. LeRoy (eds.), Colorado School of Mines, Golden, Colorado (1977).

Keith, D.R., Wilson, D.C., and Gorsuch, D.P.: "Reserve Definitions—An Attempt at Consistency," paper SPE 15865 presented at the 1986 SPE European Petroleum Conference, London, 20–22 October.

Kesler, M.G. and Lee, B.I.: "Improve Prediction of Enthalpy of Fractions," *Hydrocarbon Proc.* (March 1976) 153.

Khairy, M., El-Tayeb, S., and Hamdallah, M.: "PVT Correlations Developed for Egyptian Crudes," *Oil & Gas J.* (4 May 1998) 114.

Khan, A.M.: "An Empirical Approach to Waterflood Predictions," *JPT* (May 1971) 565.

Kimmel, J.D. and Dalati, R.N.: "Potential Tests of Oil Wells," *Pet. Eng. Handbook,* H.B. Bradley (ed.), SPE, Richardson, Texas (1987) Chap. 32.

King, G.: "4-D Seismic Improves Reservoir Management Decisions," *World Oil* (March 1996).

Kingston, P.E. and Niko, H.: "Development Planning of the Brent Field," *JPT* (October 1975) 1190.

Klins, M.A., Bouchard, A.J., and Cable, C.L.: "A Polynomial Approach to the van Everdingen-Hurst Dimensionless Variables for Water Encroachment," *SPERE* (February 1988) 320.

Kilins, M.A.: *Carbon Dioxide Flooding: Basic Mechanisms and Project Design*, IHRDC, Boston (1984)

Klotz, J.A.: "The Gravity Drainage Mechanism," *JPT* (April 1953) 19.

Knutsen, C.F.: "Definition of a Water Table," (Geologic Note), *AAPG Bull.* (September 1954) **38,** 2020.

Knutsen, C.F., Maxwell, E.L., and Millheim, K.K.: "Sandstone Continuity in the Mesaverde Formation, Rulison Field Area, Colorado," *JPT* (August 1971) 911.

Kristiansen, P. and Christie, P.: "Monitoring Foinaven Reservoir: Advances in 4-D Seismic," *World Oil* (November 1999) 71.

Kukal, G.C. *et al.:* "Critical Problems Hindering Accurate Log Interpretation of Tight Gas Sand Reservoirs," paper SPE 11620 presented at the 1983 SPE/DOE Symposium on Low Permeability, Denver, Colorado, 14–16 March.

Kulander, B.R., Dean, S.L., and Ward, B.J. Jr.: *Fractured Core Analysis,* AAPG, Tulsa, OK (1990).

Kuo, M.C.T. and DesBrisay, C.L.: "A Simplified Method for Water Coning Predictions," paper SPE 12067 presented at the 1983 SPE ATC&E, San Francisco, 5–8 October.

Kuo, M.C.T.: "Correlations Rapidly Analyze Water Coning," *Oil & Gas J.* (2 October 1989) 77.

LaChance, D.P. and Winston, R.T.: "Recognition of Waterflood Sweep and Formation Lithology in a Giant Egyptian Oil Field by Applied Petrophysics," *SPEFE* (June 1987) 150.

Lahee, F.H.: "Standardization in Compiling Data on Exploratory Drilling and Crude Oil Reserves," *Drill. & Prod. Prac.*, API, New York City (1945).

Lake, L.W. and Jensen, J.L.: "The Influence of Sample Size and Permeability Distribution on Heterogeneity Measures," *SPERE* (May 1988) 629.

Lake, L.W., Waggoner, J.R., and Castillo, J.L.: "Simulation of EOR Processes in Stochastically Generated Permeable Media," paper SPE 21237 presented at the 1991 SPE Symposium on Reservoir Simulation, Anaheim, California, 17–20 February.

Lamont, N.: "Gas Reservoir Study Promises Accurate Recovery Estimate," *Oil & Gas J.* (14 January 1963) 78.

Lasater, J.A.: "Bubble Point Pressure Correlation," *Trans.,* AIME (1958) **213,** 379.

Laustsen, D.: "Practical Decline Curve Analysis, Parts 1 and 2," *J. Cdn. Pet. Tech.* (November and December 1996) 34, 14.

Law, J.: "A Statistical Approach to the Interstitial Characteristics of Sand Reservoirs," *Trans.,* AIME (1944) **155,** 202.

Le Blanc, R.J. Sr.: "Distribution and Continuity of Sandstone Reservoirs," *JPT* (July 1977) 776.

Lee, J.: *Well Testing,* SPE Text Book Series, Vol. 1, SPE of AIME, Dallas (1982).

Lee, P.J. and Wang, P.C.: "Evaluation of Petroleum Resources From Pool Size Distributions," *Oil and Gas Assessment—Methods and Applications*, Dudley Rice (ed.), American Assn. of Petroleum Geologists (1986).

Lefkovits, H.C. and Matthews, C.S.: "Application of Decline Curves to Gravity-Drainage Reservoirs in the Stripper Stage," *Trans.,* AIME (1958) **213,** 275.

Lelek, J.L.: "Geologic Factors Affecting Reservoir Analysis, Anschutz Ranch East Field, Utah/Wyoming," *JPT* (August 1983) 1539; *Trans.,* AIME, **275.**

Leshikar, A.G.: "How To Estimate Equivalent Gas Volume of Stock Tank Condensate," *World Oil* (1961) **152,** 1.

Leverett, M.C.: "Capillary Behavior in Porous Solids," *Trans.* AIME (1941) **142,** 152

Levorsen, A.I.: *Geology of Petroleum,* W.H. Freeman and Co., San Francisco (1967).

Lewis, D.J. and Perrin, J.D.: "Wilcox Formation Evaluation: Improved Procedures for Tight Gas-Sand Evaluation," *JPT* (June 1992) 724.

Lewis, J.O. and Beal, C.H.: "Some New Methods for Estimating the Future Production of Oil Wells," *Trans.,* AIME (1918) **207,** 265.

Lewis, R.C.: "Reserve Determination Using Pseudosteady-State Techniques in a High-Pressure Sour Gas Field: A Case History," paper SPE 14363 presented at the 1985 SPE Annual Technical Conference and Exhibition, Las Vegas, Nevada, 22–25 September.

Lia, O. *et al.*: "Uncertainties in Reservoir Production Forecasts," *AAPG Bull.* (May 1997) 775.

Liu, R.J.: "Discussion of a Prediction Technique for Immiscible Processes Using Field Performance Data," *JPT* (May 1989) 558.

Long, D.R and Davis, M.J.: "A New Approach to the Hyperbolic Curve," *JPT* (July 1988) 909.

Losh, S.: "Oil Migration in a Major Growth Fault: Structural Analysis of the Pathfinder Core, South Eugene Island Block 330, Offshore Louisiana," *AAPG Bull.* (September 1998) 1694.

Lowell, J.D.: *Structural Styles in Petroleum Exploration,* Oil & Gas Consultants Intl., Tulsa, Oklahoma (1985).

Lumley, D.E. and Behrens, R.A.: "Practical Issues of 4D Seismic Reservoir Monitoring: What an Engineer Needs To Know," *SPEREE* (December 1998) 528.

Luo, M., Baker, M.R., and LeMone, D.V.: "Distribution and Generation of the Overpressure System, Eastern Delaware Basin, Western Texas and Southern New Mexico," *AAPG Bull.* (September 1994) 1386.

Macary, S. and Al Hamid, W.A.: "Technique Predicts Oil Recovery From Waterfloods," *Oil & Gas J.* (25 January 1999) 84.

Mack, J.C. and Duvall, M.L.: "Performance and Economics of Minnelusa Polymer Floods," paper SPE 12929 presented at the 1984 SPE Rocky Mountain Regional Meeting, Casper, Wyoming. 21–23 May.

Mack, J.C. and Warren, J.: "Performance and Operation of a Crosslinked Polymer Flood at Sage Spring Creek Unit A, Natrona County, Wyoming," *JPT* (July 1984) 1145; *Trans.*, AIME, **277.**

Magellan Petroleum Australia Limited: "Palm Valley Field Gas Reserves," Press Release (15 March 1996).

Mannon, R.W. and Porter, R.L.: "Decline Curve Analysis Using Computerized Type Curves," *SPECA* (November–December 1989) 13.

Marek, B.F.: "Predicting Pore Volume Compressibility of Reservoir Rocks," *SPEJ* (December 1971) 340.

Markum, B.L. *et al.*: "Development of the Beryl 'A' Field," paper EUR 97 presented at the 1978 European Offshore Petroleum Conference & Exhibition, London, 24–27 October.

Martinez, A.R.: "The Orinoco Oil Belt, Venezuela," *J. Pet. Geol.* (1987) **10,** 125.

Martinez, A.R. *et al.*: "Classification and Nomenclature Systems for Petroleum and Petroleum Reserves," *Proc.,* 12th World Petroleum Congress, Houston (1987).

Mattax, C.C. *et al.*: "Core Analysis of Unconsolidated and Friable Sands," *JPT* (December 1975) 1423.

Matthews, C.S., Brons, F., and Hazebroek, P.: "A Method for Determination of Average Pressure in a Bounded Reservoir," *Trans.,* AIME (1954) **201,** 182.

Matthews, C.S. and Lefkovits, H.C.: "Gravity Drainage Performance of Depletion-Type Reservoirs in the Stripper Stage," *Trans.,* AIME (1956) **207,** 265.

Matthews, C.S. and Russell, D.G.: *Pressure Buildup and Flow Tests in Wells,* Monograph Series, SPE, Richardson, Texas (1967) **1.**

Matthews, C.S.: "Steamflooding," *JPT* (March 1983) 465.

McCain, W.D. Jr., Rollins, J.B., and Lanzi, A.J.V.: "The Coefficient of Isothermal Compressibility of Black Oils at Pressures Below the Bubblepoint," *SPEFE* (September 1988) 659.

McCain, W.D. Jr.: *The Properties of Petroleum Fluids,* Penn Well Publishing Co., Tulsa, Oklahoma (1990).

McCain, W.D. Jr.: "Reservoir-Fluid Property Correlations—State of the Art," *SPERE* (May 1991) 266.

McCain, W.D. Jr. and Alexander, R.A.: "Sampling Gas-Condensate Wells," *SPERE* (August 1992) 358; *Trans.*, AIME, **293.**

McCain, W.D. Jr.: "Revised Gas-Oil Ratio Criteria Key Indicators of Reservoir Fluid Type," *Pet. Eng.* (April 1994) 57.

McCain, W.D. Jr. and Piper, L.D.: "Reservoir Gases Exhibit Subtle Differences," *Pet. Eng.* (March 1994) 45.

McCain, W.D. Jr. and Bridges, B.: "Volatile Oils and Retrograde Gases—What's the Difference?" *Pet. Eng.* (January 1994) 35.

McCaleb, J.A. and Willingham, R.W.: "Influence of Geologic Heterogeneities on Secondary Recovery From Permian Phosphoria Reservoir, Cottonwood Creek Field, Wyoming," *AAPG Bull.* (1967) **51,** 212.

McCray, A.W.: *Petroleum Evaluations and Economic Decisions,* Prentice Hall, Englewood Cliffs, New Jersey (1975).

McCubbin, D.G.: "Barrier-Island and Strand-Plain Facies," *Sandstone Depositional Environments,* P.A. Scholle and D. Spearing (eds.), American Assn. of Petroleum Geologists, Tulsa, Oklahoma (1982).

McEwen, C.R. Jr.: "Material Balance Calculations With Water Influx in the Presence of Uncertainty in Pressures," *SPEJ* (June 1962) 120; *Trans.,* AIME, **225.**

McKay, B.G. and Taylor, N.F.: "Definition of Petroleum Reserves Using Probability Analysis," *APEA J.* (1979) **1,** 197.

McKelvey, V.E.: "Mineral Resource Estimates and Public Policy," *American Scientist* (1972) **60,** 32.

McMichael, C.L.: "Use of Reservoir Simulation Models in the Development Planning of the Statfjord Field," paper EUR 89 presented at the 1978 European Offshore Petroleum Conference & Exhibition, London, 24–27 October.

McMichael, C.L. and Young, E.D.: "Reserve Recognition Under Production Sharing and Other Nontraditional Agreements," *SPE/WPC Supplemental Vol.,* SPE, Richardson, Texas (2001).

McNally, R.: "EOR Paying Off in Permian Basin," *Pet. Eng.* (July 1988) 34.

Mearns, E. and McBride, J.J.: "Hydrocarbon Filling History and Reservoir Continuity of Oil Fields Evaluated Using $^{87}Sr/^{86}Sr$ Isotope Ratio Variations in Formation Water, With Examples From the North Sea," *Pet. Geoscience* (May 1999).

Megill, R.E.: *Exploration Economics,* second edition, Petroleum Publishing, Tulsa, Oklahoma (1979).

Megill, R.E.: *An Introduction to Risk Analysis,* second edition, Petroleum Publishing, Tulsa, Oklahoma (1984).

Mills, A.A.: "Reservoir Monitoring in Low-Salinity Environments With Pulsed-Neutron-Capture and Gamma Ray Logs," *SPEFE* (September 1993) 177.

Mireault, R.: "Probability Analysis for Estimates of Hydrocarbons in Place," *Determination of Oil and Gas Reserves,* Petroleum Soc.-CIM, Calgary (1994) Chap. 6.

Mireault, R.: "Ca$hpot Documentation," Fekete & Assocs., Calgary (1998).

Moore, D. (ed.): *Productive Low Resistivity Well Logs of the Offshore Gulf of Mexico,* New Orleans Geological Soc., New Orleans (1993).

Morgan, J.T. and Gordon, D.T.: "Influence of Pore Geometry on Water-Oil Relative Permeability," *JPT* (October 1970) 1199.

Moritis, G.: "EOR Oil Production Up Slightly," *Oil & Gas J.* (20 April 1998) 60.

Moritis, G.: "EOR Weathers Low Oil Prices," *Oil & Gas J.* (20 March 2000) 39.

Moses, P.L.: "Engineering Applications of Phase Behavior of Crude Oil and Condensate Systems," *JPT* (July 1986) 715.

Mudford, B.S. and Best, M.E.: "Venture Gas Field, Offshore Nova Scotia: Case Study of Overpressuring in Region of Low Sedimentation Rate," *AAPG Bull.* (November 1989) 1383.

Murtha, J.: "Incorporating Historical Data in Monte Carlo Simulation," paper SPE 26245 presented at SPE Petroleum Computer Conference, New Orleans, 11–14 July 1993.

Murtha, J.: *Risk Analysis as Applied to Petroleum Investment Decisions,* Intl. Human Resources Development Corp., Boston, Massachusetts (1995).

Muskat, M. and Wyckoff, R.D.: "An Approximate Theory of Water-coning in Oil Production," *Trans.,* AIME (1935) **114,** 144.

Muskat, M.: "The Production Histories of Oil Producing Gas-Drive Reservoirs," *J. Applied Physics* (1945) **16,** 147.

Muskat, M. and Taylor, M.O.: "Effect of Reservoir Fluid and Rock Characteristics on Production Histories of Gas-Drive Reservoirs," *Trans.,* AIME (1946) **165,** 78.

Muskat, M.: "The Performance of Bottomwater-Drive Reservoirs," *Trans.,* AIME (1947) **170,** 81.

Muskat, M.: *Physical Principles of Oil Production,* McGraw-Hill Book Co., New York City (1949).

Myhill, N.A. and Stegemeier, G.L.: "Steam-Drive Correlation and Prediction," *JPT* (February 1978) 173.

Nadir, F.T. and Hay, J.T.C.: "Geological and Reservoir Modeling of the Thistle Field," paper EUR 88 presented at the 1978 European Offshore Petroleum Conference & Exhibition, London, 24–27 October 1978).

Nadir, F.T.: "Thistle Field Development," *JPT* (October 1981) 1828.

Nangea, A.G. and Hunt, E.J.: "An Integrated Deterministic/Probabilistic Approach to Reserve Estimation: An Update," paper SPE 38803 presented at the 1997 SPE Annual Technical Conference and Exhibition, San Antonio, 5–8 October.

Natl. Petroleum Council (NPC): *Enhanced Oil Recovery,* NPC, Washington, D.C. (1984).

Natrella, M.G.: *Experimental Statistics,* Natl. Bureau of Standards Handbook 91, U.S. Dept. of Commerce, Washington, D.C. (1963).

Neidell, N.S. and Beard, J.H.: "Progress in Stratigraphic Seismic Exploration and the Definition of Reservoirs," *JPT* (May 1984) 709.

Nelson, P.H.H.: "Role of Reflection Seismic in Development of Nembe Creek Field, Nigeria," *Giant Oil and Gas Fields of the Decade 1968-1978,* M.T. Halbouty (ed.), American Assn. of Petroleum Geologists, Tulsa, Oklahoma (1980).

Nemchenko, N. *et al.:* "Distinction in the Oil and Gas Reserves and Resources Classifications Assumed in Russia and the USA—Source of Distinctions," *Energy Exploration & Exploitation* (1995) **13,** 6.

Nemeth, L.K. and Kennedy, H.T.: "A Correlation of Dewpoint Pressure With Fluid Composition and Temperature," *SPEJ* (June 1967) 99; *Trans.,* AIME, **240.**

Neveux, A.R., Sakthikumer, S., and Nolray, J.M.: "Delineation and Evaluation of a North Sea Reservoir Containing Near-Critical Fluids." *SPERE* (August 1988) 842.

Newendorp, P.D.: *Decision Analysis for Petroleum Exploration,* Petroleum Publishing Co., Tulsa, Oklahoma (1975).

Newman, G.H.: "Pore-Volume Compressibility of Consolidated, Friable, and Unconsolidated Reservoir Rocks Under Hydrostatic Loading," *JPT* (February 1973) 129.

Nichols, E.A.: "Geothermal Gradients in Mid-Continent and Gulf Coast Oil Fields," *Trans.,* AIME (1947) **170,** 44.

Novosad, Z.: "Exploring the Fascinating World of Reservoir Fluid Phase Behavior," *J. Cdn. Pet. Tech.,* (January 1996) 10.

Norwegian Petroleum Directorate (NPD): *Classification of Petroleum Resources on the Norwegian Continental Shelf,* NPD, Norway (1997).

Obomanu, D.A. and Okpobiri, G.A.: "Correlating the PVT Properties of Nigerian Crudes," *J. Energy Resources Technology,* (December 1987) 214.

Oil & Gas J.: "Mobil Plans Exploration Program in Deepwater Area Off Malaysia," *Oil & Gas J.* (22 November 1993) 26.

Omar, M.I. and Todd, A.C.: "Development of New Modified Black Oil Correlations for Malaysian Crudes," paper SPE 25338 presented at the 1993 SPE Asia Pacific Oil and Gas Conference, Singapore, 8–10 February.

Ontario Securities Commission: "Natl. Policy 2-B, Guide for Engineers and Geologists Submitting Oil and Gas Reports to Canadian Provincial Securities Administrations," *Consolidated Ontario Securities Acts and Regulation,* Carswell Publishing, Toronto, Ontario (1993).

Osif, T.L.: "The Effects of Salt, Gas, Temperature, and Pressure on the Compressibility of Water," *SPERE* (February 1988) 175.

Ostermann, R.D., Ehlig-Economides, C.A., and Owolabi, O.O.: "Correlations for the Reservoir Fluid Properties of Alaskan Crudes," paper SPE 11703 presented at the 1983 SPE California Regional Meeting, Ventura, California, 23–25 March.

Otis, R.M. and Schneidermann, N.: "A Process for Evaluating Exploration Prospects," *AAPG Bull.* (July 1997) 1087.

P'an, C.H.: "Petroleum in Basement Rocks," *AAPG Bull.* (October 1982) 1597.

Parker, C.A.: "Geopressures in the Deep Smackover of Mississippi," *JPT* (August 1973) 971; *Trans.,* AIME, **255.**

Patricelli, J.A. and McMichael, C.L.: "An Integrated Deterministic/Probabilistic Approach to Reserve Estimation," paper SPE 28239 presented at the 1994 SPE Annual Technical Conference and Exhibition, New Orleans, 25–28 September.

Pearson, E.S. and Tukey, J.W.: "Approximate Means and Standard Deviations Based on Distances Between Percentage Points of Frequency Curves," *Biometrika* (1965) **52,** 3–4, 533.

Pearson, J.C.: "Estimating Oil Reserves in Russia," *Pet. Eng.* (September 1997).

Peeters, M. and Visser, R.: "A Comparison of Petrophysical Evaluation Packages: LOGIC, FLAME, ELAN, OPTIMA, and ULTRA," *Log Analyst* (July–August 1991).

Petroleum Soc.-CIM (PS-CIM): *Determination of Oil and Gas Reserves,* PS.-CIM, Calgary (1994).

Petrosky, G.E. and Farshad, F.: "Pressure-Volume-Temperature Correlations for Gulf of Mexico Crude Oils," *SPERE* (October 1998) 416.

Petzet, A.: "Pemex Defines Reserves," *Oil & Gas J.* (25 October 1999).

Pittman, E.D.: "Relationship of Porosity and Permeability to Various Parameters Derived from Mercury Injection Capillary Pressure Curves for Sandstone," AAPG *Bull.* (February 1992) 191.

Pletcher, J.L.: "A Practical Capillary Pressure Correlation Technique," *JPT* (July 1994) 556.

Plisga, G.J.: "Bottomhole Pressures," *Pet. Eng. Handbook*, H.B. Bradley (ed.), SPE, Richardson, Texas (1987) Chap. 30.

Plisga, G.J.: "Temperatures in Wells," *Pet. Eng. Handbook*, H.B. Bradley (ed.), SPE, Richardson, Texas (1987) Chap 31.

Poston, S.W., Lubojacky, R.W., and Aruna, M.: "Meren Field—An Engineering Review," *JPT* (November 1983) 2105; *Trans.*, AIME, **275.**

Poston, S.W. and Berg, R.W.: *Overpressured Gas Reservoirs,* SPE, Richardson, Texas (1997).

Prasad, R.K. and Rogers, L.A.: "Superpressured Gas Reservoirs: Case Studies and a Generalized Tank Model," paper SPE 16861 presented at the 1987 SPE Annual Technical Conference and Exhibition, Dallas, 27–30 September.

Prats, M.: *Thermal Recovery,* Monograph Series, SPE of AIME, Dallas (1982) **7.**

Pruit, J.D.: "Statistical and Geological Evaluation of Oil and Gas Production From the 'J' Sandstone, Denver Basin, Colorado, Nebraska and Wyoming," *Energy Resources of the Denver Basin*, J. Pruit and P. Coffin (eds.), Rocky Mountain Assn. of Geologists, Denver, Colorado (1978).

Purvis, R.A. and Bober, W.G.: "A Reserves Review of the Pembina Cardium Oil Pool," *J. Cdn. Pet. Tech.* (July–September 1979) 20.

Purvis, R.A.: "Analysis of Production-Performance Graphs," *J. Cdn. Pet. Tech.* (July–August 1985) 44.

Purvis, R.A.: "Further Analysis of Production Performance Graphs," *J. Cdn. Pet. Tech.* (July–August 1987) 74.

Purvis, R.A.: "Pool Production and Well Count Forecasts," *J. Cdn. Pet. Tech.* (November–December 1990) 80.

Purvis, R.A. and Dick, A.B.: "Whither Alberta's Production?" paper *CIM/AOSTRA* 91-23 presented at the 1991 CIM/AOSTRA Conference, Banff, Alberta 21–24 April.

Purvis, R.A.: "Decline Curve Methods," *Determination of Oil and Gas Reserves,* Petroleum Soc.-CIM. Calgary (1994).

Ramagost, B.P. and Farshad, F.F.: "P/Z Abnormally Pressured Gas Reservoirs," paper SPE 10125 presented at the 1981 SPE Annual Technical Conference and Exhibition, San Antonio, 5–7 October.

Rayes, D.G., Piper, L.D., McCain, W.D. Jr., and Poston, S.W.: "Two-Phase Compressibility Factors for Retrograde Gases," *SPEFE* (March 1992) 87; *Trans.*, AIME, **293.**

Reiss, L.H.: *The Reservoir Engineering Aspects of Fractured Formations,* Editions Technip, Paris (1980).

Reitzel, G.A. and Callow, G.O.: "Pool Description and Performance Analysis Leads to Understanding Golden Spike's Miscible Flood," *JPT* (July 1977) 867.

Reudelhuber, F.O. and Hinds, R.F.: "A Compositional Material Balance Method for Prediction of Recovery From Volatile Oil Depletion Drive Reservoirs," *Trans.,* AIME (1957) **210,** 19.

Richardson, J.G., Harris, D.G., Rossen, R.H., and Van Hee, G.: "The Effect of Small, Discontinuous Shales on Oil Recovery," *JPT* (November 1978) 1531.

Richardson, J.G. and Blackwell, R.J.: "Use of Simple Mathematical Models for Predicting Reservoir Behavior," *JPT* (September 1971) 1145; *Trans.*, AIME, **251.**

Richardson, J.G.: "Appraisal and Development of Reservoirs," *Leading Edge* (February 1989) **8,** 2, 42.

Riddler, G.P.: "Towards an International Classification of Reserves and Resources," *Australasian Inst. of Mining & Metallurgy Bull.* (February 1996) 31.

Roberts, T.G. and Ellis, E.H.: "Correlations of Gas-Oil Ratio History in a Solution-Gas-Drive Reservoir," *JPT* (June 1962) 595.

Robertson, J.D.: "Reservoir Management Using 3-D Seismic Data," *Leading Edge* (February 1989) **08,** 2, 25.

Robertson, J.D.: "Seismic Applications," *SPE/WPC Supplemental Vol.,* SPE, Richardson, Texas (2001).

Robinson, B.M., Holditch, S.A., and Lee, W.J.: "A Case Study of the Wilcox (Lobo) Trend in Webb and Zapata Counties, TX," *JPT* (December 1986) 1355; *Trans.*, AIME, **281.**

Robinson, J.G.: "Determination of Reserves and Value and Application of Risk," *J. Cdn. Pet. Tech.* (November 1990) 62.

Robinson, J.G.: "Standardizing Risk Analysis for the Evaluation of Oil and Gas Properties," *J. Cdn. Pet. Tech.* (June 1999) 37.

Rocky Mountain Assoc. of Geologists: "Hydrocarbon Production from Low Contrast, Low Resistivity Reservoirs, Rocky Mountain and Mid-Continent Regions, Log Examples of Subtle Plays," E.D. Dolly and J.C. Mullarky (eds.) 1996.

Rollins, J.B., McCain, W.D. Jr., and Creeger, T.J: "Estimation of Solution GOR of Black Oils," *JPT* (January 1990) 92; *Trans.*, AIME, **289.**

Rollins, J.B., Holditch, S.A., and Lee, W.J.: "Characterizing Average Permeability in Oil and Gas Formations," *SPEFE* (March 1992) 99.

Root, P.J., Warren, J.E., and Hartsock, J.H.: "Implications of Transient Flow Theory: The Estimation of Gas Reserves," *Drill. & Prod. Prac.,* API, Dallas (1966).

Rose, J.D.: "Case History—Installation of High Volume Pumping Equipment in Talco Field, Texas," paper SPE 11040 presented at the 1982 SPE Annual Technical Conference and Exhibition, New Orleans, 26–29 September.

Rose, P.R.: "Uncertainties Impacting Reserves Revenue and Costs," *Development Geology Reference Manual,* American Assn. of Petroleum Geologists, Tulsa, Oklahoma (1992).

Rose, P.R.: *Risk Analysis of Petroleum Exploration Ventures,* American Assn. of Petroleum Geologists, Tulsa, Oklahoma (In press).

Rose, W.: "Relative Permeability," *Pet. Eng. Handbook,* H.B. Bradley (ed.), SPE, Richardson, Texas (1987) Chapter 28.

Ross, J.: "Resource Definitions," *SPE/WPC Supplemental Vol.,* SPE, Richardson, Texas (2001).

Rumble, R.C., Spain, H.H., and Stamm, H.E.: "A Reservoir Analyzer Study of the Woodbine Basin," *Trans.,* AIME (1951) **192,** 331.

Russell, D.G. *et al.:* "Methods for Predicting Gas Well Performance," *JPT* (January 1966) 99; *Trans.*, AIME, **237.**

Sabet, M. and Franks, L.: "Palm Valley—A Case for Interpretation, 1965-1984," *APEA J.* (1985) **1,** 329.

Saeland G.T. and Simpson, G.S.: "Interpretation of 3-D Data in Delineating a Subunconformity Trap in Block 34/10, Norwegian North Sea," *The Deliberate Search for the Subtle Trap,* M.T. Halbouty (ed.), American Assn. of Petroleum Geologists, Tulsa, Oklahoma (1982) Memoir 32.

Saleri, N.G. and Toronyi, R.M.: "Engineering Control in Reservoir Simulation: Part I," paper SPE 18305 presented at the 1988 SPE Annual Technical Conference and Exhibition, Houston, 2–5 October.

Schafer, P.S., Hower, T., and Owens, R.W.: *Managing Water Drive Gas Reservoirs,* Gas Research Inst., Chicago (1993).

Schilthuis, R.J.: "Active Oil and Reservoir Energy," *Trans.,* AIME (1936) **118,** 33.

Schlagenhauf, M. and Jaynes, E.: "Geophysical Reserve Classification—Proved, Probable, and Possible," paper SPE 30039 presented at the 1995 SPE Hydrocarbon Economics and Evaluation Symposium, Dallas, 26–28 March.

Schlumberger Well Services Corp.: "Log Interpretation Strategies in Gas Wells," *Oilfield Review* (April 1991).

Schuyler, J.R.: "Probabilistic Reserves Lead to More Accurate Assessments," paper SPE 49032 prepared for presentation at the 1998 SPE Annual Technical Conference and Exhibition, New Orleans, 27–30 September.

Scientific Software Corp.: *Reservoir Engineering Manual*, NTIS PB-247-806, Natl. Technical Information Service, Washington, DC (1975).

Seal, W.L. and Gilreath, J.A.: "Vermillion Block 16 Field—A Study of Irregular Gas Reservoir Performance Related to Non-Uniform Sand Deposition," *Trans.*, Gulf Coast Assn. of Geological Societies (1975) **25,** 63.

Sears, G.F. and Phillips, N.Y.: "Fractured Reservoir Evaluation Using Monte Carlo Techniques," *JPT* (January 1987) 71.

Seines, K., Lien, S.C., and Haug, B.T.: "Troll Horizontal Well Tests Demonstrate Large Production Potential From Thin Oil Zones," *SPERE* (May 1994) 133; *Trans.*, AIME, **297.**

Semenovich, V.V. *et al.*: "Methods Used in the USSR for Estimating Potential Petroleum Resources," *The Future Supply of Nature Made Petroleum and Gas,* Pergamon Press, New York (1976).

Serra, O.: *Sedimentary environments from wireline logs,* Schlumberger, Houston (1985).

Shaw, M.N.: "Development of the Bream Field," *Proc.*, Offshore Australia Conference, Melbourne Australia (1991).

Shyeh, J.J. *et al.*: "Interpretation and Modeling of Time-Lapse Seismic Data: Lena Field, Gulf of Mexico," paper SPE 56731 presented at the 1999 SPE Annual Technical Conference and Exhibition, Houston, 3–6 October.

Simpson, D.E., Huffman, C.H., and Thompson, R.S.: "Net Present Value Probability Distributions From Decline Curve Reserves," paper SPE 30044 presented at the 1995 SPE Hydrocarbon Economics and Evaluation Symposium, Dallas, 26–28 March.

Sims, W.P. and Frailing, W.G.: "Lakeview Pool, Midway-Sunset Field," *Trans.*, AIME (1950) **189,** 7.

Skaar, K.J., Heiberg, S., and Veding, H.M.: *Resource Classification in Statoil,* Statoil A.S., Stavanger, Norway (1996).

Skarpnes, O., Briseid, E., and Milton, D.I.: "The 34/10 Delta Prospect of the Norwegian North Sea: Exploration Study of an Unconformity Trap," *The Deliberate Search for the Subtle Trap,* M.T. Halbouty (ed.), American Assn. of Petroleum Geologists, Tulsa, Oklahoma (1982) Memoir 32.

Slider, H.C.: "A Simplified Method of Hyperbolic Decline Curve Analysis," *JPT* (March 1968) 235.

Smith, C.R., Tracy, G.W., and Farrar, R.L.: *Applied Reservoir Engineering,* Vol. 1, Oil & Gas Consultants International, Tulsa, Oklahoma (1992) Chap. 2.

Smith, P.J. and Buckee, J.W.: "Calculating In-Place and Recoverable Hydrocarbons: A Comparison of Alternate Methods," paper SPE 13776 presented at the SPE 1985 Hydrocarbon Economics and Evaluation Symposium, Dallas, 14–15 March.

Smith, R.V.: "Open Flow of Gas Wells," *Petroleum Production Handbook*, T.C. Frick and R.W. Taylor (eds.), SPE of AIME, Dallas, Texas (1962) Chap. 30.

Smith, R.V.: "Open Flow of Gas Wells," *Pet. Eng. Handbook*, H.B. Bradley (ed.), SPE, Richardson, Texas (1987) Chap. 33.

Snedecor, G.W. and Cochran, W.G.: *Statistical Methods*, seventh ed., Iowa State University Press, Ames, Iowa (1980).

Sneider, R.M. *et al.*: "Predicting Reservoir Rock Geometry and Continuity in Pennsylvanian Reservoirs, Elk City Field, Oklahoma," *JPT* (July 1977) 851.

Sneider, R.M., Tinker, C.N., and Meckel, L.D.: "Deltaic Environment Reservoir Types and Their Characteristics," *JPT* (November 1978) 1538.

Snyder, R.H.: "A Review of the Concepts and Methodology of Determining 'Net Pay,'" paper SPE 3609 presented at the 1971 SPE Annual Technical Conference and Exhibition, New Orleans, 3–6 October.

Soc. of Petroleum Engineers: "Standards Pertaining to the Estimating and Auditing of Oil and Gas Reserve Information," *JPT* (December 1979) 1557.

Soc. of Petroleum Engineers: "Reserve Definitions Approved," *JPT* (May 1987) 576.

Soc. of Petroleum Engineers: "SPE Letter and Computer Symbols Standard for Economics, Well Logging and Formation Evaluation, Natural Gas Engineering, and Petroleum Reservoir Engineering," *Pet. Eng. Handbook*, H.B. Bradley (ed.) SPE, Richardson, Texas (1987); 1993 Revision.

Soc. of Petroleum Engineers: "SPE/WPC Reserves Definitions Approved," *JPT* (May 1997) 527.

Soc. of Petroleum Evaluation Engineers (SPEE): *Survey of Economic Parameters Used in Property Evaluation,* SPEE, Houston (1991 through 1998).

Soc. of Petroleum Evaluation Engineers: *Guidelines for Application of Petroleum Reserve Definitions,* second edition, Monograph 1, SPEE, Houston (1998).

Sognesand, S.: "Reservoir Management of the Oseberg Field During Eight Years' Production," paper SPE 38555 presented at the 1997 SPE Offshore Europe Conference, Aberdeen, U.K., 9–10 September.

Spears, R. and Dromgoole, P.: "Managing Uncertainty in Oilfield Reserves," *Middle East Well Evaluation Review* (November 1992).

Spencer, A.: "Operational Issues," *SPE/WPC Supplemental Vol.,* SPE, Richardson, Texas (2001).

Spencer, C.W.: "Geologic Aspects of Tight Gas Reservoirs in the Rocky Mountain Region," *JPT* (July 1985) 1308; *Trans.,* AIME **279.**

Spencer, C.W. and Mast, R.F. (eds.): *Geology of Tight Gas Reservoirs, Studies in Geology,* American Assn. of Petroleum Geologists, Tulsa (1986) **24.**

Sprunt, E.S.: "Arun Core Analysis: Special Procedures for Vuggy Carbonates," *The Log Analyst,* (September–October 1989) 353.

Sprunt, E.S., Maute, R.E., and Rackers, C.L.: "An Interpretation of the SCA Electrical Resistivity Study," *The Log Analyst,* (1990) **31,** 2, 75.

Squire, S.G.: "Reservoir and Pool Parameter Distributions From the Cooper/Euromanga Basin, Australia," paper SPE 37365 presented at the 1996 SPE Asia Pacific Oil and Gas Conference and Exhibition, Adelaide, Australia, 28–31 October.

Stalkup, F.I.: *Miscible Displacement,* Monograph Series, SPE, Richardson, Texas (1983) **8.**

Standing, M.B. and Katz, D.L.: "Density of Natural Gases," *Trans.,* AIME (1942) **146,** 140.

Standing, M.B.: "A Pressure-Volume-Temperature Correlation for Mixtures of California Oils and Gases, *Drill. & Prod. Prac.,* API (1947) 275.

Standing, M.B.: *Volumetric and Phase Behavior of Oilfield Hydrocarbon Systems,* Reinhold, New York City (1952); eighth printing, SPE, Richardson, Texas (1977).

Standing, M.B.: "Oil-System Correlations," *Pet. Production Handbook,* T.C. Frick (ed.), SPE, Dallas (1962).

Stanley, L.T.: *Practical Statistics for Petroleum Engineers,* Petroleum Publishing Co., Tulsa (1973).

Startzman, R.A. and Wu, C.H.: "Discussion of Empirical Prediction Technique for Immiscible Processes," *JPT* (December 1984) 2192.

Stewart, F.M., Garthwaite, D.L., and Krebill, F.K.: "Pressure Maintenance by Inert Gas Injection in the High Relief Elk Basin Field," *Trans.,* AIME (1955) **204,** 49.

Stewart, L.: "Piper Field: Reservoir Engineering," paper EUR 152 presented at the 1980 European Offshore Petroleum Conference, London, 21–24 October.

Stewart, P.R.: "Evaluation of Individual Gas Well Reserves," *Pet. Eng.* (May 1966) 85.

Stewart, P.R.: "Low-Permeability Gas Well Performance at Constant Pressure," *JPT* (September 1970) 1149.

Stiles, L.H. *et al.:* "Performance of Sharon Ridge Canyon Unit with Water Injection," *JPT* (April 1972) 431.

Stiles, L.H.: "Optimizing Waterflood Recovery in a Mature Waterflood, the Fullerton Clearfork Unit," paper SPE 6198 presented at the 1976 SPE Annual Technical Conference and Exhibition, New Orleans, 3–6 October.

Stiles, J.H. Jr. and Mc Kee, J.W.: "Cormorant: Development of a Complex Field," *SPEFE* (December 1991) 427; *Trans.,* AIME, **291.**

Stoian, E. and Telford, A.S.: "Determination of Natural Gas Recovery Factors," *J. Cdn. Pet. Tech.* (July–September 1966) 115.

Stright, D.H. Jr. and Gordon, J.I.: "Decline Curve Analysis in Fractured Low Permeability Gas Wells in the Piceance Basin," paper SPE 11640 presented at the 1983 SPE Symposium on Low Permeability, Denver, Colorado, 14–16 March.

Strobel, C.J., Gulati, M.S., and Ramey, H.J.: "Reservoir Limit Tests in a Naturally Fractured Reservoir—A Field Case Study Using Type Curves," *JPT* (September 1976) 27, 1097.

Stuart, C.A. and Kozik, H.G.: "Geopressuring Mechanism of Smackover Gas Reservoirs, Jackson Dome Area, Mississippi," *JPT* (May 1977) 579; *Trans.,* AIME (1977) **263.**

Sulaiman, S. and Bretherton, T.A.: "Thin Oil Development in the Bream Field," paper SPE 19493 presented at the 1989 SPE Asia-Pacific Conference, Sydney, 13–15 September.

Sullivan, R.E.: *Handbook of Oil and Gas Law,* Prentice Hall, Englewood Cliffs, New Jersey (1955).

Sutton, R.P.: "Compressibility Factors for High-Molecular-Weight Reservoir Gases," paper SPE 14265 presented at the 1985 SPE Annual Technical Conference and Exhibition, Las Vegas, Nevada, 22–25 September.

Sutton, R.P. and Farshad, F.: "Evaluation of Empirically Derived PVT Properties for Gulf of Mexico Crude Oils," *JPT* (August 1990) 79.

Swanson, B.F.: "A Simple Correlation Between Permeabilities and Mercury Capillary Pressures," *JPT* (December 1981) 2498.

Swinkels, W.: "Aggregation of Reserves," *SPE/WPC Supplemental Vol.,* SPE, Richardson, Texas (2001).

Sydora, L.J.: "The Hibernia Oil Field: Integration of Geophysical, Geological, and Production Data Into Reservoir Characterization and Monitoring," paper OTC 10741 presented at the 1999 Offshore Technology Conference, Houston, 3–6 May.

Takacs, G.: "Comparisons Made for Computer Z-Factor Calculations," *Oil and Gas J.* (20 December 1976) 64.

Talash, A.W.: "An Overview of Waterflood Surveillance and Monitoring," *JPT* (December 1988) 1539.

Tarner, J.: "How Different Size Gas Caps and Pressure Maintenance Programs Affect Amount of Recoverable Oil," *Oil Weekly* (1944) **144,** 32.

Tarner, J., Evans, W.R., and Kaveler, H.H.: "The Shuler Jones Sand Pool; Nine Years of Unitized Pressure Maintenance Operations," *Trans.,* AIME (1951) **192,** 121.

Tarner, J. and Curzon, J.E.: Supplement, 1975, "The Shuler Field and Unit Operations," *Field Case Histories, Oil and Gas Reservoirs,* Reprint Series, SPE of AIME, Dallas (1975) **4a,** 258.

Tearpock, D.J. and Bischke, R.E.: *Applied Subsurface Geological Mapping,* Prentice-Hall Inc. Englewood Cliffs, New Jersey (1991).

Teeuw, D.: "Prediction of Formation Compaction from Laboratory Compressibility Data," *SPEJ* (1971) 263.

Testerman, J.D.: "A Statistical Reservoir-Zonation Technique," *JPT* (August 1962) 889; *Trans.,* AIME, **225.**

Thakur, G.C.: *Reservoir Management of Mature Fields,* Intl. Human Resources Development Corp., Boston, Massachusetts (1992).

Thomas, G.W.: "The Role of Reservoir Simulation in Optimal Reservoir Management," paper SPE 14129 presented at the 1986 SPE International Meeting on Petroleum Engineering, Beijing, 17–20 March.

Thomeer, J.H.M.: "Introduction of a Pore Geometrical Factor Defined by the Capillary Pressure Curve," *JPT* (March 1960) 354; *Trans.,* AIME, **219.**

Thompson, R.S. and Wright, J.D.: *Oil Property Evaluation,* Thompson-Wright Assocs., Golden, Colorado (1984).

Thornton, R.C.N. *et al.:* "Fortesque Field—An Unconformity Trap in the Gippsland Basin, Australia," *The Deliberate Search for the Subtle Trap,* M.T. Halbouty (ed.), American Assn. of Petroleum Geologists, Tulsa, Oklahoma (1982) Memoir 32.

Thurber, J.L.: "Little Creek Field CO_2 Miscible Pilot," *Enhanced Oil Recovery Field Reports,* SPE, Richardson, Texas (March 1978) **3,** No. 4, 673.

Till, M.V., Sauer, C.W., and Smith, J.E.: "Field Development Planning for the Heidrun Field," paper SPE 15880 presented at the 1986 SPE European Petroleum Conference, London, 20–22 October.

Timur, A.: "Acoustic Logging," *Pet. Production Handbook*, H.B. Bradley (ed.), SPE, Richardson, Texas (1987).

Timmerman, E.H.: *Practical Reservoir Engineering*, Penn Well Publishing Co., Tulsa, Oklahoma (1982) **1 and 2**.

Tixier, M.P.: "Electrical Logging," *Pet. Production Handbook*, H.B. Bradley (ed.), SPE, Richardson, Texas (1987).

Tollas, J.M. and McKinney, A.: "Brent Field 3D Reservoir Simulation, " *JPT* (May 1991) 580.

Tracy, G.W.: "Simplified Form of the Material Balance Equation," *Trans.*, AIME (1955) **204**, 243.

Trewin, N. and Bramwell, M.G.: "The Auk Field, Block 30/16, UK North Sea," *United Kingdom Oil and Gas Fields, 25 Years Commemorative Vol.*, I.L. Abbotts (ed.), Geological Soc. of London, London (1991).

Trube, A.S.: "Compressibility of Undersaturated Hydrocarbon Fluids," *JPT* (December 1957) 341; *Trans.*, AIME, **210**.

Tyler, N. and Gholston, J.C.: "Reserve Growth Through Geological Characterization of Heterogeneous Reservoirs—An Example From Mud-Rich Submarine Fans Reservoirs of Permian Spraberry Trend, West Texas," *AAPG Bull.* (1987) **71**, 623.

U.K. Financial Services Authority (U.K. FSA): "The Listing Rules," Chap. 19– Mineral Companies, U.K. FSA, London (May 2000).

U.K. Oil Industry Accounting Committee (U.K. OIAC): *SORP, Statement of Recommended Practice, Accounting for Oil and Gas Exploration, Development, Production and Decommissioning Activities,* Inst. of Petroleum, London (January 2000).

U.S. Federal Securities Laws: Reg. S-X (17 CFR Part 210), Reg. 210.4-10 (a) Definitions (1982).

U.S. Geological Survey (USGS): "Geothermal Gradient Map of North America," Scale 1:500,000, USGS, Reston, Virginia (1976).

Valenti, N.P. and Buckles, R.S.: "Numerical Simulation of the North Sea Fulmar Oil Field: Evaluating Reservoir Depletion Strategies," paper SPE 15871 presented at the 1986 SPE European Petroleum Conference, London, 20–22 October.

van der Knaap, W.: "Nonlinear Behavior of Elastic Porous Media," *Trans.*, AIME (1959) **216**, 179.

van Elk, J.F., Vijayan, K., and Gupta, R.: "Probabilistic Addition of Reserves—A New Approach," paper SPE 64454 presented at the SPE Asia Pacific Oil and Gas Conference and Exhibition, Brisbane, Australia 16–18 October 2000.

van Everdingen, A.F. and Hurst, W.: "Application of the Laplace Transformation to Flow Problems in Reservoirs," *Trans.*, AIME (1949) **186**, 305.

van Lookeren, J.: "Oil Production From Reservoirs With an Oil Layer Between Gas and Bottom Water in the Same Sand,"*JPT,* (March 1965) 354; *Trans,* AIME (1965) **234**, 354.

van Meurs, P. and van der Poel, C. "A Theoretical Description of Water-Drive Processes Involving Viscous Fingering," *Trans.*, AIME (1958) **213**, 103.

van Rijswijk, J.J. *et al.:* "The Dunlin Field, a Review of Field Development and Reservoir Performance to Date," *JPT* (September 1981) 1713.

Vasquez, M.: *Correlations for Fluid Physical Property Correlation,* MS thesis, U. of Tulsa, Oklahoma Tulsa, Oklahoma (1976).

Vasquez, M. and Beggs, H.D.: "Correlations for Fluid Property Prediction," *JPT* (June 1980) 968; *Trans.*, AIME, **269**.

Veatch, R.W. Jr.: "Overview of Current Hydraulic Fracturing Design and Treatment Technology— Part 1," *JPT,* (April 1983) 677; *Trans.*, AIME, **275**.

Ventura, J. and Temansja, A.D.: "Increasing Withdrawal Rates Add Oil Reserves in Java Sea Area," *Oil & Gas J.* (18 November 1991).

von Gonten, W.D. and Choudhary, B.K.: "The Effect of Pressure and Temperature on Pore Volume Compressibility," paper SPE 2526 presented at the 1969 SPE Annual Fall Meeting, Denver, Colorado, September 28–October 1.

Wahl, W.L., Mullins, L.D., and Elfrink, E.B.: "Estimation of Ultimate Recovery from Solution Gas-Drive Reservoirs," *Trans.*, AIME (1958) **213**, 132.

Wallace, W.E.: "Water Production from Abnormally Pressured Gas Reservoirs in South Louisiana," *JPT* (August 1969) 969.

Wall Street Journal (WSJ): "Hondo Oil & Gas Company Shares no Longer Trade on the American Exchange," *WSJ* (2 July 1998) A10.

Walsh, M.P.: "A Generalized Approach to Reservoir Material Balance Equations," *J. Cdn. Pet. Tech.* (January 1995) 55.

Walstrom, J.E., Mueller, T.D., and MacFarlane, R.C.: "Evaluating Uncertainty in Engineering Calculations," *JPT* (December 1967) 1595; *Trans.*, AIME, **240.**

Warren, J.E. and Price, H.S.: "Flow in Heterogeneous Porous Media," *SPEJ* (September 1961) 153, *Trans.*, AIME, **222.**

Warren, J.E. and Root, P.J.: "The Behavior of Naturally Fractured Reservoirs," *SPEJ* (September 1963) 245; *Trans.*, AIME, **228.**

Wayhan, D.A. and McCaleb, J.A.: "Elk Basin Madison Heterogeneity—Its Influence on Performance," *JPT* (February 1969) 153.

Wayland, D.A. *et al.:* "Estimating Waterflood Recovery in Sandstone Reservoirs," *Drill. & Prod. Prac.*, API, Dallas (1971) **37, 252.**

Weimer, R.J., Howard, J.D., and Lindsay, D.R.: "Tidal Flats and Associated Tidal Channels," *Sandstone Depositional Environments,* American Assn. of Petroleum Geologists, Tulsa, Oklahoma (1982).

Wellings, R.W.: Supplement, 1975, "Recent Performance of the Leduc D-3 Reservoir," *Field Case Histories, Oil and Gas Reservoirs,* Reprint Series, SPE, Richardson, Texas (1975) **4a,** 263.

West, S.L. and Cochran, P.J.R.: "Reserves Determination Using Type Curve Matching and Extended Material Balance Methods in the Medicine Hat Shallow Gas Field," paper SPE 28609 presented at the 1994 SPE Annual Technical Conference and Exhibition, New Orleans, 25–28 September.

Wharton, J.B. Jr.: "Isopachous Maps of Sand Reservoirs," *AAPG Bull.* (1948) **32,** 7.

Whitson, C.H.: "Effect of Physical Properties Estimation on Equation of State Predictions," *SPEJ* (December 1984) 685; *Trans.*, AIME, **277.**

Wichert, E. and Aziz, K.: "Calculate Z's for Sour Gases," *Hydrocarbon Processing* (May 1972) 119.

Wieland, D.R. and Kennedy, H.T.: "Measurements of Bubble Frequency in Cores," *Trans.*, AIME (1957) **210,** 122.

Willmon, G.J.: "A Study of Displacement Efficiency in the Redwater Field," *JPT* (April 1967) 449.

Witherbee, L.J., Godfrey, R.D., and Dimelow, T.E.: "Predicting Turbidite-Contourite Reservoir Intervals in Tight Gas Sands: A Case Study From the Mancos B Sandstone," paper SPE 11609 presented at the 1983 SPE Symposium on Low Permeability, Denver, Colorado, 13–16 March.

Wong, K.H. and Ambastha, A.K.: "Decline Curve Analysis for Canadian Oil Reservoirs Under Waterflood Conditions," paper 95-08 presented at 46th Annual Technical Meeting of The Petroleum Society-CIM, Banff, Alberta, 14–17 May 1995.

Woods, R.W.: "Case History of Reservoir Performance of a Highly Volatile Type Oil Reservoir," *Trans.*, AIME (1955) **204,** 156.

World Petroleum Congress: "Classification and Nomenclature Systems for Petroleum and Petroleum Reserves," presented at 12th World Petroleum Congress, Houston 1987, John Wiley & Sons, Chichester.

Wu, C.H., Loughlin, B.A., and Jardon, M.: "Infill Drilling Enhances Waterflood Recovery," *JPT* (October 1989) 1088.

Wyllie, M.R.J.: "Relative Permeability," *Petroleum Production Handbook*, T.C. Frick (ed.), SPE, Dallas (1962), Chap. 25.

Yale, D.P. *et al.:* "Application of Variable Formation Compressibility for Improved Reservoir Analysis," paper SPE 26647 presented at the 1993 SPE Annual Technical Conference and Exhibition, Houston, 3–6 October.

Yarus, J.M. and Chambers, R.L. (eds.): *Stochastic Modeling and Geostatistics,* American Assn. of Petroleum Geologists, Tulsa, Oklahoma (1994).

Additional References

Geology/Mapping

Geology of Giant Petroleum Fields, M.T. Halbouty (ed.), American Assn. of Petroleum Geologists, Tulsa, Oklahoma (1970).

Giant Oil and Gas Fields of the Decade: 1968-1978, M.T. Halbouty (ed.), American Assn. of Petroleum Geologists, Tulsa, Oklahoma (1980).

Giant Oil and Gas Fields of the Decade: 1978-1988, M.T. Halbouty (ed.), American Assn. of Petroleum Geologists, Tulsa, Oklahoma (1992).

North American Oil and Gas Fields, J. Braunstein (ed.), American Assn. of Petroleum Geologists, Tulsa, Oklahoma (1976).

Stochastic Modeling and Geostatistics, Principles, Methods, and Case Studies, J.M. Yarus and R.L. Chambers (eds.), American Assn. of Petroleum Geologists, Tulsa, Oklahoma (1994).

Tearpock, D.J. and Bischke, R.E.: *Applied Subsurface Geological Mapping,* Prentice Hall, Englewood Cliffs, New Jersey (1991).

Treatise of Petroleum Geology—Atlas of Oil and Gas Fields of the World, N.H. Foster and E.A. Beaumont (eds.), American Assn. of Petroleum Geologists, Tulsa, Oklahoma (1990, 1991, 1992, 1993).

Log Analysis

Asquith, G. and Gibson, C.: *Basic Well Log Analysis for Geologists,* American Assn. of Petroleum Geologists, Tulsa, Oklahoma (1982).

Asquith, G.: *Handbook of Log Evaluation Techniques for Carbonate Reservoirs,* American Assn. of Petroleum Geologists, Tulsa, Oklahoma (1985).

Dewan, J.T.: *Essentials of Modern Open-Hole Log Interpretation,* Penn Well Publishing Co., Tulsa, Oklahoma (1983).

Doveton, J.H.: *Geologic Log Analysis Using Computers,* American Assn. of Petroleum Geologists, Tulsa, Oklahoma (1994).

Ellis, D.V.: "Nuclear Logging Techniques," *Pet. Eng. Handbook,* H.B. Bradley (ed.), SPE, Richardson, Texas (1987) Chap. 50.

Hilchie, D.W.: *Applied Openhole Log Interpretation,* Douglas W. Hilchie Inc., Golden, Colorado (1982).

Jorden, J.R. and Campbell, F.L.: *Well Logging I—Rock Properties, Borehole Environment, Mud and Temperature Logging,* Monograph Series, SPE, Richardson, Texas (1984) **9.**

Jorden, J.R. and Campbell, F.L.: *Well Logging II—Electric and Acoustic Logging,* Monograph Series, SPE, Richardson, Texas (1986) **10.**

Log Interpretation: Principles and Applications, third edition, Schlumberger Educational Services, Houston (1987).

Well Logging and Interpretation Techniques, Dresser Industries, Houston (1982).

Improved Recovery

Craig, F.F. Jr.: *The Reservoir Engineering Aspects of Waterflooding,* Monograph Series, SPE, Richardson, Texas (1971) **3.**

Klins, M.A.: *Carbon Dioxide Flooding: Basic Mechanisms and Project Design,* Intl. Human Resources Development Corp., Boston, Massachusetts (1984).

Petroleum Soc.-CIM (PS-CIM): "Estimation of Recovery Factors and Forecasting of Recoverable Hydrocarbons," *Determination of Oil and Gas Reserves,* PS-CIM, Calgary (1994).

Prats, M.: *Thermal Recovery,* Monograph Series, SPE, Richardson, Texas (1982) **7.**

Secondary and Tertiary Oil Recovery Processes, Interstate Oil Compact Commission, Oklahoma City (1974).

Smith, C.R.: *Mechanics of Secondary Oil Recovery,* Litton, Huntington, New York (1975).

Stalkup, F.I.: *Miscible Displacement,* Monograph Series, SPE, Richardson, Texas (1983) **8.**

Waggoner, J.R., Castillo, J.L., and Lake, L.W.: "Simulation of EOR Processes in Stochastically Generated Permeable Media," *SPEFE* (June 1992) 173; *Trans.,* AIME, **293.**

Wilhite, G.P.: *Waterflooding*, Textbook Series, SPE, Richardson, Texas (1986) **3.**

AUTHOR INDEX

A

B

SUBJECT INDEX

A

Petroleum Reserves & Resources Definitions

Oil and gas that have been discovered, but not yet produced, cannot be readily measured. Trapped in the pore spaces of rock, thousands of feet below the surface, the amount of oil or gas in a reservoir cannot be measured with precision. But it is very important to have a good estimate of the amount of oil or gas that may lie in the reservoir. A company cannot evaluate whether a discovered field will be economic to develop without estimating the amount of production that it may obtain over time to balance against the investment required. Oil and gas reserves are a substantial asset on a company's balance sheet. Without a common approach to classification and estimation of oil and gas, it would be impossible to know whether those assets were comparable from one company to another.

Petroleum Resources Management System
http://www.spe.org/industry/reserves.php

Approved by the SPE Board of Directors in March 2007, this new system for defining reserves and resources was developed over more than two years, working with WPC, AAPG, and SPEE. An archive of the prior definitions is also available.

PRMS Guide for Non-Technical Users
http://www.spe.org/industry/docs/PRMS_guide_non_tech.pdf

The guide provides the concepts in the PRMS in a four-page document that is intended to provide a quick overview for non-technical professionals.

PRMS in Spanish
http://www.spe.org/industry/docs/spanish_PRMS_2009.pdf

This translation from the English version was developed under the guidance of the SPE Oil and Gas Reserves Committee. The English version remains the official version and takes precedence over the translation in cases where there is any question of meaning or interpretation.

Mapping of Reserves Definitions
http://www.spe.org/industry/docs/OGR_Mapping.pdf

Around the world, government agencies and other organizations use slightly different definitions. This mapping provides a comparison of many of these definitions.

Estimating and Auditing Standards for Reserves
http://www.spe.org/industry/docs/Reserves_Audit_Standards_2007.pdf

To assist those responsible for estimating reserves, or auditing those estimates, a standard approach has been outlined, along with minimum qualifications for those involved in reserves auditing.

Committee Activities
http://www.spe.org/industry/ogrc.php

Learn more about the SPE Oil & Gas Reserves Committee's ongoing activities to work with other groups on a common set of definitions. The Reserves Education Committee provides outreach in communicating changes in the definitions and working toward a common definition.